Laser-Generated Periodic Nanostructures

Laser-Generated Periodic Nanostructures

Editors

Peter Simon
Jürgen Ihlemann
Jörn Bonse

MDPI • Basel • Beijing • Wuhan • Barcelona • Belgrade • Manchester • Tokyo • Cluj • Tianjin

Editors

Peter Simon
Institut für Nanophotonik
Göttingen e.V.,
Hans-Adolf-Krebs-Weg 1
Germany

Jürgen Ihlemann
Institut für Nanophotonik
Göttingen e.V.,
Hans-Adolf-Krebs-Weg 1
Germany

Jörn Bonse
Bundesanstalt für
Materialforschung
und-prüfung (BAM)
Germany

Editorial Office
MDPI
St. Alban-Anlage 66
4052 Basel, Switzerland

This is a reprint of articles from the Special Issue published online in the open access journal *Nanomaterials* (ISSN 2079-4991) (available at: https://www.mdpi.com/journal/nanomaterials/special_issues/laser-generated_periodic).

For citation purposes, cite each article independently as indicated on the article page online and as indicated below:

LastName, A.A.; LastName, B.B.; LastName, C.C. Article Title. *Journal Name* **Year**, *Volume Number*, Page Range.

ISBN 978-3-0365-2027-8 (Hbk)
ISBN 978-3-0365-2028-5 (PDF)

Contents

About the Editors

Peter Simon received his diploma and Ph.D. in Physics from the University of Szeged in 1982 and 1986, respectively. In 1988, he joined the Institute for Nanophotonics (formerly the Laser-Laboratorium Göttingen), where he participated in research associated with the generation, amplification, and characterization of femtosecond laser pulses and their application for material processing. In 1992, he was appointed as a group leader of the High Intensity Laser Technology Group, and in 2005, he was appointed as the Head of the Department of Ultrashort Pulse Photonics. He has been the head of the Department Short Pulses/Nanostructures since 2015. The subjects of his current research include the generation and amplification of ultrashort laser pulses, the compression of energetic few-cycle pulses, and the submicron-scale surface texturing of technical materials. Dr. Simon has authored more than 100 refereed journal publications and holds more than 10 patents related to his research activities.

Jürgen Ihlemann received his diploma in physics (1984) and Ph.D. in physical chemistry (1987) from the University of Göttingen. From 1984 to 1988 he was with the Max Planck Institute for biophysical chemistry, Göttingen, where he worked on picosecond laser spectroscopy. In 1989, he joined the Institute for Nanophotonics Göttingen (at that time, the Laser-Laboratorium Göttingen), Germany, where he is now head of the nanostructure technology group. His research interests include UV-laser micro- and nanomachining, patterning of surfaces and thin films, ultrashort pulse laser ablation, and the laser-based generation and manipulation of nanocrystals and nanoparticles.

Jörn Bonse is a tenured scientist at the German Federal Institute for Materials Research and Testing (BAM) in Berlin, Germany. His research interests include the fundamentals and applications of laser–matter interaction, especially with respect to ultrashort laser pulses, laser-induced periodic nanostructures, time-resolved optical techniques, laser processes in photovoltaics, surface functionalization, and laser safety. He received a doctoral degree in Physics from the Technical University of Berlin, Germany, in 2001 and a diploma degree in Physics from the University of Hannover, Germany, in 1996. Dr. Bonse has occupied various research positions at institutions such as the Max Born Institute for Nonlinear Optics and Short Pulse Spectroscopy (MBI) in Berlin, the Spanish Research Council (CSIC) in Madrid, and the Laser Zentrum Hannover (LZH) in Hannover. He was appointed as a senior laser application specialist at Newport's Spectra-Physics Lasers Division in Stahnsdorf, Germany. In 1999, he was the recipient of an award for applied research, presented by the federal German state of Thuringia, for the development of high-power fiber lasers. He received an OSA Outstanding Reviewer Award from the Optical Society of America in 2013. Between 2014 and 2017, he served as an Associate Editor for the journal Optics Express of the Optical Society of America. Since 2020, he has been a member of the Editorial Board of the journal Advanced Optical Technologies. He has authored more than 160 refereed journal publications and 2 patents related to his research activities.

nanomaterials

Editorial

Editorial: Special Issue "Laser-Generated Periodic Nanostructures"

Peter Simon [1,*], Jürgen Ihlemann [1,*] and Jörn Bonse [2,*]

[1] Institut für Nanophotonik Göttingen e.V., Hans-Adolf-Krebs-Weg 1, D-37077 Göttingen, Germany
[2] Bundesanstalt für Materialforschung und -prüfung (BAM), Unter den Eichen 87, D-12205 Berlin, Germany
* Correspondence: peter.simon@ifnano.de (P.S.); juergen.ihlemann@ifnano.de (J.I.); joern.bonse@bam.de (J.B.)

Citation: Simon, P.; Ihlemann, J.; Bonse, J. Editorial: Special Issue "Laser-Generated Periodic Nanostructures". *Nanomaterials* **2021**, *11*, 2054. https://doi.org/10.3390/nano11082054

Received: 21 July 2021
Accepted: 6 August 2021
Published: 12 August 2021

Publisher's Note: MDPI stays neutral with regard to jurisdictional claims in published maps and institutional affiliations.

The study of laser-fabricated periodic nanostructures is one of the leading topics of today's photonics research. Such structures on the surface of metals, semiconductors, dielectrics, or polymers can generate new material properties with special functionalities. Depending on the specific material parameters and the morphology of the structures, new devices such as microlasers, optical nanoswitches, optical storage devices, sensors or antifraud features can be realized. Furthermore, laser-generated surface textures can be used to improve the tribological properties of surfaces in contact and in relative motion—to reduce friction losses or wear [1–3], to modify the wettability or the cell and biofilm growth properties of surfaces through bioinspired laser engineering [4], for emerging medical applications [5,6], or as decoration elements for the refinement of precious goods [7].

This Special Issue "Laser-Generated Periodic Nanostructures" focuses on the latest experimental and theoretical developments and practical applications of laser-generated periodic structures that can be generated in a "self-organized" way (laser-induced periodic surface structures, LIPSS, ripples) [8–11] or via laser interference-based direct ablation (often referred to as direct laser interference patterning, DLIP) [12–14]. We aimed to attract both academic and industrial researchers in order to collate the current knowledge of nanomaterials and to present new ideas for future applications and new technologies. By 8 August 2021, 22 scientific articles have been published in the Special Issue [15–36], see www.mdpi.com/journal/nanomaterials/special_issues/laser-generated_periodic.

Three "Feature Papers" [15–17] were invited by the Guest Editors for the adequate framing of this Special Issue. An additional 19 papers were published as regular contributions [18–36]. The publications in this Special Issue can generally be divided into three groups: (1) manuscripts related to self-organized laser-induced periodic surface structures [15,18,19,21,23–26,28,30–36], (2) manuscripts related to interference-based periodic surface structuring [17,22,27,29], or (3) manuscripts combining both approaches [16,20].

Several publications focus on the relevance of chemical surface alterations during or after laser-processing, including studies of superficial oxidation on the formation of specific types of LIPSS [18,23–25,30], chemo-capillary effects in DLIP [29], or even post-laser irradiation effects due to reactions with the environment [30]. Other publications study the fundamental mechanisms of structure formation on the basis of experimental and theoretical plasmonic approaches [15,21,26], and also in combination with a two-temperature model [28], advanced molecular dynamics simulations [22], or multi-physical hydrodynamic continuum considerations [31,34]. Two articles investigate in vitro the growth of different types of biological cells (either scar-forming fibroblasts or bone-forming osteoblasts) on laser-generated micro- and nanostructures for medical applications [19,25]. Other applications qualify the real-life anti-icing properties of large-area laser-structured surfaces for aeronautic applications [17], and the alteration of surface superconductivity is demonstrated with a view toward future energy applications [24].

In the following, all publications in this Special Issue are briefly summarized, ordered by their date of publication.

Florian et al. [18] reported on a strong absorbing, oxidation-prone material (chromium nitride, CrN), an unusual type of self-organized near-wavelength-sized nanostructure—so-called low spatial frequency LIPSS (LSFL-∥), being parallel to the femtosecond laser (790 nm, 30 fs, 1 kHz, or 1030 nm, 350 fs, 100 kHz) beam polarization used for irradiation in air. This is in contrast to the common LSFL (LSFL-⊥) that usually form perpendicular to the laser beam polarization on strong absorbing materials. Numerical simulations of the intrasample optical beam propagation by means of finite-difference time-domain (FTDT) calculations revealed the scattering rough oxide layer surface along with an ~100 nm thick oxide layer as the origin of the LSFL-∥ being "imprinted" at the buried oxide/sample interface.

Klos et al. [19] created tailored surface micropatterns on titanium alloy (Ti6Al4V) that were covered by different multiscale surface morphologies (nanometric LIPSS (LSFL) or hierarchical micro/nanospikes) upon fs-laser (1030 nm, 400 fs, 10 kHz or 100 kHz) scan-processing in air. The authors observed that human mesenchymal stem cells (hMSCs) can be spatially controlled and mechanically strained by these laser-processed topographies. Additionally, the surface nanostructures affected the surface wettability and adsorption of proteins. As consequence, the focal cell adhesion was influenced and finally induced surface shape-related mechanical constraints on cells that can specifically promote osteogenic differentiation.

Mezera et al. [20] employed a two-step approach for creating hierarchical micro/nanostructures on polycarbonate polymer surfaces by combining DLIP—using a UV ns-laser (266 nm, 3 ns, 2 kHz) as the first step—with the subsequent processing of LIPSS using a UV ps-laser (350 nm, 7 to 10 ps, 100 kHz) as the second step. An experimental focus was laid on the influence of the direction of the laser beam polarization on the resulting surface structures, and was theoretically complemented by finite-difference time-domain (FDTD) calculations. Attenuated total reflection Fourier-transform infrared spectroscopy (ATR-FTIR) and micro-Raman spectroscopy (μ-RS) provided additional information on the structural and chemical polymer alteration upon laser-processing, revealing an enhanced polymer degradation for increasing accumulated fluence levels.

Kunz et al. [15] demonstrated in their Feature Paper that an additional gold layer, less than 300 nm thin, deposited on a wide band gap dielectric (fused silica) can significantly improve the regularity and reproducibility of so-called high spatial frequency LIPSS (HSFL) on the dielectric during large-area ultrashort laser scan-processing (1025 nm, 300 fs, 100 kHz) in air. Simultaneously, an increased total transmission of the silica glass samples, accompanied with a change in the surface wetting behavior, was proven. The beneficial effect of the Au-layer is attributed to a more homogeneous coupling of the electromagnetic laser radiation to the sample material, which reduces the influence of material properties, nonlinearities, and laser energy fluctuations. Moreover, the highly reflective gold film may act as a protective layer for those areas of the fluence profile that are insufficient for HSFL formation.

Takaya et al. [21] irradiated silicon suboxide (SiO_x, $x \approx 1$) with femtosecond laser pulses (800 nm, 100 fs, 10 Hz) in air. Upon multipulse irradiation at moderate fluences around 700 mJ/cm^2, the formation of periodic nanogratings with a period in the range of 200–300 nm was observed. The laser-induced formation of a thin surface layer with a high electron density, followed by nanoablation by plasmonic near-fields, were identified to lead to the observed patterns. Silicon suboxide is a material of high interest as it can be transferred to silica glass by thermal oxidation; the proposed process provides a useful approach for the rapid generation of nanostructures on glass.

Blumenstein et al. [22] presented a direct comparison of experimental and theory-based simulation results of UV laser-induced surface nanostructuring of gold. They investigated the structure formation after ultrashort pulse laser (248 nm, 1.6 ps) irradiation. The gold surface was exposed to a spatially periodic intensity profile with a sinusoidal shape and periods of 270 nm, 350 nm, and 500 nm (two-beam interference). For the simulations on an identical spatial scale, a hybrid atomistic-continuum model (molecular dynamics two-temperature model, MD-TTM) was used to model the interaction of the laser pulse with the gold target and the subsequent time evolution of the system. An excellent agreement between the modeling results and experimental data was found; especially the formation of narrow nanoridges—composed of two colliding side walls of neighboring grooves—was found in the simulation as well as in the experiment.

Sinev et al. [23] demonstrated the formation of oxidative LIPSS on ~10 nm thick metallic titanium films on BK7 glass upon scan-processing with an IR ns-fiber laser (1070 nm, 4–50 ns, 10–100 kHz). Regular LSFL with periods around 720 nm, modulation depths between 70 nm and 110 nm, and an orientation perpendicular to the linear laser beam polarization were characterized by optical and atomic force microscopy (AFM). According to the authors' complementary calculations, the laser-induced peak temperatures stayed below the melting temperature of the titanium for the chosen irradiation conditions.

Cubero et al. [24] provided the first experimental proof that the surface superconductive properties (e.g., the surface critical field strengths H_{c3}) of 25 μm thick niobium foils can be controlled through fs-laser generated LSFL featuring spatial periods between 570 nm and 775 nm, modulation depths around 200 nm, and an orientation perpendicular to the linear laser beam polarization (790 nm, 30 fs, 1 kHz, or 1030 nm, 280 fs, 1 kHz; irradiation in air, nitrogen, or argon). Moreover, a clear correlation was found between the relative orientation of the magnetic field and the periodic surface patterns.

Petrović et al. [25] studied the response of NIH 3T3 fibroblast cells on LIPSS (LSFL) processed with a Yb:KGW-laser system (1030 nm, 160 fs, 1 kHz) on a multilayer system consisting of 15 (Ti/Zr)-bilayers on a silicon substrate. The micro/nanopatterns of altered chemical/structural composition (qualified by energy dispersive X-ray spectroscopy (EDX), X-ray photoelectron spectroscopy (XPS), depth-profiling Auger electron spectroscopy (AES), scanning electron microscopy (SEM), and transmission electron microscopy (TEM)) were characterized regarding the influences on the cell morphology and cell proliferation and with respect to the possibility of cell orientation along the LSFL ridges. The authors reported that the Ti/Zr-multilayer thin films supported a high degree of cell proliferation and oriented cell growth and should be suitable for implants.

Liu et al. [26] investigated the formation of LIPSS (LSFL) on tilted metal surfaces (steel, titanium, aluminum, copper) upon line scanning irradiation with s-polarized fiber laser pulses (1030 nm or 515 nm, 420 fs, 100 kHz). The authors identified a characteristically different LIPSS orientation between the group of steel and titanium, where the LSFL patterns moderately rotated by some tens of degrees at elevated angles of incidences ($\theta = 30°$ or 45°), and the other materials Al and Cu, where the LSFL stayed always perpendicular to the laser beam polarization at all angles of incidence ($\theta = 0°, 30°, 45°$). The authors proposed that this finding is based on material-specific differences in the electron–phonon coupling factor, which should indirectly affect the propagation length of surface plasmon polaritons (SPPs) involved in the formation of the LSFL.

Soldera et al. [16] presented in their Feature Paper the two-beam DLIP structuring of soda lime glass with visible ps-laser pulses (532 nm, 12 ps or 70 ps, 10 kHz or 1 kHz). Periodic spot- and line-like surface textures were processed featuring spatial distances between 2.3 and 9.0 μm along with depth-to-diameter aspect ratios up to 0.29. Additionally, within these microstructures, LIPSS with spatial periods of ~300 nm were present—resulting in hierarchical micro/nanostructures. The line-patterned samples with periods between 2.3 and 3.9 μm diffracted the white light of a tungsten lamp with efficiencies of approximately 30% into specific diffraction modes. Simultaneously, all laser-processed samples altered their surface wetting behavior with water. In some cases, even the super-hydrophilic state with contact angles CA < 5° manifested.

In another Feature Paper, Milles et al. [17] studied the icephobic performance of hierarchical multiscale laser-textured aluminum alloy (AL2024) surfaces for aeronautic applications. Using ps- and ns-laser sources, the authors employed direct laser writing (DLW) and direct laser interference patterning (DLIP) to process different periodic surface topographies with feature sizes ranging between 2.6 and 50 μm. Some samples were additionally spray-coated using a perfluoropolyether (PFPE) monolayer to enforce super-hydrophobicity. Subsequently, the authors investigated the adhesion strength of ice on the microtextured surfaces. Moreover, the icephobic performance was tested in an icing wind tunnel, simulating real-world icing conditions. The authors demonstrated that optimized surface textures, consisting of cross-like DLIP patterns with a size of 2.6 µm, can lead to a relative reduction of the ice adhesion strength by ~90% when compared with a polished aluminum alloy surface. The authors proposed that specific types of surface microstructures, i.e., line-like and cross-like topographies, and feature sizes smaller than 5 µm are especially suitable for creating icephobic surfaces.

Nakata et al. [27] made use of complex multi laser beam interference to establish a laser-induced dot transfer technique (a variant of the laser-induced forward transfer, LIFT) that is able to generate periodic arrangements of several-hundred-nanometer-large nanodots on a receiving substrate via a solid–liquid–solid mechanism. Employing a single fs-laser pulse (785 nm, 240 fs) to a platinum donor film on a glass substrate in direct contact with an acceptor substrate in a vacuum chamber setup at a pressure <1.3 kPa, regular and periodic arrangements of almost spherical platinum sub-micrometer-sized nanodots were transferred onto a gold-coated silica glass acceptor substrate.

Kuznietsov et al. [28] manufactured metallic Ti-Fe multilayers, consisting of 15 (Ti/Fe)-bilayers on a steel substrate, with total stack thicknesses between 60 and 80 µm and studied the formation of LIPSS (LSFL) on these coatings upon irradiation with fs-laser pulses (1030 nm, 213 fs, 600 kHz) in a scan-processing approach. The LSFL-covered surfaces were characterized by SEM, XPS, and X-ray diffraction (XRD). Special attention was drawn to the depth-resolved laser-induced modification by using scanning electron microscopy on a previously prepared cross-sectional cut with a focused ion beam (FIB), and to the formation of the intermetallic compound TiFe. The resulting thermal laser-induced effects were simulated by a 1D two-temperature model (TTM) to predict the extent of the heat-affected region, as well as to predict the period of the LIPSS patterns.

Jähnig et al. [29] explored the application of two-beam DLIP for the structuring of steel surfaces containing different amounts of the surface-active element sulfur—allowing them to influence the laser-induced melt flow patterns. Employing single ns-laser pulses (1053 nm, 12 ns), this included the variation of the sulfur content in steel (30 ppm, 100 ppm, ~300 ppm, or 1500–3000 ppm) and the laser fluence. The authors demonstrated that, via thermocapillary melt flows (Marangoni melt convection), single-surface peak geometries with amplitudes up to ~50 nm can be generated for steel with sulfur content below 300 ppm, whereas split-peak topographies were formed for higher sulfur content steel or at high laser fluences. The experiments were complemented by numerical simulations on the basis of a smoothed particle hydrodynamics (SPH) model elucidating the influence of the sulfur content in steel on the melt pool convection during single-pulse ns-DLIP.

Wood et al. [30] systematically explored the influence of the post-laser-processing environment (CO_2 atmosphere, boiling water) on the water wettability of LIPSS(LSFL)-covered stainless steel (AISI 316) surfaces that were previously scan-processed by fs-laser irradiation (800 nm, <100 fs, 1 kHz) in air. It was found that exposure to a CO_2-rich environment after micromachining leads to an increased surface hydrophobicity, while residence in a boiling water bath instead leads to a hydrophilic surface. Moreover, important conclusions were drawn regarding the post-laser irradiation surface cleaning for removing nonsintered loosely attached nanoparticles and agglomerates from the surface; if not removed, these nanoparticle agglomerates become hydrophobic, creating a Cassie–Baxter air-trapping layer on the surface that may feature water contact angles up to 180°.

Nakhoul et al. [31] employed sequences of multiple cross-polarized double-fs-laser pulses (800 nm, 25 fs, 1 kHz) to generate a plethora of nanostructures at the surface of polished single-crystalline <100> nickel surfaces, including random patterns with concentrated nanoreliefs, 1D nanostripes, or 2D hexagonal arrays, forming disordered, labyrinthine, and bumpy nanopatterns. The surface morphologies were characterized by SEM and AFM. Some of the nanostructures featured diameters of only a few tens of nanometers, and heights up to 100 nm. The authors demonstrated that the surface structures can be controlled by the specific laser irradiation conditions, i.e., particularly the laser fluence and number of irradiating double-pulse sequences. Different ablation phenomena and hydrodynamic instabilities were discussed as potential mechanisms driving the formation of the surface nanostructures.

Shavdina et al. [32] studied the spot- and scan-processing of ~180 nm thick polystyrene (PS) films on silicon substrates upon UV fs-laser (266 nm, 100 fs, 50 Hz or 1 kHz) irradiation in air using large numbers of effective pulses per beam spot area ranging between 10^3 and 10^7. The regimes for optimum LIPSS formation were systematically explored and the influence of the thermal heating was tested for different substrate temperatures below the materials' softening point (T_g), revealing an increase in the LIPSS modulation depth from ~12 nm (RT) to ~20 nm (97 °C).

Dominic et al. [33] investigated the formation of subwavelength high spatial fr quency LIPSS (HSFL) on a tungsten surface upon fs-laser irradiation (800 nm, 60 fs, 1 kHz) in different oxidizing or inert environments (air, nitrogen, argon, 10^{-7} mbar high vacuum) and in combination with or without Ar-ion sputter removal of covering oxides prior to the laser irradiation. Their work confirms that surface oxidation is not a prerequisite for HSFL formation on tungsten.

Prudent et al. [34] experimentally studied the fluence- and pulse number-dependent evolution of surface morphology and the influence of interpulse feedback effects on the formation of LIPSS (LSFL and HSFL) on thin metallic glass ($Zr_{65}Cu_{35}$ alloy) films on silicon—starting with different film-specific nanosized precursor structures (referred to as coarse and tight columns) and subsequently irradiating them with multiple (up to 50) sequences of single or parallel-polarized double-fs laser pulses (800 nm, 60 fs) in the interpulse delay range up to 70 ps. The authors suggest that the high viscosity of an amorphous alloy can be bypassed through double-fs-pulse irradiation that drives the material towards higher temperatures and lower viscosities—a state that is more favorable to it undergoing capillary melt flows. As a consequence, the temporally tailored deposition of the optical energy allows promotion of the development of homogeneous and regular HSFL, while widely preserving the surface integrity.

Gutiérrez-Fernández et al. [35] used an experimental approach combining laser pulses (8 ns, 10 Hz) at different wavelengths (266 nm, 532 nm) and sequential irradiation with different polarization orientations to generate different nanostructure topographies, in particular linear gratings, two-dimensional grids, and arrays of nanodots on 130 nm thick films of semiconducting polymer poly(3-hexylthiophene) (P3HT) and 20 nm thick films of ferroelectric copolymer poly[vinylidenefluoride-co-trifluoroethylene] (P(VDF-TrFE)), both spin-coated on silicon wafers. Grazing incidence small- and wide-angle X-ray scattering (GISAXS/GIWAXS) performed at a synchrotron radiation beamline allowed them to qualify the crystallinity of the polymeric material in the irradiated regions. Sub-micrometric 2D periodic ferroelectric nanodot structures were laser-processed on a functional polymer bilayer consisting of P(VDF-TrFE) (top)/P3HT (bottom). Its ferroelectric nature was proven through piezoresponse force microscopy (PFM).

Werner and Chowdhury [36] investigated the formation of LIPSS on single-crystalline <111> and <100> silicon wafer surfaces following irradiation with multiple (8 to 10,000) mid-infrared fs-laser pulses (3.6 μm, 200 fs, 500 kHz). The irradiated surface spots were characterized by SEM, AFM, EDX, cross-sectional TEM, and electron beam diffraction (EBD). Extreme subwavelength structures with 50–100 nm sized features arranged parallel to the linear laser beam polarization direction and with a quasi-periodicity of 700 nm were reported. However, the size of the individual surface features depended strongly on the distance from the center of the irradiated spot and approached values down to ~λ/180 nm only. Moreover, TEM revealed that characteristic outgrowing nanospheroidal clusters (NSC) were quenched in an amorphous state from the laser-induced superficial melt/amorphous layer (40 nm).

Four items of this Special Issue "Laser-Generated Periodic Nanostructures" [18,31,33,35] were selected as "Editor's Choice" articles by the Editorial Office and the international board of Editors of Nanomaterials—based on readers' interests, academic editors' comments, and reviewers' suggestions. It is supposed that such articles will be particularly interesting or important, while simultaneously aiming to provide a snapshot of some of the work published in the various research areas of the journal.

Finally, the Guest Editors would like to express their sincere gratitude to all authors and reviewers of this Special Issue for their intense efforts and to the editorial staff of Nanomaterials for their professional support and guidance.

Author Contributions: Conceptualization, P.S., J.I. and J.B.; methodology, P.S., J.I. and J.B.; software, not applicable; validation, P.S., J.I. and J.B.; formal analysis, P.S., J.I. and J.B.; investigation, not applicable; resources, not applicable; data curation, not applicable; writing—original draft preparation, J.B.; writing—review and editing, P.S., J.I. and J.B.; visualization, not applicable; supervision, P.S., J.I. and J.B.; project administration, P.S., J.I. and J.B.; funding acquisition, not applicable. All authors have read and agreed to the published version of the manuscript.

Funding: This Editorial article received no external funding.

Institutional Review Board Statement: Not applicable.

Informed Consent Statement: Not applicable.

Data Availability Statement: Data sharing not applicable as no datasets were generated or analyzed in the current article.

Acknowledgments: The authors thank Erika Zhao and the editorial staff of Nanomaterials (Editorial Office) for their professional support and guidance.

Conflicts of Interest: The authors declare no conflict of interest.

References

1. Bonse, J.; Kirner, S.V.; Griepentrog, M.; Spaltmann, D.; Krüger, J. Femtosecond laser texturing of surfaces for tribological applications. *Materials* **2018**, *11*, 801. [CrossRef]
2. Grützmacher, P.G.; Profito, F.J.; Rosenkranz, A. Multi-Scale Surface Texturing in Tribology—Current Knowledge and Future Perspectives. *Lubricants* **2019**, *7*, 95. [CrossRef]
3. Rung, S.; Bokan, K.; Kleinwort, F.; Schwarz, S.; Simon, P.; Klein-Wiele, J.-H.; Esen, C.; Hellmann, R. Possibilities of Dry and Lubricated Friction Modification Enabled by Different Ultrashort Laser-Based Surface Structuring Methods. *Lubricants* **2019**, *7*, 43. [CrossRef]
4. Stratakis, E.; Bonse, J.; Heitz, J.; Siegel, J.; Tsibidis, G.D.; Skoulas, E.; Papadopoulos, A.; Mimidis, A.; Joel, A.C.; Comanns, P.; et al. Laser engineering of biomimetic surfaces. *Mater. Sci. Eng. R* **2020**, *141*, 100562. [CrossRef]
5. Heitz, J.; Plamadeala, C.; Muck, M.; Armbruster, O.; Baumgartner, W.; Weth, A.; Steinwender, C.; Blessberger, H.; Kellermair, J.; Kirner, S.V.; et al. Femtosecond laser-induced microstructures on Ti substrates for reduced cell adhesion. *Appl. Phys. A* **2017**, *123*, 734. [CrossRef]
6. Böker, K.O.; Kleinwort, F.; Klein-Wiele, J.-H.; Simon, P.; Jäckle, K.; Taheri, S.; Lehmann, W.; Schilling, A.F. Laser Ablated Periodic Nanostructures on Titanium and Steel Implants Influence Adhesion and Osteogenic Differentiation of Mesenchymal Stem Cells. *Materials* **2020**, *13*, 3526. [CrossRef] [PubMed]
7. Liu, H.; Lin, W.; Hong, M. Surface coloring by laser irradiation of solid substrates. *APL Photonics* **2019**, *4*, 051101. [CrossRef]
8. Vorobyev, A.Y.; Guo, C. Direct femtosecond laser surface nano/microstructuring and its applications. *Laser Photonics Rev.* **2013**, *7*, 385–407. [CrossRef]
9. Bonse, J.; Höhm, S.; Kirner, S.V.; Rosenfeld, A.; Krüger, J. Laser-induced periodic surface structures—A scientific evergreen. *IEEE J. Sel. Top. Quantum Electron.* **2017**, *23*, 9000615. [CrossRef]
10. Gräf, S. Formation of laser-induced periodic surface structures on different materials: Fundamentals, properties and applications. *Adv. Opt. Technol.* **2020**, *9*, 11–39. [CrossRef]
11. Bonse, J.; Gräf, S. Maxwell meets Marangoni—A review of theories on laser-induced periodic surface structures. *Laser Photonics Rev.* **2020**, *14*, 2000215. [CrossRef]
12. Nakata, Y. Interference laser processing. *Adv. Opt. Technol.* **2016**, *5*, 29–38. [CrossRef]
13. Lasagni, A.F. Laser interference patterning methods: Possibilities for high-throughput fabrication of periodic surface patterns. *Adv. Opt. Technol.* **2017**, *6*, 265–275. [CrossRef]
14. Klein-Wiele, J.-H.; Blumenstein, A.; Simon, P.; Ihlemann, J. Laser interference ablation by ultrashort UV laser pulses via diffractive beam management. *Adv. Opt. Technol.* **2020**, *9*, 41–52. [CrossRef]

15. Kunz, C.; Engel, S.; Müller, F.A.; Gräf, S. Large-Area Fabrication of Laser-Induced Periodic Surface Structures on Fused Silica Using Thin Gold Layers. *Nanomaterials* **2020**, *10*, 1187. [CrossRef] [PubMed]

16. Soldera, M.; Alamri, S.; Sürmann, P.A.; Kunze, T.; Lasagni, A.F. Microfabrication and Surface Functionalization of Soda Lime Glass through Direct Laser Interference Patterning. *Nanomaterials* **2021**, *11*, 129. [CrossRef]

17. Milles, S.; Vercillo, V.; Alamri, S.; Aguilar-Morales, A.I.; Kunze, T.; Bonaccurso, E.; Lasagni, A.F. Icephobic Performance of Multi-Scale Laser-Textured Aluminum Surfaces for Aeronautic Applications. *Nanomaterials* **2021**, *11*, 135. [CrossRef] [PubMed]

18. Florian, C.; Déziel, J.-L.; Kirner, S.V.; Siegel, J.; Bonse, J. The Role of the Laser-Induced Oxide Layer in the Formation of Laser-Induced Periodic Surface Structures. *Nanomaterials* **2020**, *10*, 147. [CrossRef]

19. Klos, A.; Sedao, X.; Itina, T.E.; Helfenstein-Didier, C.; Donnet, C.; Peyroche, S.; Vico, L.; Guignandon, A.; Dumas, V. Ultrafast Laser Processing of Nanostructured Patterns for the Control of Cell Adhesion and Migration on Titanium Alloy. *Nanomaterials* **2020**, *10*, 864. [CrossRef]

20. Mezera, M.; Alamri, S.; Hendriks, W.A.P.M.; Hertwig, A.; Elert, A.M.; Bonse, J.; Kunze, T.; Lasagni, A.F.; Römer, G.-w.R.B.E. Hierarchical Micro-/Nano-Structures on Polycarbonate via UV Pulsed Laser Processing. *Nanomaterials* **2020**, *10*, 1184. [CrossRef] [PubMed]

21. Takaya, T.; Miyaji, G.; Takahashi, I.; Richter, L.J.; Ihlemann, J. Fabrication of Periodic Nanostructures on Silicon Suboxide Films with Plasmonic Near-Field Ablation Induced by Low-Fluence Femtosecond Laser Pulses. *Nanomaterials* **2020**, *10*, 1495. [CrossRef]

22. Blumenstein, A.; Garcia, M.E.; Rethfeld, B.; Simon, P.; Ihlemann, J.; Ivanov, D.S. Formation of Periodic Nanoridge Patterns by Ultrashort Single Pulse UV Laser Irradiation of Gold. *Nanomaterials* **2020**, *10*, 1998. [CrossRef] [PubMed]

23. Sinev, D.A.; Yuzhakova, D.S.; Moskvin, M.K.; Veiko, V.P. Formation of the Submicron Oxidative LIPSS on Thin Titanium Films During Nanosecond Laser Recording. *Nanomaterials* **2020**, *10*, 2161. [CrossRef]

24. Cubero, Á.; Martínez, E.; Angurel, L.A.; de la Fuente, G.F.; Navarro, R.; Legall, H.; Krüger, J.; Bonse, J. Surface Superconductivity Changes of Niobium Sheets by Femtosecond Laser-Induced Periodic Nanostructures. *Nanomaterials* **2020**, *10*, 2525. [CrossRef]

25. Petrović, S.; Peruško, D.; Mimidis, A.; Kavatzikidou, P.; Kovač, J.; Ranella, A.; Novaković, M.; Popović, M.; Stratakis, E. Response of NIH 3T3 Fibroblast Cells on Laser-Induced Periodic Surface Structures on a 15×(Ti/Zr)/Si Multilayer System. *Nanomaterials* **2020**, *10*, 2531. [CrossRef] [PubMed]

26. Liu, Y.-H.; Kuo, K.-K.; Cheng, C.-W. Femtosecond Laser-Induced Periodic Surface Structures on Different Tilted Metal Surfaces. *Nanomaterials* **2020**, *10*, 2540. [CrossRef]

27. Nakata, Y.; Tsubakimoto, K.; Miyanaga, N.; Narazaki, A.; Shoji, T.; Tsuboi, Y. Laser-Induced Transfer of Noble Metal Nanodots with Femtosecond Laser-Interference Processing. *Nanomaterials* **2021**, *11*, 305. [CrossRef] [PubMed]

28. Kuznietsov, O.V.; Tsibidis, G.D.; Demchishin, A.V.; Demchishin, A.A.; Babizhetskyy, V.; Saldan, I.; Bellucci, S.; Gnilitskyi, I. Femtosecond Laser-Induced Periodic Surface Structures on 2D Ti-Fe Multilayer Condensates. *Nanomaterials* **2021**, *11*, 316. [CrossRef]

29. Jähnig, T.; Demuth, C.; Lasagni, A.F. Influence of Sulphur Content on Structuring Dynamics during Nanosecond Pulsed Direct Laser Interference Patterning. *Nanomaterials* **2021**, *11*, 855. [CrossRef]

30. Wood, M.J.; Servio, P.; Kietzig, A.-M. The Tuning of LIPSS Wettability during Laser Machining and through Post-Processing. *Nanomaterials* **2021**, *11*, 973. [CrossRef]

31. Nakhoul, A.; Maurice, C.; Agoyan, M.; Rudenko, A.; Garrelie, F.; Pigeon, F.; Colombier, J.-P. Self-Organization Regimes Induced by Ultrafast Laser on Surfaces in the Tens of Nanometer Scales. *Nanomaterials* **2021**, *11*, 1020. [CrossRef] [PubMed]

32. Shavdina, O.; Rabat, H.; Vayer, M.; Petit, A.; Sinturel, C.; Semmar, N. Polystyrene Thin Films Nanostructuring by UV Femtosecond Laser Beam: From One Spot to Large Surface. *Nanomaterials* **2021**, *11*, 1060. [CrossRef] [PubMed]

33. Dominic, P.; Bourquard, F.; Reynaud, S.; Weck, A.; Colombier, J.-P.; Garrelie, F. On the Insignificant Role of the Oxidation Process on Ultrafast High-Spatial-Frequency LIPSS Formation on Tungsten. *Nanomaterials* **2021**, *11*, 1069. [CrossRef]

34. Prudent, M.; Bourquard, F.; Borroto, A.; Pierson, J.-F.; Garrelie, F.; Colombier, J.-P. Initial Morphology and Feedback Effects on Laser-Induced Periodic Nanostructuring of Thin-Film Metallic Glasses. *Nanomaterials* **2021**, *11*, 1076. [CrossRef]

35. Gutiérrez-Fernández, E.; Ezquerra, T.A.; Nogales, A.; Rebollar, E. Straightforward Patterning of Functional Polymers by Sequential Nanosecond Pulsed Laser Irradiation. *Nanomaterials* **2021**, *11*, 1123. [CrossRef]

36. Werner, K.; Chowdhury, E. Extreme Sub-Wavelength Structure Formation from Mid-IR Femtosecond Laser Interaction with Silicon. *Nanomaterials* **2021**, *11*, 1192. [CrossRef]

 *nanomaterials*

Article

Large-Area Fabrication of Laser-Induced Periodic Surface Structures on Fused Silica Using Thin Gold Layers

Clemens Kunz, Sebastian Engel, Frank A. Müller and Stephan Gräf *

Otto Schott Institute of Materials Research (OSIM), Friedrich Schiller University Jena, Löbdergraben 32, 07743 Jena, Germany; clemens.kunz@uni-jena.de (C.K.); sebastian.engel@uni-jena.de (S.E.); frank.mueller@uni-jena.de (F.A.M.)

* Correspondence: stephan.graef@uni-jena.de; Tel.: +49-3641-947754

Received: 26 May 2020; Accepted: 17 June 2020; Published: 18 June 2020

Abstract: Despite intensive research activities in the field of laser-induced periodic surface structures (LIPSS), the large-area nanostructuring of glasses is still a challenging problem, which is mainly caused by the strongly non-linear absorption of the laser radiation by the dielectric material. Therefore, most investigations are limited to single-spot experiments on different types of glasses. Here, we report the homogeneous generation of LIPSS on large-area surfaces of fused silica using thin gold layers and a fs-laser with a wavelength $\lambda = 1025$ nm, a pulse duration $\tau = 300$ fs, and a repetition frequency $f_{rep} = 100$ kHz as radiation source. For this purpose, single-spot experiments are performed to study the LIPSS formation process as a function of laser parameters and gold layer thickness. Based on these results, the generation of large-area homogenous LIPSS pattern was investigated by unidirectional scanning of the fs-laser beam across the sample surface using different line spacing. The nanostructures are characterized by a spatial period of about 360 nm and a modulation depth of around 160 nm. Chemical surface analysis by Raman spectroscopy confirms a complete ablation of the gold film by the fs-laser irradiation. The characterization of the functional properties shows an increased transmission of the nanostructured samples accompanied by a noticeable change in the wetting properties, which can be additionally modified within a wide range by silanization. The presented approach enables the reproducible LIPSS-based laser direct-writing of sub-wavelength nanostructures on glasses and thus provides a versatile and flexible tool for novel applications in the fields of optics, microfluidics, and biomaterials.

Keywords: nanostructuring; femtosecond laser; laser-induced periodic surface structures; thin gold layer; transmission; wettability; silanization; functional surface properties

---

1. Introduction

The advantageous mechanical, physical, and chemical properties of technical and optical glasses, such as borosilicate glass, soda-lime-silicate glass, and fused silica, make them the material of choice for a large variety of applications in the fields of optics, microfluidics, photovoltaics, and biomaterials. The optimal performance of the utilized materials often requires a well-defined adjustment of the surface properties. These include the tribological behavior, specific wetting states, the optical response, as well as the behavior of the surface in contact with living cells. In addition to a modified surface chemistry and the usage of functional coatings, the surface properties are influenced by a well-defined adjustment of the surface topography. In this context, nature developed numerous outstanding skills and structures. These include the probably most known example of the lotus leaf with its superhydrophobic and self-cleaning properties [1] and the antireflective properties of the moth's eye [2], to name only a few examples. In order to make these functional principles also available to technical applications, processes are required that allow the generation of the appropriate structures on the surface of different types of materials. In this context, lasers emerged as a versatile and flexible tool. In particular, the generation of so-called laser-induced periodic surface structures (LIPSS) using ultra-short pulse lasers (fs-lasers) attracted increasing importance in recent years [3–5]. LIPSS result from the irradiation of the surface with linearly polarized fs-laser radiation close to the ablation threshold and they can be characterized as a modulation of the surface topography having spatial periods Λ typically smaller than the utilized fs-laser wavelength λ [6].

Available studies on LIPSS formation on glasses are mainly limited to single-spot experiments demonstrating that different types of LIPSS can be generated in dependence on the fs-laser peak fluence F and the number of pulses N [7–13]. According to Λ, they are divided into high-spatial frequency LIPSS (HSFL) and low-spatial frequency LIPSS (LSFL). While LSFL on glasses are characterized by $\Lambda \sim \lambda/n$, with n being the refractive index of the material, and an orientation parallel to the electrical field (E-field) vector, HSFL exhibit remarkably smaller periods ($\Lambda << \lambda$) and an orientation perpendicular to E [8,9]. For theoretical approaches to explain the formation of both LIPSS types, the reader is referred to the literature [3]. Briefly, the most-widely accepted theory to explain LSFL formation is based on interference of the incident laser radiation with surface electromagnetic waves that are generated by scattering at the rough surface [14]. As an alternative approach, self-organization of the irradiated material was used to explain the formation of HSFL and LSFL [7,15]. In this framework, LIPSS formation is described based on processes occurring during the relaxation of the surface from a laser-induced state of thermodynamic instability. Rudenko et al. recently performed numerical studies for fused silica suggesting that the formation of both LIPSS types is driven by the interference of the incident laser field with the scattered (non-radiative) near-field (HSFL) and the (radiative) far-field (LSFL), respectively, below the surface [16]. This requires the presence of initial inhomogeneities, electron defects, or scattering centers.

Despite the numerous investigations and promising results, only a few studies focused on a line-like or even area-like generation of LIPSS on glasses by a relative movement (scanning) of the laser beam [17–19]. Here, Papadopoulos et al. should be mentioned, who systematically investigated the fabrication of large areas on fused silica containing pillar-like laser-induced nanostructures by means of circularly polarized laser radiation [19]. The low number of studies is caused by some challenges during LIPSS formation on glasses, which are mainly related to their amorphous chemical structure and the relatively large band gap energy, when compared to metals and semiconductors. The latter results in strongly non-linear (multi-photon) absorption processes at the typical laser wavelengths. Consequently, glasses are more sensitive to changing irradiation properties, defects, and incubation effects.

Alternative approaches of large-area structuring are therefore based on dot-structures, hierarchical surface structures, and single-spot matrixes [20–23]. Furthermore, the application of metallic and semiconducting thin films, deposited on the substrate surface prior to laser structuring, was demonstrated to be advantageous for the formation of LIPSS on large bandgap materials including sapphire [24,25], glasses [26,27], and silicon [28]. Here, we report on the use of thin gold layers ($\leq$300 nm) to enable the fabrication of homogenous, large-area LIPSS pattern on fused silica substrates. The investigations focus on HSFL as the spatial periods Λ of these structures are typically in the order of 200–400 nm, which is why they are expected to be advantageous for optical applications in the visible spectral range. Based on a detailed single-spot analysis of the formation process as a function of the laser parameters and the layer thickness, the large-area fabrication of LIPSS was studied by means of an unidirectional scanning of the focused fs-laser beam across the substrate surface. The nanostructured surfaces were subsequently analyzed with regard to their topography and surface chemistry and characterized with respect to selected functional properties that might open up new applications in fields such as optics, microfluidics, and photovoltaics.

2. Materials and Methods

2.1. Sample Preparation

Samples of commercial fused silica (GVB, Herzogenrath, Germany) with a size of ($20 \times 10 \times 1 \ mm^3$) were used as substrate material (Figure 1b). Thin layers of gold with a thickness ≤300 nm were deposited on the substrate surface by sputtering (Sputter Coater S150B, Edwards, Irvine, CA, USA). The corresponding film thickness was determined using a white-light interference microscope (CCI HD, Taylor Hobson, Leicester, UK) equipped with a 50× objective. All samples were ultrasonically cleaned in acetone and isopropanol before sputtering and after fs-laser processing (Figure 1b). The wettability of the sample surfaces before and after fs-laser irradiation was modified by silanization with trichloro(1H,1H,2H,2H-perfluorooctyl)silane (Alfa Aesar, Haverhill, MA, USA). For this purpose, the samples were placed in a desiccator close to a 100 µL drop of the silane, which was deposited on the surface by gas phase condensation using a vacuum pump for 15 min and further storage of the samples for 15 min in the desiccator. After the deposition, the samples were stored for 2 h at 75 °C in a furnace.

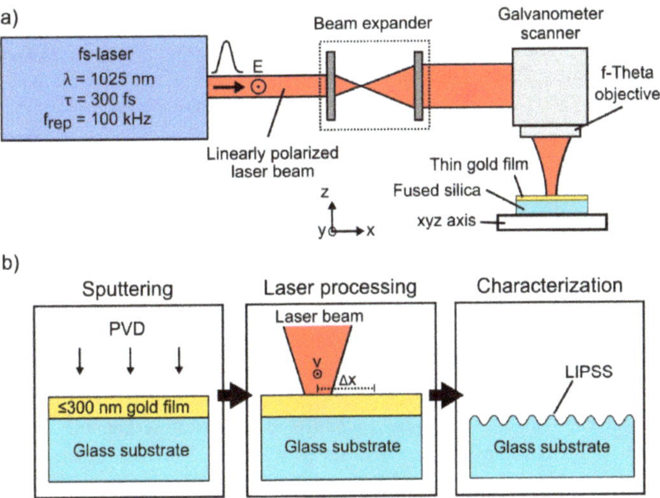

Figure 1. (**a**) Scheme of the experimental setup used for the generation of laser-induced periodic surface structures (LIPSS) on fused silica and (**b**) processing steps required for the large-area structuring with LIPSS.

2.2. Laser Processing

A diode pumped Yb:KYW thin disc fs-laser (JenLas D2.fs, Jenoptik, Jena, Germany) emitting linearly polarized fs-laser pulses (pulse duration τ = 300 fs, repetition frequency f_{rep} = 100 kHz) at a central wavelength λ = 1025 nm was used as radiation source (Figure 1a). The fs-laser beam was focused by a galvanometer scanner (IntelliScan14, Scanlab, Puchheim, Germany) including a f-Theta objective (JENar, Jenoptik, Jena, Germany) with a focal length f_L = 100 mm. The resulting focal spot diameter of the Gaussian beam ($1/e^2$ intensity) on the substrate surface was determined to be $2w_f$ ~ (22 ± 0.5) µm using the method proposed by Liu [29]. Single-spot experiments were used to evaluate the formation process of LIPSS as a function of the gold layer thickness t. For this purpose, the substrate surface was irradiated at ambient air atmosphere by N = 5 successive pulses with single-pulse energies $E_{imp} \leq 31$ µJ resulting in a maximum fs-laser peak fluence $F = 2 \cdot E_{imp}/(\pi \cdot w_f^2) = 16.3$ J/cm^2. In this context, the threshold fluence of LSFL formation (F_{LSFL}) was determined, as it indicates the transition from HSFL to LSFL and thus clearly identifies the fluence range suitable for HSFL formation. Large surface areas structured with LIPSS were realized by scanning the focused fs-laser beam ($F = 4.5$ J/cm^2) with a velocity $v = 0.15$ m/s unidirectionally across the sample surface using different values of the line spacing Δx and thin gold layers. In this context, the effective number of laser pulses N per focal spot area is given by $N_{eff} = (\pi \cdot w_f^2 \cdot f_{rep})/(v \cdot \Delta x)$.

2.3. Characterization

The morphology of the sample surfaces was evaluated by scanning electron microscopy (SEM) (Sigma VP, Zeiss, Oberkochen, Germany) at an accelerating voltage of 5 kV using the secondary electron detector. The surface topography including the modulation depth h of the LIPSS and the resulting roughness factor r was characterized by atomic force microscopy (AFM) (NanoWizard 4, JPK, Berlin, Germany) using a silicon tip (SNL-B, Bruker, Billerica, MA, USA) with a spring constant of 0.23 N/m and a resonant frequency of 23 kHz in contact mode. The spatial periods of the LIPSS were quantified by 2D-Fast-Fourier transform (2D-FFT) analyses of the SEM micrographs and verified by AFM. The chemical composition of the sample surfaces before and after fs-laser irradiation was analyzed by Raman spectroscopy (Senterra, Bruker, Billerica, MA, USA). The measurements were performed at a wavelength of 785 nm with an intensity of 100 mW and a 100× objective in a wavenumber range from 200 to 1800 cm^{-1} with a resolution of 3–5 cm^{-1}. The influence of the LIPSS formation on the functional properties was evaluated in terms of the optical properties of the sample surfaces and their wettability behavior. For this purpose, the transmission of the samples was measured in a wavelength range from 400 to 1000 nm using an integrating sphere (IS236A-4, Thorlabs, Newton, NJ, USA) and a halogen–deuterium lamp (Tidas, J&M Analytik, Essingen, Germany) as radiation source. The integration time was set to 1 s and the signal was accumulated over 30 single measurements. The wettability of the surfaces with distilled water was analyzed by contact angle (CA) measurements (Drop-shape analyzer 10 Mk2, Krüss, Hamburg, Germany) at a minimum droplet volume of 4 µL in the sessile drop mode.

3. Results and Discussion

3.1. Single-Spot Investigation

In order to evaluate the impact of the thin gold layer on LIPSS formation, the layer thickness t was varied from 20 to 300 nm and compared to the LIPSS formation on the initial fused silica surface as reference. Figure 2 shows ablation spots irradiated with $N = 5$ fs-laser pulses as a function of the fs-laser peak fluence F (top to bottom) and the gold layer thickness t (left to right). Without a gold layer (Figure 2a–d), the typical dependence of fused silica on the fs-laser peak fluence can be observed in the focal spot area [8,9]. At a specific threshold, a transition from HSFL (Figure 2b) generated at low fluences ($F = 4.5$ J/cm^2) to LSFL (Figure 2c) occurs. For the initial fused silica substrate, the threshold of LSFL formation was determined to be $F_{LSFL} = 5.1$ J/cm^2 (Figure) at the given processing conditions in line with our former investigations [9]. This transition is characterized by a change in the orientation of the LIPSS from perpendicular (HSFL) to parallel (LSFL) with respect to the direction of the linear beam polarization (see Figure 2a). At the highest fluence $F = 7.9$ J/cm^2 (Figure 2d), the ablation spot is characterized by well-pronounced, homogeneous LSFL in the intense center of the Gaussian beam profile, which are surrounded by an annular region containing HSFL. These experimental observations correlate very well with the numerical studies on LIPSS formation on fused silica reported by study, the Rudenko et al. [16]. In the corresponding authors revealed the interference below the surface between the incident laser field and the scattered near-field (HSFL) and the far-field (LSFL), respectively, to be the fundamental mechanisms of formation. Both LIPSS types are therefore generated at different sub-surface regions. Consequently, LSFL require sufficient high fluences for their "exposure" by an increased ablation of the material surface.

The ablation spots fabricated using the 20 nm thick gold layer (Figure 2e–h) reveal a shift of the fluence range of LIPSS formation towards lower values of F. This is indicated by the onset of HSFL formation already at $F = 3.4$ J/cm^2. Furthermore, the comparison with the initial fused silica surface at $F = 4.5$ J/cm^2 shows a significantly increased annular region containing HSFL as well as the formation of well-pronounced LSFL in the intensive center of the Gaussian beam profile. The corresponding value of the formation threshold was determined to be $F_{LSFL} = 4.0$ J/cm^2, which is about 80% of the value of the uncoated fused silica surface (Figure 3). On the contrary, the SEM micrographs of the sample surfaces with a larger layer thickness show LIPSS formation only at the largest investigated fluence $F = 7.9$ J/cm^2. Obviously, a thin gold layer leads to a lower formation threshold F_{LSFL} compared to the uncoated substrate, which increases almost linearly with increasing layer thickness t within the investigated thickness range (Figure 3).

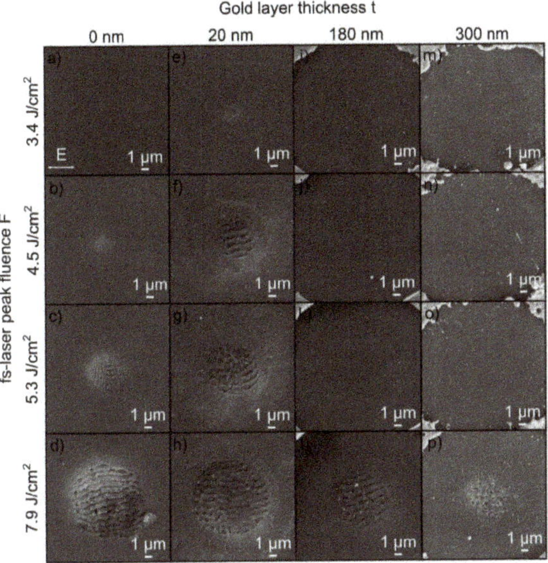

Figure 2. SEM micrographs of the fused silica surface upon the irradiation with $N = 5$ fs-laser pulses as a function of fs-laser peak fluence F and gold layer thickness t: (**a–d**) uncoated fused silica surface, (**e–h**) $t = 20$ nm, (**i–l**) $t = 180$ nm and (**m–p**) $t = 300$ nm. The direction of the linear beam polarization (*E*-field vector) is illustrated in (**a**).

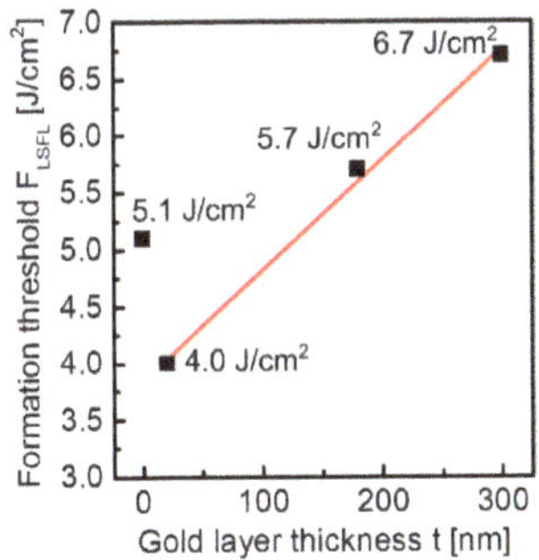

Figure 3. Threshold fluence F_{LSFL} for the formation of LSFL as a function of the gold layer thickness t determined from the corresponding SEM micrographs (see Figure 2) using the method proposed by Liu [29]. The red solid line guides the eye.

In order to explain this impact of thin metallic films on LIPSS formation, the optical properties of the gold layer and its ablation behavior have to be discussed more detailed. At $\lambda = 1025$ nm, the complex refractive index of gold is $n* = n + ik = 0.2277 + i6.4731$ [30] corresponding to a complex dielectric permittivity $\varepsilon = n*^2 = -41.85 + i2.95$. The optical penetration depth $l_\alpha = \alpha^{-1} = \lambda/(4\pi k)$ can therefore be estimated to be $l_\alpha \sim 12$ nm. According to Fresnel's formulas, the reflectance at perpendicular incidence is about 0.98. The ablation of thin gold layers has already been investigated in detail using different pulse durations with a particular focus on the influence of the layer thickness t [31–34]. In the corresponding studies, the authors reported two different regimes upon fs-laser irradiation: For layer thicknesses below a characteristic penetration depth L_c of the pulse energy into the material, the ablation threshold linearly increases with increasing t and remains constant at its bulk value for $t \geq L_c$. During the fs-laser treatment ($\tau = 28$ fs, $\lambda = 793$ nm) of gold films deposited on BK7 substrates, Krüger et al. determined L_c to be ~180 nm [32]. L_c is a measure for the heat penetration depth within the electron gas before electron–phonon relaxation occurs [32]. Therefore, it depends on the strength of electron-phonon coupling, which in the case of gold is relatively low [33]. Consequently, hot electrons excited by the deposited laser energy can penetrate deep into the material before an interaction with the lattice takes place, which is why L_c can noticeably exceed l_α. As the energy transport across the metallic–dielectric interface can occur via electron–phonon and phonon–phonon coupling [35,36], the increased interfacial electron density resulting from the diffusion of the hot electrons leads to an increased interfacial energy transfer [24,28]. In contrast to the energy coupling to the uncoated glass surface, which is solely based on strongly non-linear multiphoton processes, this leads to a stronger, more uniform deposition of the energy to the fused silica substrate and thus to the significant reduction of the LIPSS formation threshold. This reduction of F_{LSFL} is more pronounced in the case of small layer thicknesses ($t = 20$ nm) in the order of l_α. Plasmonic effects may also contribute to this process [37]. This aspect is emphasized by the fact that about 20% of the absorbed energy is still present after transmitting the 20 nm thick gold layer, which can therefore interact with the interfacial carriers. Furthermore, it should be noted that the increased coupling of the first pulse also influences subsequent processes such as accumulation and incubation occurring during multi-pulse irradiation. The observed increase in F_{LSFL} with increasing layer thickness is connected to the reported linear increase of the ablation threshold of the gold film and results from the reduced coupling and transmission of the incident laser radiation through the layer to the fused silica substrate (optical penetration depth) as the film thickness increases. Furthermore, thicker gold layers require more energy to ablate the respective layer [38]. Consequently, at larger film thickness, less laser energy is available on the fused silica surface for LIPSS formation.

Using metallic layers, the formation of LIPSS with significantly enhanced homogeneity was reported by several groups on different types of materials [24,25,28]. As the main reason, the authors assume the gold-layer-assisted homogeneous coupling field to be responsible, which reduces the influence of material properties (e.g., inhomogeneities and defects) and laser energy fluctuations. As a result, the first laser pulses generate a more homogenous modification of the material that acts as seeds for the formation of a well-pronounced LIPSS pattern by the subsequent laser pulses. This feedback mechanism includes various multi-pulse processes such as accumulation, incubation and grating-assisted coupling [3]. While a comparable impact on the LSFL was not observed within the present study, the annularly arranged, filigree HSFL fabricated with a 20 nm thin gold layer are more pronounced and more homogenous than on the uncoated substrate (Figure 2f–h).

3.2. Large-Area Fabrication of HSFL

Based on the fluence ranges evaluated in the single-spot experiments, the following investigations aim at the realization of large areas homogenously structured with HSFL. An important step is the generation of single scan lines, which will later be extended by an unidirectional scanning of the surface with a suitable line spacing Δx. Figure 4a shows an example of a single scan line containing HSFL fabricated with $v = 0.15$ m/s and $F = 4.9$ J/cm^2. In principle, the first part of the scan line reveals that the nanostructures generated within the focal spot can be written continuously on an uncoated fused silica surface. However, the SEM micrograph also emphasizes the typical problems occurring during LIPSS formation on glasses, which are caused by the strong influence of material inhomogeneities, surface contaminations, defects, and fluctuations of the pulse energy. As soon as the narrow process window is left, the structuring process switches to uncontrolled ablation, which leads to significant damage of the substrate surface. These problems are intensified when structuring large areas (Figure 4b). Here, the much smaller width of the ablated line (~8 µm) compared to the focal diameter (~22 µm) requires a certain overlap of the individual scan lines. The fs-laser fluence in the outer areas of the Gaussian profile is insufficient for the formation of HSFL, but large enough to alter the absorption properties by incubation effects, which might include the formation of self-trapped excitons and color centers [39–42]. The energy of a subsequent scan line then couples more strongly into the modified surface areas and also results in uncontrolled ablation. This results in an irregular roughness of the ablated surface instead of the intended HSFL. In order to reduce the influence of the versatile, non-controllable influencing parameters, the structuring process was systematically investigated using thin gold layers with a thickness of 20 nm.

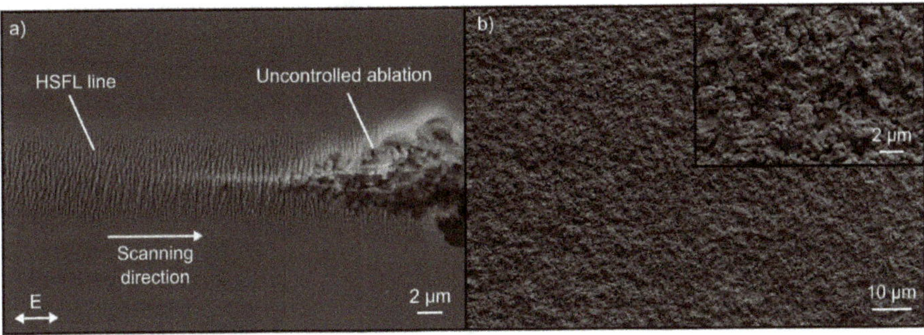

Figure 4. SEM micrographs of an uncoated fused silica surface upon fs-laser irradiation with $F = 4.9$ J/cm^2: (**a**) high-spatial frequency LIPSS (HSFL) generated by a single line scan with $v = 0.15$ m/s; (**b**) Large-area structuring by unidirectional scanning with $v = 0.15$ m/s and $\Delta x = 4$ µm. The direction of the *E*-field vector is indicated in (**a**).

For this purpose, the focused fs-laser beam was scanned with $v = 0.15$ m/s line-wise across the fused silica surface using an fs-laser peak fluence of 4.5 J/cm^2. The direction of the linear beam polarization was chosen parallel to v. Figure shows SEM micrographs of the substrate surface fabricated with line spacing Δx decreasing from 10 to 4 µm. For $\Delta x = 10$ µm, the single scan lines can clearly be distinguished. Here, a decrease of the spatial periods from the intensive center of the scan line in direction of the borders can be observed, which is related to the Gaussian intensity distribution (see also Figure 2c). Considering the ablation width of about 8 µm, $\Delta x = 4$ µm corresponds to a seamless alignment of the individual scan tracks. In this case, the overlap of the lines (regarding the Gaussian spot diameter) is about 80%. The SEM micrograph shows a homogeneous pattern with a size of $(47 \times 32$ µm$^2)$ containing HSFL with an orientation perpendicular to the *E*-field vector. From the AFM cross section measured along the white line (Figure 5c), the spatial period and the modulation depth were determined to be $\Lambda = (328 \pm 58$ nm$)$ and $h = (163 \pm 27$ nm$)$, respectively. The 2D-FFT of the SEM micrograph (Figure 5e) reveals a wider distribution of the spatial periods, as the analysis also considers deviations resulting from the fluence-dependency of the HSFL and the unidirectional scanning procedure. Consequently, HSFL with spatial periods ranging from 208 to 592 nm can be found on the fused silica surface. The most frequent period in the cross section of the 2D-FFT spectrum is about 2.75 cm^{-1}, which corresponds to a spatial period of about 362 nm. The homogeneity can be improved by using top-hat beam profiles, but this was not the subject of the current work [43,44]. Nevertheless, the results demonstrate that a thin gold layer enables the reliable and reproducible fabrication of HSFL on large surfaces areas without damaging the fused silica substrate. As already explained in the single-spot experiments, we assume in accordance with the literature that the gold layer leads to a more homogeneous coupling field, which reduces the influence of material properties and laser energy fluctuations. Furthermore, due to its very high reflectance, the gold layer may act as a protective layer for those areas of the fluence profile that are insufficient for HSFL formation.

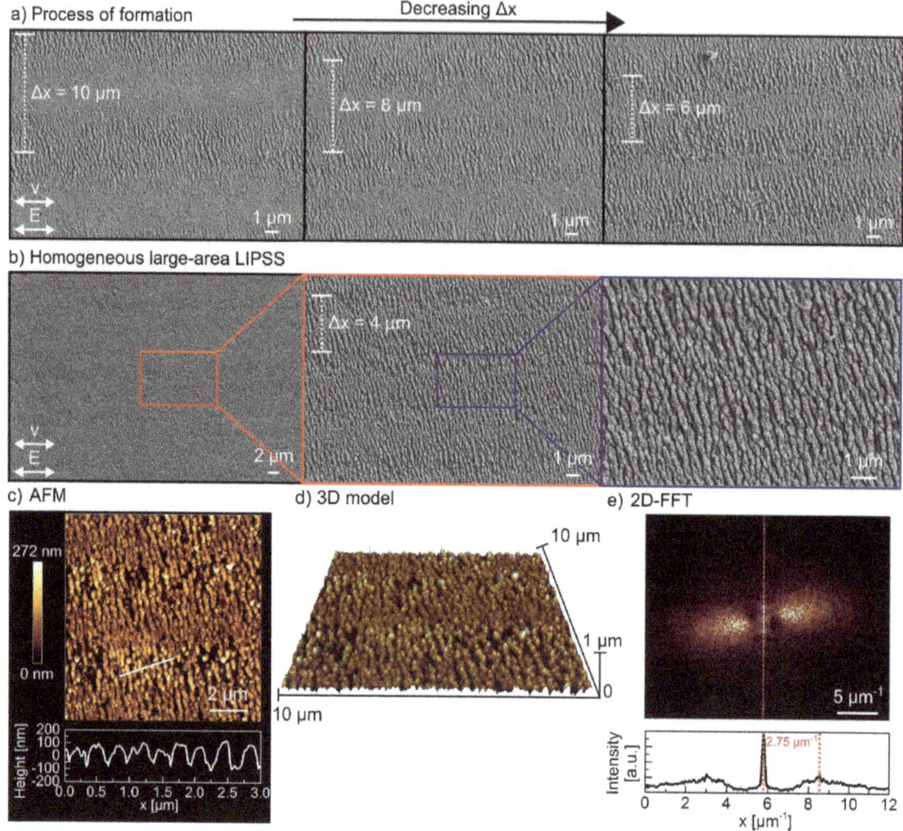

Figure 5. Large-area structuring of fused silica substrates homogeneously with HSFL using a 20 nm thin gold layer: (**a**) SEM micrographs of line scans performed with $v = 0.15$ m/s, $F = 4.5$ J/cm^2, and different line spacing Δx ranging from 10 to 6 μm; (**b**) SEM micrographs of different magnification showing a homogenous HSFL pattern fabricated with $v = 0.15$ m/s ($N = 63$), $F = 4.5$ J/cm^2, and $\Delta x = 4$ μm; (**c,d**) atomic force microscopy (AFM) image, cross section; and 3D model of the HSFL shown in (**b**); (**e**) 2-D Fourier transform of the SEM micrograph shown in (**b**); Please note the direction of the E-field vector displayed in (**a,b**).

3.3. Characterization of Surface Properties

3.3.1. Surface Chemistry

Figure 6 shows the analysis of the surface chemistry by using Raman spectroscopy. The initial fused silica surface is characterized by a main broad peak at 435 cm^{-1} (w_1), which corresponds to Si–O–Si bonds oscillating and bending in the SiO$_4$ tetrahedrons [45–47]. Peaks detected at 488 and 604 cm^{-1} can be assigned to four- (D_1) and three-membered (D_2) rings of the SiO$_4$ tetrahedrons, respectively [45,48]. At 800 cm^{-1}, the spectrum reveals a peak of the symmetric stretching mode of the Si–O–Si (w_3). Finally, peaks observed at 1078 and 1250 cm^{-1} correspond to the Si–O–Si asymmetric stretching mode (w$_4$) with transversal and longitudinal optical modes, respectively [22]. After sputtering of a 20 nm thin gold layer, the spectrum is dominated by a broad peak consisting of two maxima at 1337 and 1530 cm^{-1}. Both are related to organic residues, which might be caused by cleaning agent residues (acetone, isopropanol) as well as by carbon compounds adsorbed from the surrounding atmosphere [49]. The remarkably increased Raman intensity can be explained by the surface enhanced Raman scattering (SERS) effect reported for gold [50]. After LIPSS formation, the Raman spectrum equals to the initial fused silica surface. This confirms an almost complete ablation of the gold layer during fs-laser irradiation and excludes an impact of possible gold residues on the functional surface properties. Furthermore, a significant laser-induced modification of the chemical surface composition caused by LIPSS formation can be excluded [22,48].

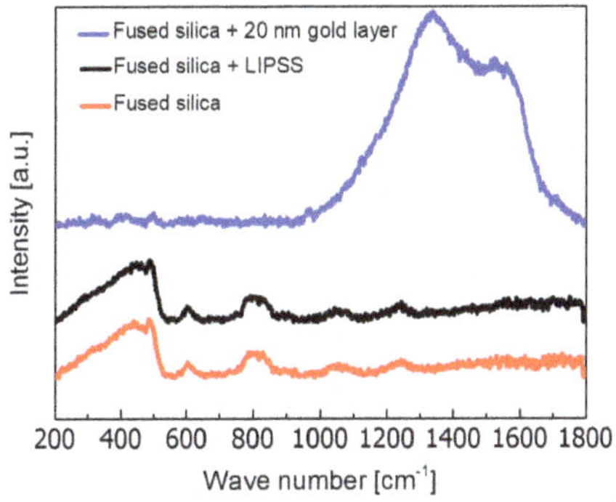

Figure 6. Raman spectrum of the initial, non-irradiated fused silica surface in comparison to fused silica surfaces covered with a thin gold layer before and after the formation of HSFL by fs-laser irradiation ($v = 0.15$ m/s, $F = 4.5$ J/cm^2, $\Delta x = 4$ μm) using a 20 nm thin gold layer. The spectra of initial fused silica and fused silica with LIPSS are magnified by a factor of 3.

3.3.2. Wettability

Figure 7 illustrates the analysis of the water contact angle θ_C of fused silica before and after fs-laser irradiation. The initial, non-irradiated fused silica surface is characterized by $\theta_C = (39.5 \pm 4.0)°$, which corresponds to a hydrophilic behavior of the flat surface, in line with literature values (Figure 7a) [21]. After fs-laser irradiation, the resulting homogeneous HSFL pattern leads to a significant reduction of θ_C to almost 0°, i.e., to a superhydrophilic wetting state. Consequently, a water droplet applied to the nanostructured surface spreads over the entire surface and forms a thin water film (top part of Figure 7b). The decreasing contact angle can be explained by the Wenzel model, which considers the impact of the surface roughness on the contact angle θ of an initially flat surface by the roughness factor r according to $\cos \theta_W = r \cdot \cos \theta$ [51]. r represents the ratio of the real surface area to its horizontal projection. Using the Wenzel model, a theoretical value of $r = 1.29$ can therefore be calculated, which is in good agreement with $r = 1.35$ determined by AFM analysis for the fused silica surface covered with HSFL in this work. The small deviation can be attributed to fact that the Wenzel model only considers topographical aspects. In addition, several studies revealed a remarkable influence of the surface chemistry on the wetting behavior [52]. By means of a silanization of the material surface, the contact angle of hydrophilic materials can additionally be altered in a well-defined manner. Using trichloro(1H,1H,2H,2H-perfluorooctyl)silane, the contact angle of the non-irradiated fused silica surface was switched to hydrophobic with $\theta = (120.0 \pm 5.0)°$, which is caused by the hydrophobic fluoro-groups of the silane. As illustrated in Figure 7b, the silanization of the surfaces structured with HSFL led to a further increase of the hydrophobicity with a contact angle of $\theta = (139.0 \pm 5.0)°$. The comparison of the silanized samples before and after fs-laser irradiation directly reveals the influence of the LIPSS and their topography and therefore confirms that a monolayer was deposited during silanization. However, it has to be noted that a superhydrophobic behavior with contact angles exceeding 150° cannot be achieved by HSFL due to the small modulation depth of the structures. According to Cassie–Baxter, such superhydrophobic surfaces would require air pockets provided by the surface topography below the water droplet [53]. Nevertheless, our experimental results demonstrate that the wettability of fused silica surfaces can be tailored in a wide range from superhydrophilic to strongly hydrophobic based on the formation of HSFL. As shown below, this is achieved without a negative impact on the transparency of the substrates.

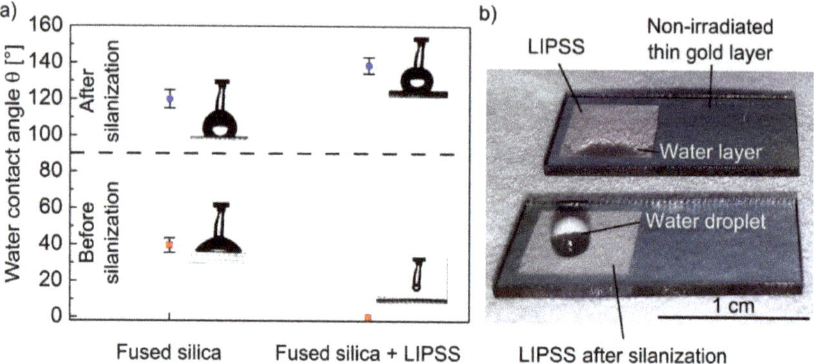

Figure 7. Contact angle analysis on fused silica with distilled water: (**a**) Wettability before and after silanization of non-irradiated substrates and surfaces homogeneously structured with HSFL. The inlets show a side-view image of the droplets during contact angle measurement; (**b**) Photographs of a droplet with a volume of 10 µL applied to the surface structured with HSFL without (top) and with (bottom) silanization (compare the corresponding inlets in (**a**)).

3.3.3. Optical Properties

The analysis of the optical properties of the fused silica substrates in terms of their transmittance shows that the initial substrate is characterized by an almost constant value of about 93% in the investigated wavelength range from 400 to 1000 nm (Figure 8a). The remaining 7% to total transmittance is mainly caused by reflection at both substrate surfaces. For light irradiating a single-interface (air/fused silica) at normal incidence, the specular reflectance can be calculated using Fresnel's formulas according to $R = ((n - 1)/(n + 1))^2$, whereby n refers to refractive index of the transmitting material. The latter varies only slightly from $n = 1.4696$ at 400 nm incident wavelength to $n = 1.4502$ at 1000 nm in the case of fused silica [54]. This results in $R \sim 3.5\%$ for a single ideal interface in the investigated spectral range. It becomes evident from the inlet in Figure 8a that the fused silica substrate structured with HSFL on its top surface is characterized by an increased transmission when compared to the non-irradiated substrate for wavelengths exceeding 550 nm. At 800 nm wavelength, a maximum transmittance of about 96% was achieved. The increase in transmittance is based on a reduced reflectivity of the structured material surface, as also indicated in Figure 8b. This can be explained by the anti-reflective effect of the periodically structured material surface, which was discussed in several studies [55,56] and which is also well-known from natural examples such as the moth's eye [2]. In addition to the specific grating profile and the depth of the generated structures, a key aspect is related to the period-to-wavelength ratio Λ/λ. For $\Lambda/\lambda < 1$, i.e., for wavelengths exceeding the spatial periods of the HSFL, the incident light is not diffracted. Instead, the incident radiation perceives effective optical properties at the material surface, which result from a structural mixing between incident and substrate material [55]. The interaction of light with such sub-λ-structures can be described with effective medium theories [55,56]. Below 550 nm, the wavelength equals to the periods measured for the HSFL ($\Lambda/\lambda \sim 1$). Here, diffraction of the incident radiation causes losses, which lead to a decrease in transmission of the substrate to ~88% at 400 nm. An improvement of the anti-reflective effect, especially by optimizing the structural geometry and structuring both substrate surfaces, will be the subject of future investigations.

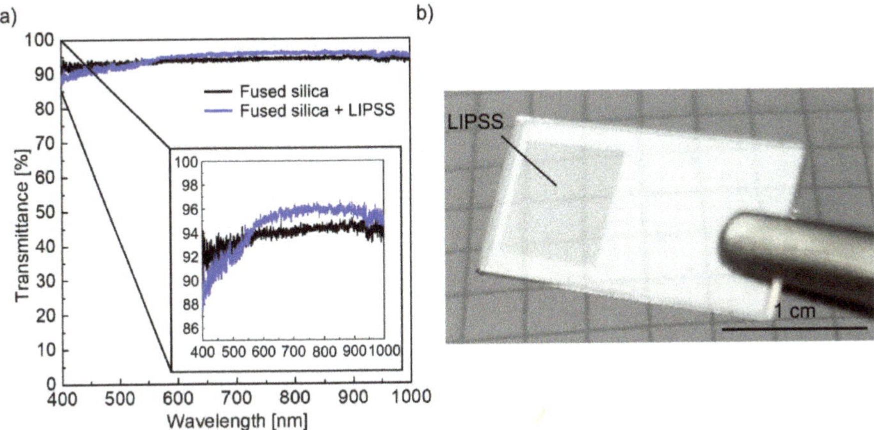

Figure 8. Optical characterization of fused silica samples before and after fs-laser irradiation: (**a**) transmittance measured using an integrating sphere and (**b**) photograph demonstrating the reduced reflectivity of the surface area homogenously structured with HSFL.

4. Conclusions

The use of thin gold layers enables the reproducible generation of homogeneous LIPSS patterns on large surface areas of fused silica. A film thickness in the order of the optical penetration depth of gold results in a remarkable reduction of the threshold fluence of LIPSS formation. The nanostructuring of the fused silica surface leads to an increased transmission of the glass samples accompanied by a change in the wetting behavior of the substrate surfaces. The present study enables the utilization of LIPSS-based nanostructuring on glasses and might therefore open up new applications in the fields of optics, photovoltaics, anti-fogging, anti-icing, and biomaterials.

Author Contributions: The authors contributed equally to the experiments, their analysis and the writing of the paper. All authors have read and agreed to the published version of the manuscript.

Funding: This research received no external funding.

Acknowledgments: The SEM facilities of the Jena Center for Soft Matter (JCSM) were established with a grant from the German Research Council (DFG). Enrico Gnecco is greatly acknowledged for providing access to the AFM facilities and Christoph Wenisch for supporting the optical characterization.

Conflicts of Interest: The authors declare no conflict of interest.

References

1. Barthlott, W.; Neinhuis, C. Purity of the sacred lotus, or escape from contamination in biological surfaces. *Planta* **1997**, *202*, 1–8. [CrossRef]
2. Bernhard, C.G. Structural and functional adaptation in a visual system. *Endeavour* **1967**, *26*, 79–84.
3. Bonse, J.; Höhm, S.; Kirner, S.V.; Rosenfeld, A.; Krüger, J. Laser-induced periodic surface structures—A scientific evergreen. *IEEE J. Select. Top. Quant. Electron.* **2017**, *23*, 9000615. [CrossRef]
4. Vorobyev, A.Y.; Guo, C. Direct femtosecond laser surface nano/microstructuring and its applications. *Laser Photon. Rev.* **2013**, *7*, 385–407. [CrossRef]
5. Liu, H.; Lin, W.; Hong, M. Surface coloring by laser irradiation of solid substrates. *APL Photonics* **2019**, *4*, 051101. [CrossRef]

6. Bonse, J.; Krüger, J.; Höhm, S.; Rosenfeld, A. Femtosecond laser-induced periodic surface structures. *J. Laser Appl.* **2012**, *24*, 042006. [CrossRef]

7. Reif, J.; Costache, F.; Henyk, M.; Pandelov, S.V. Ripples revisited: Non-classical morphology at the bottom of femtosecond laser ablation craters in transparent dielectrics. *Appl. Surf. Sci.* **2002**, *197*, 891–895. [CrossRef]

8. Höhm, S.; Rosenfeld, A.; Krüger, J.; Bonse, J. Femtosecond laser-induced periodic surface structures on silica. *J. Appl. Phys.* **2012**, *112*, 014901. [CrossRef]

9. Gräf, S.; Kunz, C.; Müller, F.A. Formation and Properties of Laser-Induced Periodic Surface Structures on Different Glasses. *Materials* **2017**, *10*, 933. [CrossRef] [PubMed]

10. Gräf, S.; Kunz, C.; Engel, S.; Derrien, T.J.Y.; Müller, F.A. Femtosecond laser-induced periodic surface structures on fused silica: The impact of the initial substrate temperature. *Materials* **2018**, *11*, 1340. [CrossRef]

11. Rosenfeld, A.; Rohloff, M.; Höhm, S.; Krüger, J.; Bonse, J. Formation of laser-induced periodic surface structures on fused silica upon multiple parallel polarized double-femtosecond-laser-pulse irradiation sequences. *Appl. Surf. Sci.* **2012**, *258*, 9233–9236. [CrossRef]

12. Schwarz, S.; Rung, S.; Hellmann, R. Generation of laser-induced periodic surface structures on transparent material-fused silica. *Appl. Phys. Lett.* **2016**, *108*, 181607. [CrossRef]

13. Seifert, G.; Kaempfe, M.; Syrowatka, F.; Harnagea, C.; Hesse, D.; Graener, H. Self-organized structure formation on the bottom of femtosecond laser ablation craters in glass. *Appl. Phys. A* **2005**, *81*, 799–803. [CrossRef]

14. Sipe, J.E.; Young, J.F.; Preston, J.S.; van Driel, H.M. Laser-induced periodic surface structure. I. Theory. *Phys. Rev. B* **1983**, *27*, 1141–1154. [CrossRef]

15. Varlamova, O.; Costache, F.; Reif, J.; Bestehorn, M. Self-organized pattern formation upon femtosecond laser ablation by circularly polarized light. *Appl. Surf. Sci.* **2006**, *252*, 4702–4706. [CrossRef]

16. Rudenko, A.; Colombier, J.P.; Höhm, S.; Rosenfeld, A.; Krüger, J.; Bonse, J.; Itina, T.E. Spontaneous periodic ordering on the surface and in the bulk of dielectrics irradiated by ultrafast laser: A shared electromagnetic origin. *Sci. Rep.* **2017**, *7*, 12306. [CrossRef]

17. Schwarz, S.; Rung, S.; Hellmann, R. One-dimensional low spatial frequency LIPSS with rotating orientation on fused silica. *Appl. Surf. Sci.* **2017**, *411*, 113–116. [CrossRef]

18. Schwarz, S.; Rung, S.; Esen, C.; Hellmann, R. Surface Plasmon Polariton Triggered Generation of 1D-Low Spatial Frequency LIPSS on Fused Silica. *Appl. Sci.* **2018**, *8*, 1624. [CrossRef]

19. Papadopoulos, A.; Skoulas, E.; Mimidis, A.; Perrakis, G.; Kenanakis, G.; Tsibidis, G.D.; Stratakis, E. Biomimetic omnidirectional antireflective glass via direct ultrafast laser nanostructuring. *Adv. Mater.* **2019**, *31*, 1901123. [CrossRef]

20. Kostal, E.; Stroj, S.; Kasemann, S.; Matylitsky, V.; Domke, M. Fabrication of biomimetic fog-collecting superhydrophilic-superhydrophobic surface micropatterns using femtosecond lasers. *Langmuir* **2018**, *34*, 2933–2941. [CrossRef]

21. Kunz, C.; Müller, F.A.; Gräf, S. Multifunctional Hierarchical Surface Structures by Femtosecond Laser Processing. *Materials* **2018**, *11*, 789. [CrossRef] [PubMed]

22. Xu, S.-Z.; Dou, H.-Q.; Sun, K.; Ye, Y.-Y.; Li, Z.; Wang, H.-J.; Liao, W.; Liu, H.; Miao, X.-X.; Yuan, X.-D.; et al. Scan speed and fluence effects in femtosecond laser induced micro/nano-structures on the surface of fused silica. *J. Non-Cryst. Solids* **2018**, *492*, 56–62. [CrossRef]

23. Lin, Y.; Han, J.; Cai, M.; Liu, W.; Luo, X.; Zhang, H.; Zhong, M. Durable and robust transparent superhydrophobic glass surfaces fabricated by a femtosecond laser with exceptional water repellency and thermostability. *J. Mater. Chem. A* **2018**, *6*, 9049–9056. [CrossRef]

24. Yin, K.; Wang, C.; Duan, J.; Guo, C.L. Femtosecond laser-induced periodic surface structural formation on sapphire with nanolayered gold coating. *Appl. Phys. A Mater. Sci. Process.* **2016**, *122*, 834. [CrossRef]

25. Wen, Q.; Wang, H.; Cheng, G.; Jiang, F.; Lu, J.; Xu, X. Improvement of ablation capacity of sapphire by gold film-assisted femtosecond laser processing. *Opt. Laser Eng.* **2020**, *128*, 106007. [CrossRef]

26. Farid, N.; Nieto, D.; O'Connor, G.M. Thin film enabling sub-250 nm nano-ripples on glass by low fluence IR picosecond laser irradiation. *Opt. Laser Technol.* **2018**, *108*, 26–31. [CrossRef]

27. Wang, L.; Cao, X.-W.; Abid, M.I.; Li, Q.-K.; Tian, W.-J.; Chen, Q.-D.; Juodkazis, S.; Sun, H.-B. Nano-ablation of silica by plasmonic surface wave at low fluence. *Opt. Lett.* **2017**, *42*, 4446–4449. [CrossRef]

28. Feng, P.; Jiang, L.; Li, X.; Rong, W.L.; Zhang, K.H.; Cao, Q. Gold-film coating assisted femtosecond laser fabrication of large-area, uniform periodic surface structures. *Appl. Opt.* **2015**, *54*, 1314–1319. [CrossRef]

29. Liu, J.M. Simple technique for measurements of pulsed Gaussian-beam spot sizes. *Opt. Lett.* **1982**, *7*, 196–198. [CrossRef]

30. Johnson, P.B.; Christy, R.W. Optical constants of the noble metals. *Phys. Rev. B* **1972**, *6*, 4370–4379. [CrossRef]

31. Matthias, E.; Reichling, M.; Siegel, J.; Kading, O.W.; Petzoldt, S.; Skurk, H.; Bizenberger, P.; Neske, E. The influence of thermal-diffusion on laser-ablation of metal-films. *Appl. Phys. A-Mater. Sci. Process.* **1994**, *58*, 129–136. [CrossRef]

32. Krüger, J.; Dufft, D.; Koter, R.; Hertwig, A. Femtosecond laser-induced damage of gold films. *Appl. Surf. Sci.* **2007**, *253*, 7815–7819. [CrossRef]

33. Güdde, J.; Hohlfeld, J.; Müller, J.G.; Matthias, E. Damage threshold dependence on electron-phonon coupling in Au and Ni films. *Appl. Surf. Sci.* **1998**, *127*, 40–45. [CrossRef]

34. Stuart, B.C.; Feit, M.D.; Herman, S.; Rubenchik, A.M.; Shore, B.W.; Perry, M.D. Optical ablation by high-power short-pulse lasers. *J. Opt. Soc. Am. B-Opt. Phys.* **1996**, *13*, 459–468. [CrossRef]

35. Lee, J.B.; Lee, S.H. Thermal boundary resistance effect on non-equilibrium energy transport in metal-dielectric thin films heated by femtosecond pulse lasers. *Mater. Trans.* **2011**, *52*, 1492–1499. [CrossRef]

36. Majumdar, A.; Reddy, P. Role of electron-phonon coupling in thermal conductance of metal-nonmetal interfaces. *Appl. Phys. Lett.* **2004**, *84*, 4768–4770. [CrossRef]

37. Derrien, T.J.Y.; Koter, R.; Krüger, J.; Höhm, S.; Rosenfeld, A.; Bonse, J. Plasmonic formation mechanism of periodic 100-nm-structures upon femtosecond laser irradiation of silicon in water. *J. Appl. Phys.* **2014**, *116*, 074902. [CrossRef]

38. Nieto, D.; Cambronero, F.; Flores-Arias, M.T.; Farid, N.; O'Connor, G.M. Aluminum thin film enhanced IR nanosecond laser-induced frontside etching of transparent materials. *Opt. Laser Eng.* **2017**, *88*, 233–242. [CrossRef]

39. Höhm, S.; Rosenfeld, A.; Krüger, J.; Bonse, J. Femtosecond diffraction dynamics of laser-induced periodic surface structures on fused silica. *Appl. Phys. Lett.* **2013**, *102*, 054102. [CrossRef]

40. Petite, G.; Guizard, S.; Martin, P.; Quéré, F. Comment on Ultrafast electron dynamics in femtosecond optical breakdown of dielectrics. *Phys. Rev. Lett.* **1999**, *83*, 5182. [CrossRef]

41. Richter, S.; Heinrich, M.; Doring, S.; Tunnermann, A.; Nolte, S.; Peschel, U. Nanogratings in fused silica: Formation, control, and applications. *J. Laser Appl.* **2012**, *24*. [CrossRef]

42. Richter, S.; Jia, F.; Heinrich, M.; Doring, S.; Peschel, U.; Tunnermann, A.; Nolte, S. The role of self-trapped excitons and defects in the formation of nanogratings in fused silica. *Opt. Lett.* **2012**, *37*, 482–484. [CrossRef] [PubMed]

43. Schwarz, S.; Rung, S.; Esen, C.; Hellmann, R. Homogeneous Low Spatial Frequency LIPSS on Dielectric Materials Generated by Beam-Shaped Femtosecond Pulsed Laser Irradiation. *J. Laser Micro Nanoeng.* **2018**, *13*, 90–94. [CrossRef]

44. Möhl, A.; Kaldun, S.; Kunz, C.; Müller, F.A.; Fuchs, U.; Gräf, S. Tailored focal beam shaping and its application in laser material processing. *J. Laser Appl.* **2020**, *31*, 042019.

45. Varkentina, N.; Dussauze, M.; Royon, A.; Ramme, M.; Petit, Y.; Canioni, L. High repetition rate femtosecond laser irradiation of fused silica studied by Raman spectroscopy. *Opt. Mater. Express* **2016**, *6*, 79–90. [CrossRef]

46. Salleo, A.; Taylor, S.T.; Martin, M.C.; Panero, W.R.; Jeanloz, R.; Sands, T.; Génin, F.Y. Laser-driven formation of a high-pressure phase in amorphous silica. *Nat. Mater.* **2003**, *2*, 796–800. [CrossRef] [PubMed]

47. Reichman, W.; Chan, J.W.; Krol, D.M. Confocal fluorescence and Raman microscopy of femtosecond laser-modified fused silica. *J. Phys. Condens. Matter* **2003**, *15*, S2447–S2456. [CrossRef]

48. De Michele, V.; Royon, M.; Marin, E.; Alessi, A.; Morana, A.; Boukenter, A.; Cannas, M.; Girard, S.; Ouerdane, Y. Near-IR- and UV-femtosecond laser waveguide inscription in silica glasses. *Opt. Mater. Express* **2019**, *9*, 4624–4633. [CrossRef]

49. Rhinow, D.; Weber, N.E.; Turchanin, A. Atmospheric pressure, temperature-induced conversion of organic monolayers into nanocrystalline graphene. *J. Phys. Chem. C* **2012**, *116*, 12295–12303. [CrossRef]

50. Sharma, B.; Frontiera, R.R.; Henry, A.I.; Ringe, E.; Van Duyne, R.P. SERS: Materials, applications, and the future. *Mater. Today* **2012**, *15*, 16–25. [CrossRef]

51. Wenzel, R.N. Resistance of solid surfaces to wetting by water. *Ind. Eng. Chem.* **1936**, *28*, 988–994. [CrossRef]

52. Kietzig, A.M.; Hatzikiriakos, S.G.; Englezos, P. Patterned superhydrophobic metallic surfaces. *Langmuir* **2009**, *25*, 4821–4827. [CrossRef] [PubMed]

53. Cassie, A.B.D.; Baxter, S. Wettability of porous surfaces. *Trans. Faraday Soc.* **1944**, *40*, 546–551. [CrossRef]

54. Malitson, I.H. Interspecimen Comparison of the Refractive Index of Fused Silica. *J. Opt. Soc. Am.* **1965**, *55*, 1205–1209. [CrossRef]
55. Raguin, D.H.; Morris, G.M. Antireflection structured surfaces for the infrared spectral region. *Appl. Opt.* **1993**, *32*, 1154–1167. [CrossRef]
56. Grann, E.B.; Moharam, M.G.; Pommet, D.A. Optimal design for antireflective tapered two-dimensional subwavelength grating structures. *J. Opt. Soc. Am. A* **1995**, *12*, 333–339. [CrossRef]

nanomaterials

MDPI

Article

Microfabrication and Surface Functionalization of Soda Lime Glass through Direct Laser Interference Patterning

Marcos Soldera [1,2,*], Sabri Alamri [3], Paul Alexander Sürmann [3], Tim Kunze [3] and Andrés Fabián Lasagni [1,3]

[1]   Institute of Manufacturing Science and Engineering, Technische Universität Dresden, George-Bähr-Str. 3c, 01069 Dresden, Germany; andres_fabian.lasagni@tu-dresden.de
[2]   PROBIEN-CONICET, Universidad Nacional del Comahue, Buenos Aires 1400, Neuquén 8300, Argentina
[3]   Fraunhofer Institute for Material and Beam Technology IWS, Winterbergstr. 28, 01277 Dresden, Germany; sabri.alamri@iws.fraunhofer.de (S.A.); paul.suermann@iws.fraunhofer.de (P.A.S.); kunze.tim@iws.fraunhofer.de (T.K.)
*   Correspondence: marcos.soldera@mailbox.tu-dresden.de

Abstract: All-purpose glasses are common in many established and emerging industries, such as microelectronics, photovoltaics, optical components, and biomedical devices due to their outstanding combination of mechanical, optical, thermal, and chemical properties. Surface functionalization through nano/micropatterning can further enhance glasses' surface properties, expanding their applicability into new fields. Although laser structuring methods have been successfully employed on many absorbing materials, the processability of transparent materials with visible laser radiation has not been intensively studied, especially for producing structures smaller than 10 µm. Here, interference-based optical setups are used to directly pattern soda lime substrates through non-lineal absorption with ps-pulsed laser radiation in the visible spectrum. Line- and dot-like patterns are fabricated with spatial periods between 2.3 and 9.0 µm and aspect ratios up to 0.29. Furthermore, laser-induced periodic surface structures (LIPSS) with a feature size of approximately 300 nm are visible within these microstructures. The textured surfaces show significantly modified properties. Namely, the treated surfaces have an increased hydrophilic behavior, even reaching a super-hydrophilic state for some cases. In addition, the micropatterns act as relief diffraction gratings, which split incident light into diffraction modes. The process parameters were optimized to produce high-quality textures with super-hydrophilic properties and diffraction efficiencies above 30%.

Keywords: glass micro-structuring; direct laser interference patterning; laser-induced periodic surface structures; multi-photon absorption; wettability; diffraction gratings

Citation: Soldera, M.; Alamri, S.; Sürmann, P.A.; Kunze, T.; Lasagni, A.F. Microfabrication and Surface Functionalization of Soda Lime Glass through Direct Laser Interference Patterning. *Nanomaterials* **2021**, *11*, 129. https://doi.org/10.3390/nano11010129

Received: 17 November 2020
Accepted: 6 January 2021
Published: 8 January 2021

Publisher's Note: MDPI stays neutral with regard to jurisdictional claims in published maps and institutional affiliations.

1. Introduction

Glass offers outstanding properties for a wide range of applications where high optical transparency, chemical stability, heat resistance, hardness, or biological compatibility are needed. Moreover, in recent years, technical glasses have become an essential material for emerging markets such as biomedical devices, micro-optics and photonics, biotechnology, and microfluidics [1–6]. Functionalization of glasses via surface modification has opened an even broader scope for these materials, showing attractive features for products with a high-added value [7–10]. For instance, surfaces with tuned hydrophilicity can enhance the filling of microchannels in microfluidics chips without the need of added pumps or valves [11], whereas large area super-hydrophobic or super-hydrophilic glasses can be used in anti-fogging and self-cleaning applications [12,13]. Super-hydrophilic surfaces immersed in water can repel oil contaminants due to the differences in the surface tensions of the liquids, paving the way for novel oil/water separation applications [14,15]. In addition, such underwater self-cleaning surfaces can also prevent biofouling, which could be attractive for the food processing sector, biomedical devices, and marine industry [16]. As another example, surface relief gratings etched in glasses have shown the capability to

23

enhance light absorption in thin film Si solar cells by diffraction and scattering, resulting in an increased power conversion efficiency up to 48% [17–19]. Moreover, the combination of super-hydrophilicity with high optical transparency and scattering on textured glass has shown the potential for photovoltaic modules with self-cleaning and increased energy harvesting abilities [20,21]. These examples demonstrate, among others, the endless possibilities for functionalized glass materials and their potential impact in many growing economical markets.

In order to functionalize the surface of these materials, special techniques able to induce chemistry-based and topography-based surface changes are necessary. For instance, plasma treatments have been used to tailor the wettability of glass substrates by depositing polymeric coatings with a controlled stoichiometry [22–24]. From the topographical point of view, chemical treatments created, in most cases, a thin coating with nano-roughness that can be controlled by adjusting the process parameters. However, these methods offer limited possibilities to fabricate deterministic surface microstructures with tunable geometrical parameters, which can be useful to induce light diffraction [25–27].

Among the numerous methods and approaches developed to fabricate deterministic microstructures on glass surfaces, those based on lithographic techniques combined with wet chemical etching or reactive ion etching are the widest spread [28–32]. Lithography is a well-established method in both lab-scale prototyping and industrial manufacturing able to pattern features down to the nanoscale with high aspect ratios and excellent texture uniformity. However, it is a multiple step process that requires a prefabricated mask and uses hazardous chemicals [33].

Laser-based structuring methods have appeared in recent decades as an attractive alternative to lithography-based techniques due to the continuously decreasing costs of the laser sources, combined with their increasing output power and high flexibility [34–36]. The laser-based micro-structuring of glass can be achieved by direct or indirect means. As an example of the latter case, the laser-induced micro-plasma ablation (LIMPA) method relies on the irradiation of a metallic absorber placed on the backside of a transparent material, e.g., fused silica. The absorbed radiation in the metal generates a plasma plume in the confined space between metal and dielectric with enough energy to ablate the materials [37–40]. In contrast, the direct laser-based processing of glasses usually requires the use of either ultraviolet laser radiation with fluences higher than 10 J/cm^2 or laser sources providing ultra-short laser pulses (USP, in the ps or fs regime), whose high-peak intensities initiate nonlinear absorption processes [41–46]. This topic has received considerable attention in recent years and the main processes can be summarized as follows.

- Multiphoton ionization: an electron in the valence band can absorb several visible or near-infrared photons and gain enough energy to cross the band gap [47].
- Tunneling photoionization: the strong electric field suppresses the Coulomb barrier and allows the electron to tunnel through. The free electrons created by nonlinear photoionization can then absorb more energy from the laser pulse by inverse bremsstrahlung. If the energy of the free carriers becomes high enough, they can also promote an electron from the valence to the conduction band by impact ionization, leading to an avalanche process [48].

One of the most distinguishing features of two-photon absorption is that the amount of absorbed power in a thin layer of the medium is proportional to the square of the light intensity (or power), while, in one-photon absorption, the ratio of absorption depends linearly with respect to intensity [49]. Therefore, considering a Gaussian beam, the absorption rate drops quadratically, moving from the center toward the periphery of the beam and any linear variation in the power would produce a quadratic variation of the absorption coefficient and, therefore, of the ablation depth [50]. As a result, the process window for the two-photon microfabrication in non-absorbing materials is very narrow, compared with linear absorption.

Besides, the interaction of USP with the material gives rise to laser-induced periodic surface structures (LIPSS), which were recently observed in transparent ceramic materials like fused silica [51], quartz [52], sapphire [53,54], soda-lime-silicate, and borosilicate glasses [55]. Although the micro-texturing based on LIPSS is a relatively simple process, features with high aspect ratios are hard to achieve and their periodicity is mainly dictated by the used radiation wavelength and its polarization state [56].

A more flexible approach for the fabrication of periodic structures is the Direct Laser Interference Patterning (DLIP) method, by which two or more coherent beams are overlapped at the material surface to generate an interference pattern. If the local intensity is high enough to initiate the ablation, a periodic pattern can be directly engraved on the material [57]. Furthermore, controlling the number of interfering beams and their overlapping angle, the radiation polarization, and deposited fluence, as well as different textures shapes, periodicities, and aspect ratios are achievable. In addition, the material properties can also be tuned [58–61]. Particularly, using two laser beams, line-like patterns can be created, whereas overlapping four beams with the same polarization at the same incidence angle as well as a dot-like array with square symmetry can be produced. Although a few pioneering works on direct structuring by interfering ultra-short laser pulses on transparent ceramics, such as amorphous SiO_2, sapphire, SiC, TiO_2, diamond, among others, have been reported [62–64], no systematic study has been done yet to optimize the laser processing parameters and link them with controlled surface properties.

In this contribution, the processability of soda lime glass substrates using DLIP with USP and visible radiation (532 nm) is addressed. In a single DLIP step, line-like and dot-like arrays with different periodicities and aspect ratios are, thus, structured. The textured glasses showed multifunctional properties, as the surface developed a super-hydrophilic characteristic combined with the ability to diffract white light into well-defined viewing angles.

2. Materials and Methods

2.1. Glass Substrates

Soda lime glass sheets (100 mm × 100 mm × 5 mm) fabricated by the float-glass method (Aachener Quarzglas-Technologie Heinrich GmbH & Co. KG, Aachen, Germany) were used in all the experiments. This glass is resistant to almost all acids, salts, and their solutions and, despite its low cost, it is characterized by excellent thermal, optical, and mechanical properties, making it a commonly used all-purpose glass material. This soda lime glass has a transmission of approximately 90% at a wavelength of 532 nm, which is the wavelength of the used laser systems (as described in Section 2.2).

2.2. Direct Laser Interference Patterning Setups

Two different DLIP systems were used to pattern either the line-like or dot-like textures. To pattern the former textures, a 12 ps pulsed laser source (EdgeWave GmbH, Würselen, Germany) with an output power of 61.4 W at a wavelength of 532 nm and at a repetition rate of 10 kHz was used. In turn, a laser source (neoLASE GmbH, Hanover, Germany) with a maximum output power of 2.7 W emitting 70 ps pulses at a wavelength of 532 nm at a repetition rate of 1 kHz was employed for structuring dot-like textures. Both DLIP systems are based on the flexible DLIP approach [65] shown schematically in Figure 1. The primary beam coming from the laser source is split into four beams by a diffractive optical element (DOE), which are then parallelized by a prism. Afterward, the secondary beams are focused by a convex lens on the sample surface. Since only two beams need to be overlapped on the sample to produce the line-like patterns, a mask is introduced before the lens to block two opposite secondary beams. By adjusting the distance between the DOE and prism, the user can easily set the overlapping angle and, thus, the texture spatial period. The samples were processed in ambient conditions at a room temperature of 22 °C and humidity of 35%.

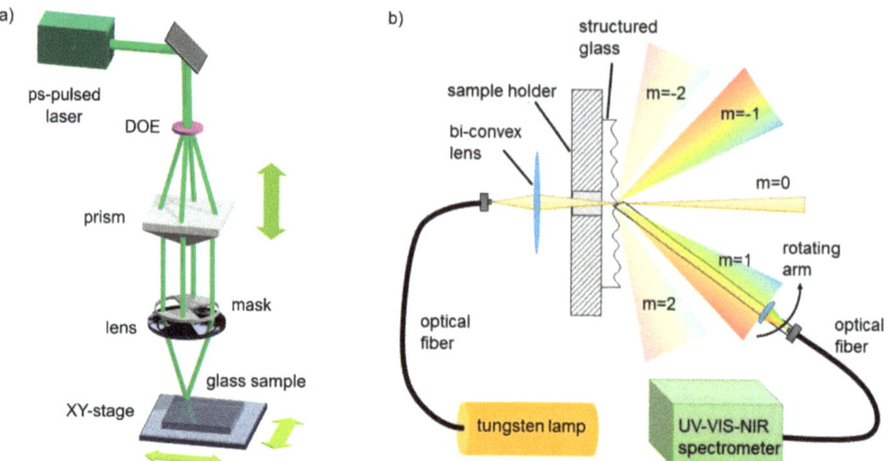

Figure 1. (**a**) Schematic setup of the four beams DLIP system used to structure glass. The mask is inserted to block two beams and letting the other two beams to interfere at the sample's surface. (**b**) Characterization setup used to measure the angular and spectral intensity of the transmitted diffracted modes by the structured glass samples.

2.3. Structuring Strategy

To pattern large areas, the followed strategies are represented in Figure 2. In the case of the two-beam setup, after firing each laser pulse, the stage where the sample is mounted moves vertically a pulse separation distance p_s, which can be tuned to deposit different amounts of an accumulated fluence dose (Figure 2a). The ratio between the spot size φ and the pulse separation distance p_s yields the effective number of applied pulses on the same area N. When a vertical line is completed, the stage moves sideways along a distance called hatch, which must be a multiple of the used spatial period to avoid destroying the periodic texture.

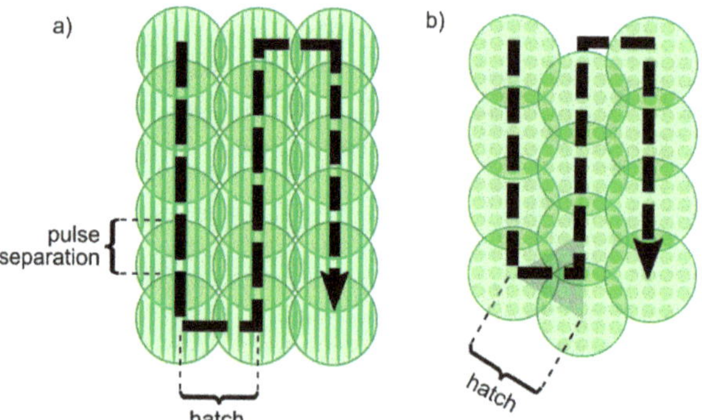

Figure 2. Patterning strategies followed to pattern (**a**) line-like textures with two-beam Direct Laser Interference Patterning (DLIP) and (**b**) dot-like textures with four-beam DLIP.

In turn, Figure 2b shows the patterning strategy used for fabricating the dot-like textures using four-beam DLIP. In this case, the sample is placed at the specified position, then it is irradiated N times until it moves to the next position. A compact hexagonal array of irradiated spots was textured with the hatch distance displaced diagonally, as shown in Figure 2b. All the structured areas had a size of 5 mm × 5 mm. Table 1 lists the fixed and variable process parameters used in this work. With the used DLIP setups, the maximum patterning throughputs for the two-beam and four-beam configurations were 3.3 cm^2/min and 2.4 cm^2/min, respectively.

Table 1. Process parameters for structuring line-like and dot-like textures.

Parameter	Line-Like Textures			Dot-Like Textures	
Spatial period [μm]	2.3	3.9	9.0	2.3	4.7
Spot size [μm]	74	52	54	45	
Hatch [μm]	23.0	15.6	18.0	16.5	13.4
Fluence [J/cm^2]	2.16–2.44	1.77–2.36	4.88–5.14	3.77–3.90	2.67–3.71
Number of pulses	10–15	8–13	7–12	5–7	2–6
Repetition rate [kHz]	10			1	

2.4. Topography Characterization

The fabricated DLIP structures were characterized by confocal microscopy (LeicaS-CAN DCM 3D, Leica Camera, Wetzlar, Germany) with a 150X magnification objective, providing a lateral and vertical resolution of 140 nm and 2 nm, respectively. The recorded images were analyzed by the software Leica Map Premium 6.0 (Leica, Wetzlar, Germany) to evaluate the depth and the roughness factor. The textured surfaces were also characterized using a scanning electron microscope (ZEISS Supra 40VP, Jena, Germany) at an operating voltage of 8.0 kV, after coating the samples with a 30-nm thick gold layer.

2.5. Wettability Characterization

The static water contact angle (CA) of the laser-treated surfaces were recorded with a contact angle system (Krüss DSA 100 S, Hamburg, Germany). The system automatically controlled the deposition of 2 μL droplets of deionized water onto the samples and calculated the CA using the tangent drop profile fitting method. The ambient temperature was 22 °C during all the measurements.

2.6. Optical Characterization

To measure the intensity of the diffracted light by the structured glass samples, the setup shown schematically (top view) in Figure 1b was used. White light emitted by a tungsten lamp is directed to the sample holder by an optical fiber and focused on the sample by a bi-convex lens. Then, the transmitted light in the plane of the movement of the goniometer (angular resolution of 2°) is collected by a plano-convex lens and coupled to an optical fiber, which directs the light toward a UV-vis-NIR spectrometer (OceanOptics GmbH, HR2000+, Ostfildern, Germany). The measured intensities were normalized to the intensity corresponding to the incident light coming from the fiber.

3. Results

3.1. Direct Fabrication of Microstructures on Glass

In order to verify the possibility to induce non-linear absorption mechanisms, a 5-mm thick soda lime glass has been irradiated with green (532 nm) laser radiation, employing a pulsed laser system (12 ps) with pulse energy E_p of at least 615 μJ and a compact optical setup for generating interference with spatial periods between 2.3 μm and 9.0 μm (see Figure 1). In a preliminary experiment, the laser fluence and spatial period were varied between 1.77 and 5.14 J/cm^2 and 2.3 and 9.0 μm, respectively, to determine the influence of these parameters on the resulting topography. The selected SEM images of Figure 3 illustrate the influence of the accumulated fluence dose on the samples patterned by two-beam DLIP with spatial periods of 2.3 μm (Figure 3a–c), 3.9 μm (Figure 3d–f), and 9.0 μm (Figure 3g–i). The number of applied pulses N and fluence F are also shown on the images. In the first row of Figure 3, it can be seen that, as F increased from 2.22 to 2.44 J/cm^2 (left to right), at a constant pulse number $N = 11$, the topography changed from a uniform texture with well-defined periodic grooves to a surface with severely damaged zones. A similar topography change was observed in the second row (Figure 3d–f), where, in this case, F was increased from 2.00 to 2.36 J/cm^2, while N was held at 9. Particularly, for the lowest fluence (Figure 3d), the trenches are relatively shallow and their depth presents a modulation across the grooves direction, which may be attributed to a non-homogeneous ablation due to the Gaussian profile of the laser radiation. As the fluence increased to $F = 2.24$ J/cm^2 (Figure 3e), the trenches became deeper but still a depth modulation can be observed. In the third row (Figure 3g–i), the fluence was kept constant at $F = 4.99$ J/cm^2, and N was increased from 7 to 11. In this case, the trend is also repeated, i.e., as the accumulated fluence dose increased, the grooves' depth increased as well and surface damage became more apparent. In the extreme case shown in Figure 3i, the surface was highly damaged, making the DLIP features hard to identify. The high-magnification images shown in the insets of Figure 3g,h allow the identification of ordered nanostructures within the micro-trenches, which are aligned perpendicularly to the radiation polarization direction (indicated with the arrows in the insets) and exhibit a spatial period of approximately ~300 nm. In agreement with previous works, the observed LIPSS spatial period is close to the ratio $\lambda/n = 350$ nm [55], where $n = 1.52$ is the refractive index of soda lime at a laser wavelength $\lambda = 532$ nm [66]. Considering their alignment and lateral feature size, these nanostructures can be identified as low spatial frequency LIPSS [67,68]. Moreover, the appearance of such LIPSS on the glass samples suggests that the absorption mechanism is mainly dictated by multi-photon absorption, as observed elsewhere [54,69–71].

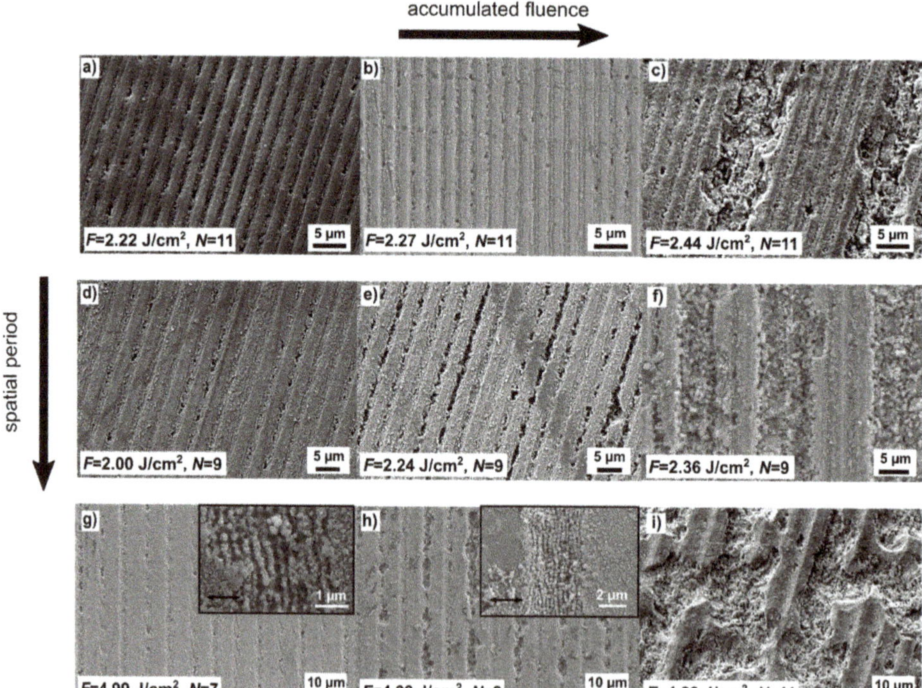

Figure 3. SEM images (secondary electron detector) of glass samples structured with line-like patterns and different spatial periods: (**a–c**) 2.3 µm, (**d–f**) 3.9 µm, and (**g–i**) 9.0 µm. The accumulated fluence dose increases from left to right. The inset of figures (**g,h**) show low spatial frequency LIPSS aligned perpendicularly to the laser polarization direction (arrows).

Although DLIP ablation of transparent materials has not been extensively investigated so far, several works have modeled and discussed the multi-photon absorption processes in transparent materials, assuming a single incident beam. Particularly, Sun et al. have modeled ablation in soda lime glass considering the ultrafast dynamics of free-electrons using a rate equation for free-electron density including multi-photon ionization, avalanche ionization, and loss terms [72]. This model predicts that using pulse durations in the picosecond regime and laser fluences in the 1.7–5 J/cm^2 range, as employed in our work, the process lies in the multi-photon ablation regime, in accordance with the results presented in this section.

Further information can be retrieved through a topography analysis by confocal microscopy. Varying the laser fluence and the pulse overlap (expressed as an equivalent number of pulses per laser spot) three contour plots (Figure 4) corresponding to samples structured with the spatial periods of 2.3 µm (a), 3.9 µm (b), and 9.0 µm (c) have been produced, summarizing the variation of the structure depth together with a qualitative interpretation of the surface quality. Namely, overlapped to the contour plots, different symbols are placed on the process variables that correspond to the actual samples, representing surfaces with either overall high homogeneity and quality (black circles) with significantly damaged areas (red diamonds) or showing an additional depth modulation due to the low intensity at the tails of the Gaussian beam (black/red squares), which are also known as a second modulation [73]. The general trend observed in the depth diagrams of Figure 4 is that, when the accumulated fluence is low, the microstructures are relatively shallow and a second depth modulation is noticed, as shown exemplarily in the profile of Figure 4a (the red line serves as a guide to the eye). As the accumulated fluence increased, the structures were deeper (i.e., up to 2.2 µm) and the topography homogeneity improved, as shown in the profile of Figure 4b. Finally, when the accumulated fluence dose was high enough, the surface became so damaged, that the DLIP texture lost its regularity (see the profile in Figure 4c). The maximum structure depths of those textures with a satisfactory texture quality were 0.53 µm, 1.12 µm, and 2.2 µm, for the patterns with spatial periods of 2.3 µm, 3.9 µm, and 9.0 µm, respectively, yielding maximum aspect ratios (depth to period ratio) of 0.23, 0.29, and 0.24, respectively. These diagrams also reveal that the process window for patterning glass with DLIP via non-linear absorption is very narrow and finding the optimum set of parameters can be challenging. As mentioned in Section 1, since the onset of non-linear absorption mechanisms strictly depends on the laser fluence dose, slight variations in pulse energy or pulse overlap may lead to large differences in the texturing results.

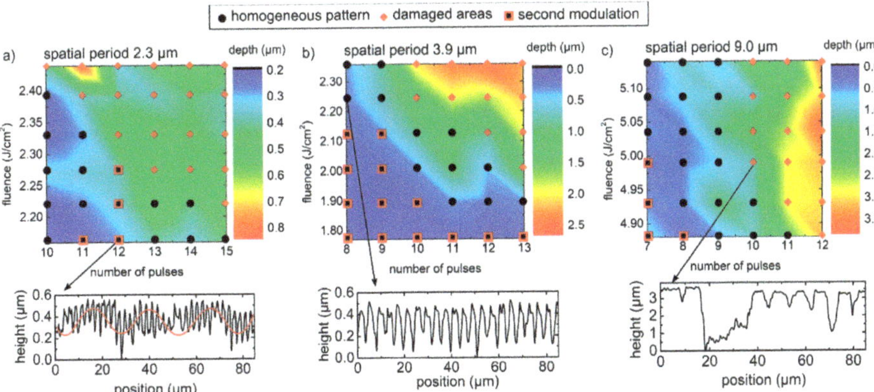

Figure 4. Depth diagrams of the line-like patterned samples with spatial periods of (**a**) 2.3 µm, (**b**) 3.9 µm, and (**c**) 9.0 µm, together with a representative extracted profile showing surfaces with (**a**) second modulation, (**b**) good overall uniformity and quality, and (**c**) significantly damaged areas.

3.2. Wettability of Textured Glass Surfaces

The wettability of the laser-treated glass surfaces was characterized by measuring the static water contact angle (CA) using deionized water (2 µL of drop volume). The CA of the flat reference was 52°, indicating a moderate hydrophilic character of the used soda lime glass substrate. After the laser texturing, all the patterned surfaces showed a strong hydrophilic behavior, and some samples reached even a super-hydrophilic state, characterized by a CA lower than 5°, according to the definition by Drelich and coworkers [74].

The inset in Figure 5 shows a box chart with the statistical distribution of the CA measured along and across the direction of the grooves (lengthwise and crosswise, respectively) of all structured samples, i.e., 108 patterned fields. The top, middle, and bottom lines of the boxes represent the 75%, 50% (or median), and 25% percentiles, respectively, whereas the filled squares indicate the mean value and the filled (empty) stars are the maximum (minimum) values. Both CAs showed similar mean values and statistical distribution, even though the standard deviation (whiskers) of the crosswise CA was 35% higher than the lengthwise measurement. A measure of the wetting anisotropy is also indicated in the chart, which was calculated as the difference between the lengthwise and crosswise CA. Analyzing this value, it can be observed that the mean value of the CA difference was only 1.3° and, therefore, it can be concluded that, despite the well-defined orientation of the DLIP grooves, the hydrophilic or super-hydrophilic states of the laser-treated samples were mainly isotropic. This behavior can be explained by taking into account the balance between capillary forces, surface tension, and gravity. For the textured glass surface, it can be assumed that the adhesion forces to the rough hydrophilic surface dominate and the droplet spread takes place in all directions regardless of the texture orientation [75–77].

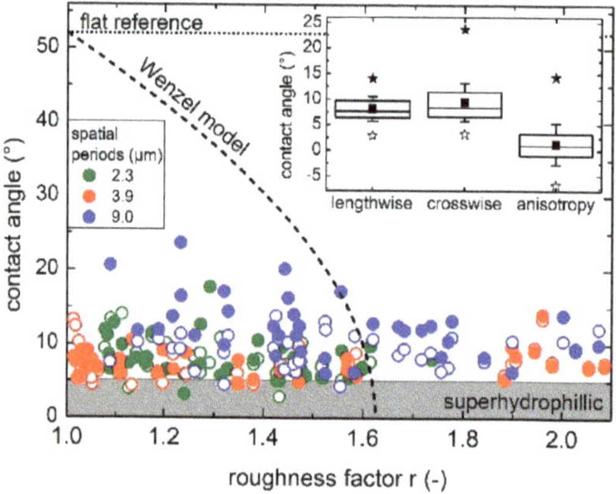

Figure 5. Water contact angle (CA) as a function of the roughness factor of all the line-like patterns. Open and filled symbols represent CA measured lengthwise and crosswise, respectively, to the direction of the grooves. The Wenzel model (dashed line) calculated with an equilibrium contact angle of 52° measured on the flat glass (dotted line) cannot explain the super-hydrophilic state reached for those samples with roughness factors below 1.6. The inset shows the statistical distribution of the CA measured lengthwise and crosswise as well as their difference (anisotropy). See text for details.

In Figure 5, the CA of all the samples are plotted as a function of the samples' roughness factor *r*, which is defined as the ratio of the real surface area to the projected surface area [78]. The open (filled) symbols stand for the CA measured lengthwise (crosswise) to the direction of the grooves. Strikingly, CAs close to 5° in both measuring directions were reached for roughness factors as low as 1.05. However, no clear trend was observed for the dependence of CA measured parallel or perpendicular to the DLIP grooves as the roughness factor increases.

The well-known Wenzel model can be used to explain the highly hydrophilic wetting state upon surface texturing [79]. According to this model, the liquid penetrates all the protrusions of the rough surface, wetting the entire surface of the solid. Then, the dependence of the contact angle θ_W as a function of the roughness factor r is given by $cos(\theta_W) = r \times cos(\theta_0)$, where θ_0 is the CA of the flat surface [80]. As a result of the model, the hydrophilic behavior of an initially hydrophilic flat surface increases when its roughness increases. Although many studies suggest that this model fails to accurately predict the contact angle in real structured surfaces, it usually succeeds in providing a qualitative estimate of the general CA behavior [81–84]. Figure 5 also shows that the Wenzel model (dashed curve) assuming $\theta_0 = 52°$ predicts a super-hydrophilic state if the roughness factor is higher than ~1.6. Nevertheless, in many structured samples, the super-hydrophilic state was reached for roughness factors well below that value, even approaching a roughness factor of 1. A possible explanation for this discrepancy may come from the fact that all the patterned samples have a hierarchical texture with several types of nano-scaled features that add to the DLIP trenches (Figure 3). These nanostructures, such as low and high spatial frequency LIPSS as well as defects arising from the redeposited material, chipping and non-uniform ablation, cannot be resolved by the used confocal microscope. Therefore, the actual roughness factor could be significantly higher (i.e., >1.6) than the measured values, correlating to a super-hydrophilic state according to Wenzel's model. Superhydrophilicity on nanostructured and micro-structured glass with low anisotropy was also reported in many works for different types of geometrical textures and aspect ratios [12,20,85,86].

On the other hand, most of the measured values do not reach contact angles lower than 5° even though the samples show high roughness values. This behavior can be ascribed to the presence of atmospheric organic contaminants on the samples' surface, which deposit on the surface through physisorption (Van der Waals bonds) [87]. In fact, a common source of such accidental contamination are hydrocarbons present in ambient air and this has been demonstrated to be present even in high-end nanofabrication clean rooms [88,89]. This contamination, which is very common in metal surfaces, has been confirmed on inert surfaces as SiO_2 by means of spectroscopic investigations [90]. As commonly known, this thin organic layer leads to a decrease of the surface energy of the substrate and increases its hydrophobicity (i.e., an increase of the CA) [87], which may prevent the contact angles reported in this work to reach a value close to zero.

3.3. Optical Properties

Using an in-house developed optical characterization setup (see Section 2 for details), the angular and spectral dependent intensity of the diffracted light in transmission mode was measured for all the line-like textures. Figure 6 shows the measured intensity (colors) for three selected samples with different spatial periods: (a) 2.3 µm ($F = 2.27$ J/cm^2, $N = 13$), (b) 3.9 µm ($F = 2.12$ J/cm^2, $N = 9$), and (c) 9.0 µm ($F = 5.04$ J/cm^2, $N = 11$). The black lines represent the well-known diffraction grating equation [91] assuming the corresponding grating period and different diffraction orders m, namely $m = 0$, ±1, ±2, ±3. The intensity scale bar, on a logarithmic scale, is expressed as the percentage of the light coming from the optical fiber. As can be seen in Figure 6a,b, there is a very good match between the predicted diffraction orders using the grating equation with the measured intensities, confirming that the produced line-like textures behave effectively as relief diffraction gratings. However, for the sample with a spatial period of 9.0 µm (Figure 6c), the diffraction orders are tightly distributed and they cannot be separated from scattering effects, especially for short wavelengths. Consequently, the diagram presents a blur vertical band around the zero order instead of well-defined diffraction modes. The spatial period of the LIPSS (~300 nm) is shorter than the wavelengths used in this setup (380–900 nm), ruling out the propagation of higher diffraction [91]. Therefore, all the diffraction modes seen in Figure 6 are associated with the periodic DLIP microstructures.

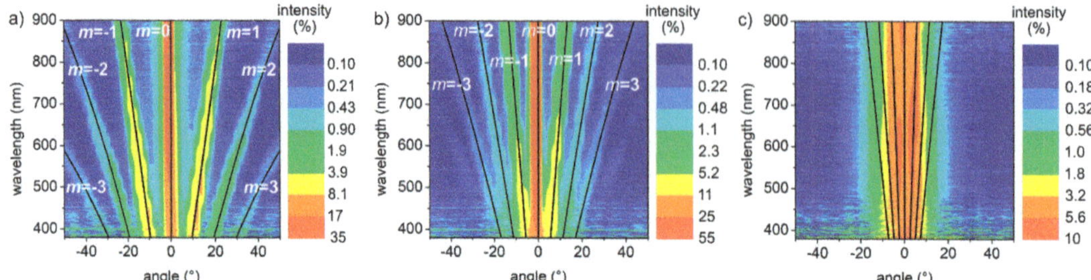

Figure 6. Spectral and angular dependence of the transmitted intensity through three selected structured samples with spatial periods of (**a**) 2.3 µm, (**b**) 3.9 µm, and (**c**) 9.0 µm. The black lines represent the diffraction grating equation for different diffraction orders m.

Next, the diffraction efficiency DE is defined as the ratio between the sum of intensities of the diffracted orders ($m = \pm1, \pm2, \pm3$) to the intensity of the incoming light from the optical fiber integrated in the vis-NIR spectral range, i.e., 380–900 nm. In Figure 7, the diffraction efficiency as a function of the number of laser pulses and fluence is shown for the line-like patterns with spatial periods of (a) 2.3 µm and (b) 3.9 µm. The same symbols displayed in the depth diagrams of Figure 4 representing the surface quality are overlapped with the diffraction efficiency data in Figure 7. It can be seen that the textures with both periods showed maximum diffraction efficiencies of approximately 30%. Since the maximum achievable DE depends strongly on the structure height and on the texture uniformity, the regions where the maximum DE in the diagrams of Figure 7 were found can be directly correlated with their topographical characteristics shown in Figure 4. For instance, the maximum DE for the textures with a spatial period of 2.3 µm were observed for low fluences and a high number of pulses that correspond to damage-free textures with relatively large structured depths of approximately ~0.5 µm (Figure 4). For a lower number of pulses, the structure depth and the diffraction efficiency became lower. As expected, in the regions with high accumulated fluence, i.e., high number of pulses with high fluence, which present a significant surface damage, the diffraction efficiency is low. A similar correlation between topography and DE was observed for the textures with a spatial period of 3.9 µm (Figure 7b), where a high diffraction efficiency region is located along the diagonal of the diagram, corresponding to the samples with a lower amount of defects. The maximum DE of 33% was found for homogeneous patterns with a structure depth of ~0.5 µm, but also DE values > 25% were found for structure depths around 1–1.3 µm. As indicated in Figures 4b and 7b, second modulation effects were observed in several samples, reducing the overall surface uniformity. This phenomenon was particularly evident for low accumulated fluences (low number of pulses with low fluence) that led to low structure depths and low DE. These differences in the dependence of the chosen process parameters with the topography characteristics explain the different locations of the regions where the DE is maximized in Figure 7a,b.

Next, the maximum DE achieved in this study of 33% by the texture with a structure depth of 0.5 µm and spatial period of 3.9 µm is compared to an ideal thin sinusoidal phase grating, according to the Fraunhofer diffraction theory [92]. For the calculations, a dispersive index of refraction was considered for the soda lime glass [66]. It was assumed that the shape of the grating was a perfect sinusoidal function with a constant peak-to-valley height of 0.5 µm and the total diffraction efficiency was calculated as the sum of the efficiency of the first three orders averaged in the wavelength range of 380–900 nm. Considering that a portion of the incoming light is reflected at the air/glass interfaces, the calculation gave a diffraction efficiency of 57.8% (see Appendix A for details). The difference between the theoretical and experimental DE might come mainly from the deviation of the experimental topography from an ideal sinusoidal function. For instance, the depth of the texture is not uniform, and the presence of LIPSS and debris promote the scattering of incoming light into random directions. However, debris and other imperfections might be controlled by further optimization of the process. LIPSS formation is an intrinsic mechanism of the interaction between USP and the material and cannot be avoided.

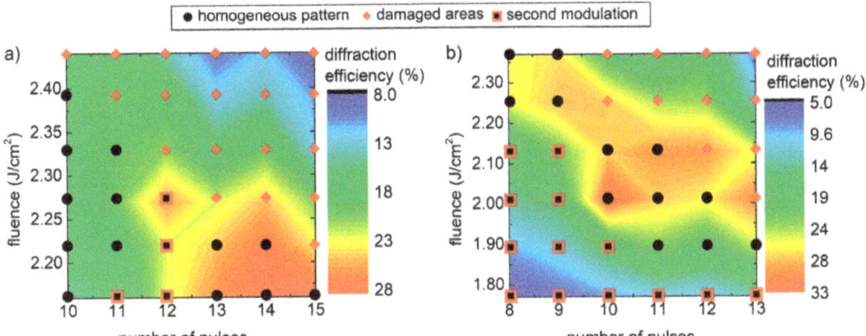

Figure 7. Diffraction efficiency determined for the line-like textures with spatial periods of (**a**) 2.3 μm and (**b**) 3.9 μm as a function of the process parameters.

In Table A1 of Appendix B, the samples with the maximum diffraction efficiencies and their process parameters are listed. In the table, the sample with a spatial period of 5.9 μm that had the deepest microstructure with satisfactory overall uniformity is also included.

3.4. Outlook on Four-Beam DLIP Structuring

As stated above, many different texture geometries can be achieved by modifying the DLIP configuration such as by using a different number of laser beams, and changing their geometrical arrangement, polarization state, or scanning strategy. As a proof of concept, four-beam DLIP was used to generate a square array of holes with a spatial period of 4.7 μm by irradiating the glass surface with $N = 5$ pulses per spot with a pulse duration of 70 ps at a fluence of 3.90 J/cm^2, as shown in the SEM images of Figure 8. LIPSS were also visible in the holes as well as a large amount of debris inside and around the holes produced by redeposited material and chipping. Confocal microscopy characterization revealed an average structure depth of 0.65 μm, representing an aspect ratio of 0.14.

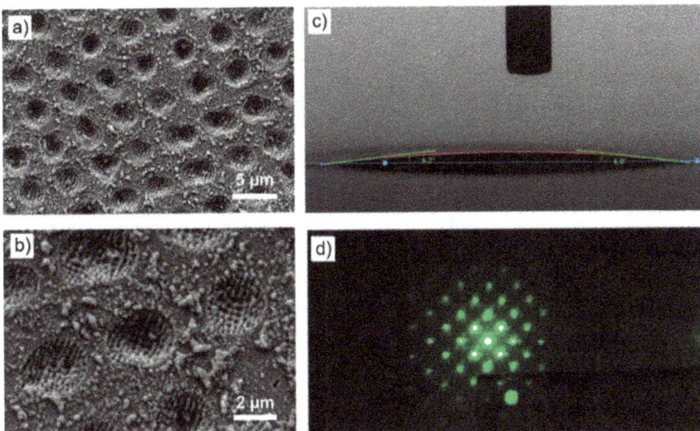

Figure 8. (**a**,**b**) SEM micrographs showing a glass surface structured by four-beam DLIP (5 applied pulses per spot at a fluence of 3.90 J/cm^2). Aside from the well-defined periodic dot-like array (spatial period of 4.7 μm), LIPSS are visible inside the craters. (**c**) Photograph of a water droplet spreading on the textured surface with a contact angle of ~6°. (**d**) Diffraction pattern on a screen upon illuminating the structured glass with a green diode laser.

The dot-like texture showed enhanced hydrophilicity, allowing water droplets to spread with contact angles of 6° (Figure 8c). Likewise, the textured glass acted as a relief diffraction grating when illuminated with a coherent light source. Figure 8d shows the two-dimensional diffraction pattern projected on a screen when the sample was illuminated with a green laser diode.

4. Conclusions

In this work, the feasibility of producing micropatterns with one and two dimensional periodicities on soda lime glass by non-linear absorption of interfering laser beams is demonstrated. The process window for obtaining homogenous line-like patterns without a significant damaged area turned to be particularly narrow and strongly dependent on the chosen spatial period. Uniform line-like textures with spatial periods of 2.3 μm, 3.9 μm, and 9.0 μm and maximum aspect ratios of 0.23, 0.29, and 0.24, respectively, were fabricated. The textures also presented low spatial frequency LIPSS with a periodicity of ~300 nm and aligned perpendicularly to the laser polarization. This effect also confirms the multi-photon absorption in the transparent soda lime glass.

The laser-treated surfaces showed enhanced functional properties. On the one hand, all the samples increased their hydrophilic characteristic, even reaching the super-hydrophilic state (CA < 5°) in some cases. On the other hand, homogenous structured samples with spatial periods of 2.3 μm and 3.9 μm diffracted white light into well-defined diffraction modes with efficiencies close to 30%. In this way, both enhanced properties can be combined in a single pattern to obtain multifunctional glass surfaces. As an example, patterning a line-like texture with a period of 2.3 μm with 14 pulses at a fluence of 2.22 J/cm^2 resulted in a super-hydrophilic surface able to diffract light with an efficiency of 25%.

It was also demonstrated the feasibility of employing four-beam DLIP to pattern periodic dot-like arrays with ultra-short laser pulses with a wavelength in the visible spectrum. Further investigations need to be done to optimize the process and to study the modified surface properties.

Although the throughputs achieved in this study (~3 cm^2/min) are relatively low to be attractive for industrial applications, the DLIP method can be scaled up to mass production, e.g., throughputs ~1 m^2/min, by using laser sources with higher power output that allows larger spot sizes, by optimizing the DLIP optics and employing faster beam guidance systems, such as galvo-scanners or polygon scanners.

Author Contributions: Conceptualization, M.S. and S.A. Methodology, M.S. and S.A. Software, S.A. Validation, M.S. and S.A. Formal analysis, M.S., S.A. and P.A.S. Investigation, M.S., S.A. and P.A.S. Resources, T.K. and A.F.L. Data curation, M.S., S.A. and P.A.S. Writing—original draft preparation, M.S. Writing—review and editing, S.A., T.K. and A.F.L. Visualization, M.S. and S.A. Supervision, A.F.L. Project administration, T.K. and A.F.L. Funding acquisition, T.K. and A.F.L. All authors have read and agreed to the published version of the manuscript.

Funding: M.S. acknowledges the support of the Alexander von Humboldt Foundation. The work of A.F.L. was supported in the framework of the Reinhart Koselleck project (323477257), which has received funding from the German Research Foundation (German: Deutsche Forschungsgemeinschaf DFG).

Institutional Review Board Statement: Not applicable.

Informed Consent Statement: Not applicable.

Data Availability Statement: The data presented in this study are available on request from the corresponding author.

Acknowledgments: The authors would like to thank S. Milles (TU Dresden) for the SEM analysis.

Conflicts of Interest: The authors declare no conflict of interest. The funders had no role in the design of the study, in the collection, analyses, or interpretation of data, in the writing of the manuscript, or in the decision to publish the results.

Appendix A. Calculation of Diffraction Efficiency of a Thin Sinusoidal Phase Grating

Following the Fraunhofer diffraction calculations presented in Reference [92], the diffraction efficiency of the diffraction mode m of an ideal thin sinusoidal phase grating can be estimated as:

$$\eta_m = J_m^2(\beta/2), \tag{A1}$$

where J is the Bessel function of the first kind and β is the maximum phase contrast between the waves emerging at the back side of the grating and is given by:

$$\beta = \frac{2\pi h}{\lambda}(n(\lambda) - 1), \tag{A2}$$

where $n(\lambda)$ is the dispersive refractive index of soda lime glass and h is the grating depth. The diffraction efficiency considering the orders $m = \pm 1, \pm 2$ and ± 3 is:

$$\eta = 2 \times \left(J_1^2 + J_2^2 + J_3^2\right), \tag{A3}$$

which is also a function of the wavelength. The reflection at the glass/air interface can be calculated by Reference [91].

$$R(\lambda) = \left(\frac{n-1}{n+1}\right)^2. \tag{A4}$$

Considering that the light interacts with the air/glass interface of the front and back sides of the grating, the transmitted intensity is reduced due to reflection by a factor $(1 - R)^2$.

Finally, the averaged diffraction efficiency across the visible-NIR spectrum is calculated as:

$$DE = \frac{1}{900\text{nm} - 380\text{nm}} \int_{380\text{nm}}^{900\text{nm}} (1 - R(\lambda))^2 \eta(\lambda) d\lambda.$$

Appendix B.

Table A1. Selected samples with their spatial periods, process parameters, structure depth, diffraction efficiency, and contact angle. The samples with the spatial periods of 2.3 µm and 3.9 µm were selected because they have shown the maximum diffraction efficiency for each period. In turn, the sample with the spatial period of 5.9 µm was selected, as it had the deepest microstructures with satisfactory homogeneity.

Spatial period [µm]	2.3	3.9	5.9
Fluence [J/cm2]	2.16	2.01	4.88
Number of pulses	14	10	11
Structure depth [µm]	0.34	0.30	2.21
Diffraction efficiency [%]	27.8	32.7	-
Contact angle [°]	7.2	5.9	10.1

References

1. Righini, G.C.; Chiappini, A. Glass Optical Waveguides: A Review of Fabrication Techniques. *Opt. Eng.* **2014**, *53*, 071819. [CrossRef]
2. Zhang, H.; Lee, Y.Y.; Leck, K.J.; Kim, N.Y.; Ying, J.Y. Recyclable Hydrophilic- Hydrophobic Micropatterns on Glass for Microarray Applications. *Langmuir* **2007**, *23*, 4728–4731. [CrossRef] [PubMed]
3. Ainslie, K.M.; Desai, T.A. Microfabricated Implants for Applications in Therapeutic Delivery, Tissue Engineering, and Biosensing. *Lab Chip* **2008**, *8*, 1864–1878. [CrossRef] [PubMed]
4. Bertin, A.; Schlaad, H. Mild and Versatile (Bio-)Functionalization of Glass Surfaces via Thiol−Ene Photochemistry. *Chem. Mater.* **2009**, *21*, 5698–5700. [CrossRef]
5. Ottevaere, H.; Cox, R.; Herzig, H.P.; Miyashita, T.; Naessens, K.; Taghizadeh, M.; Völkel, R.; Woo, H.J.; Thienpont, H. Comparing Glass and Plastic Refractive Microlenses Fabricated with Different Technologies. *J. Opt. A Pure Appl. Opt.* **2006**, *8*, S407–S429. [CrossRef]

6.   Kuna, L.; Haase, A.; Sommer, C.; Zinterl, E.; Krenn, J.R.; Wenzl, F.P.; Pachler, P.; Hartmann, P.; Tasch, S.; Leising, G. Improvement of Light Extraction from High-Power Flip-Chip Light-Emitting Diodes by Femtosecond Laser Direct Structuring of the Sapphire Backside Surface. *J. Appl. Phys.* **2008**, *104*, 074507. [CrossRef]
7.   Fuchs, A.; Kanoufi, F.; Combellas, C.; Shanahan, M.E.R. Wetting and Surface Properties of (Modified) Fluoro-Silanised Glass. *Colloid Surf. A* **2007**, *307*, 7–15. [CrossRef]
8.   Kolli, M.; Hamidouche, M.; Bouaouadja, N.; Fantozzi, G. HF Etching Effect on Sandblasted Soda-Lime Glass Properties. *J. Eur. Ceram. Soc.* **2009**, *29*, 2697–2704. [CrossRef]
9.   Wei, M.; Bowman, R.S.; Wilson, J.L.; Morrow, N.R. Wetting Properties and Stability of Silane-Treated Glass Exposed to Water, Air, and Oil. *J. Colloid Interf. Sci.* **1993**, *157*, 154–159. [CrossRef]
10.   Fang, Z.; Qiu, X.; Qiu, Y.; Kuffel, E. Dielectric Barrier Discharge in Atmospheric Air for Glass-Surface Treatment to Enhance Hydrophobicity. *IEEE Trans. Plasma Sci.* **2006**, *34*, 1216–1222. [CrossRef]
11.   Kim, S.H.; Yang, Y.; Kim, M.; Nam, S.-W.; Lee, K.-M.; Lee, N.Y.; Kim, Y.S.; Park, S. Simple Route to Hydrophilic Microfluidic Chip Fabrication Using an Ultraviolet (UV)-Cured Polymer. *Adv. Funct. Mater.* **2007**, *17*, 3493–3498. [CrossRef]
12.   Domke, M.; Sonderegger, G.; Kostal, E.; Matylitsky, V.; Stroj, S. Transparent Laser-Structured Glasses with Superhydrophilic Properties for Anti-Fogging Applications. *Appl. Phys. A* **2019**, *125*, 675. [CrossRef]
13.   Chen, Y.; Zhang, Y.; Shi, L.; Li, J.; Xin, Y.; Yang, T.; Guo, Z. Transparent Superhydrophobic/Superhydrophilic Coatings for Self-Cleaning and Anti-Fogging. *Appl. Phys. Lett.* **2012**, *101*, 033701. [CrossRef]
14.   Akhtar, N.; Thomas, P.J.; Svardal, B.; Almenningen, S.; de Jong, E.; Magnussen, S.; Onck, P.R.; Fernø, M.A.; Holst, B. Pillars or Pancakes? Self-Cleaning Surfaces without Coating. *Nano Lett.* **2018**, *18*, 7509–7514. [CrossRef] [PubMed]
15.   Xue, Z.; Wang, S.; Lin, L.; Chen, L.; Liu, M.; Feng, L.; Jiang, L. A Novel Superhydrophilic and Underwater Superoleophobic Hydrogel-Coated Mesh for Oil/Water Separation. *Adv. Mater.* **2011**, *23*, 4270–4273. [CrossRef] [PubMed]
16.   Zhang, S.; Jiang, G.; Gao, S.; Jin, H.; Zhu, Y.; Zhang, F.; Jin, J. Cupric Phosphate Nanosheets-Wrapped Inorganic Membranes with Superhydrophilic and Outstanding Anticrude Oil-Fouling Property for Oil/Water Separation. *ACS Nano* **2018**, *12*, 795–803. [CrossRef] [PubMed]
17.   Zhang, W.; Bunte, E.; Worbs, J.; Siekmann, H.; Kirchhoff, J.; Gordijn, A.; Hüpkes, J. Rough Glass by 3d Texture Transfer for Silicon Thin Film Solar Cells. *Phys. Status Solidi C* **2010**, *7*, 1120–1123. [CrossRef]
18.   Chen, T.-G.; Yu, P.; Tsai, Y.-L.; Shen, C.-H.; Shieh, J.-M.; Tsai, M.-A.; Kuo, H.-C. Nano-Patterned Glass Superstrates with Different Aspect Ratios for Enhanced Light Harvesting in a-Si: H Thin Film Solar Cells. *Opt. Express* **2012**, *20*, A412–A417. [CrossRef]
19.   Yang, G.; van Swaaij, R.A.; Tan, H.; Isabella, O.; Zeman, M. Modulated Surface Textured Glass as Substrate for High Efficiency Microcrystalline Silicon Solar Cells. *Sol. Energy Mater. Sol. Cells* **2015**, *133*, 156–162. [CrossRef]
20.   Son, J.; Kundu, S.; Verma, L.K.; Sakhuja, M.; Danner, A.J.; Bhatia, C.S.; Yang, H. A Practical Superhydrophilic Self Cleaning and Antireflective Surface for Outdoor Photovoltaic Applications. *Sol. Energy Mater. Sol. Cells* **2012**, *98*, 46–51. [CrossRef]
21.   Verma, L.K.; Sakhuja, M.; Son, J.; Danner, A.J.; Yang, H.; Zeng, H.C.; Bhatia, C.S. Self-Cleaning and Antireflective Packaging Glass for Solar Modules. *Renew. Energy* **2011**, *36*, 2489–2493. [CrossRef]
22.   Trinh, Q.H.; Hossain, M.d.M.; Kim, S.H.; Mok, Y.S. Tailoring the Wettability of Glass Using a Double-Dielectric Barrier Discharge Reactor. *Heliyon* **2018**, *4*, e00522. [CrossRef] [PubMed]
23.   Terpiłowski, K.; Rymuszka, D.; Goncharuk, O.V.; Sulym, I.Y.; Gun'ko, V.M. Wettability of Modified Silica Layers Deposited on Glass Support Activated by Plasma. *Appl. Surf. Sci.* **2015**, *353*, 843–850. [CrossRef]
24.   Kontziampasis, D.; Boulousis, G.; Smyrnakis, A.; Ellinas, K.; Tserepi, A.; Gogolides, E. Biomimetic, Antireflective, Superhydrophobic and Oleophobic PMMA and PMMA-Coated Glass Surfaces Fabricated by Plasma Processing. *Microelectron. Eng.* **2014**, *121*, 33–38. [CrossRef]
25.   Wang, S.; Deng, Y.; Yang, L.; Shi, X.; Yang, W.; Chen, Z.-G. Enhanced Antibacterial Property and Osteo-Differentiation Activity on Plasma Treated Porous Polyetheretherketone with Hierarchical Micro/Nano-Topography. *J. Biomater. Sci. Polym. Ed.* **2018**, *29*, 520–542. [CrossRef] [PubMed]
26.   Kontziampasis, D.; Trantidou, T.; Regoutz, A.; Humphrey, E.J.; Carta, D.; Terracciano, C.M.; Prodromakis, T. Effects of Ar and O2 Plasma Etching on Parylene C: Topography versus Surface Chemistry and the Impact on Cell Viability. *Plasma Process. Polym.* **2016**, *13*, 324–333. [CrossRef]
27.   Awaja, F.; Tripathi, M.; Wong, T.-T.; O'Brien, T.; Speranza, G. The Chemistry and Topography of Stabilized and Functionalized Graphene Oxide Coatings. *Plasma Process Polym.* **2018**, *15*, 1800084. [CrossRef]
28.   Hein, E.; Fox, D.; Fouckhardt, H. Glass Surface Modification by Lithography-Free Reactive Ion Etching in an Ar/CF4-Plasma for Controlled Diffuse Optical Scattering. *Surf. Coat. Tech.* **2011**, *205*, S419–S424. [CrossRef]
29.   Jain, H.; Vlcek, M. Glasses for Lithography. *J. Non Cryst. Solids* **2008**, *354*, 1401–1406. [CrossRef]
30.   Hicks, E.M.; Lyandres, O.; Hall, W.P.; Zou, S.; Glucksberg, M.R.; Van Duyne, R.P. Plasmonic Properties of Anchored Nanoparticles Fabricated by Reactive Ion Etching and Nanosphere Lithography. *J. Phys. Chem. C* **2007**, *111*, 4116–4124. [CrossRef]
31.   Hannes, B.; Vieillard, J.; Chakra, E.B.; Mazurczyk, R.; Mansfield, C.D.; Potempa, J.; Krawczyk, S.; Cabrera, M. The Etching of Glass Patterned by Microcontact Printing with Application to Microfluidics and Electrophoresis. *Sensors Actuators B Chem.* **2008**, *129*, 255–262. [CrossRef]
32.   Coltro, W.K.T.; Piccin, E.; da Silva, J.A.F.; do Lago, C.L.; Carrilho, E. A Toner-Mediated Lithographic Technology for Rapid Prototyping of Glass Microchannels. *Lab Chip* **2007**, *7*, 931–934. [CrossRef] [PubMed]

33. Garcia, R.; Knoll, A.W.; Riedo, E. Advanced Scanning Probe Lithography. *Nat. Nanotechnol.* **2014**, *9*, 577–587. [CrossRef] [PubMed]
34. Lim, M.P.; Guo, X.; Grunblatt, E.L.; Clifton, G.M.; Gonzalez, A.N.; LaFratta, C.N. Augmenting Mask-Based Lithography with Direct Laser Writing to Increase Resolution and Speed. *Opt. Express* **2018**, *26*, 7085–7090. [CrossRef] [PubMed]
35. Buividas, R.; Mikutis, M.; Juodkazis, S. Surface and Bulk Structuring of Materials by Ripples with Long and Short Laser Pulses: Recent Advances. *Prog. Quantum Electron.* **2014**, *38*, 119–156. [CrossRef]
36. Cheng, J.; Liu, C.; Shang, S.; Liu, D.; Perrie, W.; Dearden, G.; Watkins, K. A Review of Ultrafast Laser Materials Micromachining. *Opt. Laser Technol.* **2013**, *46*, 88–102. [CrossRef]
37. Lorenz, P.; Ehrhardt, M.; Zimmer, K. Laser-Induced Front Side and Back Side Etching of Fused Silica with KrF and XeF Excimer Lasers Using Metallic Absorber Layers: A Comparison. *Appl. Surf. Sci.* **2012**, *258*, 9742–9746. [CrossRef]
38. Zhang, J.; Sugioka, K.; Midorikawa, K. Laser-Induced Plasma-Assisted Ablation of Fused Quartz Using the Fourth Harmonic of a Nd+: YAG Laser. *Appl. Phys. A* **1998**, *67*, 545–549. [CrossRef]
39. Shkuratova, V.A.; Kostyuk, G.K.; Sergeev, M.M.; Zakoldaev, R.A.; Yakovlev, E.B. Speckle-Free Smoothing of Coherence Laser Beams by a Homogenizer on Uniaxial High Birefringent Crystal. *Opt. Mater. Express* **2019**, *9*, 2392–2399. [CrossRef]
40. Alamri, S.; Sürmann, P.A.; Lasagni, A.F.; Kunze, T. Interference-Based Laser-Induced Micro-Plasma Ablation of Glass. *Adv. Opt. Tech.* **2020**, *9*, 79–88. [CrossRef]
41. Niino, H.; Kawaguchi, Y.; Sato, T.; Narazaki, A.; Ding, X.; Kurosaki, R. Surface Microfabrication of Fused Silica Glass by UV Laser Irradiation. In Proceedings of the Photon Processing in Microelectronics and Photonics III, San Jose, CA, USA, 25–29 January 2004; International Society for Optics and Photonics: San Jose, CA, USA, 2004; Volume 5339, pp. 112–117.
42. Herman, P.R.; Marjoribanks, R.S.; Oettl, A.; Chen, K.; Konovalov, I.; Ness, S. Laser Shaping of Photonic Materials: Deep-Ultraviolet and Ultrafast Lasers. *Appl. Surf. Sci.* **2000**, *154–155*, 577–586. [CrossRef]
43. Schwerter, M.; Gräbner, D.; Hecht, L.; Vierheller, A.; Leester-Schädel, M.; Dietzel, A. Surface-Passive Pressure Sensor by Femtosecond Laser Glass Structuring for Flip-Chip-in-Foil Integration. *J. Microelectromech. Syst.* **2016**, *25*, 517–523. [CrossRef]
44. Ihlemann, J.; Wolff-Rottke, B. Excimer Laser Micro Machining of Inorganic Dielectrics. *Appl. Surf. Sci.* **1996**, *106*, 282–286. [CrossRef]
45. Stoian, R.; Boyle, M.; Thoss, A.; Rosenfeld, A.; Korn, G.; Hertel, I.V.; Campbell, E.E.B. Laser Ablation of Dielectrics with Temporally Shaped Femtosecond Pulses. *Appl. Phys. Lett.* **2002**, *80*, 353–355. [CrossRef]
46. Du, D.; Liu, X.; Korn, G.; Squier, J.; Mourou, G. Laser-induced Breakdown by Impact Ionization in $SiO2$ with Pulse Widths from 7 Ns to 150 Fs. *Appl. Phys. Lett.* **1994**, *64*, 3071–3073. [CrossRef]
47. Schaffer, C.B.; Brodeur, A.; Mazur, E. Laser-Induced Breakdown and Damage in Bulk Transparent Materials Induced by Tightly Focused Femtosecond Laser Pulses. *Meas. Sci. Technol.* **2001**, *12*, 1784. [CrossRef]
48. Mao, S.S.; Quéré, F.; Guizard, S.; Mao, X.; Russo, R.E.; Petite, G.; Martin, P. Dynamics of Femtosecond Laser Interactions with Dielectrics. *Appl. Phys. A* **2004**, *79*, 1695–1709. [CrossRef]
49. Jamshidi-Ghaleh, K.; Masalehdan, H. Modeling of Nonlinear Responses in BK7 Glass under Irradiation of Femtosecond Laser Pulses. *Opt. Quantum Electron.* **2009**, *41*, 47–53. [CrossRef]
50. Sun, M.; Eppelt, U.; Russ, S.; Hartmann, C.; Siebert, C.; Zhu, J.; Schulz, W. Numerical Analysis of Laser Ablation and Damage in Glass with Multiple Picosecond Laser Pulses. *Opt. Express* **2013**, *21*, 7858–7867. [CrossRef]
51. Rohloff, M.; Das, S.K.; Höhm, S.; Grunwald, R.; Rosenfeld, A.; Krüger, J.; Bonse, J. Formation of Laser-Induced Periodic Surface Structures on Fused Silica upon Multiple Parallel Polarized Double-Femtosecond-Laser-Pulse Irradiation Sequences. *Appl. Surf. Sci.* **2012**, *258*, 9233–9236. [CrossRef]
52. Ahsan, M.d.S.; Lee, M.S.; Hasan, M.K.; Noh, Y.-C.; Sohn, I.-B.; Ahmed, F.; Jun, M.B.G. Formation Mechanism of Self-Organized Nano-Ripples on Quartz Surface Using Femtosecond Laser Pulses. *Optik* **2015**, *126*, 5979–5983. [CrossRef]
53. Yin, K.; Wang, C.; Duan, J.; Guo, C. Femtosecond Laser-Induced Periodic Surface Structural Formation on Sapphire with Nanolayered Gold Coating. *Appl. Phys. A* **2016**, *122*, 834. [CrossRef]
54. Alamri, S.; Fraggelakis, F.; Kunze, T.; Krupop, B.; Mincuzzi, G.; Kling, R.; Lasagni, A.F. On the Interplay of DLIP and LIPSS Upon Ultra-Short Laser Pulse Irradiation. *Materials* **2019**, *12*, 1018. [CrossRef] [PubMed]
55. Gräf, S.; Kunz, C.; Müller, F.A. Formation and Properties of Laser-Induced Periodic Surface Structures on Different Glasses. *Materials* **2017**, *10*, 933. [CrossRef]
56. Gnilitskyi, I.; Derrien, T.J.-Y.; Levy, Y.; Bulgakova, N.M.; Mocek, T.; Orazi, L. High-Speed Manufacturing of Highly Regular Femtosecond Laser-Induced Periodic Surface Structures: Physical Origin of Regularity. *Sci. Rep.* **2017**, *7*, 1–11. [CrossRef]
57. Lasagni, A.F. Laser Interference Patterning Methods: Possibilities for High-Throughput Fabrication of Periodic Surface Patterns. *Adv. Opt. Tech.* **2017**, *6*, 265–275. [CrossRef]
58. Indrišiūnas, S.; Voisiat, B.; Gedvilas, M.; Račiukaitis, G. New Opportunities for Custom-Shape Patterning Using Polarization Control in Confocal Laser Beam Interference Setup. *J. Laser. Appl.* **2017**, *29*, 011501. [CrossRef]
59. Lasagni, A.; Mücklich, F. FEM Simulation of Periodical Local Heating Caused by Laser Interference Metallurgy. *J. Mater. Proc. Technol.* **2009**, *209*, 202–209. [CrossRef]
60. Gachot, C.; Catrin, R.; Lasagni, A.; Schmid, U.; Mücklich, F. Comparative Study of Grain Sizes and Orientation in Microstructured Au, Pt and W Thin Films Designed by Laser Interference Metallurgy. *Appl. Surf. Sci.* **2009**, *255*, 5626–5632. [CrossRef]
61. Daniel, C.; Lasagni, A.; Mücklich, F. Stress and Texture Evolution of Ni/Al Multi-Film by Laser Interference Irradiation. *Surf. Coat. Technol.* **2004**, *180–181*, 478–482. [CrossRef]

62. Kawamura, K.; Sarukura, N.; Hirano, M.; Hosono, H. Holographic Encoding of Permanent Gratings Embedded in Diamond by Two Beam Interference of a Single Femtosecond Near-Infrared Laser Pulse. *Jpn. J. Appl. Phys.* **2000**, *39*, L767. [CrossRef]
63. Kawamura, K.; Ogawa, T.; Sarukura, N.; Hirano, M.; Hosono, H. Fabrication of Surface Relief Gratings on Transparent Dielectric Materials by Two-Beam Holographic Method Using Infrared Femtosecond Laser Pulses. *Appl. Phys. B* **2000**, *71*, 119–121. [CrossRef]
64. Hosono, H.; Kawamura, K.; Matsuishi, S.; Hirano, M. Holographic Writing of Micro-Gratings and Nanostructures on Amorphous SiO2 by near Infrared Femtosecond Pulses. *Nucl. Instrum. Methods B* **2002**, *191*, 89–97. [CrossRef]
65. Lasagni, A.F.; Roch, T.; Berger, J.; Kunze, T.; Lang, V.; Beyer, E. To Use or Not to Use (Direct Laser Interference Patterning), That Is the Question. In Proceedings of the SPIE 9351, Laser-based Micro- and Nanoprocessing IX, San Francisco, CA, USA, 10–12 February 2015; p. 935115.
66. Rubin, M. Optical Properties of Soda Lime Silica Glasses. *Sol. Energy Mater.* **1985**, *12*, 275–288. [CrossRef]
67. Bonse, J.; Höhm, S.; Kirner, S.V.; Rosenfeld, A.; Krüger, J. Laser-Induced Periodic Surface Structures—A Scientific Evergreen. *IEEE J. Sel. Top. Quantum Electron.* **2017**, *23*, 9000615. [CrossRef]
68. Bonse, J.; Krüger, J.; Höhm, S.; Rosenfeld, A. Femtosecond Laser-Induced Periodic Surface Structures. *J. Laser Appl.* **2012**, *24*, 042006. [CrossRef]
69. Cheng, G.; Mishchik, K.; Mauclair, C.; Audouard, E.; Stoian, R. Ultrafast Laser Photoinscription of Polarization Sensitive Devices in Bulk Silica Glass. *Opt. Express* **2009**, *17*, 9515–9525. [CrossRef] [PubMed]
70. Mishchik, K.; Cheng, G.; Huo, G.; Burakov, I.M.; Mauclair, C.; Mermillod-Blondin, A.; Rosenfeld, A.; Ouerdane, Y.; Boukenter, A.; Parriaux, O.; et al. Nanosize Structural Modifications with Polarization Functions in Ultrafast Laser Irradiated Bulk Fused Silica. *Opt. Express* **2010**, *18*, 24809–24824. [CrossRef] [PubMed]
71. Soileau, M. Ripple Structures Associated with Ordered Surface Defects in Dielectrics. *IEEE J. Quantum Electron.* **1984**, *20*, 464–467. [CrossRef]
72. Sun, M.; Eppelt, U.; Russ, S.; Hartmann, C.; Siebert, C.; Zhu, J.; Schulz, W. Laser Ablation Mechanism of Transparent Dielectrics with Picosecond Laser Pulses. In Proceedings of the Laser-Induced Damage in Optical Materials, Boulder, CO, USA, 23–26 September 2012; International Society for Optics and Photonics: Boulder, CO, USA, 2012; Volume 8530, p. 853007.
73. Aguilar-Morales, A.I.; Alamri, S.; Kunze, T.; Lasagni, A.F. Influence of Processing Parameters on Surface Texture Homogeneity Using Direct Laser Interference Patterning. *Opt. Laser Technol.* **2018**, *107*, 216–227. [CrossRef]
74. Drelich, J.; Chibowski, E.; Meng, D.D.; Terpilowski, K. Hydrophilic and Superhydrophilic Surfaces and Materials. *Soft Matter* **2011**, *7*, 9804–9828. [CrossRef]
75. Lee, S.J.; Ha, N.; Kim, H. Superhydrophilic–Superhydrophobic Water Harvester Inspired by Wetting Property of Cactus Stem. *ACS Sustain. Chem. Eng.* **2019**, *7*, 10561–10569. [CrossRef]
76. Son, H.H.; Seo, G.H.; Jeong, U.; Kim, S.J. Capillary Wicking Effect of a Cr-Sputtered Superhydrophilic Surface on Enhancement of Pool Boiling Critical Heat Flux. *Int. J. Heat Mass Transf.* **2017**, *113*, 115–128. [CrossRef]
77. Si, Y.; Dong, Z.; Jiang, L. Bioinspired Designs of Superhydrophobic and Superhydrophilic Materials. *ACS Cent. Sci.* **2018**, *4*, 1102–1112. [CrossRef] [PubMed]
78. Ran, C.; Ding, G.; Liu, W.; Deng, Y.; Hou, W. Wetting on Nanoporous Alumina Surface: Transition between Wenzel and Cassie States Controlled by Surface Structure. *Langmuir* **2008**, *24*, 9952–9955. [CrossRef]
79. Wenzel, R.N. Resistance of Solid Surfaces to Wetting by Water. *Ind. Eng. Chem.* **1936**, *28*, 988–994. [CrossRef]
80. Wolansky, G.; Marmur, A. Apparent Contact Angles on Rough Surfaces: The Wenzel Equation Revisited. *Colloid Surf. A* **1999**, *156*, 381–388. [CrossRef]
81. Joanny, J.F.; De Gennes, P.-G. A Model for Contact Angle Hysteresis. *J. Chem. Phys.* **1984**, *81*, 552–562. [CrossRef]
82. Leroy, F.; Müller-Plathe, F. Rationalization of the Behavior of Solid- Liquid Surface Free Energy of Water in Cassie and Wenzel Wetting States on Rugged Solid Surfaces at the Nanometer Scale. *Langmuir* **2011**, *27*, 637–645. [CrossRef]
83. Marmur, A.; Bittoun, E. When Wenzel and Cassie Are Right: Reconciling Local and Global Considerations. *Langmuir* **2009**, *25*, 1277–1281. [CrossRef]
84. Erbil, H.Y.; Cansoy, C.E. Range of Applicability of the Wenzel and Cassie- Baxter Equations for Superhydrophobic Surfaces. *Langmuir* **2009**, *25*, 14135–14145. [CrossRef] [PubMed]
85. Yu, E.; Kim, S.-C.; Lee, H.J.; Oh, K.H.; Moon, M.-W. Extreme Wettability of Nanostructured Glass Fabricated by Non-Lithographic, Anisotropic Etching. *Sci. Rep.* **2015**, *5*, 9362. [CrossRef] [PubMed]
86. Kostal, E.; Stroj, S.; Kasemann, S.; Matylitsky, V.; Domke, M. Fabrication of Biomimetic Fog-Collecting Superhydrophilic– Superhydrophobic Surface Micropatterns Using Femtosecond Lasers. *Langmuir* **2018**, *34*, 2933–2941. [CrossRef] [PubMed]
87. Li, Z.; Wang, Y.; Kozbial, A.; Shenoy, G.; Zhou, F.; McGinley, R.; Ireland, P.; Morganstein, B.; Kunkel, A.; Surwade, S.P.; et al. Effect of Airborne Contaminants on the Wettability of Supported Graphene and Graphite. *Nat. Mater.* **2013**, *12*, 925–931. [CrossRef]
88. Kurokawa, A.; Odaka, K.; Azuma, Y.; Fujimoto, T.; Kojima, I. Diagnosis and Cleaning of Carbon Contamination on SiO2 Thin Film. *J. Surf. Anal.* **2009**, *15*, 337–340. [CrossRef]
89. Smith, P.J.; Lindley, P.M. Analysis of Organic Contamination in Semiconductor Processing. In Proceedings of the AIP Conference Proceedings, Bellaire, MI, USA, 23–27 June 1998; Volume 449, pp. 133–139. [CrossRef]

90. Shinozaki, A.; Arima, K.; Morita, M.; Kojima, I.; Azuma, Y. FTIR-ATR Evaluation of Organic Contaminant Cleaning Methods for SiO2 Surfaces. *Anal. Sci.* **2003**, *19*, 1557–1559. [CrossRef]
91. Hecht, E. *Optics*, 4th ed.; Addison Wesley: Essex, UK, 2001; pp. 476–479.
92. Goodman, J.W. *Introduction to Fourier Optics*, 2nd ed.; McGraw-Hill: New York, NY, USA, 1996; pp. 81–83.

Article

Icephobic Performance of Multi-Scale Laser-Textured Aluminum Surfaces for Aeronautic Applications

Stephan Milles [1,*], Vittorio Vercillo [2], Sabri Alamri [3], Alfredo I. Aguilar-Morales [3], Tim Kunze [3], Elmar Bonaccurso [2] and Andrés Fabián Lasagni [1,3]

1   Institute of Manufacturing Science and Engineering, Technische Universität Dresden, George-Bähr, Str. 3c, 01069 Dresden, Germany; andres_fabian.lasagni@tu-dresden.de

2   Airbus Central Research and Technology, Materials X, Willy-Messerschmitt, Str. 1, 82024 Taufkirchen, Germany; vittorio.vercillo@gmail.com (V.V.); elmar.bonaccurso@airbus.com (E.B.)

3   Fraunhofer Institute for Material and Beam Technology IWS, Winterbergstr. 28, 01277 Dresden, Germany; sabri.alamri@iws.fraunhofer.de (S.A.); aguilarmoralesalfredo@gmail.com (A.I.A.-M.); tim.kunze@iws.fraunhofer.de (T.K.)

*   Correspondence: stephan.milles@tu-dresden.de; Tel.: +49-351-463-40292

Abstract: Ice-building up on the leading edge of wings and other surfaces exposed to icing atmospheric conditions can negatively influence the aerodynamic performances of aircrafts. In the past, research activities focused on understanding icing phenomena and finding effective countermeasures. Efforts have been dedicated to creating coatings capable of reducing the adhesion strength of ice to a surface. Nevertheless, coatings still lack functional stability, and their application can be harmful to health and the environment. Pulsed laser surface treatments have been proven as a viable technology to induce icephobicity on metallic surfaces. However, a study aimed to find the most effective microstructures for reducing ice adhesion still needs to be carried out. This study investigates the variation of the ice adhesion strength of micro-textured aluminum surfaces treated using laser-based methods. The icephobic performance is tested in an icing wind tunnel, simulating realistic icing conditions. Finally, it is shown that optimum surface textures lead to a reduction of the ice adhesion strength from originally 57 kPa down to 6 kPa, corresponding to a relative reduction of ~90%. Consequently, these new insights will be of great importance in the development of functionalized surfaces, permitting an innovative approach to prevent the icing of aluminum components.

Keywords: multi-scale textures; aluminum; direct laser interference patterning; superhydrophobicity; icephobicity

check for updates

Citation: Milles, S.; Vercillo, V.; Alamri, S.; Aguilar-Morales, A.I.; Kunze, T.; Bonaccurso, E.; Lasagni, A.F. Icephobic Performance of Multi-Scale Laser-Textured Aluminum Surfaces for Aeronautic Applications. *Nanomaterials* **2021**, *11*, 135. https://doi.org/10.3390/nano11010135

Received: 10 December 2020
Accepted: 6 January 2021
Published: 8 January 2021

Publisher's Note: MDPI stays neutral with regard to jurisdictional claims in published maps and institutional affiliations.

1. Introduction

Atmospheric icing includes all meteorological phenomena in which ice is accreting and accumulating on surfaces. As a consequence of icing, serious issues in different industrial sectors result, such as in telecommunications where transmission lines and towers can collapse due to the increased weight, leading to power outages [1]. The efficiency and output of renewable energy sources, such as wind and solar, can be also seriously endangered by ice formation [2]. In the aviation industry, flights can be delayed or canceled due to ice build-up on aircraft, and in the worst cases, fatal accidents can occur from icing-related events [3]. In particular, icing can take place both when aircraft are on the ground and in flight. On the ground, snow or freezing rain result in aircraft surface contamination, which can interfere with the aerodynamic properties of the aircraft or even lead to damage of the engines due to dislodged ice fragments. This requires on-ground de-icing techniques, which typically consist of hot mixtures of water and glycol, which are directly sprayed on the aircraft surfaces covered with ice or snow [4]. During the flight, icing is caused by the impingement of supercooled, micrometer-sized water droplets in clouds, which may either directly adhere upon impact or flow back as a thin film of water and collect into rivulets because of surface tension [5,6].

Most of the technical surfaces today consist of aluminum alloys, which, due to their light weight and high strength, are employed, e.g., on the leading edge of the wings [7–9]. The term de-icing implies the removal of ice after its formation on a surface exposed to extreme weather conditions, whereas anti-icing describes the capacity of a surface to reduce or avoid ice formation [10]. On airplanes, de-icing is achieved by the use of thermal energy [11] through hot bleed air diverted from the compressor of the engine to the location to be de-iced [10]. Alternatively, electrothermal heaters are used for areas where thermal de-icing systems are not applicable, such as propellers, spinners, nose cones, helicopter rotors, and hubs [12]. Having a surface that avoids or retards the adhesion of ice, which is defined as icephobic, can reduce the power needed to protect the surfaces from ice accretion during the flight and revolutionize the technology implemented in large passengers' aircraft (anti-icing) [13,14]. Novel light-weight ice protection system (IPS) with decreased power requirements can then be implemented in combination with advanced surfaces, replacing finally conventional IPS.

In the last decades, increasing interest in a novel technological concept, the hybrid ice protection systems (HIPS), was observed. HIPS includes an engineered surface combined with an ice protection system (IPS) [15]. Alternatively, the ice can be ultrasonically detached from the surface; however, ultrasonic initiation requires several hundred watts of energy and is, therefore, not ideal for the practical application [16,17]. In contrast to in-flight ice protection strategies, on ground ice accumulation is generally removed by ultrasonic methods, warmed up air, or by chemical coatings [15]. These coatings, which decrease the adhesion of water to the surface by reducing the free surface energy, are applied over large areas and are also not durable in the long term [18]. The mechanism behind this approach has shown that chemically-treated substrates, which are equipped with the superhydrophobic surface, have additional anti-icing properties. Furthermore, these chemically-treated surfaces have also shown a delayed frost formation for test conditions at temperatures of −6 °C. Despite the promising results, these solutions are not environmentally friendly, are time-consuming, and are related to high costs [19].

A novel approach to target these objectives are micro-textured functional surfaces with icephobic properties. An icephobic surface has per definition the following properties: (i) they delay the solidification time of the water or reduce ice accretion while the heterogeneous nucleation is hindered [20] and (ii) ice removal is promoted due to low adhesion strength between ice [21,22]. Superhydrophobic coatings on aluminum have demonstrated a remarkable potential to tackle the challenge of icing aircraft surfaces [23]. However, it should be mentioned that functional water-repellent surfaces are not always icephobic as well [24].

A fast and cost-efficient way to directly fabricate functional surfaces is laser surface texturing (LST). In LST, microstructures are generated on a surface by using laser-based methods like direct laser writing (DLW) or direct laser interference patterning (DLIP), among others [25]. In DLW, a single laser beam is guided by a scanner and focused by an optical element on the substrate. Depending on the laser wavelength, pulse duration, and pulse energy, a large variety of materials like metals, ceramics, and polymers can be processed [26,27]. For instance, this technology has been used to create superhydrophobic microstructures on aluminum as well as on titanium. Without applying chemical treatments, superhydrophobicity is reached with a static contact angle (SCA) above 150° and a sliding contact angle below 10° [28–30]. In addition to the topography, the chemical composition of the laser-generated texture also plays a role in the wettability modification. In fact, carbon-rich molecules can be absorbed from the ambient air, and the surface chemistry can be accordingly changed [31,32], making the wetting behavior time-dependent and its functionality limited for practical applications [33]. Moreover, microstructures fabricated on aluminum with surface features in the range of 30 to 50 μm have shown ice-delaying properties for single droplets freezing on the surface [34,35].

Another laser-based technology that can be applied for producing textured surfaces is direct laser interference patterning (DLIP). This method requires the use of at least two laser beams, which are superimposed on the substrate. This superimposition leads to an interference-generated modulation of the laser intensity distribution, which allows fabricating different periodic micro-patterns. The feature size of the pattern is usually in the range between ~500 nm and 10–20 µm [36,37]. In particular, the use of pulsed laser sources (nanosecond to femtosecond) has allowed the fabrication of features in the sub-µm range, even on metals [36,38]. Furthermore, DLIP has been widely employed for fabricating microstructures for biomedical applications on ceramics to influence the wettability on steel or to reduce the friction of automotive steer rings [39–41]. It is well known that hierarchical (also called multi-scale) microstructures are advantageous compared to a single-scale structure. For instance, they can provide multiple surface properties, including wear protection, ensuring a target functionality (e.g., antibacterial) for a longer time [42,43]. In addition, the superhydrophobic condition combined with an ice-repellent functionality has been also shown in recent work by creating 7.0 µm periodic pillars on aluminum as well as a hierarchical structure based on a combination of DLIP and DLW technology [35]. Nevertheless, ice repellency through microstructures has been only demonstrated under static testing conditions. In contrast, recently published results have shown that different laser-fabricated microstructures on titanium alloy may show dynamic ice-repellent properties, and a specific micro-texture design rule for minimizing the ice adhesion on titanium has been found [44]. A trend has been observed for surfaces in which a lower surface roughness R_z adds an advantageous effect on reducing ice adhesion [44]. By using a dynamic vibration testing technique with an electromagnetic oscillator, the ice adhesion and also its de-bonding forces can be determined accurately on a plethora of substrates [44,45]. However, the combination of different laser fabrication methods, the resulting hierarchical topographies, and their impact on ice adhesion, especially for aluminum, have not been investigated so far. Furthermore, to the best of our knowledge, it is not clear to what extent the roughness can still be reduced to further minimize ice adhesion.

In this study, various single- and multiscale patterns are fabricated on aluminum 2024 by means of DLW and DLIP. By combining both laser methods, complex hierarchical structures are introduced. The wetting characteristics are analyzed by static water contact angle and sliding angle measurements. The icephobic property is analyzed in relation to the adhesion of the formed ice on the laser-textured surfaces. Therefore, the terms "ice-repellent" and "icephobic" are to be understood in such a way that it requires reduced strength to release the ice from its surface. The performance regarding the dynamic ice adhesion is examined by interfacial shear stress experiments for rime, glaze, and mixed icing conditions using a lab-scaled icing wind tunnel. Hence, in these experiments, it is also feasible to take into account the wind speed and the droplet size, which allows a significantly more realistic view of the icing and deicing processes of laser-textured aluminum substrates.

2. Materials and Methods

2.1. Materials

The samples used in all experiments consisted of Al2024 (AMAG rolling GmbH, Ranshofen, Austria) in the shape of a cantilever with dimensions of $125 \times 13 \times 1$ mm^3. They were polished to a surface roughness R_z of 150 nm $\pm$ 40 nm. Prior to the laser processing, the samples were cleaned from contamination using isopropanol in an ultrasonic bath for 10 min. After the laser process, the structured samples were spray-coated with a hydrophobic agent (Mecasurf, perfluoropolyether compound by Surfactis, Angers, France) and stored under atmospheric conditions of 22 °C room temperature and 55% relative humidity. The non-flammable agent is composed of a chemically active perfluoropolyether compound dissolved in a fluorinated solvent, which is clear colorless and has an atmospheric lifetime of 0.77 years for a temperature range between -50 °C and $+150$ °C [46]. The used spray bottle (Buerkle GmbH, Bad Bellingen, Germany) was equipped with a 0.6 mm diameter nozzle, and the spray volume was 1.2 mL $\pm$ 0.1 mL per stroke. Five strokes were applied per sample with an interval of ~20 mm on the length-scale and a distance of ~100 mm to the surface in order to coat the cantilever homogeneously. This agent creates covalently bonded fluoride groups to cover the surface, saturating the material with non-polar compounds (reducing the surface energy) with an ultrathin layer (<5 nm). It is important to mention that the chemical reaction can take place only if the surface is not passivated with organic compounds: Therefore, the samples have been chemically treated directly after the laser process [41]. The compound reacts only with the surface of the material, while the unreacted compound evaporates due to the high volatility of the molecules, leaving a deposited monolayer [47].

2.2. Laser Structuring Methods

Two different structuring methods, direct laser writing (DLW) and direct laser interference patterning (DLIP), were used for fabricating the microstructures. For DLW processing, a fiber laser was used because the pulse duration could be adjusted and optimized for aluminum processing. For the DLIP experiments, however, a solid-state laser was used since it is especially characterized by its pulse stability for ultrashort laser pulses (ps range). Moreover, under determined texturing conditions, nanosecond laser pulses can create deeper micro-tranches on metal surfaces than picosecond ones due to the displacement of molten material [48]. The DLW technology is implemented by a laser surface texturing workstation (GF machining solutions P 600, Geneva, Switzerland). The system is featured with a galvanometer scanner system (Scanlab GmbH, Puchheim, Germany), including two mirrors and an IR (1064 nm) Ytterbium fiber laser (YLPN-1-4x200-30-M, IPG Laser GmbH, Burbach, Germany), providing a maximal output power of 30 W (Figure 1a). The pulse duration can be modified from 4 up to 200 ns, and the pulse repetition rate is flexible from 2 to 1000 kHz. In our experiments, 14 ns pulse durations were applied, and the repetition rate was set to 30 kHz. The beam was focused onto the sample using a 254 mm focal length F-theta objective, obtaining a spot with a diameter of 70 μm. The laser fluence was set to 1.06 J/cm^2 by controlling the laser power. The laser beam was scanned along the surface of the sample with a constant speed of 250 mm/s. In order to achieve sufficient material removal, the structuring process was repeated 10 times in all cases. The mesh-like surface structures processed by this setup consisted of 3 sequentially formed line-like structures, whose orientation was each time rotated by 60°. The distance between the parallel lines was fixed to 50 μm, which is, therefore, defining the spatial period of the structures. Previous works using a similar texturing approach showed that this texture provided remarkable results in terms of water repellence and static icing properties and was, therefore, considered for the dynamic icing tests. A detailed description of the fabrication process was already published elsewhere [33,35,49].

For processing the samples with DLIP technology, a two-beam configuration was used. The used configuration utilizes an IR (1064 nm) solid-state Nd:YVO$_4$ laser (Edgewave PX200, Würselen, Germany) with 10 ps pulses and maximum output power of 10 W. The workstation is equipped with a commercially available DLIP optical head (manufactured by Fraunhofer IWS, Dresden, Germany), which uses a diffractive optical element (DOE) to split the initial laser beam into two sub-beams (Figure 1b). These two beams are guided to a prism where they are parallelized and then focused to a beam radius $\omega < 100$ μm on the sample surface using a 60 mm aspheric lens. Using this configuration, a line-like interference intensity profile is obtained within a circular area with a radius of ablated area rth of ~70 μm (Figure 1c). In this study, the intersection angle θ between the beams was set to 8.7° and 23.6° to obtain a spatial period Λ of 7.0 μm and 2.6° μm, respectively. The interference principle is depicted in Figure 1c, the laser fluence was set to 1.93 J/cm^2, and the repetition rate of the laser was fixed to 10 kHz. The samples were translated in x- and y-directions by using high-precision axes (Aerotech, Pittsburgh, PA, USA) with an accuracy of ±2.5 μm. By moving the sample in the y-direction, the pulse-to-pulse overlap can be controlled, which finally determines the number of laser pulses, influencing the ablated structure depth. To obtain the pillar-like structures, it is necessary to rotate the samples by 90° and repeat the process.

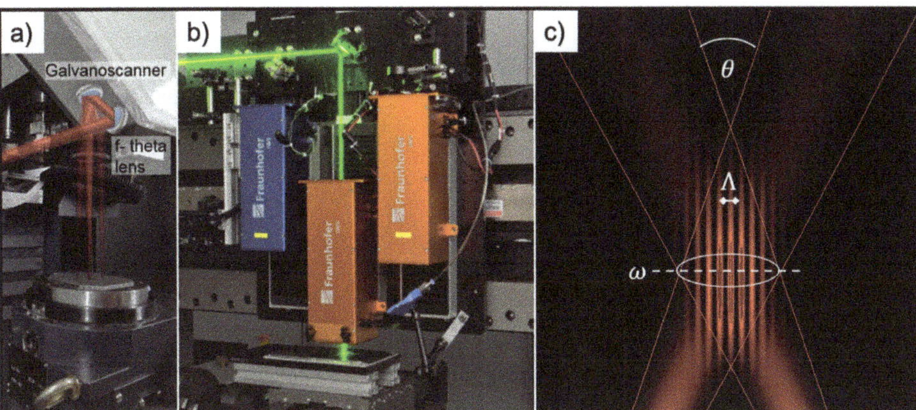

Figure 1. DLW (direct laser writing) configuration equipped with a galvanoscanner and an f-theta-lens used in the experiments (**a**), 3D-DLIP workstation (developed at the Technische Universität Dresden) with a 1064 nm wavelength laser source emitting 10 ps pulses for the processing of 2D and 3D parts ((**b**), the visible laser beam illustrates the process), graphical representation of the used two-beam direct laser interference patterning (DLIP) principle, showing the most relevant parameters describing both the pattern and the process: ω—beam diameter; θ—beams intersection angle; Λ—laser intensity profile/ablated structure period (**c**) [35,42].

The hierarchical structures, consisting of the DLW and DLIP, were manufactured by combining both strategies successively. A detailed description of the fabrication processes, the laser process windows, and the choice of the processing parameters was already published elsewhere [33,35,49].

2.3. Surface Characterization

To analyze the surface topography, scanning electron microscopy (ZEISS Supra 40VP, Jena, Germany) at an operating voltage of 15.0 kV was used. The wettability of the surfaces was characterized via static contact angle (SCA) and sliding angle (SA) measurements using the sessile drop method with a video-based optical goniometer (DSA 25, Krüss GmbH, Hamburg, Germany). This method was chosen due to its suitability for reproducible measuring. Deionized water droplets of 10 µL volume were deposited in atmospheric conditions for all measurements. The wetting tests were repeated 6 times to ensure statistical significance. Additionally, the real-to-projected area ratio Sdr was measured using a confocal microscope (Sensofar S Neox, Terrassa, Spain) with a 150× objective, resulting in vertical and lateral resolution of 2 nm and 140 nm. The surface roughness R_z was measured with a benchtop stylus profiler (DektakXT, Bruker, Billerica, MA, USA), following the standard DIN ISO 25178 and DIN EN ISO 4287, respectively [50,51]. For acquiring the results, a stylus with a 2 µm diameter was pressed on the surface with a weight of 2 mg and moved at a speed of approx. 0.05 mm/s. Every measurement point was repeated three times for a statistical purpose.

2.4. Ice Adhesion Test

To analyze the ice adhesion behavior for four different ice conditions, the interfacial shear stress was measured using a circular closed-loop wind tunnel iCORE (icing and Contamination Research facility, Airbus Defence and Space GmbH, Taufkirchen, Germany, Figure 2a) [50,52]. Therefore, the laser-structured area was mounted inside the test section, oriented against the flow direction, and the sample was fixed on one end to an electromagnetic shaker. Under previously defined conditions, ice accumulated on the structured area, which was then removed by inducing mechanical vibration (Figure 2b) [45]. The vibration caused shear stress at the interface between the accreted ice layer, and the cantilever was measured by a strain gauge and used to characterize the ice interfacial shear strength. The stress-strain curve, caused by the vibration, would be monitored during deformation, from which the failure stress could be identified microscopically [53]. The crack-propagation was afterward also visible (Figure 2c). In order to calculate the stress accurately, the ice thickness was measured and averaged at several points of the cantilever. Thus, the local waviness of the ice layer was taken into account. Furthermore, the dimensions of the cantilever were kept constant to ensure comparability of the results.

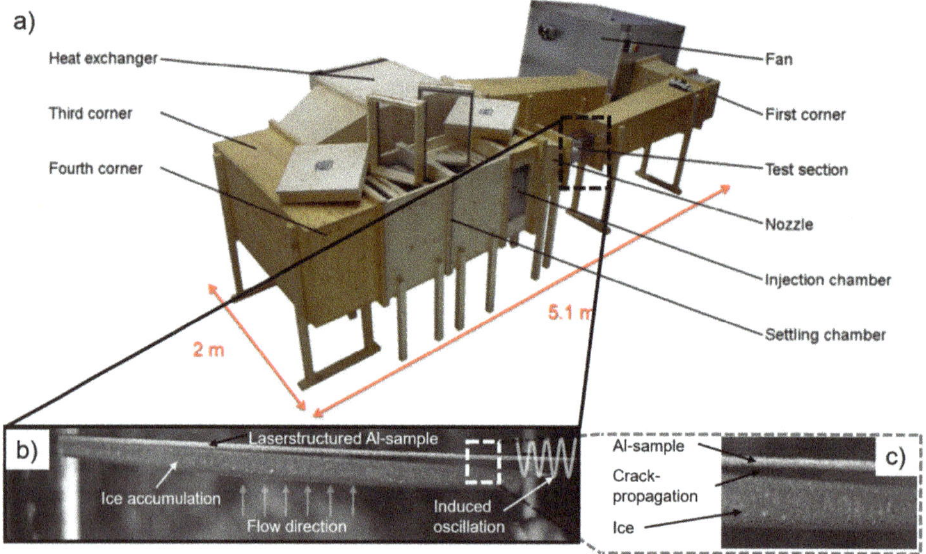

Figure 2. Icing wind tunnel iCORE (icing and Contamination Research facility), which allows adjusting the wind speed and temperature by accessing a fan and a heat exchanger. The droplet size and stream water content in the test section are set by the settling and injection chambers before the nozzle (**a**) [51]. In the test section, ice accumulates on the laser-structured sample, and the specimen is exposed to mechanical shear stress due to induced oscillation (**b**), which finally leads to crack propagation (**c**).

The calculation was performed under the assumption of a constant Young's modulus of 9 GPa for each type of ice (E_{ice}) based on several publications. It is worth mentioning that there were slight differences in the E_{ice} for the different icing conditions. Nevertheless, our assumption allowed comparability of the ice adhesion data within each ice type. For simulating atmospheric in-flight conditions, four different ice conditions (rime, mixed rime, glaze, and mixed glaze) were investigated by applying five testing parameters. These are the icing conditions to which an aircraft mostly can be exposed during a flight. They result from the varying flight altitude, which is associated with different air temperatures and ice-water mixtures. For instance, the glaze ice condition usually occurs at an altitude around 2000 m at a temperature range between 0 °C and −10 °C. Contrarily, the rime ice condition appears at altitudes above 3000 m and at a temperature of at least −20 °C. Furthermore, different ice conditions are characterized by varying densities and hardnesses [15].

The ice-producing parameters were related to the icing design envelope specifications from the Federal Aviation Administration (FAA) [54]. The influencing parameters were beside the total air temperature (−5 °C and −20 °C) and the airspeed (50 m/s and 80 m/s), the liquid water content (LWC), the mean effective volume droplet diameter (MVD), and the approximate freezing fraction (AFF). The MVD is defined as the midpoint (median) droplet size, 20 μm in our case, where half of the water volume in the cloud is in droplets smaller, and half of the volume is in droplets larger than the median [54]. The approximate freezing fraction (AFF) determines the fraction of water that freezes initially when water droplets impinge on a surface. For instance, for rime ice, the freezing fraction is 1, which means all the impinging water is freezing immediately. In contrast to that, for the glaze, the freezing fraction is close to 0, and most of the impinging water will stay liquid and run back [55]. The LWC defines the density of super-cooled water droplets in the icing cloud and is ranging between 0.3 and 1. An overview of the parameters for creating the icing regimes is given in Table A1 in the Appendix A.

3. Results and Discussion

3.1. Laser Structuring of Aluminum Substrates

For the structuring process of the aluminum substrates, DLW and DLIP methods were used to create different topographies, with feature sizes ranging from 2.6 μm to 50 μm, employing ps and ns laser sources, respectively. In contrast to the untextured Al reference (Figure 3a), the DLW-treated substrates showed a morphology dominated by melting processes due to the used nanosecond pulse duration. In particular, the mesh-like structure with a repetitive feature size of 50 μm consisted of molten and re-solidified material, as shown in Figure 3b. Furthermore, the topography of the DLW-treated samples was assessed through confocal imaging, revealing a roughness R_z of ~43.1 μm and an average structure depth of 35.7 μm. Significantly less melting occurred on the DLIP-treated samples due to the ps-based micro-processing. Applying a spatial period of 7.0 μm, perfectly ordered pillar-like microstructures were fabricated, and a roughness R_z of ~4.1 μm was reached (Figure 3c). To produce a multiscale pattern, the ns-DLW and ps-DLIP technologies were combined, resulting in a hierarchical structure consisting of 7.0 μm pillars on the 50 μm meshes. The ps-pulsed laser source was selected in order to avoid further thermal treatments of the DLW structures. The added DLIP-pillars led to a slight decrease of the roughness (compared to DLW) to an R_z of ~38.1 μm (Figure 3d). It could be assumed that the additional DLIP process slightly smoothened the roughness peaks caused by the ns-DLW melting since the utilization of the ps laser sources resulted in ablation and vaporization of the treated material.

Figure 3. Scanning electron microscope images of (**a**) an untextured reference and of microstructures fabricated with (**b**) DLW for a 50 μm mesh-like structure using a laser fluence of 1.06 J/cm², a scanning speed of 250 mm/s at a repetition rate of 30 kHz; (**c**) DLIP for a 7.0 μm pillar-like structure using a laser fluence of 1.93 J/cm², a pulse-to-pulse overlap of 99% at a repetition rate of 10 kHz; (**d**) a combination of DLW and 7.0 μm DLIP for a multiscale structure; (**e**) DLIP for a 2.6 μm line-like structure and (**f**) DLIP for 2.6 μm cross-like structure using the same fluence, repetition rate, and overlap as for the 7.0 μm DLIP structures. The inserted confocal images provide an impression of the topography. The laser parameters are provided in the experimental section.

In addition to the above-mentioned patterns, also line-like and cross-like structures with lower spatial periods were produced. These had a spatial period of 2.6 μm and a surface roughness R_z of 0.61 μm for a line-like and 0.96 μm for a cross-like pattern (Figure 3e,f), showing a significant lower roughness compared to the 7.0 μm pillar-like structures.

The ice adhesion functionality of a surface is dominated by the surface's topography and chemistry, as mentioned in the introduction. It is well-known that laser processing of aluminum surfaces leads to an influence of the chemical composition on the textured surface [33,44]. Directly after laser processing, the structured substrates and the untextured reference were spray-coated using a perfluoropolyether (PFPE) monolayer coating, which reduces the surface tension and ensures constant chemical surface conditions all over the surface. PFPE is very inert, and the chemical bond has high stability due to the covalent carbon-fluorine bond [41,50,56]. After the treatment, all samples showed a superhydrophobic characteristic due to the lower surface tension combined with the produced micro-texture, regardless of the size or orientation of the microstructures. In particular, the DLIP-produced cross-like patterns showed static water contact angles (SWCA) of 164°, and the hierarchical pattern an angle of 172°. The untreated reference remained in the hydrophilic condition with an SWCA of 59°, and the PFPE coated reference led to a hydrophobic transition with an SWCA of 122°. All angles are reported in Table 1. The obtained results clearly show the influence of the surface pattern on the wetting properties of the samples and confirms that a spray-coated and untextured surface does not develop a superhydrophobic characteristic [41,44].

Table 1. Topographical and wetting characteristics of the fabricated samples treated with perfluoropolyether (PFPE), direct laser writing (DLW), direct laser interference patterning (DLIP) and an combination of DLW and DLIP (DLW/DLIP).

Parameter	Untreated Reference	Reference with PFPE	DLW 50 μm	DLIP 7.0 μm	DLW/DLIP 50/7.0 μm	DLIP Line 2.6 μm	DLIP Cross 2.6 μm
Geometry	-	-	Mesh-like	Pillar-like	Hierarchical	Line-like	Cross-like
Spatial period (μm)	-	-	50	7	50/7	2.6	2.6
Static contact angle (°)	59 ± 2	122 ± 2	171 ± 2	165 ± 1	172 ± 1	166 ± 1	164 ± 3
Sliding angle (°)	No sliding	No sliding	2 ± 1	4 ± 2	2 ± 1	3 ± 1	9 ± 4
Roughness R_z (μm)	0.15 ± 0.04	0.18 ± 0.04	43.15 ± 0.32	4.16 ± 0.39	38.16 ± 2.33	0.61 ± 0.03	0.96 ± 0.17
Real area to projected area ratio Sdr (%)	1.1	1.1	240	130	203	5.9	12.9

While the static water contact angle is commonly known as a criterion to characterize the wettability of a surface, it does not necessarily allow to determine the (super-)hydrophobic character of a surface. Therefore, the sliding angle (SA), denoted as the tilting angle where the droplet starts to slide on the surface, is an additional parameter that has to be considered [57]. In this study, all laser-fabricated structures exhibited a sliding angle below 10°, allowing the samples to be considered as fully superhydrophobic (see Table 1) [58,59]. In detail, the sliding angles for the microstructured surfaces ranged between 2° (DLW and DLW + DLIP) and 9° (DLIP cross). Both references did not show any sliding characteristics due to their hydrophilic and moderate hydrophobic surface state. The geometries and their periodicities, as well as the wetting characteristics (SCA and SA) and the roughness R_z of the considered topographies, are summarized in Table 1. A detailed description of the data acquisition is given in the experimental section.

It is important to state that the determination of the wetting contact angle is representative of the macroscopic wetting behavior of the treated surfaces, and it is not sufficient to describe the wetting phenomena involved during in-flight (icing) conditions [20]. In fact, as stated in the introductory part, supercooled microscopic droplet imping at high velocity on metallic surfaces, making the effect of surface tension, hydrostatic pressure, and surface microstructure size more relevant [4,60].

It is worth mentioning that the durability of the laser-generated structures and their chemical composition when exposed to consecutive icing/de-icing cycles was examined in a previous study [50]. It was demonstrated that the chemical functionalization with the PFPE substrate after laser processing produced surfaces that preserved their superhydrophobicity beyond 16 icing/de-icing cycles, indicating a likelihood of covalent bonds between perfluoropolyether and metal oxide. The same observation was found in the samples presented in this study. The SCA and SA did not decrease after consecutive icing/de-icing cycles.

3.2. Ice Adhesion Analyzed by Mechanically Induced Stress

It is well-known that superhydrophobic aluminum leads to delayed ice-formation due to the reduced contact area between a single water droplet and the surface [35]. However, most of the works present in literature are based on the icing of water droplets under static conditions. An alternative to the static analysis of the anti-icing properties are methods in which the surface is mechanically strained while measuring the ice adhesion [61]. It is fundamental that the surface roughness and the microstructure are optimized in order to allow proper anti-icing functionality [44,62]. In fact, it has been demonstrated that water-repellent surfaces with a spatial period of 35 µm even lead to a stronger ice adhesion compared to an untreated titanium reference [50].

The microstructures in this study were fabricated on $125 \times 13 \times 1$ mm^3 cantilevers, which were placed afterward in an icing wind tunnel in order to investigate the ice adhesion. Four icing conditions were analyzed, which simulated flight conditions encountered in different atmospheric regimes. These generated four qualities of ice were rime, mixed rime, mixed glaze, and glaze ice. After that ice was accreted on the surface of the cantilever, the icing cloud was turned off, and an oscillation was induced on the cantilever through a piezoelectric transducer. A linear frequency sweep was applied, letting this increase with time and causing an increase in the amplitude of the oscillation when the resonance frequency of the oscillating body was reached [45]. Since the resonance frequency depends on the thickness and density of the accreted ice, this differed among the investigated samples (60–70 Hz). Once a critical amplitude value was reached, the ice detached from the surface (crack initiation), and the interfacial shear stress could be retrieved and quantified. A strain gauge was placed on the backside of the cantilever in order to detect the strain during the oscillation and display the change of the mechanical properties of the composite beam (i.e., ice and metal) due to ice detachment. An example of a strain gauge reading is shown in Figure 4 for the untextured reference, DLW, and DLW + DLIP treated surfaces. In all cases, the amplitude changed drastically at 0.75 s. Such variation was related to a discontinuity in the mechanical properties of the composite beam due to an interfacial crack between ice and cantilever.

The amplitude of the strain recorded right before the interfacial crack represented the limit at which the interfacial shear stress matched the adhesion strength of the ice to the surfaces. Exceeding this value triggered a crack at the ice-surface interface. The Euler-Bernoulli beam theory could be used to correlate the strain to the maximum interfacial shear stress (i.e., ice adhesion) between ice and surface, which is expressed in Equation (1) [45]:

$$\tau_{int} = \frac{\varepsilon_{EF-al} E_{ice} \left(h_{ice}^2 + 2h_{ice}|e| \right)}{2(x-l)(h_{cl} - |e|)}, \tag{1}$$

where τ_{int} is the interfacial shear stress, ε_{EF-al} is the amplitude of the strain measured by the strain gauge, E_{ice} is the Young's Modulus of ice, h_{ice} is the thickness of the ice, h_{cl} is the thickness of the cantilever, x is the distance of the strain gauge to the fixed end of the cantilever, l is the length of the cantilever, and e is the eccentricity of the neutral axis of the ice/metal beam with respect to the ice/metal interface [53]. The eccentricity of the neutral axis could be calculated from Equation (2):

$$e = \frac{\left(h_{cl}^2 + \frac{E_{ice}}{E_{cl}}h_{ice}^2\right)}{2\left(h_{cl} + \frac{E_{ice}}{E_{cl}}h_{ice}\right)}, \tag{2}$$

where E_{cl} is the Young's Modulus of the bulk cantilever material. The Young's Modulus of ice used for shear adhesion strength calculation was taken from prior research for fully dense ice [63,64]. As a result, a low shear stress value resulted in less mechanical force necessary to remove the ice from the surface [65]. In general, the lower the maximum amplitude of the strain before the interfacial crack is, the lower the ice adhesion (with all the other parameters fixed). Therefore, based on the three shown exemplary measurements from Figure 4, it can be concluded that this method is suitable to quantitatively evaluate the adhesion of ice under different conditions [45]. It is worth to mention that other authors have also reported superhydrophobic aluminum surfaces tested in an ice wind tunnel, leading to a significant delay of the initiation of ice formation [66]. Moreover, it was demonstrated that the ice adhesion was reduced from ~360 kPa to ~80 kPa for coated and etched superhydrophobic aluminum (AA6061-T6) substrates, which corresponded to a reduction of approximately 78%. The samples were tested exclusively for the glaze ice condition in a centrifuge set-up [67].

In contrast to that, the results of the ice adhesion tests for the seven samples (two references and five micro-structured cantilevers) by means of the retrieved interfacial shear stress under four different icing conditions are summarized in Figure 5.

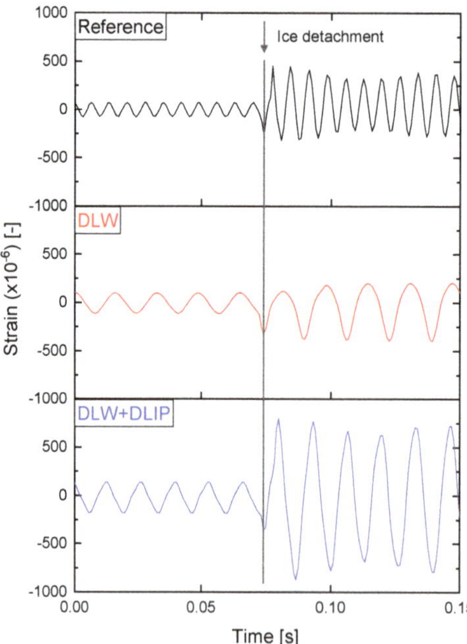

Figure 4. Example of strain gauge readings recorded during the ice adhesion tests for three different samples of an untextured reference AA2024 (**top**), DLW (**middle**), and DLW + DLIP (**bottom**) under test conditions of mixed/rime ice. The time range in which the ice detachment occurred was framed to 0.15 s to obtain a more detailed overview.

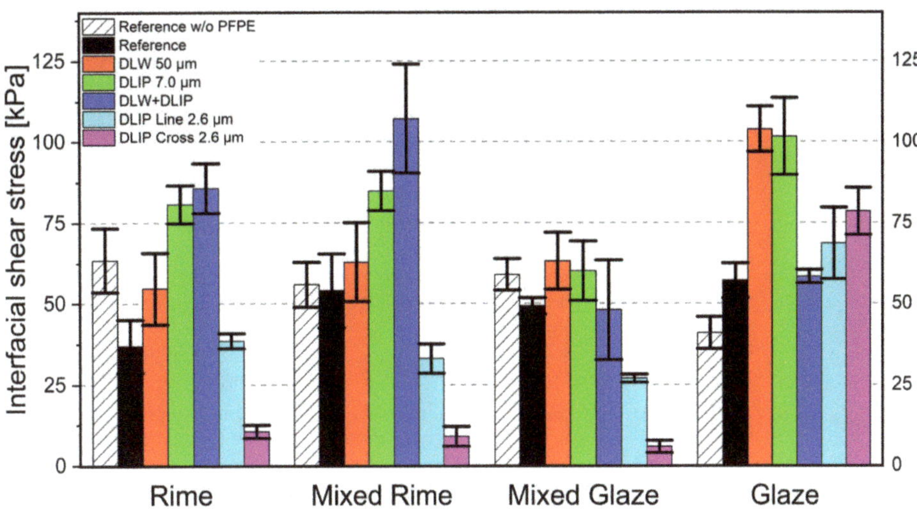

Figure 5. Histograms depicting the results of the ice adhesion tests for the four different ice conditions of rime, mixed rime, mixed glaze, and glaze with the interfacial shear stress for the five laser-structured specimens, a reference, and a PFPE untreated (w/o PFPE) reference.

Each measurement was repeated 4 times for each surface type and for each ice condition (in total, 96 measurements). The average value of one measurement series defines the height of the bar, while the error bars result from the standard deviation of each measurement series. The term reference refers to the untextured sample, and it provides average interfacial shear stress of around 36 kPa, whereas, for mixed rime, mixed glaze, and glaze conditions, the stress was slightly increased to around 50 kPa. Besides that, some trends were observed, which are discussed in the following.

The DLW structure (50 µm), the pillar-like DLIP structure (7 µm), and the hierarchical pattern showed interfacial shear stress between 48 and 64 kPa, which was similar (for mixed glaze ice) or even significantly higher (up to 107 kPa for rime and mixed rime ice) compared to the reference. It could be assumed that the roughness R_z of 43.15 µm, 4.16 µm, and 38.16 µm (DLW, DLIP, and DLW+DLIP) resulting from the coarse feature sizes of 50 µm (DLW) or 7 µm (DLIP) provided a favorable topography for the droplets to attach and interlock. Note that the mean effective volume droplet diameter (MVD) of the impinging droplets was ~20 µm and thus in the order of magnitude of the investigated microstructures (see Table A1 in the Appendix A).

In contrast, the DLIP structures with a spatial period of 2.6 µm showed a significantly lower interfacial shear stress. The stress for the smallest examined texture was between 27 and 38 kPa (DLIP line-like 2.6 µm) and even lower between 6 and 10 kPa (DLIP cross-like 2.6 µm) for the icing condition of rime, mixed rime, and mixed glaze. This amounted to a shear stress reduction of ~90% compared to the reference. This behavior could be explained with a synergic effect of high hydrophobicity and low surface roughness.

Besides that, it is worth mentioning that both references did not show superhydrophobic characteristics for all icing conditions. This property is fundamentally necessary since impacting droplets will not adhere to the surface. The droplet rolls off the surface because the anti-wetting capillary pressure is greater than the dynamic and water hammer wetting pressures [68].

In fact, the texture size produced on the "DLIP cross" sample is low enough to both repel micrometer-sized water droplets and to prevent the formation of large anchor points for the ice adhesion and accretion [69]. Although having similar roughness, the "DLIP line" sample shows instead a much higher ice adhesion. This can be explained by the surface modifications in the micro-scale, which only take place in one direction (line-like texture). Hence, the overall response to ice adhesion is, therefore, higher than for an isotropic texture (cross-like). Therefore, it can be assumed that a line-like geometry provides a preferred orientation regarding ice accumulation. Such assumption finds support in the work from Lo et al. [70]. Their research was focused on ice nucleation studies of different surfaces using environmental SEM (ESEM). The authors showed that ice nucleation was controlled and confined on V-shaped microgroove patterned surfaces. The ice is showing preferential nucleation sites along the manufactured grooves. This anchoring theory is also valid for the hierarchical DLW+DLIP structure with a lower roughness (38.16 μm) compared to the DLW structures (43.15 μm). However, this topography seems to offer more anchoring possibilities for ice formation. Due to the hierarchical surface, this hypothesis is especially claimed for lower temperatures, which are present in the rime and mixed rime conditions at −20 °C.

3.3. Influence of the Surface Topography on the Ice Adhesion

To further investigate the influence of the surface peak-to-peak-roughness, the interfacial shear stress was plotted as a logarithmic function of the real-to-projected area ratio Sdr in %, for the four ice adhesion characteristics (Figure 6). The corresponding roughness and Sdr parameters are additionally given in Table 1.

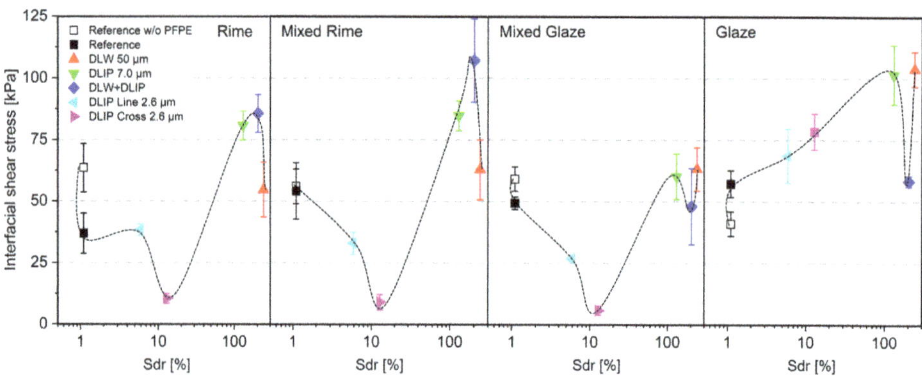

Figure 6. Interfacial shear stress of the laser-treated samples and the two references as a logarithmic function of the Sdr ratio for the icing conditions of rime, mixed rime, mixed glaze, and glaze. The black dashed spline is meant only to guide the eye of the reader.

Although the experimental data did not show a clear trend, it was evident that a defined Sdr ratio was favorable in order to reach low interfacial shear stress under rime, mixed rime, and mixed glaze conditions. The differences of the Sdr values for the 50 μm DLW and 2.6 μm DLIP textures could be ascribed in-depth and the aspect ratios of the textures. Since the 2.6 μm textures were a few hundred nm deep, the aspect ratio was consequently low and in the range of ~0.1. In contrast, the DLW textures showed a structure depth of ~35.7 μm, which resulted then in an aspect ratio of ~0.7. As a result, the DLW textures provided a significantly higher Sdr ratio. The Sdr ratio was in the range of ~13 (for the DLIP cross 2.6 μm textures), corresponding to a surface roughness R_z of 0.96 μm, and resulted in the detachment of the ice from the aluminum surface at interfacial shear stress below 10 kPa. However, a direct correlation of the individual textures regarding their ice adhesion, as shown in Figure 6, was difficult due to their significantly diverse topography. Based on the experimental results, it could be concluded that a smoother surface seemed to promote a higher adhesion [71].

These findings were supported by previous studies reporting the systematic reduction of ice adhesion strength with the decrease of surface roughness regardless of surface wettability [72,73]. In-situ icing studies indicated that ice did not anchor on smoothened metallic surfaces but anchored on the as-received metallic surfaces, which is consistent with the results in Figure 6 [73]. The larger microstructures (DLIP 7.0 µm, DLW 50 µm, and DLW + DLIP) showed significantly higher ice adhesion stress, which was related to the average effective drop diameter (MVD). When the 20 µm droplets hit the textured surface, especially the deep DLIP pillars or the DLW structures (Sdr between 130 and 240%) offered space between the microscopic features. The ice adhesion was, in this case, even significantly higher compared to the reference and the DLIP structures with the 2.6 µm structures. This hypothesis was further supported by earlier research results obtained on titanium substrates [44].

Apart from the roughness and the Sdr parameter, it was observed that not only the spatial period influenced the ice adhesion but also the pattern geometry. Two samples were fabricated with the same spatial period of 2.6 µm with line and cross-like geometries (see Figure 3e,f). Nevertheless, the cross-like geometry showed substantial lower ice adhesion. It was further demonstrated that the icing condition also had a significant influence on the shear stress. For example, it has been shown that for the glaze condition, increased stress was measured for all microstructured surfaces than for the reference (see Figure 6, glaze). This could be linked to the combination of a higher temperature (-5 °C), higher speed (80 m/s), and high liquid water content (LWC) of 1 g/m^3 (see Table A1 in the Appendix A). Under this condition, the pressure generated at the impact could cause the partial wetting of the structures, given the nature of glaze ice [68]. As a matter of fact, only a lower portion of water is freezing on impact, and the rest is able to run wet and freeze afterward. As a consequence, liquid water can penetrate within the laser-generated structures and freeze in them, increasing the mechanical interlocking between ice and surface. Subsequently, it was reported in other studies that the gravity effect on the droplet trajectories favored ice formation in this case [60,74].

In summary, the DLIP 2.6 µm line and cross are the only laser-treated surfaces, exhibiting remarkable icephobic properties in this study. The cross-like strategy is performing better, most likely due to its isotropic morphology, which is reducing the amount of anchoring points for the ice formation. The DLIP 7.0 µm and the DLW 50 µm are not able to repel supercooled water droplets, despite their superhydrophobicity [44]. However, such surfaces could still be used for repelling larger droplets, possibly at lower impact speeds. The combination of DLW and DLIP does not provide any benefit in terms of ice adhesion reduction in dynamic conditions, although previous results remarked the advantages when combining both geometries in static conditions [35,75,76].

4. Conclusions

In this work, five different periodic patterns were textured on AA2024 clad cantilever samples using direct laser interference patterning, direct laser writing, and a combination of both methods. Additionally, two untextured reference samples were considered. The spatial period of the textures ranged from 2.6 µm to 50 µm, showed a line-, cross-, pillar-, or mesh-like pattern, and covered a corresponding roughness R_z between 0.61 µm and 43.15 µm linked to an Sdr ratio between 5.9 and 240. After a chemical post-treatment, all textures showed superhydrophobic wetting characteristics with a static contact angle over 166° and a sliding angle below 10°. This combination of material processing, laser texturing, and the subsequent chemical treatment was conducted for the first time on aluminum cantilevers. Using a dynamic vibration testing technique, the shear stress between the structured sample and ice was investigated. Each structural geometry was tested under four different icing conditions, namely rime ice, glaze ice, and two mixed ice conditions. For the rime ice and mixed ice conditions, the cross-like DLIP pattern with a size of 2.6 µm showed the lowest interfacial shear stress, between 6 and 10 kPa. In contrast, the stress of the reference was between 36 and 57 kPa independently of the icing condition. This magnitude of shear stress is extraordinarily low and has not been demonstrated by any known alternative approach under these icing conditions to date to the best of our knowledge. Surface patterns with larger feature sizes, such as 7.0 µm DLIP, 50 µm DLW, or the hierarchical structure, showed even higher shear stresses (between 80 and 108 kPa) compared to the reference. Consequently, it can be stated that microstructures with feature sizes smaller than 5 µm and a particular structure type (line-like and cross-like) are especially suitable for designing icephobic surfaces. Using the results of these experiments, the field of the perfect icephobic texture for real dynamic icing conditions can be further narrowed down, and future research can build on this knowledge. The results obtained and the methodology developed in this work can be further adapted to other alloys and even polymers, unlocking completely new application examples in the challenge facing icing surfaces in the fields of aerospace technology.

Author Contributions: Conceptualization, S.M. and A.F.L.; methodology, V.V. and A.I.A.-M.; software, S.A.; validation, V.V., S.A. and E.B.; formal analysis, S.M., V.V. and S.A.; investigation, S.M., V.V., S.A., and A.I.A.-M.; resources, A.F.L. and E.B.; data curation, S.M. and V.V.; writing—original draft preparation, S.M. and V.V.; writing—review and editing, S.M., V.V., S.A., A.I.A.-M., T.K., E.B., and A.F.L.; visualization, S.M., V.V., S.A., and A.I.A.-M.; supervision, T.K., E.B., and A.F.L.; project administration, T.K., E.B., and A.F.L.; funding acquisition, T.K., E.B., and A.F.L. All authors have read and agreed to the published version of the manuscript.

Funding: This work was carried out in the framework of the Reinhart Koselleck project (Grant Agreement No. 323477257), which has received funding from the German Research Foundation (German: Deutsche Forschungsgemeinschaft DFG), and in the framework of the LASER4FUN project (www.laser4fun.eu), which has received funding from the European Union's Horizon 2020 research and innovation programme under the Marie Skłodowska-Curie Grant Agreement No. 675063. The work of A.F.L. is also supported by the German Research Foundation (DFG) under the Excellence Initiative program by the German federal and state government to promote top-level research at German universities.

Institutional Review Board Statement: Not applicable.

Informed Consent Statement: Not applicable.

Data Availability Statement: Data is contained within the article.

Conflicts of Interest: The authors declare no conflict of interest. The funders had no role in the design of the study; in the collection, analyses, or interpretation of data; in the writing of the manuscript, or in the decision to publish the results.

Abbreviations

ω	Beam radius
θ	Intersection angle
Λ	Spatial period
E_{ice}	Young's modulus of the ice layer
E_{cl}	Young's Modulus of the bulk cantilever material
R_z	Maximum peak to valley roughness
Sdr	Real area to projected area ratio
τ_{int}	Interfacial shear stress
ε_{EF-al}	Amplitude of the strain
h_{ice}	Thickness of the ice
h_{cl}	Thickness of the cantilever
x	Distance of the strain gauge to the fixed end of the cantilever
l	Length of the cantilever
e	Eccentricity of the neutral axis of the ice/metal beam

Appendix A

Table A1 summarizes the parameters for creating the four regimes, and each measurement was repeated at least 4 times for statistical purposes.

Table A1. Icing conditions of total air temperature (TAT), airspeed, liquid water content (LWC), mean effective volume droplet diameter (MVD), and an approximate freezing fraction (AFF) used for the ice adhesion tests in the icing wind tunnel iCORE (icing and Contamination Research facility) [50].

Ice Type	TAT (°C)	Airspeed (m/s)	LWC (g/m^3)	MVD (µm)	AFF (–)
Rime	−20	50	0.3	20	1.0
Mixed/Rime	−20	50	0.8	20	0.7
Mixed/Glaze	−5	50	0.3	20	0.5
Glaze	−5	80	1.0	20	0.2

References

1. Makkonen, L.; Lehtonen, P.; Hirviniemi, M. Determining ice loads for tower structure design. *Eng. Struct.* **2014**, *74*, 229–232. [CrossRef]
2. Jelle, B.P. The challenge of removing snow downfall on photovoltaic solar cell roofs in order to maximize solar energy efficiency—Research opportunities for the future. *Energy Build.* **2013**, *67*, 334–351. [CrossRef]
3. Cao, Y.; Tan, W.; Wu, Z. Aircraft icing: An ongoing threat to aviation safety. *Aerosp. Sci. Technol.* **2018**, *75*, 353–385. [CrossRef]
4. Li, Q.; Guo, Z. Fundamentals of icing and common strategies for designing biomimetic anti-icing surfaces. *J. Mater. Chem. A* **2018**, *6*, 13549–13581. [CrossRef]
5. Stankov, B.B.; Bedard, A.J.; Edard, A.J. Remote sensing observations of winter aircraft icing conditions—A case study. *J. Aircr.* **1994**, *31*, 79–89. [CrossRef]
6. Cebeci, T.; Kafyeke, F. Aircrafticing. *Annu. Rev. Fluid Mech.* **2003**, *35*, 11–21. [CrossRef]
7. Mao, Y.; Liming, K.; Chunping, H.; Fencheng, L.; Qiang, L. Formation characteristic, microstructure, and mechanical performances of aluminum-based components by friction stir additive manufacturing. *Int. J. Adv. Manuf. Technol.* **2015**, *83*, 1637–1647. [CrossRef]
8. Sasse, C.; Brühl, F.; Schäfer, M. Thermal Treatment of Aviation Aluminum Alloys in Strip Processing Lines. Available online: /KEM.746.168 (accessed on 2 April 2020).
9. Zhang, X.; Ma, J.; Jia, Z.; Song, D. Machining parameter optimisation for aviation aluminium-alloy thin-walled parts in high-speed milling. *Int. J. Mach. Mach. Mater.* **2018**, *20*, 180–192. [CrossRef]
10. Thomas, S.K.; Cassoni, R.P.; MacArthur, C.D. Aircraft anti-icing and de-icing techniques and modeling. *J. Aircr.* **1996**, *33*, 841–854. [CrossRef]
11. Masiulaniec, K.C. A Numerical Simulation of the Full Two-Dimensional Electrothermal De-Icer Pad; National Aeronautics and Space Administration, Scientific and Technical Information Division. In Proceedings of the 21st Aerospace Sciences Meeting Conference, Reno, NV, USA, 10–13 January 1983; pp. 1–7.
12. Yaslik, A.D.; De Witt, K.J.; Keith, T.G.; Boronow, W. Three-dimensional simulation of electrothermal deicing systems. *J. Aircr.* **1992**, *29*, 1035–1042. [CrossRef]
13. Work, A.; Lian, Y. A critical review of the measurement of ice adhesion to solid substrates. *Prog. Aerosp. Sci.* **2018**, *98*, 1–26. [CrossRef]
14. Rønneberg, S.; Laforte, C.; He, J.; Zhang, Z. Comparison of Icephobic Materials through Interlaboratory Studies. In *Ice Adhesion*; John Wiley & Sons: Hoboken, NJ, USA, 2020; pp. 285–324.
15. Huang, X.; Tepylo, N.; Pommier-Budinger, V.; Budinger, M.; Bonaccurso, E.; Villedieu, P.; Bennani, L. A survey of icephobic coatings and their potential use in a hybrid coating/active ice protection system for aerospace applications. *Prog. Aerosp. Sci.* **2019**, *105*, 74–97. [CrossRef]
16. Palacios, J.; Smith, E.; Rose, J.; Royer, R. Ultrasonic De-Icing of Wind-Tunnel Impact Icing. *J. Aircr.* **2011**, *48*, 1020–1027. [CrossRef]
17. Wang, Z. Recent progress on ultrasonic de-icing technique used for wind power generation, high-voltage transmission line and aircraft. *Energy Build.* **2017**, *140*, 42–49. [CrossRef]
18. Meuler, A.J.; Smith, J.D.; Varanasi, K.K.; Mabry, J.M.; McKinley, G.H.; Cohen, R.E. Relationships between Water Wettability and Ice Adhesion. *ACS Appl. Mater. Interfaces* **2010**, *2*, 3100–3110. [CrossRef]
19. Yan, W.; Liu, H.; Chen, T.; Sun, Q.; Zhu, W. Fast and low-cost method to fabricate large-area superhydrophobic surface on steel substrate with anticorrosion and anti-icing properties. *J. Vac. Sci. Technol. A* **2016**, *34*, 041401. [CrossRef]
20. Alizadeh, A.; Yamada, M.; Li, R.; Shang, W.; Otta, S.; Zhong, S.; Ge, L.; Dhinojwala, A.; Conway, K.R.; Bahadur, V.; et al. Dynamics of Ice Nucleation on Water Repellent Surfaces. *Langmuir* **2012**, *28*, 3180–3186. [CrossRef]
21. Schutzius, T.M.; Jung, S.; Maitra, T.; Eberle, P.; Antonini, C.; Stamatopoulos, C.; Poulikakos, D. Physics of Icing and Rational Design of Surfaces with Extraordinary Icephobicity. *Langmuir* **2015**, *31*, 4807–4821. [CrossRef]
22. Alamri, S.; Vercillo, V.; Aguilar-Morales, A.I.; Schell, F.; Wetterwald, M.; Lasagni, A.F.; Bonaccurso, E.; Kunze, T. Self-Limited Ice Formation and Efficient De-Icing on Superhydrophobic Micro-Structured Airfoils through Direct Laser Interference Patterning. *Adv. Mater. Interfaces* **2020**, *7*. [CrossRef]

23. De Pauw, D.; Dolatabadi, A. Effect of Superhydrophobic Coating on the Anti-Icing and Deicing of an Airfoil. *J. Aircr.* **2017**, *54*, 490–499. [CrossRef]

24. Nosonovsky, M.; Hejazi, V. Why Superhydrophobic Surfaces Are Not Always Icephobic. *ACS Nano* **2012**, *6*, 8488–8491. [CrossRef] [PubMed]

25. Song, Y.; Wang, C.; Dong, X.; Yin, K.; Zhang, F.; Xie, Z.; Chu, D.; Duan, J. Controllable superhydrophobic aluminum surfaces with tunable adhesion fabricated by femtosecond laser. *Opt. Laser Technol.* **2018**, *102*, 25–31. [CrossRef]

26. Waller, E.H.; Dix, S.; Gutsche, J.; Widera, A.; Von Freymann, G. Functional Metallic Microcomponents via Liquid-Phase Multiphoton Direct Laser Writing: A Review. *Micromachines* **2019**, *10*, 827. [CrossRef] [PubMed]

27. Alqattan, B.; Yetisen, A.K.; Butt, H. Direct Laser Writing of Nanophotonic Structures on Contact Lenses. *ACS Nano* **2018**, *12*, 5130–5140. [CrossRef] [PubMed]

28. Li, J.; Zhou, Y.; Fan, F.; Du, F.; Yu, H. Controlling surface wettability and adhesive properties by laser marking approach. *Opt. Laser Technol.* **2019**, *115*, 160–165. [CrossRef]

29. Huerta-Murillo, D.; Morales, A.I.A.; Alamri, S.; Cardoso, J.; Jagdheesh, R.; Lasagni, A.-F.; Ocaña, J.L. Fabrication of multi-scale periodic surface structures on Ti-6Al-4V by direct laser writing and direct laser interference patterning for modified wettability applications. *Opt. Lasers Eng.* **2017**, *98*, 134–142. [CrossRef]

30. Milles, S.; Soldera, M.; Kuntze, T.; Lasagni, A.F. Characterization of self-cleaning properties on superhydrophobic aluminum surfaces fabricated by direct laser writing and direct laser interference patterning. *Appl. Surf. Sci.* **2020**, *525*, 146518. [CrossRef]

31. Samanta, A.; Wang, Q.; Shaw, S.K.; Ding, H. Roles of chemistry modification for laser textured metal alloys to achieve extreme surface wetting behaviors. *Mater. Des.* **2020**, *192*, 108744. [CrossRef]

32. Li, J.; Du, F.; Zhao, Y.; Zhao, S.; Yu, H. Two–step fabrication of superhydrophobic surfaces with anti–adhesion. *Opt. Laser Technol.* **2019**, *113*, 273–280. [CrossRef]

33. Milles, S.; Voisiat, B.; Nitschke, M.; Lasagni, A.F. Influence of roughness achieved by periodic structures on the wettability of aluminum using direct laser writing and direct laser interference patterning technology. *J. Mater. Process. Technol.* **2019**, *270*, 142–151. [CrossRef]

34. Chen, K.; Sun, T. Effects of microstructure design on aluminum surface hydrophobic and ice-retarding properties. *Asia-Pacific J. Chem. Eng.* **2017**, *12*, 307–312. [CrossRef]

35. Milles, S.; Soldera, M.; Voisiat, B.; Lasagni, A.F. Fabrication of superhydrophobic and ice-repellent surfaces on pure aluminium using single and multiscaled periodic textures. *Sci. Rep.* **2019**, *9*, 1–13. [CrossRef] [PubMed]

36. Bieda, M.; Siebold, M.; Lasagni, A.F. Fabrication of sub-micron surface structures on copper, stainless steel and titanium using picosecond laser interference patterning. *Appl. Surf. Sci.* **2016**, *387*, 175–182. [CrossRef]

37. Indrišiūnas, S.; Voisiat, B.; Žukauskas, A.; Račiukaitis, G. Direct laser beam interference patterning technique for fast high aspect ratio surface structuring. *Proc. SPIE* **2015**, *9350*, 935003. [CrossRef]

38. Fang, S.; Hsu, C.-J.; Klein, S.; Llanes, L.; Bähre, D.; Mücklich, F. Influence of Laser Pulse Number on the Ablation of Cemented Tungsten Carbides (WC-CoNi) with Different Grain Size. *Lubricants* **2018**, *6*, 11. [CrossRef]

39. Roitero, E.; Lasserre, F.; Anglada, M.; Mücklich, F.; Jimenez-Pique, E. A parametric study of laser interference surface patterning of dental zirconia: Effects of laser parameters on topography and surface quality. *Dent. Mater.* **2017**, *33*, e28–e38. [CrossRef] [PubMed]

40. König, F.; Rosenkranz, A.; Grützmacher, P.G.; Mücklich, F.; Jacobs, G. Effect of single- and multi-scale surface patterns on the frictional performance of journal bearings—A numerical study. *Tribol. Int.* **2020**, *143*, 106041. [CrossRef]

41. Aguilar-Morales, A.I.; Alamri, S.; Voisiat, B.; Kunze, T.; Lasagni, A.F. The Role of the Surface Nano-Roughness on the Wettability Performance of Microstructured Metallic Surface Using Direct Laser Interference Patterning. *Materials* **2019**, *12*, 2737. [CrossRef]

42. Zwahr, C.; Helbig, R.; Werner, C.; Lasagni, A.F. Fabrication of multifunctional titanium surfaces by producing hierarchical surface patterns using laser based ablation methods. *Sci. Rep.* **2019**, *9*, 1–13. [CrossRef] [PubMed]

43. Cardoso, J.; Aguilar-Morales, A.; Alamri, S.; Huerta-Murillo, D.; Cordovilla, F.; Lasagni, A.; Ocaña, J. Superhydrophobicity on hierarchical periodic surface structures fabricated via direct laser writing and direct laser interference patterning on an aluminium alloy. *Opt. Lasers Eng.* **2018**, *111*, 193–200. [CrossRef]

44. Vercillo, V.; Tonnicchia, S.; Romano, J.; García-Girón, A.; Aguilar-Morales, A.I.; Alamri, S.; Dimov, S.S.; Kunze, T.; Lasagni, A.F.; Bonaccurso, E. Design Rules for Laser-Treated Icephobic Metallic Surfaces for Aeronautic Applications. *Adv. Funct. Mater.* **2020**, *30*. [CrossRef]

45. Strobl, T.; Raps, D.; Hornung, M. Comparative Evaluation of Ice Adhesion Behavior. *Mater. Sci.* **2012**, *6*, 1622–1627.

46. Portet, D.; Lecollinet, G.; Hindre, F.; Bejanin, S.; Chappard, D.; Denizot, B. Procede de Recouvrement de Surfaces Metalliques ou Minerales par des Monocouches Autoassemblees de Composes gem-Bisphosphoniques et Leurs Utilisations. Patent WO2008017721A2, 14 February 2008.

47. Choi, J.; Kawaguchi, M.; Kato, T. Self-assembled monolayer formation on magnetic hard disk surface and friction measurements. *J. Appl. Phys.* **2002**, *91*, 7574. [CrossRef]

48. Chichkov, B.N.; Momma, C.; Nolte, S.; Von Alvensleben, F.; Tünnermann, A. Femtosecond, picosecond and nanosecond laser ablation of solids. *Appl. Phys. A* **1996**, *63*, 109–115. [CrossRef]

49. Aguilar-Morales, A.I.; Alamri, S.; Lasagni, A.F. Micro-fabrication of high aspect ratio periodic structures on stainless steel by picosecond direct laser interference patterning. *J. Mater. Process. Technol.* **2018**, *252*, 313–321. [CrossRef]

50. Vercillo, V.; Cardoso, J.T.; Huerta-Murillo, D.; Tonnicchia, S.; Laroche, A.; Guillén, J.M.; Ocaña, J.L.; Lasagni, A.F.; Bonaccurso, E. Durability of superhydrophobic laser-treated metal surfaces under icing conditions. *Mater. Lett. X* **2019**, *3*, 100021. [CrossRef]

51. DIN EN ISO 4287:2010-07, *Geometrische Produktspezifikation (GPS)_- Oberflächenbeschaffenheit: Tastschnittverfahren_- Benennungen, Definitionen Und Kenngrößen Der Oberflächenbeschaffenheit (ISO_4287:1997_+ Cor_1:1998_+ Cor_2:2005_+ Amd_1:2009); Deutsche Fassung EN_ISO_4287:1998_+ AC:2008_+ A1:2009*; Beuth Verlag GmbH: Berlin, Germany, 2009.

52. Hauk, T.; Strobl, T.; Raps, D. Implementation and Calibration of the Icing and Contamination Research Facility (ICORE). In Proceedings of the 25th European Conference on Liquid Atomization and Spray Systems (ILASS), Chania, Greece, 1–4 September 2013; p. 8.

53. Thompson, D.S.; Meng, D.; Afshar, A.; Bassou, R.; Zong, J.; Bonaccurso, E.; Laroche, A.; Vercillo, V. Initial Development of a Model to Predict Impact Ice Adhesion Stress. In Proceedings of the 2018 Atmospheric and Space Environments Conference; American Institute of Aeronautics and Astronautics (AIAA), Atlanta, GA, USA, 25–29 June 2018.

54. Jeck, R.K. *Icing Design Envelopes (14 CFR Parts 25 and 29, Appenddix C) Converted to a Distance-Based Format*; Federal Aviation Administration Technical Center: Atlantic City, NJ, USA, 2002.

55. Ruff, G.A. *Analysis and Verification of the Icing Scaling Equations*; Arnold Engineering Development Center: Tullahoma, TN, USA, 1986; Volume 1.

56. Garcia-Giron, A.; Romano, J.-M.; Batal, A.; Dashtbozorg, B.; Dong, H.; Solanas, E.M.; Angos, D.U.; Walker, M.; Penchev, P.; Dimov, S.S. Durability and Wear Resistance of Laser-Textured Hardened Stainless Steel Surfaces with Hydrophobic Properties. *Langmuir* **2019**, *35*, 5353–5363. [CrossRef]

57. Eral, H.B.; Mannetje, D.J.C.M.; Oh, J.M. Contact angle hysteresis: A review of fundamentals and applications. *Colloid Polym. Sci.* **2013**, *291*, 247–260. [CrossRef]

58. Su, C.; Li, Y.; Dai, Y.-Z.; Gao, F.; Tang, K.; Cao, H. Fabrication of three-dimensional superhydrophobic membranes with high porosity via simultaneous electrospraying and electrospinning. *Mater. Lett.* **2016**, *170*, 67–71. [CrossRef]

59. Subramanyam, S.B.; Kondrashov, V.; Rühe, J.; Varanasi, K.K. Low Ice Adhesion on Nano-Textured Superhydrophobic Surfaces under Supersaturated Conditions. *ACS Appl. Mater. Interfaces* **2016**, *8*, 12583–12587. [CrossRef]

60. Dehkordi, H.B.; Farzaneh, M.; Van Dyke, P.; Kollár, L.E. The effect of droplet size and liquid water content on ice accretion and aerodynamic coefficients of tower legs. *Atmos. Res.* **2013**, 362–374. [CrossRef]

61. Hassan, M.F.; Lee, H.P.; Lim, S.P. The variation of ice adhesion strength with substrate surface roughness. *Meas. Sci. Technol.* **2010**, *21*, 075701. [CrossRef]

62. Bharathidasan, T.; Kumar, S.V.; Bobji, M.; Chakradhar, R.; Basu, B.J. Effect of wettability and surface roughness on ice-adhesion strength of hydrophilic, hydrophobic and superhydrophobic surfaces. *Appl. Surf. Sci.* **2014**, *314*, 241–250. [CrossRef]

63. Eskandarian, M.; Jennings, R.; Coté, M.; Arsenault, B. Fracture Behavior of Typical Structural Adhesive Joints Under Quasi-Static and Cyclic Loadings. *SAE Int. J. Mater. Manuf.* **2010**, *3*, 622–627. [CrossRef]

64. Ginnings, D.; Corruccini, R. An improved ice calorimeter—The determination of its calibration factor and the density of ice at 0 degrees C. *J. Res. Natl. Inst. Stand. Technol.* **1947**, *38*, 583. [CrossRef]

65. Golovin, K.; Kobaku, S.P.R.; Lee, D.H.; DiLoreto, E.T.; Mabry, J.M.; Tuteja, A. Designing durable icephobic surfaces. *Sci. Adv.* **2016**, *2*, e1501496. [CrossRef]

66. Saleema, N.; Farzaneh, M.; Paynter, R.W.; Sarkar, D. Prevention of Ice Accretion on Aluminum Surfaces by Enhancing Their Hydrophobic Properties. *J. Adhes. Sci. Technol.* **2011**, *25*, 27–40. [CrossRef]

67. Kulinich, S.A.; Farzaneh, M. On ice-releasing properties of rough hydrophobic coatings. *Cold Reg. Sci. Technol.* **2011**, *65*, 60–64. [CrossRef]

68. Deng, T.; Varanasi, K.K.; Hsu, M.; Bhate, N.; Keimel, C.; Stein, J.; Blohm, M.L. Nonwetting of impinging droplets on textured surfaces. *Appl. Phys. Lett.* **2009**, *94*, 133109. [CrossRef]

69. Liu, J.; Zhu, C.; Liu, K.; Jiang, Y.; Song, Y.; Francisco, J.S.; Zeng, X.C.; Wang, J. Distinct ice patterns on solid surfaces with various wettabilities. *Proc. Natl. Acad. Sci. USA* **2017**, *114*, 11285–11290. [CrossRef]

70. Lo, C.-W.; Sahoo, V.; Lu, M.-C. Control of Ice Formation. *ACS Nano* **2017**, *11*, 2665–2674. [CrossRef] [PubMed]

71. Varanasi, K.K.; Deng, T.; Smith, J.D.; Hsu, M.; Bhate, N. Frost formation and ice adhesion on superhydrophobic surfaces. *Appl. Phys. Lett.* **2010**, *97*, 234102. [CrossRef]

72. Kulinich, S.A.; Farzaneh, M. Ice adhesion on super-hydrophobic surfaces. *Appl. Surf. Sci.* **2009**, *255*, 8153–8157. [CrossRef]

73. Memon, H.; Liu, J.; De Focatiis, D.S.; Choi, K.-S.; Hou, X. Intrinsic dependence of ice adhesion strength on surface roughness. *Surf. Coat. Technol.* **2020**, *385*, 125382. [CrossRef]

74. Momen, G.; Jafari, R.; Farzaneh, M. Ice repellency behaviour of superhydrophobic surfaces: Effects of atmospheric icing conditions and surface roughness. *Appl. Surf. Sci.* **2015**, *349*, 211–218. [CrossRef]

75. Sarshar, M.A.; Song, D.; Swarctz, C.; Lee, J.; Choi, C.-H. Anti-Icing or Deicing: Icephobicities of Superhydrophobic Surfaces with Hierarchical Structures. *Langmuir* **2018**, *34*, 13821–13827. [CrossRef]

76. He, Y.; Jiang, C.; Cao, X.; Chen, J.; Tian, W.; Yuan, W. Reducing ice adhesion by hierarchical micro-nano-pillars. *Appl. Surf. Sci.* **2014**, *305*, 589–595. [CrossRef]

Article

The Role of the Laser-Induced Oxide Layer in the Formation of Laser-Induced Periodic Surface Structures

Camilo Florian [1,*], Jean-Luc Déziel [2], Sabrina V. Kirner [1], Jan Siegel [3] and Jörn Bonse [1,*]

[1] Bundesanstalt für Materialforschung und -prüfung (B A M), Unter den Eichen 87, 12205 Berlin, Germany; Sabrina.Kirner@gmx.net

[2] Département de Physique, Université Laval, Pavillon Alexandre-Vachon 1045, Av. de la Médecine, Québec, QC G1V0A6, Canada; jean-luc.deziel.1@ulaval.ca

[3] Laser Processing Group, Instituto de Óptica IO-CSIC, Serrano 121, 28006 Madrid, Spain; j.siegel@io.cfmac.csic.es

[*] Correspondence: Camilo.Florian-Baron@bam.de (C.F.); Joern.Bonse@bam.de (J.B.)

Received: 6 December 2019; Accepted: 9 January 2020; Published: 14 January 2020

Abstract: Laser-induced periodic surface structures (LIPSS) are often present when processing solid targets with linearly polarized ultrashort laser pulses. The different irradiation parameters to produce them on metals, semiconductors and dielectrics have been studied extensively, identifying suitable regimes to tailor its properties for applications in the fields of optics, medicine, fluidics and tribology, to name a few. One important parameter widely present when exposing the samples to the high intensities provided by these laser pulses in air environment, that generally is not considered, is the formation of a superficial laser-induced oxide layer. In this paper, we fabricate LIPSS on a layer of the oxidation prone hard-coating material chromium nitride in order to investigate the impact of the laser-induced oxide layer on its formation. A variety of complementary surface analytic techniques were employed, revealing morphological, chemical and structural characteristics of well-known high-spatial frequency LIPSS (HSFL) together with a new type of low-spatial frequency LIPSS (LSFL) with an anomalous orientation parallel to the laser polarization. Based on this input, we performed finite-difference time-domain calculations considering a layered system resembling the geometry of the HSFL along with the presence of a laser-induced oxide layer. The simulations support a scenario that the new type of LSFL is formed at the interface between the laser-induced oxide layer and the non-altered material underneath. These findings suggest that LSFL structures parallel to the polarization can be easily induced in materials that are prone to oxidation.

Keywords: laser-induced oxide layer; laser-induced periodic surface structures; LIPSS; surface chemistry; nanostructuring; femtosecond laser processing

---

1. Introduction

The fabrication of laser-induced periodic surface structures (LIPSS) on metals, semiconductors and dielectrics can be realized with linearly polarized high intensity ultrashort laser pulses [1]. These structures are parallel line-grating-like modifications formed at the sample surface in a self-ordered way either parallel ($\parallel$) or perpendicular ($\perp$) to the laser beam polarization. Their periodicity (Λ_{LIPSS}) usually ranges from hundreds of nanometers up to some micrometers and it is used to classify them into the general categories as low-spatial frequency LIPSS (LSFL), when $\Lambda_{LSFL} \sim \lambda$, and high-spatial frequency LIPSS (HSFL) for $\Lambda_{HSFL} \ll \lambda$, where λ is the laser wavelength [2]. Suitable manufacturing strategies have been identified, including the optimization of laser processing parameters (laser fluence, wavelength, repetition rate, angle of incidence, number of pulses per spot area) [3–8], material properties (optical, thermal and mechanical properties) [9–11], and the ambient medium in which they are generated (air, vacuum, reactive atmospheres) [12–15] for applications in optics, medicine, fluid transport and tribology among others [1].

The abundance of parameters and the complexity of the mechanisms involved pose difficulties to develop a model that satisfyingly describes experimental findings. However, there have been recent significant advances in models based on finite-difference time-domain calculations (FDTD) that describe the formation of ablative LSFL in order to account for the periodicity of LIPSS on various materials under realistic irradiation conditions, including the formation of random defects at the surface [16]. For metals, Rudenko et al. [17] implemented a numerical FDTD approach for calculating the electromagnetic fields scattered from ensembles of isolated dipole-like scatterers. In addition, FDTD simulations were used for the description of the processes that occur when different types of structures are formed in dielectrics under femtosecond laser irradiation [18]. In the latter work, the formation mechanisms of different types of LIPSS including LSFL parallel and HSFL perpendicular to the polarization were presented consistently with experimental demonstrations and for the LSFL in line with an analytical theory [19]. Based on the intrinsic optical properties of the irradiated materials, FDTD simulations allow to classify the periodic structures into specific categories [16,20,21], including type-d ("dissident") structures corresponding to LSFL being parallel to the polarization with periodicities $\Lambda_{LSFL} \sim \lambda/n$ (with n being the refractive index of material), type-s ("scattering") for LSFL perpendicular to the polarization with $\Lambda_{LSFL} \sim \lambda$, type-m ("mixed") for a special kind of LSFL parallel to the polarization with periodicities $\Lambda_{LSFL} \sim \lambda$, and type-r ("roughness") for HSFL structures that typically are perpendicular to the laser polarization and $\Lambda_{HSFL} << \lambda$. Fuentes-Edfuf et al. [8] recently published a study on the formation of LIPSS in metals where the key parameters analyzed were the beam incident angle and the role of surface roughness for the generation of surface plasmons that are ultimately responsible for the formation of LIPSS. The experiments were done in air atmosphere at room temperature using fluences above the ablation threshold. Thus, the temperatures reached are high enough to allow the sample oxidation on the irradiated areas. Despite that those studies include experimental results in good agreement with simulations, the laser-induced oxidation was not considered. Usually, this effect is only accounted when the final application for which the LIPSS where fabricated is compromised, such as in the case of surface wetting [22] and tribology [23].

About 20 years ago, the fs-laser ablation behavior of TiN, CrN and other hard-coating materials was studied in detail by several groups using low repetition rate (≤1 kHz) Titanium:Sapphire (Ti:Sa) laser pulses at 790 nm wavelength [24,25]. HSFL with an orientation perpendicular to the laser beam polarization and periods of 100–200 nm were found at fluences close to the ablation threshold [24]. At gradually increasing fluences, the structures showed a "smooth transition" of the HSFL towards LSFL of larger periods featuring the same orientation [25]. These structures are termed LSFL$^\perp$ in the following. Large area chemical analyzes revealed that the HSFL were formed in the regime of strong superficial oxidation [24] but an explicit link between the LIPSS topography and surface chemistry was not established. Recently, Kirner et al. presented the first spatially and depth-resolved investigation of both HSFL and LSFL$^\perp$ formed upon fs-laser scan processing of titanium surfaces in air environment [26]. These depths-profiling Auger microscopy measurements revealed that a superficial oxide layer is accompanying the formation of LIPSS in the ablative regime. Its extent of more than 150 nm into the bulk largely exceeds the topographical depth modulation of the HSFL, which typically accounts for some tens of nanometers only. However, this "classical regime" of ablative LIPSS formation has to be distinguished from other processing regimes of LIPSS that may manifest in the sub-ablative regime through localized heat-induced effects: Öktem et al. [27] presented a detailed study on the formation of superficial oxide-based periodic structures in metals by inducing the incorporation of oxygen to form oxides (specifically titanium dioxide and tungsten oxide). This process occurs when the incoming focused laser beam interferes with radiation scattered at existing nanostructures or surface defects leading to the imprint of a periodic fluence pattern of maxima and minima, allowing the incorporation of oxygen in the places where the intensity maxima is located [27]. Dostovalov et al. [28,29] very recently published several works on thermochemical formation of elevated parallel and perpendicular surface structures on metallic chromium films produced by local oxidation at the local intensity maxima produced by high-repetition rate (200 kHz) laser pulses in the non-ablative regime. One of them indicates that it is possible to control the periodicity of the LIPSS by choosing the proper Cr film thickness. A simulation allowed determining the influence of different spatial concentrations of two different Cr oxides, i.e., CrO_2 and Cr_2O_3, along one of the produced ridges. The simulated field spatial distribution demonstrates that changes in the chemistry induced thermally by the laser pulses affect the final intensity pattern imprinted into the material [28]. In another similar system of Cr films, the authors consider different beam scanning speeds to investigate the influence on the transition from HSFL perpendicular to the polarization to LSFL being parallel to it (LSFL$^{\parallel}$), involving the formation of both Cr oxides and the interaction of a surface electromagnetic wave between two dielectric materials, air and glass, in which the Cr film has been deposited [29]. In this case, the proposed model includes the optical properties of both Cr oxides and ratios between them; however, the obtained periodicities are ~1.5 times larger than the ones obtained experimentally. Only by assuming a certain porosity inside the oxide layer, the predicted periodicities are close to the measured values [29]. From all this, it is clear that the incorporation of additional effects, such as superficial oxidation, into the calculations to predict the LSFL periodicity and orientation is not a trivial task.

Therefore, in this paper we present a different approach by numerically simulating a layered system via FDTD calculations of the laser beam intensity inside a superficial laser-generated oxidation layer and at its interface to an underlying substrate of CrN as a model of an oxidation-prone strong absorbing material. Surprisingly, we found instead of the usual LSFL$^{\perp}$ some LSFL$^{\parallel}$ that were not reported yet for this type of material. These new structures are formed for static and dynamic irradiations of CrN in the ablative regime and could be generated using two different laser systems in two independent laboratories. The irradiated samples were characterized chemically by energy dispersive X-ray analyses (EDX) and μ-Raman spectroscopy, and morphologically by scanning electron microscopy (SEM) and atomic force microscopy (AFM). When comparing the numerical FDTD simulations with the experimental results, we obtain a significant agreement of their characteristics, suggesting that the formation of the laser-induced oxide layer is a key parameter for the formation of the LSFL$^{\parallel}$ structures.

2. Materials and Methods

The samples used for the experiments were provided by Miba Gleitlager GmbH (Laakirchen, Austria) within the frame of the European FET Open research project LiNaBioFluid. They consist of steel disks (16MnCr5, 1.7131) with dimensions of 35 mm diameter and 1 cm height, with a central borehole. The upper flat polished surface was covered via physical vapor deposition (sputtering) with a 2.5 μm thick layer of chromium nitride (CrN) on top of a ~200 nm thin bonding intermediate chromium (Cr) layer, as shown in Figure 1A. The samples were stored in a desiccator and were cleaned in an ultrasonic bath in isopropanol/acetone for 5 min and subsequent drying with pressurized air.

The CrN layer was characterized by X-ray diffraction (XRD). The measurements (XRD 300 TT, Seifert, Ahrensburg, Germany) were performed in Bragg-Brentano geometry using Cu Kα-radiation. Since the element iron from the underlying steel substrate caused a significant X-ray fluorescence, the acquired diffraction curves were corrected by a linear background subtraction. The XRD analysis (Figure 1B) indicates peaks of the Carlsbergite phase of stoichiometric CrN and of the underlying steel substrate. The widths of the CrN peaks and their peak intensities point toward a textured growth of CrN with crystalline grain sizes of several tens up to a few hundreds of nanometers—consistent with the SEM micrograph and AFM data shown in Figure 3A. Optical characterization was performed using a Variable Angle Spectroscopic Ellipsometry (VASE) to determine the optical constants of the CrN layer with a J.A. Woollam M-2000DI ellipsometer (Lincoln, NE, USA) and the WVase32 software (version 3.81) of the same company. The corresponding data of the refractive index n and the extinction k are presented in Figure 1C in the spectral range of 200 to 1700 nm. From these measurements, the linear absorption coefficient $\alpha = 4\pi k/\lambda$ and the associated optical penetration depth $1/\alpha$ can be calculated. At

800 nm ($n = 2.439$, $k = 1.623$) and 1030 nm ($n = 2.945$, $k = 1.7306$) wavelengths, the latter account to $1/\alpha$ ~39 nm and 47 nm, respectively. The optical penetration depth of the laser radiation in CrN is more than one order of magnitude smaller that the thickness of the CrN layer. Hence, from the optical point of view, the CrN can be considered as bulk material here.

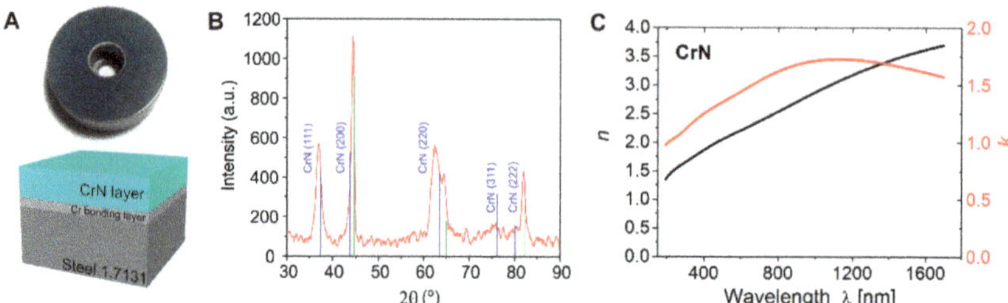

Figure 1. (**A**) Picture of the actual sample with a sketch that illustrates the steel 1.7131 substrate, covered with a 200 nm Cr film for bonding a 2.5 µm thick chromium nitride (CrN) top layer. (**B**) X-ray diffraction (XRD) data of a non-irradiated area of a CrN samples as displayed in (**A**), including peaks corresponding to iron (marked green) and a CrN form known as Carlsbergite (marked blue). (**C**) Plot of the refractive index n (left axis) and the extinction coefficient k (right axis) of the CrN layer for different wavelengths as measured by ellipsometry. In our calculations we used $n = 2.439$ and $k = 1.623$ for $\lambda = 800$ nm.

Laser irradiation setups. The irradiation experiments performed involved loosely focused femtosecond laser pulses from two different laser sources: Figure 2A shows a solid state Ti:Sa laser amplifier system (Compact Pro, Femtolasers, Vienna, Austria) operated at 1 kHz repetition rate delivering pulses of 790 nm center wavelength and duration of ~30 fs (full width at half maximum, FWHM).

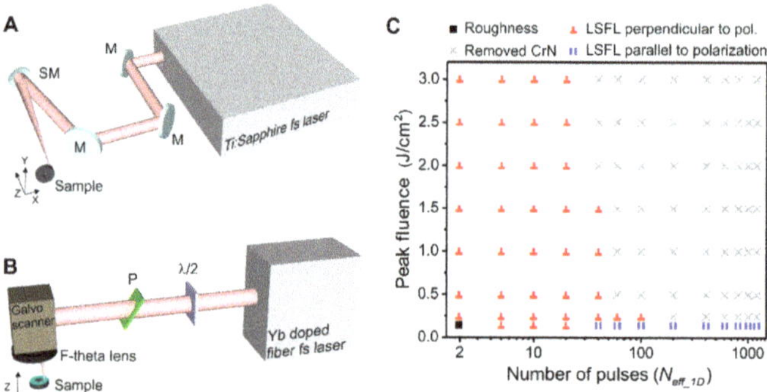

Figure 2. (**A**) Setup of the Ti:Sa laser system (BAM). The beam is directed using several mirrors (M), the polarization is defined by a half-wave plate ($\lambda/2$) and is finally focused in a static position on the sample by a spherical dielectric mirror (SM) of 500 mm focal length. The sample is positioned and translated by a motorized X-Y-Z translation stage. (**B**) Setup of the Yb-doped fiber laser (CSIC). The beam power is controlled by a combination of a half-wave plate ($\lambda/2$) and a thin film polarizer (P). The beam scans the sample by a galvanometric mirror-based scanning head, and it is focused by a 100 mm F-theta lens. The sample is mounted in a Z-axis translation stage to control the focusing of the beam. (**C**) Plot that displays the obtained type of surface structures on CrN for a combination of number of effective pulses (N_{eff_1D}) and peak fluences used for the formation of (■) roughness, (‖) indicates the presence of LSFL$^{\|}$, ($\perp$) indicates the presence of LSFL$^{\perp}$ and (×) marks the conditions where the CrN layer has been destroyed.

This setup focused the laser pulses using a spherical broadband dielectric mirror of 500 mm focal length (f/#-number ~130), obtaining a $1/e^2$-beam diameter at the focus of $2w_0 = 130$ μm. The sample was mounted normal to the incident focused laser beam and positioned with an *X-Y-Z* motorized translation stage (VT-80, MICOS, Eschbach, Germany), scanning the sample in the *X-Y*-plane and controlling the focus with the *Z*-axis. The second laser system, in Figure 2B, shows an Yb-doped fiber laser (Satsuma HP, Amplitude Systemes, Talence, France) with repetition rates ranging from 1 kHz to 2 MHz emitting pulses of 1030 nm wavelength and 350 fs duration (FWHM). For the experiments presented in this work the repetition rate was fixed at 100 kHz. The pulses were focused through a 100 mm focal length (f/#-number ~20) F-Theta lens coupled with a fast galvanometric-mirror based scanning head (SCANcube 14, Scanlab GmbH, Puchheim, Germany) keeping the beam diameter of $2w_0 = 39$ μm on a nominal square area of 70×70 mm^2. The experiments were carried out in the 3 $\times$ 1 mm^2 central region of this working space to avoid any possible modifications due to the beam angle of incidence. In both laser processing systems, the sample was fixed on a holder that allowed the setting of the perpendicularity of the sample surface to the incident laser beam. In addition, during the irradiation, an air suction system was used to extract actively the ablated debris and produced gases in both setups. Figure 2C provides a "morphological map" of the structures that were produced on a CrN sample including LSFL parallel to the laser polarization ($\parallel$) and perpendicular to it ($\perp$). Under a combination of high laser fluences and high number of pulses, the CrN layer is removed from the steel substrate, indicated by the symbol ($\times$).

Irradiation procedure. For the generation of LIPSS on CrN, we have performed a series of static (spot only) and dynamic (spatially overlapped pulses to form lines) irradiations using different incident peak laser fluences (ϕ_0) and number of effective pulses per spot diameter (N_{eff_1D}). Peak fluences were determined according to the method of Liu [30]. The calculation of N_{eff_1D} follows the equations presented in [31] that take into account the scanning speed (v), laser beam waist radius at the focus (w_0) and the laser pulse repetition rate (f) as $N_{eff\_1D} = 2w_0f/v$. For the processing of LIPSS-covered areas, the line separation distance (Δ) is considered as an additional parameter. All laser irradiations were performed in air at room temperature.

Surface characterization. In order to characterize the sample surface morphology, we use two characterization techniques: SEM (Gemini Supra 40, Carl Zeiss, Oberkochen, Germany) and AFM (Dimension 3100, Digital Instruments, Santa Barbara, CA, USA). The SEM micrographs were acquired in SE-InLens mode at 5 and 10 kV electron acceleration voltage. This imaging mode acquires the so-called SE1 secondary electrons produced mainly at the impact point of the primary electron beam allowing for a high spatial resolution. Two-dimensional Fourier transform (2D-FT) analyses were performed using the free software ImageJ (version 1.52a, National Institutes of Health, Bethesda, MD, USA) to get information on the periodicity of the structures. In order to account the periodicity variation, we performed a radial average around the data cloud of interest in the Fourier space, then we fitted the data into a Gaussian distribution from which we obtained sigma (σ) as the periodicity range displayed in all the 2D-FTs [32]. AFM topographic data of specific samples were acquired in tapping mode using silicon probes with a nominal tip radius of 10 nm. Images contained 512 lines in *X*- and 256 lines in *Y*-direction, allowing a nominal resolution in *Z* of 70 pm (4.7 μm/16 bit in the full *Z*-scale range). The data are displayed as 3D representation of the topography, from which the root-mean-squared (RMS) roughness of the surfaces is obtained and a cross-section topography profile is extracted, parallel to the *X*-axis.

Chemical and structural characterization of laser-irradiated samples. Energy dispersive X-ray analyses (EDX) using a Thermo NSS 3.1 detector (Thermo Fisher Scientific, Waltham, MA, USA) were performed on selected samples to obtain the compositional information of the fs-laser irradiated samples. The acceleration voltage in this case was 5 kV to ensure the necessary surface sensitivity. Single point μ-Raman spectra were recorded as five accumulated acquisitions in backscattering geometry to characterize the local material structure and composition in the fs-laser modified regions. The continuous wave 532 nm wavelength laser beam was focused by a 100× microscope objective (numerical aperture NA = 0.85) to a probing spot diameter of ~1 μm, resulting in a power of 0.1 mW at the sample surface.

3. Results and Discussion

Laser irradiations. The scanning electron micrograph displayed in Figure 3A corresponds to a non-irradiated sample, revealing the CrN grains at the surface. According to the information extracted from the 2D-FT, there is neither a clear periodicity nor orientation of the structures of the pristine surface. From the AFM data, it is possible to obtain the RMS-roughness of $R_{CrN} = 25$ nm. The fs-laser irradiations follow the procedure described in the "Materials and Methods" section. There are two different irradiation strategies: areas processed by dynamically scanning the sample surface with the laser beam or static irradiation of fixed spot locations.

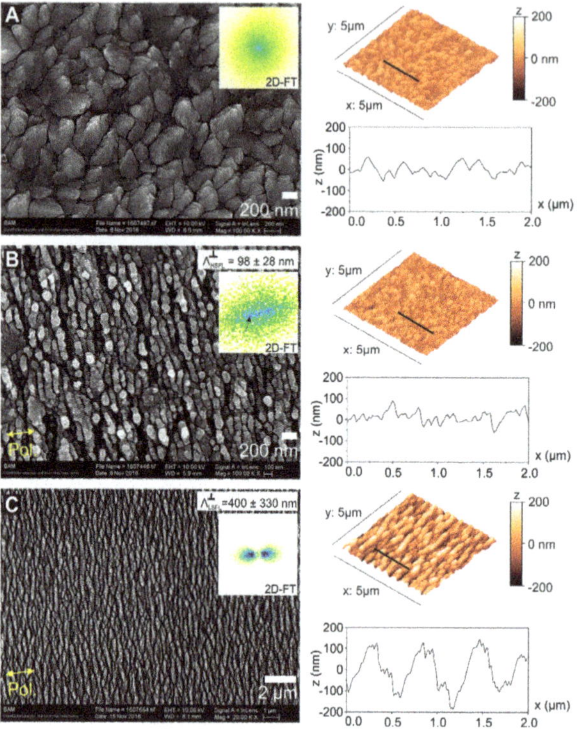

Figure 3. Top-view scanning electron microscopy (SEM) micrographs and 3D reconstructed atomic force microscopy (AFM) data acquired in different areas of the sample with its corresponding two-dimensional Fourier transform to estimate the structures periodicity (Λ). (**A**) Non-irradiated surface showing the randomly oriented CrN grains with no clear periodicity in the Two-dimensional Fourier transform (2D-FT). The AFM data shows a profile taken at the position of the indicated black line, featuring a surface root-mean-squared roughness of $R_{CrN} = 25$ nm. (**B**) HSFL produced on an irradiated area at $\phi_0 = 0.15$ J/cm^2, $N_{eff\_1D} = 200$ and line separation distance $\Delta = 50$ μm. The periodicity in this case is $\Lambda^{\perp}_{HSFL} = 98 \pm 14$ nm, the modulation depth is d_{HSFL} ~50 nm and the root-mean-squared-roughness is $R_{HSFL} = 30$ nm. (**C**) LSFL$^{\perp}$ processed at $\phi_0 = 0.5$ J/cm^2, $N_{eff\_1D} = 10$ and line separation distance $\Delta = 50$ μm. The periodicity is $\Lambda^{\perp}_{LSFL} = 400 \pm 330$ nm, the modulation depth d_{HSFL} ~200 nm and the root-mean-squared-roughness is $R_{LSFL} = 73$ nm. Both structures (Figure 3B,C) were produced with the Ti:Sa system ($\lambda = 790$ nm) and the structures orientation is perpendicular to the laser beam polarization, indicated by yellow double arrows.

The first set of experiments contained dynamic irradiations to process 2×1 mm^2 areas using a Ti:Sa laser system ($\phi_0 = 0.15$ J/cm^2, $N_{\text{eff\_1D}} = 200$, $\Delta = 50$ μm). These parameters are defined in "Materials and Methods" section. At this fluence, the final morphology of the laser-induced structures corresponds to HSFL oriented perpendicular to the laser beam polarization, as displayed in Figure 3B. The periodicity obtained in this case from the 2D-FT of the SEM micrograph is $\Lambda_{\text{HSFL}}^{\perp} = 98 \pm 14$ nm, and the RMS roughness $R_{\text{HSFL}} = 30$ nm. From the AFM cross-sectional profile, it can be seen that the HSFL are superimposed to the roughness of the initial topography of the CrN surface. Note that the measurement of the RMS roughness takes into account the whole surface data including both, the CrN grain roughness with the superimposed laser-generated HSFL, the topography of which can be better observed in the profile of Figure 3B.

Another laser-processed area is produced using an approx. three-times larger laser fluence allowing the fabrication of LSFL perpendicular to the laser polarization ($\phi_0 = 0.5$ J/cm^2, $N_{\text{eff\_1D}} = 10$, $\Delta = 50$ μm). The corresponding surface morphology is shown in Figure 3C, including an SEM micrograph and AFM data. The periodicity range of the structures extracted from the 2D-FT of the SEM micrograph is $\Lambda_{\text{LSFL}}^{\perp} = 400 \pm 330$ nm. The large period variation of about 83% is partly attributed to the effects of period splitting via localized electromagnetic field-enhancement effects [33]. Moreover, the consequences of laser scanning must be considered, where the LSFL$^{\perp}$ generated at the high fluence part of the Gaussian laser beam are subsequently "overwritten" by HSFL formed by the following low-fluence wing of the spatial beam profile. From the AFM cross-section, it is possible to estimate a periodicity of the footprint of the bigger $\Lambda_{\text{LSFL}}^{\perp}$ structures of ~600 nm and an RMS roughness $R_{\text{LSFL}} = 73$ nm, in good agreement with the work of Yasumaru et al. studying the characteristics of fs-laser generated LIPSS on CrN [25].

Surprisingly, when observing the irradiated areas for both laser systems under a regular optical microscope, as shown in Figure 4, it is possible to detect another type of regular LSFL (LSFL$^{\parallel}$), structures (indicated by the sets of red parallel lines in Figure 4) being parallel to the polarization (as indicated by the yellow double-arrows.) Importantly, from the AFM data shown in Figure 3B the only detected structures at the sample surface are HSFL perpendicular to the polarization, which clearly confirms that the LSFL$^{\parallel}$ are not located at the surface but constitute in a sub-surface layer.

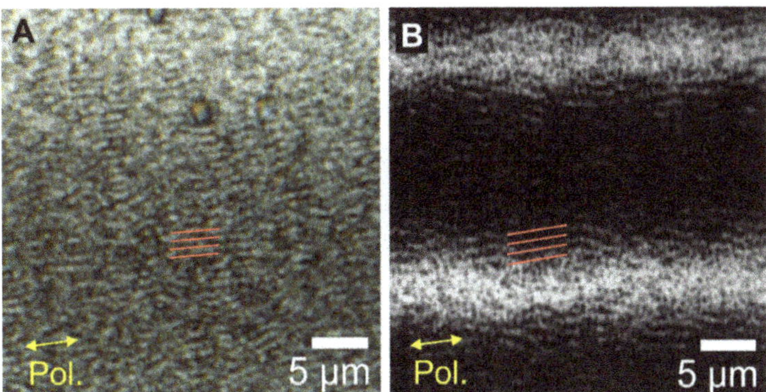

Figure 4. Morphological characterization via top-view optical microscopy (OM) of the samples irradiated with (**A**) Ti:Sa laser system ($\lambda = 790$ nm), using the same parameters of Figure 3B, featuring the structures parallel to the laser polarization, highlighted with a set of four parallel red lines, and (**B**) Yb-doped fiber laser system ($\lambda = 1030$ nm, $\phi_0 = 0.15$ J/cm^2, $N_{\text{eff\_1D}} = 1000$), featuring LSFL$^{\parallel}$ structures parallel to the laser polarization highlighted with a set of four parallel red lines. In both cases, the yellow double arrows indicate the polarization direction.

More in detail, Figure 4A shows the morphologies detected in areas processed with the Ti:Sa laser system, where the topography corresponds to LSFL$^{\parallel}$ with periodicities of $\Lambda^{\parallel}_{LSFL}$ = 820 ± 30 nm. Since the topography corresponds to the same region shown in Figure 3B, it can be assumed that HSFL are present at the surface despite that is not possible to resolve them via optical microscopy. The LSFL$^{\parallel}$ fabricated with the Yb-doped fiber laser system (ϕ_0 = 0.15 J/cm^2, N_{eff_1D} = 1000, Δ = 20 μm) are shown in Figure 4B, presenting a periodicity of $\Lambda^{\parallel}_{LSFL}$ = 1.26 ± 0.08 μm.

In order to discard possible effects due to the laser beam scanning on the LSFL$^{\parallel}$, we performed an additional static (*N*-on-1) irradiation experiment in a single spot with the Ti:Sa laser system under the same experimental conditions as in Figures 3B and 4A (ϕ_0 = 0.15 J/cm^2, N = 200). The corresponding SEM micrograph is included in Figure 5A, taken with an acceleration voltage of 5 kV. Similarly, LSFL structures parallel to the laser beam polarization are present with a periodicity of $\Lambda^{\parallel}_{LSFL}$ = 780 ± 65 nm, which confirms that the phenomenon does not depend on the scanning overlap of the laser pulses. A second micrograph is acquired at higher magnification and acceleration voltage of 10 kV (Figure 5B), showing HSFL similar to those of Figure 3B, with periodicities of $\Lambda^{\perp}_{HSFL}$ = 102 ± 12 nm. In order to confirm that the LSFL$^{\parallel}$ are not located at the sample surface, AFM data displaying the topography is included in Figure 5C, where again only HSFL structures superimposed to the CrN grains can be observed.

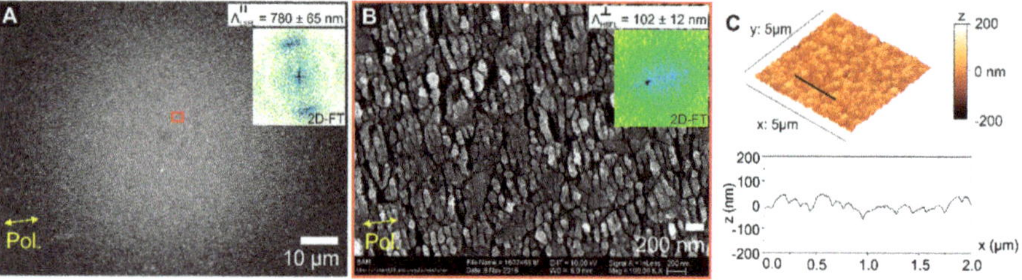

Figure 5. (**A**). Top-view SEM micrograph of a spot made under the same irradiation conditions of the area shown in Figures 3B and 4A (ϕ_0 = 0.15 J/cm^2 and N = 200), acquired in InLens mode at 5 kV. It is also possible to detect the formation LSFL parallel to the beam polarization with periodicity $\Lambda^{\parallel}_{LSFL}$ = 780 ± 65 nm. (**B**) SEM micrograph of the red square inset in (**A**). In this case, the magnification factor is 100.000× and the acceleration voltage was 10 kV, showing the presence of HSFL perpendicular to the polarization with a periodicity $\Lambda^{\perp}_{HSFL}$ = 102 ± 12 nm. (**C**) AFM 3D-data including a profile taken on the indicated black line showing an RMS roughness R_{HSFL} = 32 nm.

Chemical characterization. Additional analyses were performed for the material modification accompanying the formation of the LSFL$^{\parallel}$ structures. A first chemical characterization of the surface was conducted by EDX analysis on the spot shown in Figure 5A using an acceleration voltage of 5 kV to ensure the necessary surface sensitivity. The results are compiled in Figure 6, where the spatial distribution of oxygen (O), chromium (Cr), carbon (C) and nitrogen (N) are encoded in false colors. The EDX elemental map of O confirms an increased concentration of oxygen in the irradiated area keeping the same size and shape of the irradiated spot. Contrary to chromium and carbon, where the concentration is practically the same in the irradiated area and its surroundings, the nitrogen concentration slightly decreases in the irradiated area. This indicates that the oxygen is located close to the sample surface and is, thus, reducing the signal of the nitrogen from underneath due to the formation of a superficial oxidation layer on the CrN. Those results further support the idea that the formation of the LSFL$^{\parallel}$ structures may take place at the interface between the laser-induced oxidation layer and the underlying CrN material, in line with the findings obtained from the SEM/AFM characterizations.

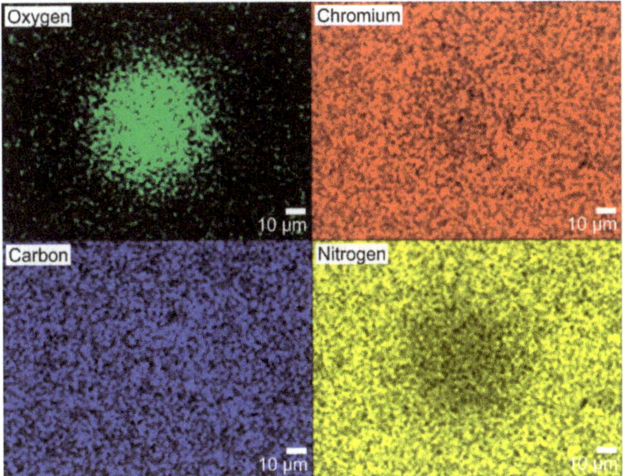

Figure 6. EDX analysis of the fs-laser irradiated spot shown in Figure 5A, where the concentrations of oxygen, chromium, carbon and nitrogen are displayed in false colors. Brighter colors indicate larger signals. Note that the spatial distribution of oxygen is consistent with the irradiated spot shown in Figure 5A.

Complementary structural information was gained from μ-Raman spectroscopy. Figure 7 presents results of the structures produced on a line track produced with the Yb-fiber laser system under the same conditions as the one presented in Figure 4B. In this case, the SEM micrograph of Figure 7A was taken using 5 kV electron acceleration voltage. The LSFL structures are oriented parallel to the polarization direction, indicated by the yellow double-arrows. However, the morphology is dominated by HSFL structures perpendicular to the polarization, presenting a periodicity of $\Lambda_{HSFL}^{\perp} = 213 \pm 35$ nm. The value is larger than the one corresponding to Figure 3B mainly due to the increased laser wavelength ($\lambda = 1030$ nm) and the somewhat different irradiation conditions used in this case.

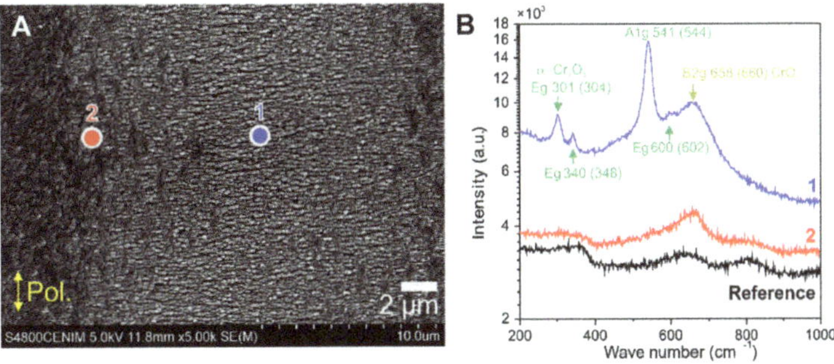

Figure 7. (**A**) Top-view SEM micrograph taken at 5 kV of an irradiated line on the CrN surface using the Yb-doped fiber laser system. The whitish regions indicate that the laser-processing generates LSFL‖ visible in the borders of the line (red dot) parallel to the beam polarization and HSFL in the center (blue dot) perpendicular to the laser beam polarization (indicated by the yellow double-arrows). Irradiation conditions are the same as those of Figure 4B. (**B**) μ-Raman spectra acquired at the positions indicated by the red and blue dots in (**A**) with an additional spectrum that corresponds to a non-irradiated area (black line—dot not shown in (**A**) for comparison. Note the logarithmic scale for the intensity axis.

In order to get detailed information of the composition of the oxide layer, μ-Raman spectra were acquired on different positions across the fs-laser processed line track. For experimental details of the μ-Raman characterization, see the "Materials and Methods" section. The μ-Raman spectra plotted in Figure 7B were recorded on a non-irradiated location of the sample (black line), at the border of the fs-laser track (red line) and at its center (blue line) in the two positions marked in Figure 7A. When comparing to the spectra from the non-irradiated area, both the border of the laser track and its center present an increased Raman signal around 660 cm^{-1}. According to [34], it corresponds to the B_{2g} mode of CrO_2. The Raman spectrum from the border of the track does not present any other identifiable peaks attributed to oxides. In the case of the center of the laser track (blue line), the Raman spectrum shows the presence of several pronounced peaks on a broad background that confirm the presence of the corundum structure of Cr_2O_3 with modes A_{1g} (541 cm^{-1}) and E_g (301, 340 and 600 cm^{-1}) [35,36].

FDTD Simulations. In order to check the hypothesis whether the LSFL$^{\parallel}$ may be formed at the chromium oxide/chromium nitride interface, and more specifically, to identify the role of the oxide layer and its thickness, we have employed the finite-difference time-domain (FDTD) method to calculate the spatial intensity distribution induced by the fs-laser beam at the interface between the CrN layer and the covering oxide layer. It was assumed that the oxide layer is composed mainly of Cr_2O_3 to simplify the calculations to a single material for the oxide layer. For these simulations, we numerically solved the Maxwell's Equations (1) and (2), where $\vec{H}$ is the magnetizing field, $\vec{E}$ the electric field, ϵ_0 the electric permittivity of free space and μ_0 permeability of free space,

$$\vec{\nabla} \times \vec{H} = \epsilon_0 \epsilon_r \frac{\partial \vec{E}}{\partial t} + \sigma \vec{E} \tag{1}$$

$$\vec{\nabla} \times \vec{E} = -\mu_0 \frac{\partial \vec{H}}{\partial t} \tag{2}$$

considering the relative permittivity ϵ_r defined as:

$$\epsilon_r = \mathrm{Re}(\widetilde{n})^2 - \mathrm{Im}(\widetilde{n})^2 \tag{3}$$

in addition, the conductivity σ as:

$$\sigma = 2\epsilon_0 \omega \mathrm{Re}(\widetilde{n})\,\mathrm{Im}(\widetilde{n}) \tag{4}$$

where $\widetilde{n} = n + ik$ is the complex refractive index of the medium and ω is the angular frequency of the laser radiation ($\omega = 2.35 \times 10^{15}$ s^{-1} for $\lambda = 800$ nm irradiation wavelength). The optical constants n and k for the CrN layer were measured experimentally for the substrate system using ellipsometry measurements of a non-irradiated sample, see Figure 1C in the "Materials and Methods" section: $n = 2.439$ and $k = 1.623$ for $\lambda = 800$ nm. For the oxidized layer of Cr_2O_3, $n = 2$ and $k = 0$ were found in [37], and for vacuum $n = 1$ and $k = 0$.

The laser beam considered had Gaussian temporal and spatial distributions. At the surface of the sample it is considered as a plane wave, polarized along the X-axis. The temporal envelope of the electric field at the sample surface is described by:

$$\vec{E}(z = 0, t) = E_0 \exp\left[-\left(\frac{t}{\tau}\right)^2\right] \sin(\omega t)\hat{e}_x \tag{5}$$

The 3D-geometry used to compute the results is included in Figure 8A. The simulation domain sizes were fixed at 3.2 μm in X- and Y-directions. The Z-dimension was adjusted in each case depending on the thickness of each simulated element. The 3D-domain was subdivided into 3D-space cells which sizes were 5 nm for X and Y and 2 nm for Z. From bottom to top, there is a layer of CrN that starts at $Z = -\infty$ and finishes at the interface with the oxide layer ($Z = 0$). In our simulation, the real CrN layer thickness of 2.5 μm is thick enough to be considered optically as an infinite layer (see the values of the optical penetration depth of some tens of nanometers provided in the "Materials and Methods" section). Therefore, the steel substrate and the chromium-bonding layer were excluded for the simulations since they do not have any relevant optical contribution. On top of the CrN layer, there is an oxide film that has a specific thickness T_{ox} ranging from 0 to 500 nm. On top of the oxide layer, and made up of the same oxidized materials, lay the grating-like HSFL structures with periodicity $\Lambda_{HSFL} = 100$ nm for all the simulations. This value is consistent with the HSFL periodicity found via 2D-FT of Figures 3B and 5B, $\Lambda_{HSFL}^{\perp} = 98 \pm 14$ nm and 102 ± 12 nm, respectively. At the sample surface, there are two variables to consider: the modulation depth of the HSFL d_{HSFL} ranging from 0 to 100 nm, and a non-organized roughness R_{HSFL} ranging from 0 to 50 nm. The rest of the domain is always large enough to contain at least 10 nm of vacuum above the HSFL region. The simulations ran from $t = -2\tau$ to $t = 2\tau$, with $\tau = 25$ fs (FWHM $= 2\tau\ln2 \approx 35$ fs pulse duration) using time steps of 5.516 attoseconds. We computed the time-averaged intensity in arbitrary units during the whole interaction, following:

$$I = \int_{-2\tau}^{2\tau} \left(E_x^2 + E_y^2 + E_z^2\right)dt \tag{6}$$

All simulated intensity distributions that were evaluated at the interface between the oxide layer and the CrN sample are compiled in Figure 8. Figure 8B presents a simulation of the intensity in a sample modelled with a HSFL modulation depth $d_{HSFL} = 100$ nm, an oxide layer of thickness $T_{ox} = 100$ nm and an additional non-organized roughness $R_{HSFL} = 20$ nm. Under these conditions, that are very close to the real HSFL structures observed in the experiments of Figures 3 and 5, it is already possible to recognize the signature of the LSFL$^{\|}$ with period $\Lambda_{LSFL}^{\|} = 580 \pm 98$ nm. Some remnants of the interaction between the laser and the HSFL grating at the surface are still present and manifest as lines oriented along the Y-axis with a periodicity of 100 nm in the X-direction. The other simulations shown in Figure 8 address systematic variations of the HSFL modulation depth d_{HSFL} (Figure 8C), the roughness parameter R_{HSFL} (Figure 8D), and the oxide layer thickness T_{ox} (Figure 8E), respectively. In order to observe its influence on the formation of the intensity pattern responsible of the LSFL$^{\|}$ formation, the FDTD simulations were performed changing one parameter each time, while keeping the others constant, as it is indicated in each set of sub-figures.

The simulations obtained upon changing the modulation depth (d_{HSFL}) of the HSFL gratings formed on the oxide layer, are shown in Figure 8C. Without HSFL structures on the oxide layer ($d_{HSFL} = 0$ nm), the intensity distribution corresponds to a non-organized pattern, only affected by the initial roughness present at the oxide surface set to $R_{HSFL} = 20$ nm here. For metals it has been demonstrated that the roughness at the surface is an important parameter that may influence the final periodicity of the LIPSS [8,17]. However, this is only valid when the structures are produced at the first irradiated interface (the surface), in our case, the oxide layer, and not at the interface between two different materials. For all the other modulation depths shown ($d_{HSFL} > 0$ nm), there is a two-fold modulation in the calculated intensity patterns at the oxide/nitride interface: in the Y-axis direction, the HSFL remnants are still visible and are practically constant for all the studied d_{HSFL} values. In the X-axis direction, there is certain modulation in the form of LSFL parallel to the laser polarization, significantly more pronounced when $d_{HSFL} \gtrsim 40$ nm. This result is consistent with the experimentally measured modulation depths for the HSFL shown in Figures 3B and 5C (AFM data), having values of d_{HSFL} ~50 nm for both cases.

The second parameter varied in the simulation was the non-organized roughness superimposed to the HSFL covered oxide layer, R_{HSFL}, as shown in Figure 8D. Note that this parameter is different from the root-mean-squared roughness obtained from the AFM data in Figures 3B and 5C. When $R_{HSFL} = 0$ nm, the sample surface corresponds to a smooth HSFL grating. Since the symmetry is not broken due to the absence of any surface defects, it is expected that the intensity along the Y-axis is not modulated in the corresponding FDTD simulations. As R_{HSFL} increases, the relative signature of the HSFL is drastically reduced and the intensity modulation parallel to the laser polarization (X-axis) featuring the

LSFL$^{\parallel}$ characteristics becomes evident, reaching average values of $\Lambda^{\parallel}_{LSFL} = 650 \pm 50$ nm when $R_{HSFL} = 40$ nm, in reasonable agreement to the experimental periodicities found in Figure 5A for the interfacial LSFL $\Lambda^{\parallel}_{LSFL} = 780 \pm 65$ nm.

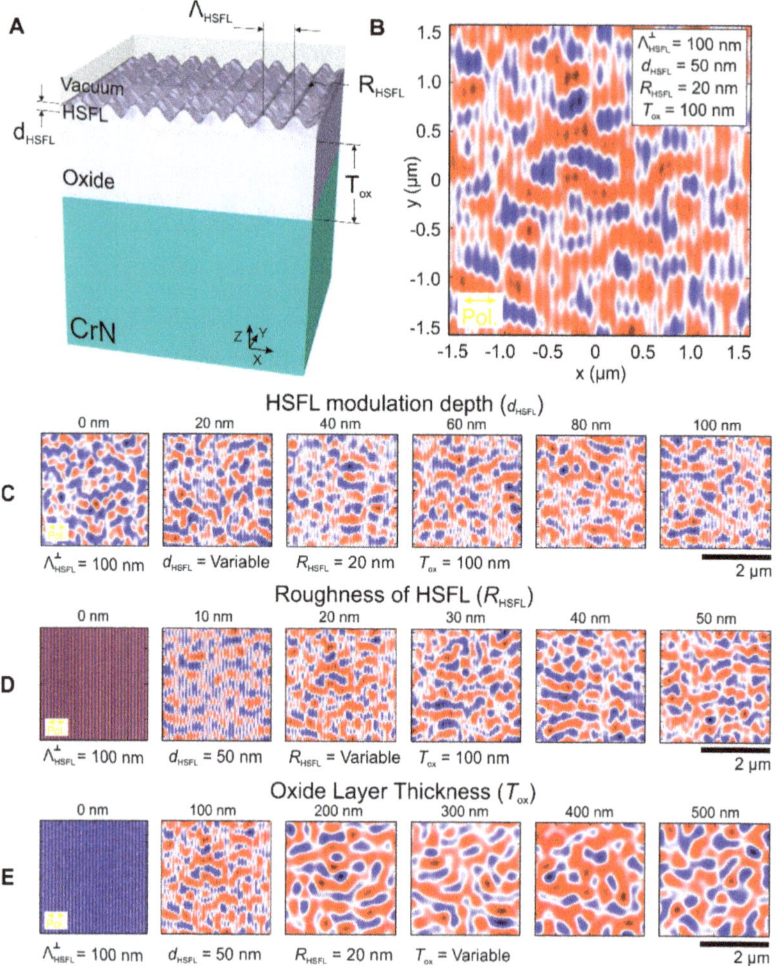

Figure 8. (**A**) 3D-representation (not to scale) of the simulated domain with dimensions in X and Y of 3.2 µm and variable Z including the different layers, visualizing the relevant parameters studied in the finite-difference time-domain (FDTD) simulations. (**B**) Intensity pattern obtained at the interface of the oxide layer and the CrN sample for a sample with HSFL periodicity $\Lambda^{\perp}_{HSFL} = 100$ nm, HSFL modulation depth $d_{HSFL} = 50$ nm, roughness of the HSFL $R_{HSFL} = 20$ nm, and oxide layer thickness $T_{ox} = 100$ nm. Subsequent simulations take into account the same conditions as in (**B**) changing in (**C**) d_{HSFL}, from 0 to 100 nm, (**D**) R_{HSFL} from 0 to 50 nm and in (**E**) T_{ox} from 0 to 500 nm. The laser beam ($\lambda = 800$ nm) polarization direction is horizontal. The colors of the intensity plots are encoded in a false color scheme where red indicated deviations above and blue below the average intensity value of the image, represented in white.

The final parameter varied is the oxide layer thickness, T_{ox}, as shown in Figure 8E. When $T_{ox} = 0$ nm, the incident laser beam has propagated only through vacuum, and no dispersive or reflecting interface elements modify the wave front upon propagation. Hence, apart from the influence of the non-organized surface roughness and remnants of the HSFL grating, there is no significant modulation on the laser beam intensity observed here. The periodicity that can be seen in this case, corresponds to the presence of the HSFL that are present at the surface of the CrN layer. When T_{ox} = 100 nm, the intensity is already modulated in the Y-direction while still exhibiting the signature of the HSFL in the X-direction. For even thicker oxide layers, the regularity of the intensity pattern is reduced and even the contribution from the HSFL disappears due to propagation effects in the oxide and the superposition of the incident waves with that partially reflected from the CrN interface.

More in detail, in order to see how the beam is propagating inside this layered system, Figure 9 provides intensity distributions at different positions Z along the beam propagation axis. Each plot show insets with the corresponding Z-positions at which the intensity distribution is calculated. The position Z = 0 nm correspond to the interface between the oxide layer and the CrN sample underneath. In general, positive Z-positions indicate that the intensity distribution is located inside the oxide layer (or above it) and negative Z-positions correspond to intensities inside the CrN material. The results for a layered system with a modulation depth of d_{HSFL} = 50 nm, roughness R_{HSFL} = 20 nm and oxide layer thickness T_{ox} = 100 nm and 200 nm are compared in Figures 9A and 9B, respectively.

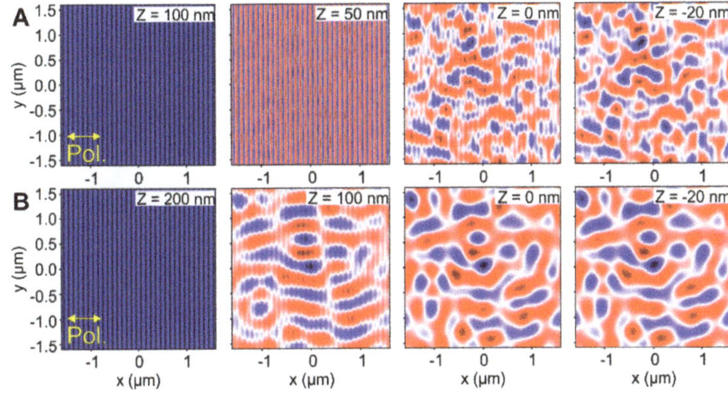

Figure 9. FDTD-intensity distribution at different Z-positions along the beam axis for two different layered systems with HSFL modulation depth d_{HSFL} = 50 nm and. In (**A**) the oxide layer thickness is T_{ox} = 100 nm and in (**B**) T_{ox} = 200 nm. The interface between the oxide layer and the CrN is located at Z = 0 nm. The plots at Z = −20 nm in both cases correspond to intensities calculated 20 nm inside the CrN from the oxide-nitride interface. The laser beam (λ = 800 nm) polarization direction is horizontal. The colors of the intensity plots are encoded in a false color scheme where red indicated deviations above and blue below the average intensity value of the image, represented in white.

In Figure 9A (top row), where the oxide layer thickness is 100 nm, the intensity distribution is practically unchanged at Z = 100 nm, corresponding to the position at the base of the HSFL structures. In the middle of the oxide layer (Z = 50 nm) the signature of the HSFL is strongly visible and at this small propagation distance along the Z-axis, the beam pattern symmetry is not completely broken yet along the Y-axis. At the interface between the oxide layer and the CrN sample (Z = 0 nm) the signature of the HSFL is still present but slightly reduced, presenting a clear variation along the Y-axis with the signature of LSFL parallel to the laser polarization (LSFL$^{\parallel}$). This pattern is maintained inside the CrN sample reducing significantly the signs of the HSFL, as it is shown in the plot at Z = −20 nm.

In Figure 9B (bottom row) a similar system with a just thicker oxide layer of T_{ox} = 200 nm is analyzed. The intensity distribution at Z = 200 nm is calculated at the base of the HSFL where the intensity distribution remains unchanged (similarly as the one at Z = 100 nm in Figure 9A). In the middle of the oxide layer (Z = 100 nm), the modulation of the intensity already shows the very pronounced imprint of LSFL$^{\parallel}$ with a contribution from the HSFL top grating. At the oxide/nitride interface $Z = 0$ nm, the signature of the HSFL is vanished presenting a smooth intensity distribution that corresponds to the LSFL$^{\parallel}$. This distribution is preserved inside the CrN layer ($Z = -20$ nm). From these results, it is possible to conclude that for a layered system with HSFL of 100 nm periodicity, HSFL modulation depth $d_{HSFL} = 50$ nm, roughness $R_{HSFL} = 20$ nm, (all in good agreement with the experimental parameters found in Figures 3B and 5B), the oxide layer thickness value should be considered between $100 < T_{ox} < 200$ nm to produce effectively LSFL$^{\parallel}$.

For addressing a less idealized simulation scenario, Figure 10 includes the results of a similar layered system considering a more realistic topography for the HSFL structures. The parameters for this simulation are HSFL real periodicity $\Lambda_{HSFL}^{\perp} = 100$ nm, an artificial HSFL modulation depth $d_{HSFL} = 75$ nm based on a false color grayscale image of the SEM micrograph shown in Figure 10A, no additional roughness ($R_{HSFL} = 0$ nm) and an oxide layer thickness of $T_{ox} = 100$ nm.

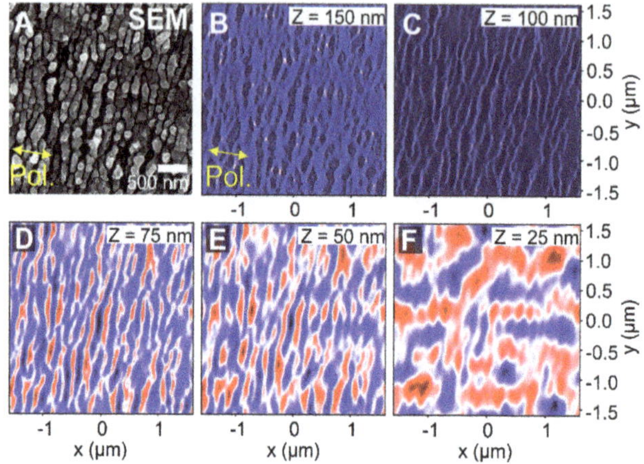

Figure 10. FDTD simulation of a layered system that includes periodic HSFL $\Lambda_{HSFL}^{\perp} = 100$ nm given by the real structures shown in the SEM micrograph on the top left (**A**), HSFL modulation depth $d_{HSFL} = 75$ nm, no added roughness ($R_{HSFL} = 0$ nm) and oxide layer thickness $T_{ox} = 100$ nm. The intra oxide film position varies between $Z = 150$ nm (**B**), $Z = 100$ nm (**C**), $Z = 75$ nm (**D**), $Z = 50$ nm (**E**), and $Z = 25$ nm (**F**). The laser beam ($\lambda = 800$ nm) polarization direction is horizontal. The colors of the intensity plots are encoded in a false color scheme where red indicated deviations above and blue below the average intensity value of the image, represented in white.

In this case, the intensity distribution was acquired at different positions only inside the oxide layer (at positive Z-positions). The top surface (not shown) is located at $Z = 175$ nm, ($T_{ox} = 100$ nm $+ d_{HSFL} = 75$ nm). At $Z = 150$ nm (Figure 10B), it is possible to recognize the morphology shown in the SEM micrograph. The intensity distribution remains essentially unchanged due to the short distance that the beam has propagated inside the oxide layer, similarly as in the case of the HSFL structures ($Z = 100$ nm, Figure 10C). From there, once the beam propagates 25 nm more along the beam axis ($Z = 75$ nm, Figure 10D), the signature of the HSFL is strong and slowly fades away as the beam propagates towards the oxide/nitride interface. At $Z = 25$ nm (Figure 10F), the LSFL$^{\parallel}$ characteristics with an orientation parallel to the laser beam polarization and larger periods are evident. Remarkably, considering the real topography, the intensity distributions along different positions along the Z-axis present a good agreement with the more idealized simulations presented in Figures 8 and 9. This ultimately confirms the relevance of the oxidation at the surface for the fabrication of such buried (interfacial) LSFL structures parallel to the laser beam polarization on oxidation prone materials.

In view of their orientation parallel to the laser beam polarization, their spatial periods of the order of λ/n, and their sub-surface presence in a transparent material, it is very likely that the physical origin of the simulated intra-film and interfacial intensity patterns carrying the LSFL$^{\parallel}$ characteristics here lies in the type-d ("dissident") electromagnetic field structures identified earlier in FDTD simulations [16,18,20,38]. These intensity patterns are formed via the far-field interference between the electromagnetic field scattered at the rough surface and the laser beam refracted into the film and propagating in the sub-surface dielectric material [18,38]. A visualization of the electromagnetic fields emitted from single and multiple type-d scattering centers on dielectric surfaces predominantly in the direction perpendicular to the polarization can be found in [18,39]. Note that the type-m ("mixed") features observed in FDTD (associated with LSFL parallel to the polarization and periods close to the wavelength λ) can be ruled out here as origin of the LSFL$^{\parallel}$ since the necessary condition of $n = \text{Re}(\widetilde{n}) \approx \text{Im}(\widetilde{n}) = k$ [20] on the optical properties of the oxide material is not fulfilled in our case here ($n = 2$, $k = 0$ [37]). Interestingly, the spatial periods of the sub-surface LSFL$^{\parallel}$ patterns simulated by FDTD rather lie between λ and λ/n as it was previously reported in [21]. The deviation between the experimentally observed and numerically simulated periods $\Lambda^{\parallel}_{\text{LSFL}}$ may arise from the fact that in the experiments the laser-generated oxide layer may not a have a sharp interface to the bulk but may exhibit a graded composition [26] between the sample surface and the CrN bulk material. The latter would be associated then with optical constants varying at depth. Moreover, high-intensity laser-induced transient changes of the refractive index n may result in somewhat increased spatial periods (for details refer to the basic Drude model presented in [19]).

4. Conclusions

From the presented results, four conclusions can be drawn: (i) it can be clearly observed that the oxide layer is an important parameter that influences strongly the formation of a regular intensity pattern at the interface to the underlying material that presumably leads to the formation of the LSFL structures parallel to the laser polarization. (ii) in order to allow the formation of a regular intensity pattern that can be imprinted at the interface between the superficial oxide layer and the underlying material (CrN), the thickness of the oxide layer should be in the order of ~100 nm, as it has been also suggested in [28]. (iii) the presence of superficial sub-wavelength HSFL on the oxide layer is essentially required to form the near-wavelength LSFL$^{\parallel}$. (iv) the intensity pattern carrying the spatial LSFL$^{\parallel}$ characteristics is formed via the joint action of electromagnetic scattering, propagation, and interference effects and does not require the presence of a partially metallic oxide layer (via laser-excited conduction band electrons) or that of nano-plasmas scattering from localized defects [18], as visualized in the summarizing scheme presented in Figure 11. The findings suggest that these effects can be present on strong absorbing materials that are prone to oxide formation.

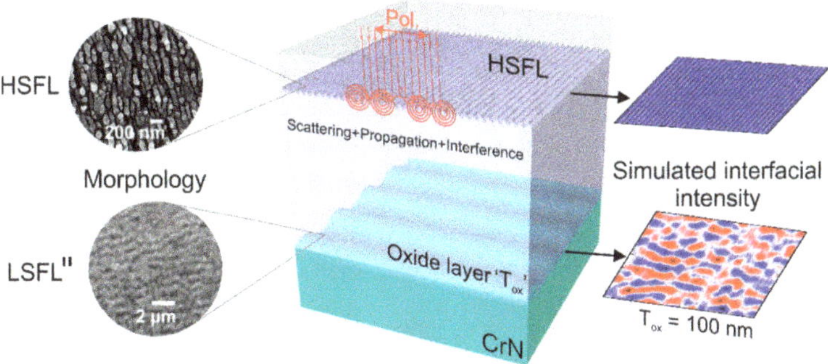

Figure 11. Scheme of the interfacial LSFL$^{\parallel}$ formation, featuring scattering, propagation and interference effects.

Author Contributions: J.S. and J.B. designed the experiments. C.F. and S.V.K. conducted the experiments. C.F., J.S. and J.B. analyzed the experimental results. J.-L.D. implemented, performed, and analyzed the FDTD simulations. C.F. wrote the manuscript. All authors contributed to the scientific discussion and revision of the article. J.B. directed the project. All authors have read and agree to the published version of the manuscript.

Funding: This work was supported through the European Horizon 2020 FET Open programme by the following projects: "CellFreeImplant" (Grant Agreement No. 800832), "LiNaBioFluid" (Grant Agreement No. 665337), and the Spanish Ministry of Science, Innovation by the project "UDiSON" (TEC2017–82464–R) and the collaborative project i-Link+2018 (Reference LINKA20044).

Acknowledgments: The authors would like to thank S. Benemann (BAM 6.1) for SEM and EDX characterizations, V.-D. Hodoroaba (BAM 6.1) for valuable discussions of the SEM/EDX data, including Monte Carlo simulations (Casino), A. Hertwig (BAM 6.7) for the ellipsometric measurements of the CrN optical constants, M. Sahre and T. Lange (both BAM 6.7) for the XRD analysis, M. Sahre for the AFM data, S. Sanchez-Cortez (IO-CSIC) for μ-Raman spectroscopy and A. Lopez and A. García (CENIM-CSIC) for SEM and EDX characterizations.

Conflicts of Interest: The authors declare no conflict of interest.

References

1. Bonse, J.; Höhm, S.; Kirner, S.V.; Rosenfeld, A.; Krüger, J. Laser-induced periodic surface structures—A scientific evergreen. *IEEE J. Sel. Top. Quantum Electron.* **2017**, *23*, 9000615. [CrossRef]

2. Bonse, J.; Krüger, J.; Höhm, S.; Rosenfeld, A. Femtosecond laser-induced periodic surface structures. *J. Laser Appl.* **2012**, *24*, 042006. [CrossRef]

3. Bonse, J.; Baudach, S.; Krüger, J.; Kautek, W.; Lenzner, M. Femtosecond laser ablation of silicon–modification thresholds and morphology. *Appl. Phys. A Mater. Sci. Process.* **2002**, *74*, 19–25. [CrossRef]

4. Aguilar, A.; Mauclair, C.; Faure, N.; Colombier, J.-P.; Stoian, R. In-situ high-resolution visualization of laser-induced periodic nanostructures driven by optical feedback. *Sci. Rep.* **2017**, *7*, 16509. [CrossRef]

5. Gnilitskyi, I.; Derrien, T.J.Y.; Levy, Y.; Bulgakova, N.M.; Mocek, T.; Orazi, L. High-speed manufacturing of highly regular femtosecond laser-induced periodic surface structures: physical origin of regularity. *Sci. Rep.* **2017**, *7*, 8485. [CrossRef]

6. Höhm, S.; Herzlieb, M.; Rosenfeld, A.; Krüger, J.; Bonse, J. Dynamics of the formation of laser-induced periodic surface structures (LIPSS) upon femtosecond two-color double-pulse irradiation of metals, semiconductors, and dielectrics. *Appl. Surf. Sci.* **2016**, *374*, 331–338. [CrossRef]

7. Sedao, X.; Lenci, M.; Rudenko, A.; Faure, N.; Pascale-Hamri, A.; Colombier, J.P.; Mauclair, C. Influence of pulse repetition rate on morphology and material removal rate of ultrafast laser ablated metallic surfaces. *Opt. Lasers Eng.* **2019**, *116*, 68–74. [CrossRef]

8. Fuentes-Edfuf, Y.; Sánchez-Gil, J.A.; Florian, C.; Giannini, V.; Solis, J.; Siegel, J. Surface plasmon polaritons on rough metal surfaces: Role in the formation of laser-induced periodic surface structures. *ACS Omega* **2019**, *4*, 6939–6946. [CrossRef]

9. Sipe, J.E.; Young, J.F.; Preston, J.S.; van Driel, H.M. Laser-induced periodic surface structure. I. Theory. *Phys. Rev. B* **1983**, *27*, 1141–1154. [CrossRef]

10. Bonse, J.; Rosenfeld, A.; Krüger, J. On the role of surface plasmon polaritons in the formation of laser-induced periodic surface structures upon irradiation of silicon by femtosecond-laser pulses. *J. Appl. Phys.* **2009**, *106*, 104910. [CrossRef]

11. Gräf, S.; Kunz, C.; Engel, S.; Derrien, T.; Müller, F. Femtosecond laser-induced periodic surface structures on fused silica: the impact of the initial substrate temperature. *Materials* **2018**, *11*, 1340. [CrossRef] [PubMed]

12. Miyaji, G.; Miyazaki, K.; Zhang, K.; Yoshifuji, T.; Fujita, J. Mechanism of femtosecond-laser-induced periodic nanostructure formation on crystalline silicon surface immersed in water. *Opt. Express.* **2012**, *20*, 14848–14856. [CrossRef] [PubMed]

13. Derrien, T.J.Y.; Koter, R.; Krüger, J.; Höhm, S.; Rosenfeld, A.; Bonse, J. Plasmonic formation mechanism of periodic 100-nm-structures upon femtosecond laser irradiation of silicon in water. *J. Appl. Phys.* **2014**, *116*, 074902. [CrossRef]

14. Kuladeep, R.; Dar, M.H.; Deepak, K.L.N.; Rao, D.N. Ultrafast laser induced periodic sub-wavelength aluminum surface structures and nanoparticles in air and liquids. *J. Appl. Phys.* **2014**, *116*, 113107. [CrossRef]

15. Gesuele, F.; Nivas, J.J.J.; Fittipaldi, R.; Altucci, C.; Bruzzese, R.; Maddalena, P.; Amoruso, S. Multi-imaging analysis of nascent surface structures generated during femtosecond laser irradiation of silicon in high vacuum. *Appl. Phys. A Mater. Sci. Process.* **2018**, *124*, 204. [CrossRef]

16. Skolski, J.Z.P.; Römer, G.R.B.E.; Obona, J.V.; Ocelik, V.; Huis in 't Veld, A.J.; De Hosson, J.T.M. Laser-induced periodic surface structures: Fingerprints of light localization. *Phys. Rev. B* **2012**, *85*, 075320. [CrossRef]

17. Rudenko, A.; Mauclair, C.; Garrelie, F.; Stoian, R.; Colombier, J.-P. Self-organization of surfaces on the nanoscale by topography-mediated selection of quasi-cylindrical and plasmonic waves. *Nanophotonics* **2019**, *8*, 459–465. [CrossRef]

18. Rudenko, A.; Colombier, J.-P.; Höhm, S.; Rosenfeld, A.; Krüger, J.; Bonse, J.; Itina, T.E. Spontaneous periodic ordering on the surface and in the bulk of dielectrics irradiated by ultrafast laser: A shared electromagnetic origin. *Sci. Rep.* **2017**, *7*, 12306. [CrossRef]

19. Höhm, S.; Rosenfeld, A.; Krüger, J.; Bonse, J. Femtosecond laser-induced periodic surface structures on silica. *J. Appl. Phys.* **2012**, *112*, 014901. [CrossRef]

20. Déziel, J.-L.; Dumont, J.; Gagnon, D.; Dubé, L.J.; Messaddeq, S.H.; Messaddeq, Y. Toward the formation of crossed laser-induced periodic surface structures. *J. Opt.* **2015**, *17*, 075405. [CrossRef]

21. Déziel, J.-L.; Dubé, L.J.; Messaddeq, S.H.; Messaddeq, Y.; Varin, C. Femtosecond self-reconfiguration of laser-induced plasma patterns in dielectrics. *Phys. Rev. B* **2018**, *97*, 205116. [CrossRef]

22. Kietzig, A.M.; Hatzikiriakos, S.G.; Englezos, P. Patterned superhydrophobic metallic surfaces. *Langmuir* **2009**, *25*, 4821–4827. [CrossRef] [PubMed]

23. Kirner, S.V.; Slachciak, N.; Elert, A.M.; Griepentrog, M.; Fischer, D.; Hertwig, A.; Sahre, M.; Dörfel, I.; Sturm, H.; Pentzien, S.; et al. Tribological performance of titanium samples oxidized by fs-laser radiation, thermal heating, or electrochemical anodization. *Appl. Phys. A Mater. Sci. Process.* **2018**, *124*, 326. [CrossRef]

24. Bonse, J.; Sturm, H.; Schmidt, D.; Kautek, W. Chemical, morphological and accumulation phenomena in ultrashort-pulse laser ablation of TiN in air. *Appl. Phys. A Mater. Sci. Process.* **2000**, *71*, 657–665. [CrossRef]

25. Yasumaru, N.; Miyazaki, K.; Kiuchi, J. Fluence dependence of femtosecond-laser-induced nanostructure formed on TiN and CrN. *Appl. Phys. A Mater. Sci. Process.* **2005**, *81*, 933–937. [CrossRef]

26. Kirner, S.V.; Wirth, T.; Sturm, H.; Krüger, J.; Bonse, J. Nanometer-resolved chemical analyses of femtosecond laser-induced periodic surface structures on titanium. *J. Appl. Phys.* **2017**, *122*, 104901. [CrossRef]

27. Öktem, B.; Pavlov, I.; Ilday, S.; Kalaycıoğlu, H.; Rybak, A.; Yavaş, S.; Erdoğan, M.; Ilday, F.Ö. Nonlinear laser lithography for indefinitely large-area nanostructuring with femtosecond pulses. *Nat. Photonics* **2013**, *7*, 897–901. [CrossRef]

28. Dostovalov, A.V.; Korolkov, V.P.; Okotrub, K.A.; Bronnikov, K.A.; Babin, S.A. Oxide composition and period variation of thermochemical LIPSS on chromium films with different thickness. *Opt. Express.* **2018**, *26*, 7712–7723. [CrossRef]

29. Dostovalov, A.V.; Derrien, T.J.-Y.; Lizunov, S.A.; Přeučil, F.; Okotrub, K.A.; Mocek, T.; Korolkov, V.P.; Babin, S.A.; Bulgakova, N.M. LIPSS on thin metallic films: New insights from multiplicity of laser-excited electromagnetic modes and efficiency of metal oxidation. *Appl. Surf. Sci.* **2019**, *491*, 650–658. [CrossRef]

30. Liu, J.M. Simple technique for measurements of pulsed Gaussian-beam spot sizes. *Opt. Lett.* **1982**, *7*, 196. [CrossRef]

31. Bonse, J.; Mann, G.; Krüger, J.; Marcinkowski, M.; Eberstein, M. Femtosecond laser-induced removal of silicon nitride layers from doped and textured silicon wafers used in photovoltaics. *Thin Solid Films* **2013**, *542*, 420–425. [CrossRef]

32. Fuentes-Edfuf, Y.; Sánchez-Gil, J.A.; Garcia-Pardo, M.; Serna, R.; Tsibidis, G.D.; Giannini, V.; Solis, J.; Siegel, J. Tuning the period of femtosecond laser induced surface structures in steel: From angled incidence to quill writing. *Appl. Surf. Sci.* **2019**, *493*, 948–955. [CrossRef]

33. Hou, S.; Huo, Y.; Xiong, P.; Zhang, Y.; Zhang, S.; Jia, T.; Sun, Z.; Qiu, J.; Xu, Z. Formation of long- and short-periodic nanoripples on stainless steel irradiated by femtosecond laser pulses. *J. Phys. D Appl. Phys.* **2011**, *44*, 505401. [CrossRef]

34. Baranov, A.V.; Bogdanov, K.V.; Fedorov, A.V.; Yarchuk, M.V.; Ivanov, A.I.; Veiko, V.P.; Berwick, K. Micro-Raman characterization of laser-induced local thermo-oxidation of thin chromium films. *J. Raman Spectrosc.* **2011**, *42*, 1780–1783. [CrossRef]

35. Barshilia, H.C.; Rajam, K.S. Raman spectroscopy studies on the thermal stability of TiN, CrN, TiAlN coatings and nanolayered TiN/CrN, TiAlN/CrN multilayer coatings. *J. Mater. Res.* **2004**, *19*, 3196–3205. [CrossRef]

36. Gomes, A.S.O.; Yaghini, N.; Martinelli, A.; Ahlberg, E. A micro-Raman spectroscopic study of Cr(OH)$_3$ and Cr$_2$O$_3$ nanoparticles obtained by the hydrothermal method. *J. Raman Spectrosc.* **2017**, *48*, 1256–1263. [CrossRef]

37. Gesheva, K.A.; Ivanova, T.; Szekeres, A.M.; Trofimov, O. Surface characterization of chromium oxide thin films in dependence on CVD growth process parameters. *ECS Trans.* **2007**, *2*, 229–236.

38. Zhang, H.; Colombier, J.-P.; Li, C.; Faure, N.; Cheng, G.; Stoian, R. Coherence in ultrafast laser-induced periodic surface structures. *Phys. Rev. B* **2015**, *92*, 174109. [CrossRef]

39. Déziel, J.-L. Ablation Laser et Croissance de Réseaux de Surface. M.Sc. Thesis, Université Laval, Quebec City, QC, Canada, September 2015. [CrossRef]

Article

Ultrafast Laser Processing of Nanostructured Patterns for the Control of Cell Adhesion and Migration on Titanium Alloy

Antoine Klos [1], Xxx Sedao [2,3], Tatiana E. Itina [2], Clémentine Helfenstein-Didier [4], Christophe Donnet [2], Sylvie Peyroche [1], Laurence Vico [1], Alain Guignandon [1,†] and Virginie Dumas [4,†,*]

[1] SAINBIOSE Laboratory INSERM U1059, University of Lyon, Jean Monnet University, F-42270 Saint Priest en Jarez, France; antoine.klos@3sr-grenoble.fr (A.K.); sylvie.peyroche@univ-st-etienne.fr (S.P.); vico@univ-st-etienne.fr (L.V.); alain.guignandon@univ-st-etienne.fr (A.G.)

[2] Hubert Curien Laboratory, University of Lyon, Jean Monnet University, UMR 5516 CNRS, F-42000 Saint-Etienne, France; xxx.sedao@univ-st-etienne.fr (X.S.); tatiana.itina@univ-st-etienne.fr (T.E.I.); christophe.donnet@univ-st-etienne.fr (C.D.)

[3] GIE Manutech-USD, 20 rue Benoit Lauras, F-42000 Saint-Etienne, France

[4] Laboratory of Tribology and Systems Dynamics, National School of Engineers of Saint-Etienne, University of Lyon, UMR 5513 CNRS, F-42100 Saint-Etienne, France; clementine.didier@enise.fr

* Correspondence: virginie.dumas@enise.fr

† These authors shared senior authorship.

Received: 26 March 2020; Accepted: 28 April 2020; Published: 30 April 2020

Abstract: Femtosecond laser texturing is a promising surface functionalization technology to improve the integration and durability of dental and orthopedic implants. Four different surface topographies were obtained on titanium-6aluminum-4vanadium plates by varying laser processing parameters and strategies: surfaces presenting nanostructures such as laser-induced periodic surface structures (LIPSS) and 'spikes', associated or not with more complex multiscale geometries combining micro-pits, nanostructures and stretches of polished areas. After sterilization by heat treatment, LIPSS and spikes were characterized to be highly hydrophobic, whereas the original polished surfaces remained hydrophilic. Human mesenchymal stem cells (hMSCs) grown on simple nanostructured surfaces were found to spread less with an increased motility (velocity, acceleration, tortuosity), while on the complex surfaces, hMSCs decreased their migration when approaching the micro-pits and preferentially positioned their nucleus inside them. Moreover, focal adhesions of hMSCs were notably located on polished zones rather than on neighboring nanostructured areas where the protein adsorption was lower. All these observations indicated that hMSCs were spatially controlled and mechanically strained by the laser-induced topographies. The nanoscale structures influence surface wettability and protein adsorption and thus influence focal adhesions formation and finally induce shape-based mechanical constraints on cells, known to promote osteogenic differentiation.

Keywords: femtosecond laser; multiscale-patterning; wettability; human mesenchymal stem cell; cell adhesion; cell spreading; cell motility; protein adsorption

---

1. Introduction

Because of their excellent biocompatibility and superior mechanical strength, titanium and related alloys, such as titanium-6aluminum-4vanadium (Ti6Al4V), are widely used for biomedical applications [1,2]. However, it remains a challenge to achieve a homogenized bone–implant interface for decent osseointegration of dental and orthopedic implants in human bodies. Several attempts to solve this issue have been reported. In particular, it was shown to be possible to modify surface bio-functionalities, such as improved osseointegration at the implant surface, through surface engineering [3,4]. The process of osseointegration involves a complex chain of events, from protein adsorption to recruitment of mesenchymal stem cells (MSCs) and osteoblasts, finally leading to bone formation at the interface of implant.

Human MSCs (hMSCs) are multipotent bone-marrow derived stem cells that can differentiate in a wide diversity of cell types such as osteoblasts, adipocytes, chondrocytes or myoblasts and they represent a model of choice in the field of biointegration tissue–implants. It is widely accepted that surface properties and, in particular, topography and wettability affect protein adsorption and the subsequent cell adhesion [5,6] and fate [7]. Surface micropatterning can be applied to modify the microenvironment of cells and modulate their behavior. Most micropatterning techniques used in order to apply shape-based mechanical constraint on cells [8] derive from surface grafting of pro-adherent proteins surrounded by an anti-adhesion coating. Others (e.g., photolithography) are mostly applied to polymers [9]. Laser texturing is the only technique that can, in a single step, provide controlled multiscaled design on metallic surfaces. Previous studies showed that by inducing sub-micrometric roughness, laser texturing could enhance osteoblastic differentiation of hMSCs [10,11] and in vivo osseointegration [12,13]. Many efforts have been made in order to enhance the osteogenic potential but engraftment of hMSCs is a critical step in the process of osseointegration. Hence, ultrafast laser texturing appears to be a promising surface biofunctionalization approach to control hMSCs adhesion, spreading and migration. It was widely demonstrated that multiscale surfaces can be achieved by using femtosecond laser texturing, such as nano-textured grooves [14,15], grids [11] or pits [10] within the micrometric scale. Moreover, laser texturing allows one to obtain various types of laser-induced surface nanostructures, such as laser-induced periodic surface structures (LIPSS), grooves, spikes, columns, etc. [16–21]. Of peculiar bio-engineering importance, these nanostructures have different wetting properties [19,22], which are time-dependent and age towards hydrophobicity [23,24].

Here, our goal was to create finely designed patterns on a Ti6Al4V surface, with both micro- and nanometric dimensions. It is of interest to evaluate the influence of these engineered double-scaled surface patterns on cell behavior. It is known that hMSC osteoblastic differentiation could be directed by the state of intracellular tension [8,25–28]. In order to engineer titanium surfaces with optimal properties for improved osteogenesis, one must understand the complex interrelationships among material surface properties, adsorbed proteins, and early cellular responses. In this in vitro study, we focused on the short-term cellular responses which are known to be crucial in regulating events that lead to cell differentiation. The aim was to investigate the influence of the laser nanostructures on cell adhesion, spreading and migration and compare it to conventional smooth titanium. We analyzed the wetting properties of laser nanostructures before and after sterilization, then we investigated the effect on the adsorption of proteins on these areas and the establishment of focal contacts. The multi-scale complex design was conceived in order to spatially control cell adhesion and multicellular organization. Specific design with carefully arranged pro- and anti-adhesive areas could mechanically constrain the cellular morphology, which is critical in numerous biomedical and tissue-engineering applications.

We first investigated the properties, such as the topography and wettability, of surfaces fully covered by simple laser-induced nanostructures, and their influences on cell spreading and motility were also quantified. Based on the knowledge acquired from the aforementioned studies, we designed and laser-texturized multiscale surface patterns with combinations of nanostructures, micro-pits, and polished surface areas. Such surface patterns were developed in order to position cell nuclei and stimulate the cells to adopt a desired cell shape, thanks to anti-adhesive nanostructures. The key performance factors, such as the distribution of vinculin adhesion points, surface protein adsorption rates and the nuclei location were characterized.

2. Materials and Methods

2.1. Titanium Alloy Samples

Mirror-polished titanium alloy grade 5 (Ti6Al4V) samples were purchased from Goodfellow (Huntingdon, UK). The samples were square with dimensions of 10 mm by 10 mm with a thickness of 1 mm.

2.2. Laser Surface Texturing

Surface texturing and patterning were performed by a femtosecond laser with a galvanometer scanner system within the GIE Manutech-USD platform (Saint-Etienne, France). As a laser source, Tangor HP (Amplitude Systems) was used delivering femtosecond laser pulses with a duration of about 400 femtosecond (fs) at a wavelength of 1030 nm. After passing through a 56 mm f-theta lens, the laser beam converged to a focus spot with diameter $2\omega_0$ = 16.3 µm ($1/e^2$) in the focal plane. The laser fluence was adjusted by varying the laser power at a given laser repetition rate. The laser was linearly polarized.

Two different types of laser-processed surfaces were realized:

- two entirely laser-nanostructured samples fully covered by either LIPSS or spikes, termed as such in the following;
- two multiscale laser-patterned (LP) samples covered with a combination of micro-pits and laser-nanostructured patterns, together with stretches of polished surface areas (LIPSS + polished or spikes + polished), termed LP_LIPSS and LP_spikes, respectively.

LP_LIPSS and LP_spikes were realized through a multiple-step patterning strategy: the stretches of LIPSS or spikes areas were first laser-patterned and micro-pits were then machined in the next step. A spiral trajectory from the periphery to the center of the micro-pits was used to obtain a homogeneous photon energy distribution and, therefore, a homogeneous depth. A superposition of the features made up the complex surface patterns. The principle laser parameters for generating the aforementioned structures are summarized in Table 1. The laser-processed samples were then sonicated in demineralized water to clean them after which they were dried at room temperature.

Table 1. Laser parameters.

Topographies	Pulse Energy (E)	Fluence Peak	Pulse Rate	Distance between Pulses	Hatch Distance	Number of Pass
Micro-Pits	32×10^{-8} J	0.31 J·cm^{-1}	10 kHz	2 µm	4 µm	5
LIPSS	32×10^{-8} J	0.31 J·cm^{-1}	100 kHz	4 µm	4 µm	3
Spikes	270×10^{-8} J	2.59 J·cm^{-1}	100 kHz	4 µm	4 µm	1

The fluence peak is defined as: $F_{peak} = \dfrac{2\,E}{\pi\,\omega_0^2}$.

2.3. Surface Topography

Topographies and roughness (S_a) were analyzed using an optical 3D microscope (InfiniteFocus G4, ALICONA, Graz, Austria) which uses the focus variation microscopy technique. This allows 3D surface topography images to be obtained in which each pixel is at its maximum focus and from which the altitude variations can be analyzed. S_a (Arithmetical mean height of the surface) is a 3D-roughness parameter defined by the ISO 25178 standard. For further surface analysis, scanning electron microscopy (SEM) (TESCAN VEGA3 SB, Brno, Czech Republic) was carried out at 20.0 kV with a secondary electron detector.

2.4. Surface Wettability

Wettability measurements were carried out with a laboratory-developed multiscale and multifunctional system within the GIE MANUTECH-USD consortium. They were performed in a controlled atmosphere (Temperature (T) = 23 ± 0.6 °C, Relative humidity (RH) = 40 ± 4%). Cell basal medium was used as the testing medium (MSCBM, Prod. No. PT-3238, Lonza, Basel, Switzerland). Approximately 2-μL droplets of culture medium were deposited on the surfaces, and the evolution of the droplet shape was visualized with a camera and a sample rotation stage enabling 360° contact angle measurements. The platform moved at a speed of 0.1 rad·s^{-1}. The droplet profile and especially the contact angle (CA) were extracted from a complete droplet 360° rotation leading to approximately 50 measurements per droplet. The first CA measurements were carried out 38 days after laser processing. The second step of CA measurements was performed on the same samples, 24 h after a sterilization procedure in which the samples received a dry heat treatment at 180 °C for 2 h. The data represent the measurements from 4 consecutive droplets on each surface.

2.5. Cell Culture

Human bone marrow-derived mesenchymal stem cells (Lonza) at passage 6 were maintained in a T75-flask for a week in a growth medium (MSCGM, Prod. No. PT-3001, Lonza) before seeding. Cells were seeded on samples at 4000 cells·cm^{-2} in 6-well plates for 24 h, except in the case of the adhesion plaques study for which the growth time was 48 h.

2.6. Fluorescent Cell Labeling

Seeded cells were fixed in 4% paraformaldehyde for 30 min. Permeabilization was performed with 0.1% Triton X-100 in phosphate-buffered saline (PBS) for 3 min. Samples were incubated with rhodamine-conjugated phalloidin diluted at 1:50 in PBS, at 37 °C for 1.5 h, then with vinculin antibody, FITC (Fluorescein isothiocyanate) conjugated (Prod. No. F7053, Sigma-Aldrich, St. Louis, MO, USA) diluted at 1:50 in PBS, at 4 °C for 12 h. Afterwards, nuclei labeling was performed with 1 μg·mL^{-1} DAPI (4′,6-diamidino-2-phenylindole) diluted in PBS at 37 °C for 20 min. Washes were performed using PBS between each step of the experiment. Specimens were observed by confocal microscopy (ZEISS LSM 800 Airyscan, Oberkochen, Germany).

2.7. Living-Cell Labeling

For living-cell observations, DIL (1′-dioctadecyl-3,3,3′,3′,3′-tetramethylindocarbocyanine perchlorate, Prod. No. D282, Thermofisher, Waltham, MA, USA) previously diluted at 1000 μM in DMSO (Dimethyl sulfoxide) was used. DIL has a high affinity for the phospholipids that make up the cellular membrane and becomes fluorescent when binding with them. Cells were labeled in suspension at 1:500 with the previous solution in growth medium, and incubated (temperature (T) = 37 °C, relative humidity (RH) = 95%, 5% CO_2) for 1 h, following the protocol of Dumas et al. [10]. After centrifugation (5 min, 2000 rpm), the culture medium was changed. Then, DIL-labeled cells were seeded on samples at a density of 4000 cells·cm^{-2}. After letting the cells adhere to the surface for 1 h, the samples were put upside-down on biocompatible silicon pillars before the time-lapse recording started. The confocal microscope (ZEISS LSM 800 Airyscan) was equipped with an incubator 37 °C, 5% CO_2. Images were recorded every 30 min for 63 h.

2.8. Simultaneous Visualization of Surfaces and Cells through Confocal Microscopy

Surfaces and mesenchymal stem cells labelled with different fluorescent-coupled antibodies and fluorescent dyes were observed simultaneously using confocal microscopy. Labeling detection was performed by using up to three different laser wavelengths: 405 nm for DAPI, 488 nm for anti-vinculin-FITC antibody and 561 nm for rhodamine-conjugated phalloidin. Surfaces observation was based on the laser reflection on the substrate. For this, images were acquired in a 560–700 nm wavelength detection range, including the fourth laser emission wavelength at 640 nm, here used as a spotlight. This is why polished surfaces, which are highly reflective, appear brighter than laser-nanostructured areas.

2.9. Protein Adsorption Assay

This experiment was based on the work of Miao et al. [29]. FITC-albumin (Prod. No. A9771, Sigma–Aldrich) was dissolved in PBS at 15 µM and then diluted at 1:100 in basal culture medium. Afterwards, 70 µL of FITC-albumin solution was deposited on sample surfaces and incubated overnight at 4 °C. Samples were carefully rinsed in PBS before observation by fluorescent microscopy (ZEISS LSM 800 Airyscan). The albumin adsorption rate is related to the mean fluorescence intensity. In order to measure such intensity on specific zones, the laser-patterned surface was thresholded to identify polished and nanostructured areas. Thus, a Boolean operation allowed us to individually measure the mean fluorescence intensity of both polished and nanostructured areas. The difference between both intensities was then statistically compared to the theoretical value $\mu_0 = 0$, which represents the no intensity difference. The protein adsorption assay was made on sterilized samples to consider the surface properties changes.

2.10. Image Analysis for Cell Behavior

Images were analyzed using the ImageJ [30] freeware with Fiji [31] package.

Spreading measurements were obtained after automatic image processing including binarization preceded by 8-bit conversion, background removal and filtering. Measurements were carried out on 4 different areas (1.25×1.25 mm^2) and in duplicate.

For cell motility analysis, the "Manual Tracking" plug-in and IBIDI's plug-in "Chemotaxis and Migration Tool" (Gräfelfing, Germany) were used with ImageJ on contrast-adjusted videos. Tortuosity was defined as the ratio between the accumulated distance and the Euclidean distance. It was computed every 3 frames (1.5 h) and averaged on the whole recording for each cell. Cell velocity was defined from the ratio of the accumulated distance and the total recording duration, for each cell. Mean acceleration was computed from two consecutive cell velocity values, which were calculated from 2 consecutive cell position measurements within a time step between frames of 30 min. The absolute value of acceleration/de-acceleration was then averaged cell by cell for the whole migration path (63 h).

Nuclei location measurements were performed after binarization preceded by 8-bit conversion, background removal and filtering, with an automatic image processing for the counting of the whole nuclei. The counting of nuclei inside micro-pits was done manually, considering a nucleus inside a pit when half or more of the projected surface of the nucleus was inside. Data corresponding to 300 nuclei analyzed for LP_LIPSS and 494 for LP_spikes was normalized according to the surface proportion of micro-pits.

The proportion of focal adhesion points on polished surfaces was measured with consideration taken to both their number and their surface. The number of focal adhesion points was determined manually whereas the measurement by surface was semi-automatic. Indeed, focal adhesion points were binarized after an image processing following the steps: (i) coarse blurring (Gaussian, radius = 40); (ii) successive subtraction of the blurred image to the raw image followed by local contrast enhancement; (iii) successive slight blurring (median, radius = 1) followed by local contrast enhancement. Moreover, the laser-patterned surface was filtered to identify polished and nanostructured areas. Then, a Boolean operation was used to only measure the surface of focal adhesion points on polished areas. Data were normalized according to the proportion of polished areas.

2.11. Statistical Analysis

Data were analyzed through the Rstudio [32] freeware. Shapiro–Wilk normality tests and Bartlett homoscedasticity tests were performed for following statistical test selection. The Kruskal–Wallis test followed by a Mann–Whitney pairwise comparison with Bonferroni correction was used for statistical analysis ($n < 15$, or homoscedasticity or normality of data not respected).

3. Results

3.1. Surface Characterization

3.1.1. Multiscale Surface Topography

Two simple laser-induced nanostructures based on a literature survey [16,17,19] were obtained by adjusting femtosecond laser parameters (Figure 1a left panel). LIPSS and spikes were chosen because of their relative differences in terms of topography, small roughness and hydrophobic behavior. LIPSS are, as their name suggests, periodic (about 600 nm) and highly anisotropic whereas spikes are poorly periodic. The average areal surface roughness (S_a) was measured to be 76 ± 5 nm for the LIPSS and 85 ± 5 nm for the spikes, which is consistent with other studies [33,34]. The S_a of the polished surface was 65 ± 4 nm.

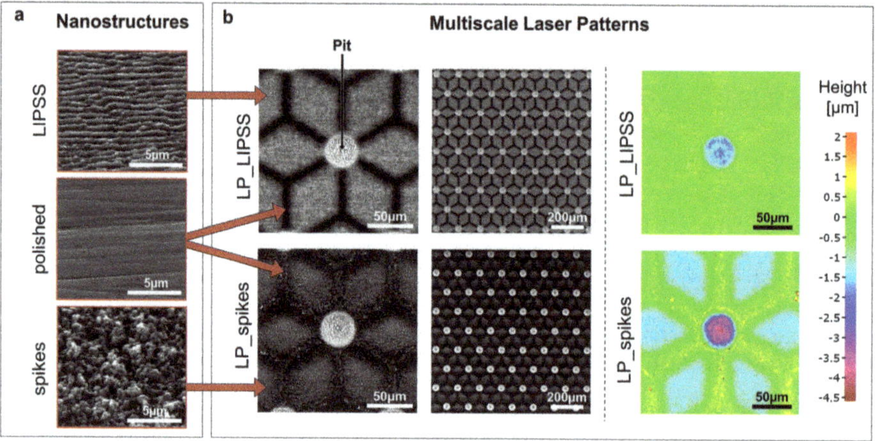

Figure 1. Nanostructures (LIPSS or spikes) and multiscale laser patterns (LP) produced by a femtosecond laser. (**a**) Scanning electron microscope (SEM) micrographs of polished surfaces and laser-nanostructured surfaces with LIPSS or spikes (tilt 45°). (**b**) SEM micrographs and focus variation microscopy of laser-patterned surfaces. LP_LIPSS and LP_spikes with a mix of polished and nanostructured areas. The micro-pits located in the center of the patterns were deeper and covered with LIPSS.

By adapting laser trajectories, multiscale laser-patterned surfaces can be created. The laser-untouched areas remain polished whereas their laser-processed counterparts become either textured with LIPSS or with spikes. Furthermore, finely increasing the number of laser scan passes and laser fluence allows an etching of the surface to several micrometers of depth. Hence, it is possible to obtain microscale surface modifications like laser-etched pits. Multiscale laser-patterned LP_LIPSS and LP_spikes surfaces (Figure 1) were thus formed first by microscale etching followed by imprinting of superficial laser-induced nanostructures. The patterns had a star-like shape with polished branches 60 µm long and 10 µm wide for an optimal adhesion of the focal contacts in these areas. The center of each pattern displayed pits approximately 3 µm deep and 40 µm in diameter. The geometry of the µm-pits was carefully designed for the nuclei to reside in. These parameters were the optimized values derived from a set of pre-selection experiments (data not shown). The distance between the micropatterns was adjusted to create a contiguous network of periodically repeated patterns. It is worth mentioning that, by using focus variation microscopy, we observed that the higher laser fluence used to generate spikes caused a difference in surface depth of about 1.5 μm relative to the polished areas (Figure 1b).

3.1.2. Surface Wettability

In order to characterize surface wettability by cell culture medium, contact angles were measured on both totally laser-nanostructured and polished surfaces. Figure 2 and Video S1 in Supplementary Materials show that contact angles were modified by laser texturing: they were either increased in the case of LIPSS, or notably decreased reaching the superhydrophilicity limit for spikes. It was observed that sterilization deeply affected the surface wettability. Indeed, we measured a significant increase in contact angle after such a treatment for all surfaces. This increase was remarkable for laser-nanostructured samples, reaching superhydrophobicity. The contact angle of the polished surface, however, did not change as much as the laser-nanostructured ones and remained within the same order of magnitude. Clearly, sterilization (180 °C, 2 h) made the nanostructured surfaces superhydrophobic while the smooth surfaces remained moderately hydrophilic.

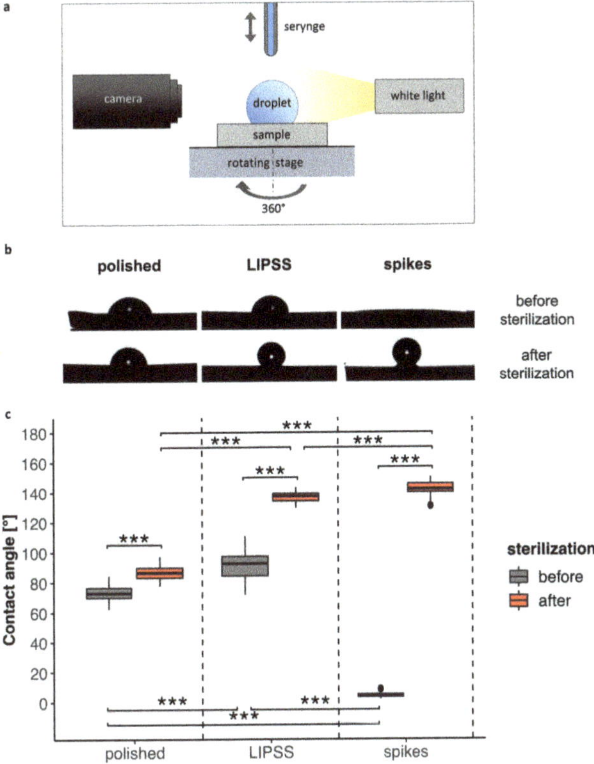

Figure 2. Comparison of the contact angles on polished and laser-nanostructured surfaces before and after sterilization. (**a**) Schematic of experimental setup for wettability characterization. (**b**) Representative droplet profile according to surfaces and sterilization process (**c**) 360° measurements from four consecutive droplets (at least 200 measurements per boxplot) of culture medium (2 μL) deposited on each surface (one sample). The Kruskal–Wallis test followed by Mann–Whitney pairwise comparison with Bonferroni correction was performed for data before and after sample sterilization. *** *p*-value < 0.001.

3.2. Cell Behavior

3.2.1. Cell Spreading

Cell spreading was investigated to quantify cell responses to nano-textured samples with different surface properties; 24 h after cell seeding, cytoskeletons and nuclei were stained to measure an average of cell areas. Figure 3 shows a significant decrease in cell spreading areas on laser-nanostructured surfaces compared with polished ones. Moreover, spreading on spikes occurred to a significantly smaller degree than spreading on LIPSS or LP_LIPSS and less by almost half compared with that on a polished surface. On LP_spikes, spreading increased compared with spikes, but did not reach the level observed on the polished specimens. Basically, cell spreading decreased in the presence of nanostructures (with or without smooth areas).

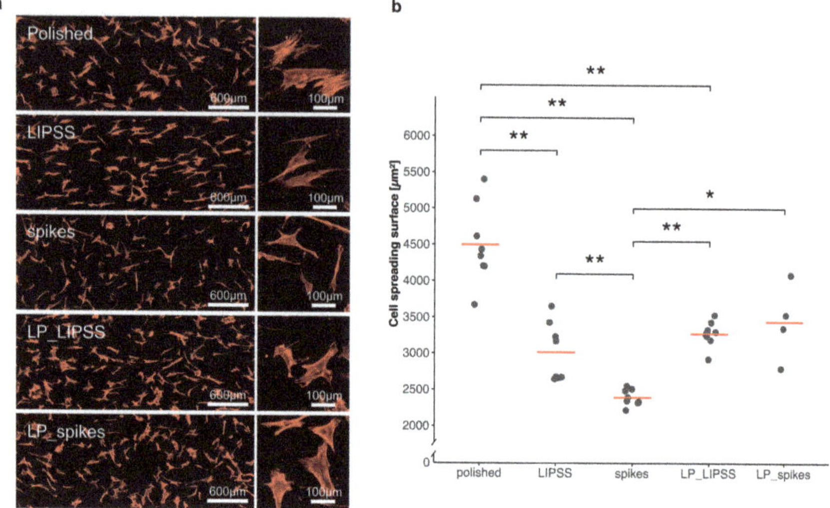

Figure 3. Quantification of cell spreading areas on polished, laser-nanostructured and laser-patterned surfaces. (**a**) Cytoskeleton staining images representing human mesenchymal stem cells (hMSCs) spreading 24 h after seeding on the five different surfaces previously sterilized. The rectangular graphs on the left are global views at a low magnification, and the square graphs on the right are representative cells at a high magnification; (**b**) four different areas of each surface were analyzed and, except for LP_spikes, the displayed data corresponds to duplicates of each condition (each dot represents the mean of more than 30 cells per area). The red lines denote mean values. The Kruskal–Wallis test followed by Mann–Whitney pairwise comparison with Bonferroni correction was performed. * p-value < 0.05, ** p-value < 0.01.

3.2.2. Cell Motility

The cell velocity, acceleration and the tortuosity of trajectories were measured by manual tracking of nuclei (Figure 4b) on totally laser-nanostructured (LIPSS and spikes) and laser-patterned samples (LP_LIPSS and LP_spikes). The cell velocity was significantly increased (approximatively +100%) on LIPSS and spikes as compared with polished and laser-patterned surfaces (Figure 4a left panel). Furthermore, no difference in cell migration velocity appeared between polished and laser-patterned samples. Moreover, no cell velocity difference was noted between LIPSS and spikes nor between LP_LIPSS and LP_spikes patterns. The tortuosity of the trajectories was enhanced on all laser-treated surfaces compared with polished ones. Cell acceleration was increased on all the laser-treated samples whereas the mean velocity was only improved on LIPSS and spikes. This result strongly corroborated our observations from the time-lapse videos that the cells tended to stop their migration when crossing a micro-pit (see Video S2 in Supplementary Materials). Altogether, the results from a 63-h video strongly suggested that in presence of nanostructures the speed variations (cell acceleration or cell deceleration) increased and cell trajectories became more tortuous compared with the polished surface.

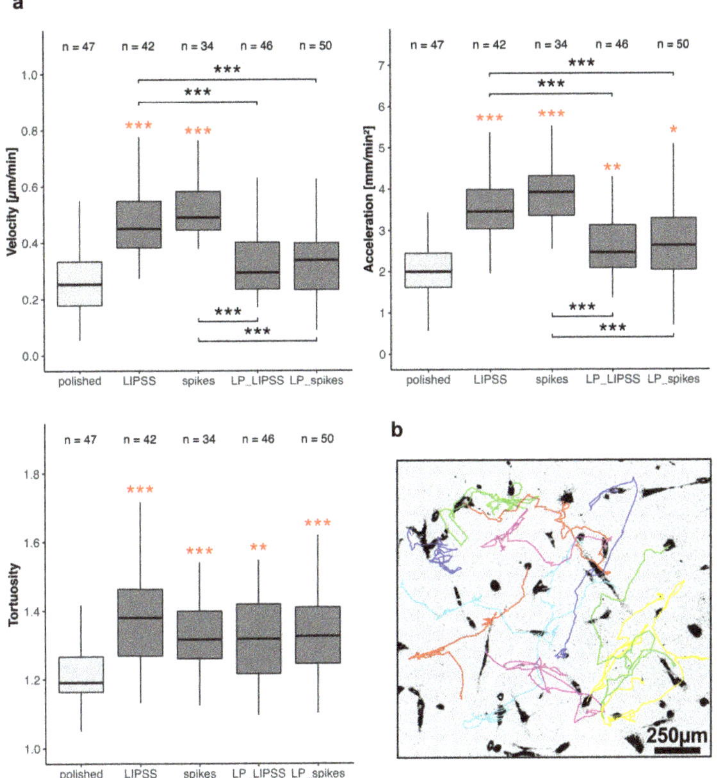

Figure 4. Cell motility on polished and laser-treated surfaces. hMSCs were observed 1 h post-seeding for 63 h and every 30 min on sterilized Ti6Al4V samples. (**a**) Mean velocity and acceleration (absolute value) were measured for each cell individually. Mean tortuosity was defined as the ratio between the accumulated distance and the Euclidean distance every 1.5 h and averaged on the whole record for each cell. Three areas per sample were analyzed, the exact number of analyzed cells is indicated above each boxplot. The Kruskal–Wallis test followed by Mann–Whitney pairwise comparison with Bonferroni correction was performed on data. * p-value < 0.05, ** p-value < 0.01, *** p-value < 0.001 Red asterisks denote significant differences versus polished surface. (**b**) Representative migration trajectories of cells on spikes by manual tracking of nuclei.

3.2.3. Nuclei Location

The nuclei location inside micro-pits was quantified. Figure 5 shows that the nuclei were preferentially located inside the micro-pits on LP_LIPSS surfaces, whereas this tendency was not observed on LP_spikes.

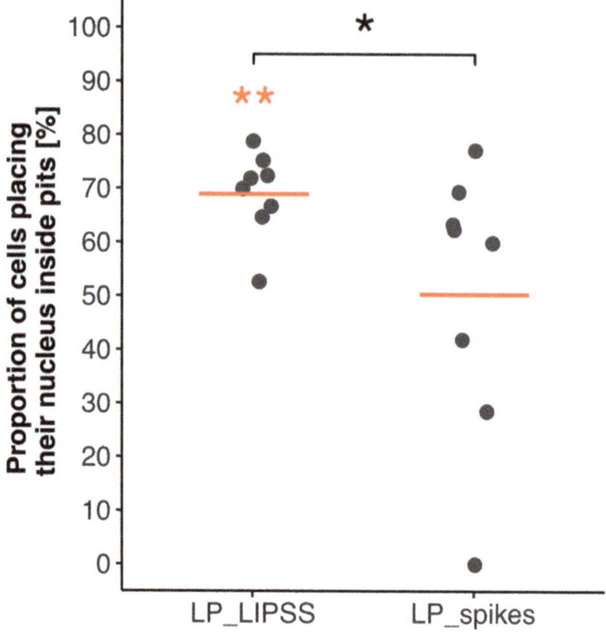

Figure 5. Quantification of nuclei proportion inside pits on laser-patterned surfaces 24 h after cell seeding. Values were normalized according to the proportion of pits area. Four different areas of each surface were analyzed and the displayed data correspond to duplicates of each condition (300 nuclei analyzed for LP_LIPSS and 494 for LP_spikes). Red lines denote mean values. The Mann–Whitney statistic test was performed. * p-value < 0.05, ** p-value < 0.01. Red asterisks denote significant differences with the random theoretical value ($\mu_0 = 50\%$).

3.2.4. Cell Adhesion

In order to determine hMSC interactions with multiscale patterns, cell adhesion was investigated at the sub-cellular scale. For this, hMSCs were seeded on sterilized laser-patterned samples (LP_LIPSS and LP_spikes) and were analyzed by immunofluorescence assays for vinculin detection, a protein involved in cell adhesion. hMSCs were found to place significantly their focal adhesion points on polished areas rather than on laser-nanostructured ones (Figure 6a). These observations were confirmed by measurements after image processing (Figure 6b). The number and area of adhesion plaques were quantified, and data were normalized according to surface proportions. Both number and area measurements led to the finding that focal adhesions were mostly and preferentially located on the polished areas. Moreover, a significant difference was measured between laser-patterned LP_LIPSS and LP_spikes for both factors. Vinculin staining was also performed on polished samples, and samples fully covered by simple laser-induced nanostructures. It revealed that the focal adhesion clusters were well established on polished surface but no focal adhesion was observed on LIPSS or spikes (data not shown). This observation was consistent with the results concerning the preferential distribution of focal contacts on smooth areas of complex patterns.

Collectively, the migration versus adhesion data suggest that the complex patterns reduce the migration but increase the adhesion compared to simple nanostructures. This latter fact could be an important determinant for osteoblast differentiation.

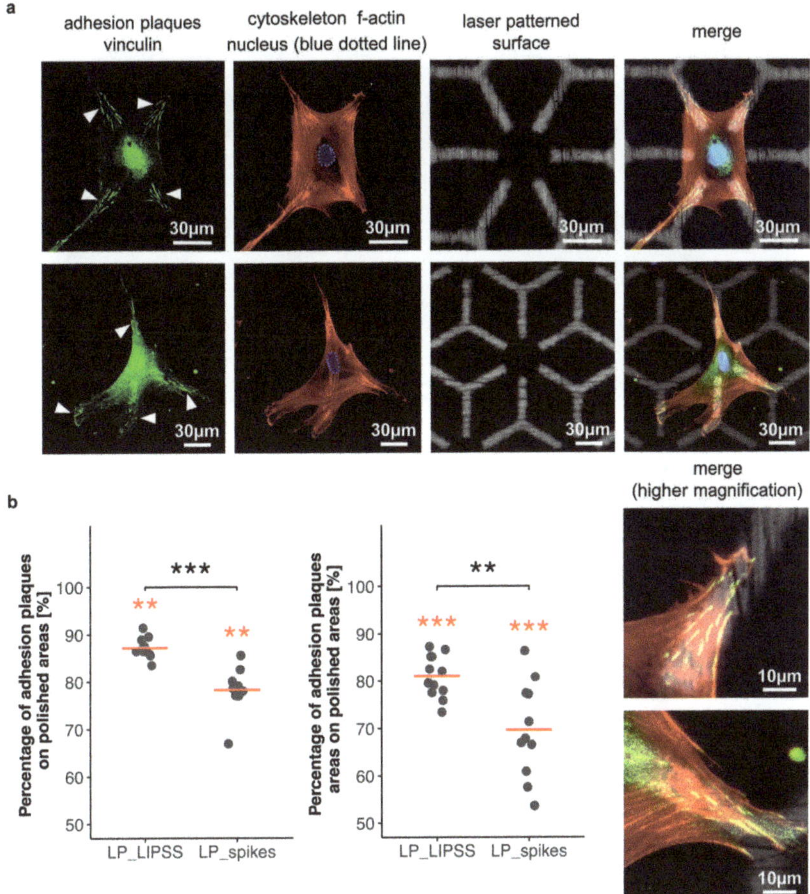

Figure 6. Quantification of focal adhesion repartition between polished areas and laser-nanostructured ones on laser-patterned surfaces. (**a**) Two representative cells 48 h after seeding on laser-patterned surfaces. Focal adhesions (white arrows) mostly located on polished areas. Nucleus in blue dotted line located inside the micro-pit. Cytoskeleton with a star like shape in red. Laser patterned surface with polished areas in white and nanostructured areas in black. (**b**) Ratio of adhesion plaques on polished regions compared to adhesion plaques on laser-nanostructured ones was obtained either by counting the number of adhesion plaques per cell (left panel) or by measuring the areas of the adhesion plaques per cell (right panel). A total of 760 focal adhesions were analyzed on LP_LIPSS and 665 on LP_spikes. Values were normalized according to the proportion of polished and laser-nanostructured regions on each sample. The Mann–Whitney U statistic tests were performed. Red lines denote mean values. ** p-value < 0.01, *** p-value < 0.001. Red asterisks denote significant differences versus the random theoretical value ($\mu_0 = 50\%$).

3.3. Protein Adsorption

To observe preferential adsorption of protein an assay with fluorescent albumin was performed (Figure 7). During this experiment, laser-patterned samples LP_LIPSS and LP_spikes were investigated. The mean fluorescence intensity was measured on polished and on laser-nanostructured areas to quantify albumin adsorption on these surfaces. The mean fluorescence was found to be higher (p-value

< 0.1) on polished areas than on their laser-nanostructured counterparts. It can be noted that the difference in mean fluorescence intensity was stronger for LP_LIPSS than on LP_spikes.

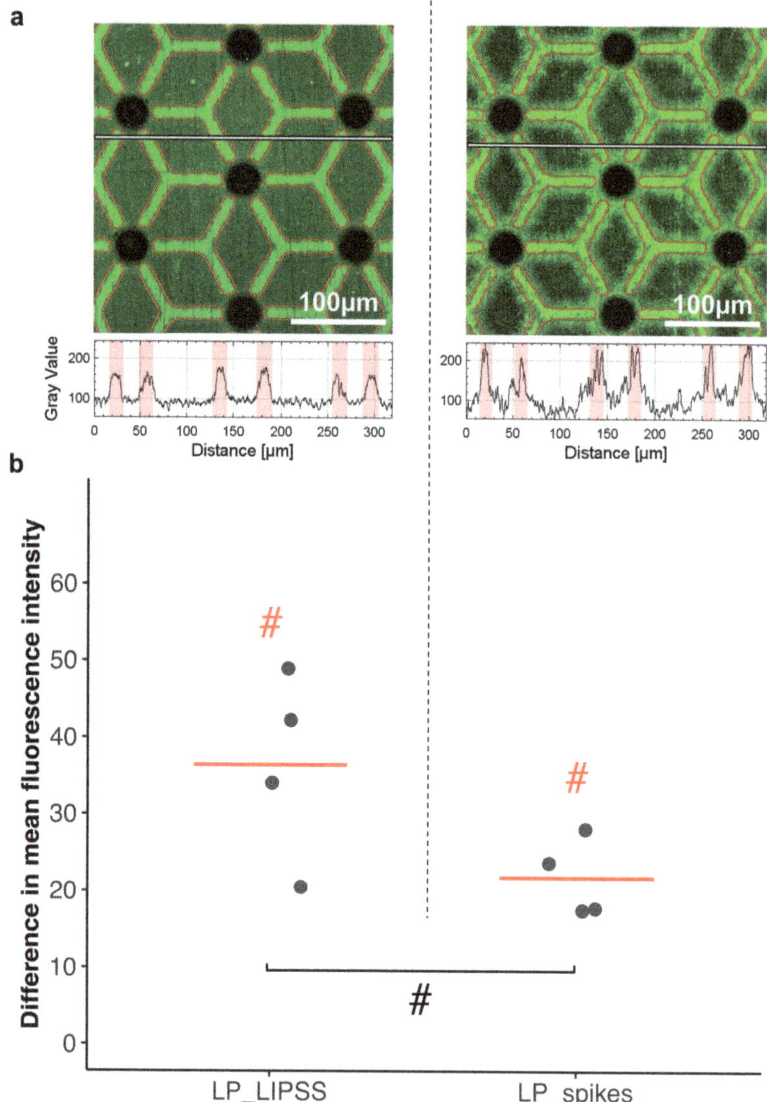

Figure 7. Comparison of albumin adsorption on polished areas and nanostructured areas. Fluorescent albumin was detected on laser-patterned surfaces with LIPSS (LP_LIPSS) or spikes (LP_spikes). (**a**) Representative image of fluorescence intensity. Polished areas are outlined in red, textured ones are the remainder (**top panel**). Respective profiles of the fluorescence intensity, red regions represent polished areas along the profile (**bottom panel**). (**b**) The difference in mean fluorescence intensity between polished and textured regions is displayed according to an 8-bit grayscale. Four measurements on areas of 320 µm by 320 µm per sample were analyzed. The Mann–Whitney one-tailed statistic test was performed. Red lines denote mean values. # p-value < 0.1. Red hash signs denote differences with the random theoretical value ($\mu_0 = 0$).

4. Discussion

4.1. Uniform Surface Topography (Polished Surface, Laser-Induced Periodic Surface Structures (LIPSS), and Spikes)

It is well known that the wettability of laser-nanostructured surfaces changes with time [23,24] and can be modified by storage in different media just after laser processing [35]. In fact, most laser-induced nanostructures show a total wetting (CA < 5°) shortly after the laser treatments. These structures then become more and more hydrophobic with the lapse of time. The first hydrophilic state could be explained by the laser removal of non-polar and organic surface contaminants [36]. Furthermore, according to the Wenzel model, the roughness of a hydrophilic surface enhances even more the hydrophilicity, hence the complete wetting observed on the sample with spikes. Moreover, it was shown that the laser irradiation increased the thickness of the natural oxide layer at the very surface of the laser-processed titanium samples [36,37] and this was especially demonstrated on LIPSS topographies. Such an oxide layer leads to high surface polarity, in turn related to hydrophilic properties due to the polarity of the water molecule [36]. More recently, surface aging towards hydrophobicity has also been shown to be connected with the progressive adsorption of non-polar and organic compounds [38]. It was proven that this trend could be accelerated through a low heat treatment (<200 °C) [37,39] comparable to the sterilization process used in this study.

Here, we show that total wetting could be reached in a relatively time-stable (38 days) manner by using laser-texturing (spikes). In particular, we emphasize that the sterilization process is a critical step for a more stable and reliable prediction of the surface properties for laser-nanostructured samples. Indeed, heat treatment appears to significantly promote hydrophobicity. The fact that these changes were not observed on polished surfaces contrary to their laser-nanostructured counterparts can be explained by the better-established chemical stability of the passivation layer of the polished surface. It remains challenging to analyze the independent effects of the surface topography and chemistry on wettability. Both are without doubt involved, but the cross effects are predominant. Because laser-induced nanostructures are generated by adjusting the laser fluence, a set of electronic, thermal and structural variations considerably affects the surface chemistry.

Some studies have already demonstrated the great influence of regular micro- and/or nano-structures on wettability [40,41] and that laser induced nanostructures such as LIPSS leads to anisotropy in wetting [19,42]. Hence, the large variation in contact angles on LIPSS (see the box plot in Figure 2) denoted a droplet anisotropy with an elongation perpendicular to the LIPSS waviness, which caused a wider contact angle range. The phenomenon became less evident after sterilization because of the shorter droplet baseline due to non-wetting. It is important to notice that a cell culture medium with an unknown surface tension was used as the testing medium for the contact angle measurements. This is why the results cannot be compared in absolute values with data from the literature; only relative trends can be proposed. This choice was made to mimic as well as possible real cellular experimental conditions.

Cells respond to their environment by adapting their behavior through changes in their morphology but also by adjusting their spreading, adhesion and motility. Such behavior modifications can influence cellular proliferation, direct cell fate, and trigger cellular differentiation, in particular when it comes to stem cells. Cellular morphology relies on the structure of the cytoskeleton, and changes in cell shape and contractility are, therefore, often a consequence of its (re)-organization. Moreover, cells display a mechano-sensing ability triggering the rearrangement of the cytoskeleton upon external mechanical stimuli. Thus, any morphological or mechanical deformation induced by the substrate topography are experienced by the intrinsic nuclear mechano-transduction pathways and induce modifications of the tension and the structure of the cytoskeleton as well as changes in the nucleus as a response. Since the nucleus, the nuclear membrane and the chromatin are directly connected to the cytoskeleton, any mechanical stimuli cause changes in the chromatin state thus leading to changes in gene expression [7,43].

Here, cell spreading was strongly reduced 24 h post-seeding on totally laser-nanostructured surfaces (Figure 3). This decrease can be explained based on a hypothesis that our textured samples might work as anti-adherent surfaces. As a result, cells would become less prone to spread. hMSCs cultured on specific patterns inducing cell shape with higher aspect ratios enhance the cytoskeletal tension and exhibit a clear tendency to differentiate towards osteoblastic lineage [25–27]. Interestingly, McBeath et al. [28] observed that hMSCs grown on pro-adherent patterns would differentiate into osteoblastic cells or adipocytes if sufficiently spread or unspread, respectively. It seems that adjusting cellular spreading would control stem cell lineage commitment [28]. Dumas et al. [10] showed that modifications of cell morphology induced by laser-nanostructured surfaces could direct MSCs differentiation into osteoblasts.

4.2. Multiscale Surface Topography (Laser-Patterned Samples with Polished Surface Areas (LP_LIPSS), and Laser-Patterned Samples with Spikes (LP_spikes))

Time-lapse fluorescence microscopy was used for cell motility measurements, 1 h post-seeding and for 63 h. An increase of the cellular velocity was observed for hMSCs seeded on totally laser-nanostructured samples, but not on laser-patterned ones (relative to polished). Bertolo et al. [44] demonstrated that cell velocity is correlated with cell spreading and that faster cells have small spreading areas. This is adequate for our results. Moreover, the augmentation in cell velocity, acceleration and tortuosity on the totally laser-nanostructured samples denote a tendency for cells to explore their environment. We assume that the exploration of the environment by the cells results from the search for more suitable adherent sites. Interestingly, the velocity and the acceleration of hMSCs seeded on laser-patterned surfaces (LP_LIPSS and LP_spikes) were significantly reduced compared with totally laser-nanostructured samples, and even reached the average velocity measured for hMSCs on polished surfaces (Figure 4b left panel). This could be explained both by the fact that the polished surface may offer more possible adherent sites and that pits can constitute a location for cell implantation. Indeed, it was widely observed that cells slowed down during pit-crossing or even settled down, placing their nucleus inside the pit (Supplementary Materials, Video S2). Further measurements showed that on LP_LIPSS, up to 70% of the cells' nuclei were located inside pits. This last result proved that the control of the cell "engraftment" could be performed using topography modifications at the microscale. The fact that this behavior might simply be the consequence of gravity was discarded since it was observed during real-time observation with upside-down samples. As shown by Pieuchot et al. [45], cells respond to cell-scale curvature variations during their migration and position themselves in concave valleys such as the pits on LP_LIPSS. It is worth noting that the nuclei were not significantly located inside pits on LP_spikes. This could be explained by the pattern topography itself (Figure 1b) which is better defined in terms of height on LP_LIPSS compared to LP_spikes. Indeed, on LP_LIPSS, only pits represent a concave zone, while on LP_spikes the higher laser fluence used to generate spikes led to slightly deeper areas (−1.5 µm), which could also be considered as concave.

To validate our previous idea connecting cell motility and the research for suitable adherent sites, focal adhesion studies were performed on laser-patterned samples. We showed that hMSCs adhered preferentially to polished areas compared to the nano-textured ones, and this result was enhanced on LP_LIPSS (Figure 6). Since the cells also adhered to totally textured samples, it is hard to conclude whether laser-nanostructured areas had intrinsic anti-adhesive properties. Nevertheless, when grown on laser-patterned surfaces (i.e., a mix of polished and laser-nanostructured areas), a significant amount of the cells placed their focal adhesion points on polished areas rather than on textured ones. Studies indicated that cell adhesion is improved on moderate hydrophilic biomaterial surfaces. Dowling et al. [46] indicated that a surface with a wettability comparable to our smooth titanium (60–80° contact angle) showed better cell adhesion compared to more hydrophilic (12°) or more hydrophobic ones (155°). However, cell adhesion is highly governed by surface properties including roughness, and chemical composition, such complex interrelationships are the possible reasons for the contradictory reported behaviors. For example, superhydrophobic surfaces were reported to be extremely cell repellent [47,48] whereas it was shown, in another work, that cells adhere and proliferate on superhydrophobic surfaces [49]. In our study, we have noted that there were fewer focal contacts on the smooth areas of the LP_spikes compared to the LP_LIPSS (Figure 6), this could be due to the fact that the polished areas are not at the same height level than the areas with nanostructures in LP_spikes. This irregular depth in LP_spikes could decrease the development of focal adhesions on smooth areas. In order to achieve a better understanding of the connection between laser-induced topographies and anti-adherent properties, we studied the protein adsorption rate of such laser-patterned surfaces. The surface wettability was shown to be an important factor in protein adhesion to biomaterial surfaces. Generally hydrophobic surfaces are considered to be more protein-adsorbent than are hydrophilic surfaces [50]. However, different studies regarding the effects of surface wettability on protein adhesion have not always been consistent, the adhesion of proteins to a surface is a time-dependent process that can involve relatively large energy scales [51]. In the present study, it was found that albumin adsorption was significantly higher on polished areas as compared to laser-nanostructured ones. Studies have shown that cells do not adhere directly to the surface but to proteins, in turn adsorbed on the first surface layer [52]. The protein layers formed within the first few minutes of contact with the culture medium consist almost exclusively of albumin [53]. This initial protein adsorption behavior that affect cell-adhesive protein such as fibronectin could subsequently regulate the cell adhesion via focal contacts [54,55]. Therefore, variations in albumin adsorption rate should be representative of favorable focal adhesion locations. These results strongly suggest that within laser-patterned surfaces, the laser-nanostructured areas behaved like anti-adhesive sites for proteins and cells. Therefore, laser patterning could be used to create local "anti-adherent sites" in order to constrain and stimulate the stem cells to form the desired shape. Indeed, in this study, star-shaped cells were observed on laser-patterned surfaces. Polished and pro-adherent branches were guiding lines for cell extensions and the remoteness of adhesion point locations promoted a high contractility of the cell cytoskeleton. Furthermore, it is known that cytoskeleton contractility and star-shaped hMSCs enhance stem cell differentiation into the osteoblastic lineage. Indeed, Kilian et al. [25] demonstrated that changing the shape of the patterns altered the osteogenesis, so stem cells grown on flower-shaped (low contractility) or star-shaped (high contractility) patterns would differentiate into adipocytes or osteoblasts, respectively.

Altogether, our results strongly suggest that surfaces with a predictive cell behavior can be developed from the understanding of cell-surface interactions. Such surfaces will constitute a major improvement for the control of the interface between the implant and the surrounding tissues. Complex multiscale surface geometries (LP_LIPSS and LP_spikes) combining micro-pits and nanostructures lead to an efficient cell positioning on specific textured areas and the control of focal contact adhesion repartition. Creating anti-adherent sites delimits appropriately the adhesive sites elsewhere, and the combination of the two leads to cell confinement for control of shape and mechanics. We expect that mechanically strained cells on a patterned surface should differentiate preferentially into osteoblastic cells. A correlation between late differentiation and early parameters, such as shape or migration, is not within the scope of the present report. A separate study will be carried out on this subject matter.

The laser texturing as a single-step process for creating micron- and nanoscale directly on metals could constitute a potential tool to improve the performance of biomedical titanium and its attributes would be useful for a range of applications in regenerative medicine including orthopedics and dentistry. Otherwise, the anti-adherent function can be applied to achieve other functionalities for other biomedical implants, for instance, the contact surfaces of nails and splints for bone fracture fixation where a minimum cell adherence is highly desired.

Supplementary Materials: The following are available online at http://www.mdpi.com/2079-4991/10/5/864/s1: Video S1: Nanostructured surface wettability before and after sterilization process (droplet 360° rotation not shown). Video S2: A representative time lapse video of cell migration for 25 h on LP_LIPSS (5 frame/s).

Author Contributions: A.K. wrote the manuscript, carried out cell experiments and performed surface characterizations (topographies and wettability). A.G. helped with video microscopy and V.D. with SEM and optical 3D surface measurements. S.P. coordinated the cell experiment. X.S. developed and performed the laser texturing. A.K., A.G. and V.D. analyzed and interpreted the results. V.D., X.S., T.E.I., A.G., L.V., C.H.-D. and C.D. contributed to the different stages of the design of the experiments and writing. V.D. and A.G. conceived the experiments and directed the research. All authors have read and agreed to the published version of the manuscript.

Funding: The authors thank the EUR MANUTECH SLEIGHT for financial support and the French ANR for the Equipex MANUTECH-USD support regarding the laser texturing and wettability measurements (project TEXLID).

Acknowledgments: The authors thank Cyril Mauclair for his expertise in femtosecond laser texturing and Alina Pascale Hamri for her precious help concerning the wettability measurements.

Conflicts of Interest: The authors declare no conflict of interest.

References

1. Hanawa, T. Zirconia versus titanium in dentistry: A review. *Dent. Mater. J.* **2019**. [CrossRef] [PubMed]
2. Jäger, M.; Jennissen, H.; Dittrich, F.; Fischer, A.; Köhling, H. Antimicrobial and osseointegration properties of nanostructured titanium orthopaedic implants. *Materials* **2017**, *10*, 1302. [CrossRef] [PubMed]
3. Lausmaa, J. Mechanical, thermal, chemical and electrochemical surface treatment of titanium. In *Titanium in Medicine: Material Science, Surface Science, Engineering, Biological Responses and Medical Applications*; Springer: Berlin, Germany, 2013; Volume 2, pp. 231–266. ISBN 978-3-642-63119-1.
4. Liu, X.; Chu, P.; Ding, C. Surface modification of titanium, titanium alloys, and related materials for biomedical applications. *Mater. Sci. Eng. R Rep.* **2004**, *47*, 49–121. [CrossRef]
5. Lotfi, M.; Nejib, M.; Naceur, M. Cell adhesion to biomaterials: Concept of biocompatibility. In *Advances in Biomaterials Science and Biomedical Applications*; Pignatello, R., Ed.; InTech: London, UK, 2013; ISBN 978-953-51-1051-4.
6. Ponsonnet, L.; Reybier, K.; Jaffrezic, N.; Comte, V.; Lagneau, C.; Lissac, M.; Martelet, C. Relationship between surface properties (roughness, wettability) of titanium and titanium alloys and cell behaviour. *Mater. Sci. Eng. C* **2003**, *23*, 551–560. [CrossRef]
7. Anselme, K.; Wakhloo, N.T.; Rougerie, P.; Pieuchot, L. Role of the nucleus as a sensor of cell environment topography. *Adv. Healthc. Mater.* **2018**, *7*, 1701154. [CrossRef]
8. Théry, M.; Pépin, A.; Dressaire, E.; Chen, Y.; Bornens, M. Cell distribution of stress fibres in response to the geometry of the adhesive environment. *Cell Motil. Cytoskeleton* **2006**, *63*, 341–355. [CrossRef]
9. Ermis, M.; Antmen, E.; Hasirci, V. Micro and Nanofabrication methods to control cell-substrate interactions and cell behavior: A review from the tissue engineering perspective. *Bioact. Mater.* **2018**, *3*, 355–369. [CrossRef]
10. Dumas, V.; Guignandon, A.; Vico, L.; Mauclair, C.; Zapata, X.; Linossier, M.T.; Bouleftour, W.; Granier, J.; Peyroche, S.; Dumas, J.-C.; et al. Femtosecond laser nano/micro patterning of titanium influences mesenchymal stem cell adhesion and commitment. *Biomed. Mater.* **2015**, *10*, 055002. [CrossRef]
11. Chen, P.; Aso, T.; Sasaki, R.; Ashida, M.; Tsutsumi, Y.; Doi, H.; Hanawa, T. Adhesion and differentiation behaviors of mesenchymal stem cells on titanium with micrometer and nanometer-scale grid patterns produced by femtosecond laser irradiation. *J. Biomed. Mater. Res. A* **2018**, *106*, 2735–2743. [CrossRef]
12. Götz, H.E.; Müller, M.; Emmel, A.; Holzwarth, U.; Erben, R.G.; Stangl, R. Effect of surface finish on the osseointegration of laser-treated titanium alloy implants. *Biomaterials* **2004**, *25*, 4057–4064. [CrossRef]
13. Lee, B.E.J.; Exir, H.; Weck, A.; Grandfield, K. Characterization and evaluation of femtosecond laser-induced sub-micron periodic structures generated on titanium to improve osseointegration of implants. *Appl. Surf. Sci.* **2018**, *441*, 1034–1042. [CrossRef]
14. Dumas, V.; Rattner, A.; Vico, L.; Audouard, E.; Dumas, J.C.; Naisson, P.; Bertrand, P. Multiscale grooved titanium processed with femtosecond laser influences mesenchymal stem cell morphology, adhesion, and matrix organization. *J. Biomed. Mater. Res. A* **2012**, *100A*, 3108–3116. [CrossRef] [PubMed]
15. Raimbault, O.; Benayoun, S.; Anselme, K.; Mauclair, C.; Bourgade, T.; Kietzig, A.-M.; Girard-Lauriault, P.-L.; Valette, S.; Donnet, C. The effects of femtosecond laser-textured Ti-6Al-4V on wettability and cell response. *Mater. Sci. Eng. C* **2016**, *69*, 311–320. [CrossRef] [PubMed]

16. Vorobyev, A.Y.; Guo, C. Femtosecond laser structuring of titanium implants. *Appl. Surf. Sci.* **2007**, *253*, 7272–7280. [CrossRef]

17. Nayak, B.K.; Gupta, M.C.; Kolasinski, K.W. Formation of nano-textured conical microstructures in titanium metal surface by femtosecond laser irradiation. *Appl. Phys. A* **2008**, *90*, 399–402. [CrossRef]

18. Ranella, A.; Barberoglou, M.; Bakogianni, S.; Fotakis, C.; Stratakis, E. Tuning cell adhesion by controlling the roughness and wettability of 3D micro/nano silicon structures. *Acta Biomater.* **2010**, *6*, 2711–2720. [CrossRef]

19. Cunha, A.; Serro, A.P.; Oliveira, V.; Almeida, A.; Vilar, R.; Durrieu, M.-C. Wetting behaviour of femtosecond laser textured Ti–6Al–4V surfaces. *Appl. Surf. Sci.* **2013**, *265*, 688–696. [CrossRef]

20. Ahmmed, K.M.T.; Ling, E.J.Y.; Servio, P.; Kietzig, A.-M. Introducing a new optimization tool for femtosecond laser-induced surface texturing on titanium, stainless steel, aluminum and copper. *Opt. Lasers Eng.* **2015**, *66*, 258–268. [CrossRef]

21. Sedao, X.; Lenci, M.; Rudenko, A.; Faure, N.; Pascale-Hamri, A.; Colombier, J.P.; Mauclair, C. Influence of pulse repetition rate on morphology and material removal rate of ultrafast laser ablated metallic surfaces. *Opt. Lasers Eng.* **2019**, *116*, 68–74. [CrossRef]

22. Bizi-Bandoki, P.; Benayoun, S.; Valette, S.; Beaugiraud, B.; Audouard, E. Modifications of roughness and wettability properties of metals induced by femtosecond laser treatment. *Appl. Surf. Sci.* **2011**, *257*, 5213–5218. [CrossRef]

23. Bizi-bandoki, P.; Valette, S.; Audouard, E.; Benayoun, S. Time dependency of the hydrophilicity and hydrophobicity of metallic alloys subjected to femtosecond laser irradiations. *Appl. Surf. Sci.* **2013**, *273*, 399–407. [CrossRef]

24. Zhang, Y.; Zou, G.; Liu, L.; Zhao, Y.; Liang, Q.; Wu, A.; Zhou, Y.N. Time-dependent wettability of nano-patterned surfaces fabricated by femtosecond laser with high efficiency. *Appl. Surf. Sci.* **2016**, *389*, 554–559. [CrossRef]

25. Kilian, K.A.; Bugarija, B.; Lahn, B.T.; Mrksich, M. Geometric cues for directing the differentiation of mesenchymal stem cells. *Proc. Natl. Acad. Sci. USA* **2010**, *107*, 4872–4877. [CrossRef] [PubMed]

26. Yao, X.; Ding, J. Effect of cell anisotropy on differentiation of stem cells on micropatterned surfaces through the controlled single cell adhesion. *Biomaterials* **2011**, *32*, 8048–8057. [CrossRef]

27. Yao, X.; Peng, R.; Ding, J. Effects of aspect ratios of stem cells on lineage commitments with and without induction media. *Biomaterials* **2013**, *34*, 930–939. [CrossRef] [PubMed]

28. McBeath, R.; Pirone, D.M.; Nelson, C.M.; Bhadriraju, K.; Chen, C.S. Cell Shape, cytoskeletal tension, and rhoa regulate stem cell lineage commitment. *Dev. Cell* **2004**, *6*, 483–495. [CrossRef]

29. Miao, Y.H.; Helseth, L.E. Adsorption of bovine serum albumin on polyelectrolyte-coated glass substrates: Applications to colloidal lithography. *Colloids Surf. B Biointerfaces* **2008**, *66*, 299–303. [CrossRef]

30. Schneider, C.A.; Rasband, W.S.; Eliceiri, K.W. NIH Image to ImageJ: 25 years of image analysis. *Nat. Methods* **2012**, *9*, 671–675. [CrossRef]

31. Schindelin, J.; Arganda-Carreras, I.; Frise, E.; Kaynig, V.; Longair, M.; Pietzsch, T.; Preibisch, S.; Rueden, C.; Saalfeld, S.; Schmid, B.; et al. Fiji: An open-source platform for biological-image analysis. *Nat. Methods* **2012**, *9*, 676–682. [CrossRef]

32. RStudio Team. *RStudio: Integrated Development for R*; RStudio. Inc.: Boston, MA, USA, 2015.

33. Lutey, A.H.A.; Gemini, L.; Romoli, L.; Lazzini, G.; Fuso, F.; Faucon, M.; Kling, R. Towards laser-textured antibacterial surfaces. *Sci. Rep.* **2018**, *8*, 10112. [CrossRef]

34. Schnell, G.; Staehlke, S.; Duenow, U.; Nebe, J.B.; Seitz, H. Femtosecond laser nano/micro textured Ti6Al4V surfaces—Effect on wetting and MG-63 cell adhesion. *Materials* **2019**, *12*, 2210. [CrossRef]

35. Kietzig, A.-M.; Hatzikiriakos, S.G.; Englezos, P. Patterned superhydrophobic metallic surfaces. *Langmuir* **2009**, *25*, 4821–4827. [CrossRef] [PubMed]

36. Huerta-Murillo, D.; García-Girón, A.; Romano, J.M.; Cardoso, J.T.; Cordovilla, F.; Walker, M.; Dimov, S.S.; Ocaña, J.L. Wettability modification of laser-fabricated hierarchical surface structures in Ti-6Al-4V titanium alloy. *Appl. Surf. Sci.* **2019**, *463*, 838–846. [CrossRef]

37. Yang, Z.; Zhu, C.; Zheng, N.; Le, D.; Zhou, J. Superhydrophobic surface preparation and wettability transition of titanium alloy with micro/nano hierarchical texture. *Materials* **2018**, *11*, 2210. [CrossRef] [PubMed]

38. Yang, Z.; Liu, X.; Tian, Y. Insights into the wettability transition of nanosecond laser ablated surface under ambient air exposure. *J. Colloid Interface Sci.* **2019**, *533*, 268–277. [CrossRef] [PubMed]

39. Ngo, C.-V.; Chun, D.-M. Control of laser-ablated aluminum surface wettability to superhydrophobic or superhydrophilic through simple heat treatment or water boiling post-processing. *Appl. Surf. Sci.* **2018**, *435*, 974–982. [CrossRef]

40. Kubiak, K.J.; Wilson, M.C.T.; Mathia, T.G.; Carval, P. Wettability versus roughness of engineering surfaces. *Wear* **2011**, *271*, 523–528. [CrossRef]

41. Li, P.; Xie, J.; Deng, Z. Characterization of irregularly micro-structured surfaces related to their wetting properties. *Appl. Surf. Sci.* **2015**, *335*, 29–38. [CrossRef]

42. May, A.; Agarwal, N.; Lee, J.; Lambert, M.; Akkan, C.K.; Nothdurft, F.P.; Aktas, O.C. Laser induced anisotropic wetting on Ti–6Al–4V surfaces. *Mater. Lett.* **2015**, *138*, 21–24. [CrossRef]

43. Kirby, T.J.; Lammerding, J. Emerging views of the nucleus as a cellular mechanosensor. *Nat. Cell Biol.* **2018**, *20*, 373–381. [CrossRef]

44. Bertolo, A.; Gemperli, A.; Gruber, M.; Gantenbein, B.; Baur, M.; Pötzel, T.; Stoyanov, J. In vitro cell motility as a potential mesenchymal stem cell marker for multipotency: Mesenchymal stem cell motility in vitro. *STEM CELLS Transl. Med.* **2015**, *4*, 84–90. [CrossRef] [PubMed]

45. Pieuchot, L.; Marteau, J.; Guignandon, A.; Dos Santos, T.; Brigaud, I.; Chauvy, P.-F.; Cloatre, T.; Ponche, A.; Petithory, T.; Rougerie, P.; et al. Curvotaxis directs cell migration through cell-scale curvature landscapes. *Nat. Commun.* **2018**, *9*, 3995. [CrossRef] [PubMed]

46. Dowling, D.P.; Miller, I.S.; Ardhaoui, M.; Gallagher, W.M. Effect of surface wettability and topography on the adhesion of osteosarcoma cells on plasma-modified polystyrene. *J. Biomater. Appl.* **2011**, *26*, 327–347. [CrossRef] [PubMed]

47. Ratner, B.D. The biocompatibility of implant materials. In *Host Response to Biomaterials*; Elsevier: Amsterdam, the Netherlands, 2015; pp. 37–51. ISBN 978-0-12-800196-7.

48. Kulinets, I. Biomaterials and their applications in medicine. In *Regulatory Affairs for Biomaterials and Medical Devices*; Elsevier: Amsterdam, the Netherlands, 2015; pp. 1–10. ISBN 978-0-85709-542-8.

49. Oliveira, S.M.; Alves, N.M.; Mano, J.F. Cell interactions with superhydrophilic and superhydrophobic surfaces. *J. Adhes. Sci. Technol.* **2014**, *28*, 843–863. [CrossRef]

50. Noh, H.; Vogler, E.A. Volumetric interpretation of protein adsorption: Competition from mixtures and the Vroman effect. *Biomaterials* **2007**, *28*, 405–422. [CrossRef]

51. Fang, F.; Satulovsky, J.; Szleifer, I. Kinetics of protein adsorption and desorption on surfaces with grafted polymers. *Biophys. J.* **2005**, *89*, 1516–1533. [CrossRef]

52. Othman, Z.; Cillero Pastor, B.; van Rijt, S.; Habibovic, P. Understanding interactions between biomaterials and biological systems using proteomics. *Biomaterials* **2018**, *167*, 191–204. [CrossRef]

53. Wargenau, A.; Fekete, N.; Beland, A.V.; Sabbatier, G.; Bowden, O.M.; Boulanger, M.D.; Hoesli, C.A. Protein film formation on cell culture surfaces investigated by quartz crystal microbalance with dissipation monitoring and atomic force microscopy. *Colloids Surf. B Biointerfaces* **2019**, *183*, 110447. [CrossRef]

54. Hindié, M.; Camand, E.; Agniel, R.; Carreiras, F.; Pauthe, E.; Van Tassel, P. Effects of human fibronectin and human serum albumin sequential adsorption on preosteoblastic cell adhesion. *Biointerphases* **2014**, *9*, 029008. [CrossRef]

55. Zhou, Q.; Chen, J.; Luan, Y.; Vainikka, P.A.; Thallmair, S.; Marrink, S.J.; Feringa, B.L.; van Rijn, P. Unidirectional rotating molecular motors dynamically interact with adsorbed proteins to direct the fate of mesenchymal stem cells. *Sci. Adv.* **2020**, *6*, eaay2756. [CrossRef]

Article

Hierarchical Micro-/Nano-Structures on Polycarbonate via UV Pulsed Laser Processing

Marek Mezera [1,*], Sabri Alamri [2], Ward A.P.M. Hendriks [3], Andreas Hertwig [4], Anna Maria Elert [4], Jörn Bonse [4], Tim Kunze [2], Andrés Fabián Lasagni [2,5] and Gert-willem R.B.E. Römer [1]

1   Department of Mechanics of Solids, Surfaces and Systems (MS3), Faculty of Engineering Technology, University of Twente, Drienerlolaan 5, 7522 NB Enschede, The Netherlands; g.r.b.e.romer@utwente.nl
2   Fraunhofer Institut für Werkstoff- und Strahltechnik (IWS), Winterbergstraße 28, 01277 Dresden, Germany; sabri.alamri@iws.fraunhofer.de (S.A.); tim.kunze@iws.fraunhofer.de (T.K.); andres_fabian.lasagni@iws.fraunhofer.de (A.F.L.)
3   Optical Science Group, MESA + Institute for Nanotechnology, University of Twente, Drienerlolaan 5, 7500 AE Enschede, The Netherlands; w.a.p.m.hendriks@utwente.nl
4   Bundesanstalt für Materialforschung und -prüfung (BAM), Unter den Eichen 87, 12205 Berlin, Germany; andreas.hertwig@bam.de (A.H.); anna-maria.elert@bam.de (A.M.E.); joern.bonse@bam.de (J.B.)
5   Institut für Fertigungstechnik, Technische Universität Dresden, Georg-Bähr-Str. 3c, 01069 Dresden, Germany
*   Correspondence: m.mezera@utwente.nl

Received: 15 May 2020; Accepted: 2 June 2020; Published: 17 June 2020

Abstract: Hierarchical micro/-nanostructures were produced on polycarbonate polymer surfaces by employing a two-step UV-laser processing strategy based on the combination of Direct Laser Interference Patterning (DLIP) of gratings and pillars on the microscale (3 ns, 266 nm, 2 kHz) and subsequently superimposing Laser-induced Periodic Surface Structures (LIPSS; 7–10 ps, 350 nm, 100 kHz) which adds nanoscale surface features. Particular emphasis was laid on the influence of the direction of the laser beam polarization on the morphology of resulting hierarchical surfaces. Scanning electron and atomic force microscopy methods were used for the characterization of the hybrid surface structures. Finite-difference time-domain (FDTD) calculations of the laser intensity distribution on the DLIP structures allowed to address the specific polarization dependence of the LIPSS formation observed in the second processing step. Complementary chemical analyzes by micro-Raman spectroscopy and attenuated total reflection Fourier-transform infrared spectroscopy provided in-depth information on the chemical and structural material modifications and material degradation imposed by the laser processing. It was found that when the linear laser polarization was set perpendicular to the DLIP ridges, LIPSS could be formed on top of various DLIP structures. FDTD calculations showed enhanced optical intensity at the topographic maxima, which can explain the dependency of the morphology of LIPSS on the polarization with respect to the orientation of the DLIP structures. It was also found that the degradation of the polymer was enhanced for increasing accumulated fluence levels.

Keywords: Direct Laser Interference Patterning; Laser-induced Periodic Surface Structures; polycarbonate; hierarchical structures; surface functionalization

---

1. Introduction

In the course of evolution, flora and fauna adapted distinct surface structures, which induced specific functionalities and therefore ensured survival and procreation. A well-known example of a functional surface found in nature is the lotus leaf, which is water repellent and self-cleaning [1]. Other examples are the wings of butterflies and cicada, which are bactericidal [2–4], or the skin of sharks, which present self-cleaning, anti-biofouling, hydrodynamic and drag reduction properties [5]. Other examples, like the tenebrionid beetle *Stenocara* collects drinking water on its integument from morning fog and transports the collected water on its skin towards its mouthparts [6]. In this way, the insect can survive to the Namibian desert climate. Similar functionalities can be also found on moisture harvesting lizards [7]. All of these different specific surface properties are the result of so-called functionalized surfaces, which often consist of regular (hierarchical) micro- and nanometer sized surface structures. Functionalized surfaces have received increased scientific attention in recent years, aiming to reproduce them (biomimetics) due to their potential for new applications, such as anti-bacterial hip implants [8,9], increased [10] or decreased [11] cell-tissue growth onto implantable materials, liquid motion flow in microfluidics [12], fluid transport in tribological systems [13], friction control [14,15], wettability control [9,16,17] or colorization of surfaces [18–21].

Two well established laser-based methods have been shown in the past to be capable of creating micro- and nanostructured surfaces directly on the materials surface and thus functionalizing them, namely *Direct Laser Interference Patterning* (DLIP) and *Laser-induced Periodic Surface Structures* (LIPSS). DLIP is a method that produces micrometer and sub-micrometer sized, regular (hierarchical) structures on various materials, such as metals [17], polymers [16,22] or ceramics [23]. The periodic structures are created due to the interference pattern, which is produced when two or more laser beams are overlapped, leading to material removal (ablation) at the interference maxima of the spatially modulated intensity distribution. In the case of two-beam interference, the spatial period of the interference pattern can be controlled by the laser wavelength (λ) and the angle of incidence of the interfering laser beams (θ) [17]. Employing modern laser and beam scanning technology, the DLIP technique can fulfill industrial demands by addressing individual patterns of several square micrometer areas only, at processing rates of 0.9 m^2/min and 0.3 m^2/min for polymers and metals, respectively [24].

The second approach is based on LIPSS. LIPSS are regular (hierarchical) micro- to nanometer sized surface ripples, which appear due to the interference of (1) the impinging laser radiation with its scattered light at the surface or (2) laser triggered surface plasmon polaritons [25–27] and can be processed on solids due to polarized, (ultra-) short pulsed laser irradiation at laser peak fluence levels close to the ablation threshold [25,28]. The direction of the LIPSS depends on the material and the (linear) laser beam polarization. Their periodicity depends on several process parameters, such as the laser wavelength (λ), the angle of incidence (α), the number of pulses processing effectively impinging one spot (N_{eff}) and the laser peak fluence (F_0) [9,25,28]. Several types of LIPSS can be distinguished, depending on the laser processing parameters and the material, e.g., common *Low Spatial Frequency LIPSS* (LSFL) with a period of about the laser wavelength ($\Lambda_{LIPSS} \sim \lambda$), *High Spatial Frequency LIPSS* (HSFL) with a period well below the laser wavelength ($\Lambda_{LIPSS} < \lambda/2$), or even *hexagonally arranged triangular nanopillars* with an overall period close to the laser wavelength ($\Lambda_{LIPSS} \sim \lambda$) [25,28]. LIPSS can be processed on metals [8–10,18–20,29], semiconductors [12,21,25,28], dielectrics [30], ceramics [31–33] and polymers [34–37]. Large area processing of LIPSS is easily achieved in a one-step approach by scanning the focused laser beam in a meandering way across the sample surface. Since the central high fluence part of the Gaussian laser beam profile can generate different types of LIPSS than its low fluence wing, hierarchical micro-/nano-structures are easily feasible [9,38]. Moreover, as the periodicity of HSFL is not constrained by the optical diffraction limit, these extremely fine nanostructures may be even superimposed to sub-micrometric LSFL [39].

While individual scientific communities have already independently explored and optimized the processing of LIPSS and DLIP-based structures in detail [25,40,41], the combination of both techniques is still widely unexplored [42]. Although hierarchical structures can be achieved by using LIPSS [9,38] or DLIP [16] methods separately, the hybrid two-step laser process can provide an enhanced flexibility control of the surface features as well as explore new geometries in view of tailored surface functionalities.

Commercially available polycarbonate is used as sample material due to its wide range of applications, such as for products in the electronic and the automotive sector, in building and construction and for optical information storage systems, because of its unique combination of properties such as excellent toughness, high electrical insulation, transparency and large heat distortion resistance [43]. However, UV radiation leads to depolymerization of the molecular structure of polycarbonate [44–47] and this material modification could impair the use for applications. Hence, the quantitative changes of the molecular structure due to the two laser based surface functionalization techniques need to be assessed.

In this work, the evolution of nanometer sized LSFL on top of different types of micrometer sized DLIP structures on polycarbonate with respect to the laser peak fluence is studied, depending on the morphology and dimensions of the DLIP structures, as well as the direction of the laser polarization relative to the orientation of the DLIP structure, to obtain hierarchical structures. Additionally, the structural molecular changes of the polycarbonate due to the laser irradiation are studied.

2. Materials and Methods

Commercially available Bisphenol-A polycarbonate (PC) plates (Makrolon™ of Covestro AG, Leverkusen, Germany) with a thickness of 5 mm and a surface roughness of $R_a \approx 2$ nm were used as samples. DLIP and LIPSS methods were used on the samples using three different laser setups; one at the University of Twente (TruMicro 5050 of Trumpf GmbH, Ditzingen, Germany) and two at the Faunhofer Institute for Material and Beam Technology IWS (DLIP-µFab, Fraunhofer IWS, Dresden, Germany; Fuego of Time-Bandwidth Products AG, Zurich, Switzerland), as listed in Table 1.

Table 1. Laser sources and process parameters to produce hierarchical structures.

Structure Type	DLIP		LIPSS	
Laser source	Laser-export *TECH-263 Advanced*		Trumpf *TruMicro 5050*	Time-Bandwidth *Fuego*
DLIP Period [µm]	1.5	10	-	-
Wavelength [nm]	266	266	343	355
Pulse Duration [s]	3×10^{-9}	3×10^{-9}	7×10^{-12}	1×10^{-11}
Pulse Frequency [kHz]	2	2	100	100
Beam Quality Factor M^2 [-]	<1.3	<1.3	<1.3	<1.3
Laser Spot Diameter [µm]	25	25	174	120
Scan Velocity [mm/s]	2.5	2.5	1000	600
Number of Overscans [-]	1	1	1000	1000
Line Pitch [µm]	7.33	10.16	10	3
Geometrical Pulse-to-Pulse Overlap [%]	85	90	94	95
Peak Fluence Levels [J/cm^2]	0.25 (ridges) + 0.2 (Pillars)	1.45 (ridges) + 1.13 (Pillars)	$2...4 \times 10^{-3}$	$2...4 \times 10^{-3}$

2.1. Direct Laser Interference Patterning Configurations

The structuring of the PC samples was conducted by a compact two-beam DLIP system (DLIP-µFab, Fraunhofer IWS, Dresden, Germany), which produces confined DLIP treated areas containing the periodic structures created per laser pulse (also called pixels), with a diameter of $d_P \approx 25$ µm. The pixel diameter was calculated using the D-squared method described elsewhere [48]. The system uses a frequency quadrupled Q-switched laser head (TECH-263 Advanced of Laser-export Co. Ltd., Moscow, Russia) with a maximum pulse energy of 50 µJ and operating at a wavelength of $\lambda = 263$ nm and a pulse duration shorter than 3 ns. The laser beam has a nearly Gaussian intensity distribution (TEM$_{00}$) with a beam quality of $M^2 < 1.3$. The setup of the used DLIP optics allows the primary beam from the laser source to split into two single beams by means of a diffractive optical element. The sub-beams are parallelized by a prism and finally overlapped at the sample surface using a focusing aspheric lens.

As can be observed in Figure 1a, an interference pattern is obtained within the volume where the two single laser beams overlap. Changing the position of the prism modifies the interfering angle

θ, and leads to a change of the spatial period Λ of the periodic structures. In the employed setup, spatial periods Λ in the range between 1.0 μm and 11.0 μm can be produced. In order to structure larger areas than the DLIP pixel, the sample is moved using a high precision computer-controlled stage system (PRO155-05, Aerotech GmbH, Fürth, Germany), resulting in square-shaped processed areas with an edge length of 30 mm full covered with a ridge-like pattern. In particular, the samples were moved in the direction parallel to the interference lines with a spatial pulse separation p and successively displaced laterally of a quantity h (hatch distance), chosen as an integer of the spatial period (see Figure 1b). Moreover, for producing micro-"pillars", the areas treated with DLIP and containing a ridge-like pattern have been rotated by 90 degrees and re-irradiated, selectively ablating the previous pattern in the interference maxima. Note, that we use the term "pillar" here in the following for simplicity, since micro-"cones" are usually referred to as self-assembling surface structures [49–51].

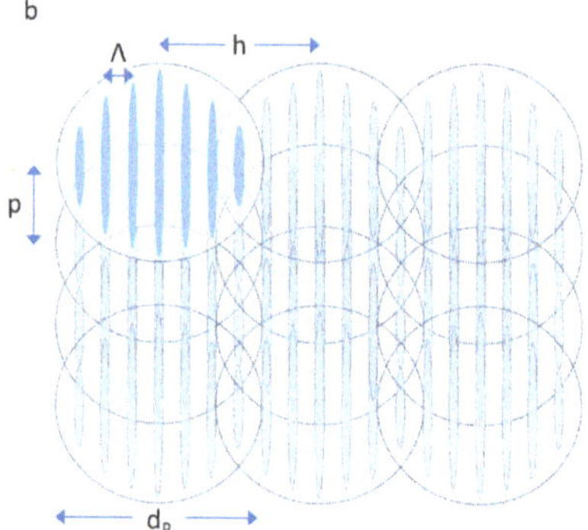

Figure 1. (**a**) Depiction of the interference phenomenon between two laser beams overlapping with an angle θ and (**b**) Scheme of the texturing approach for the displacement of several DLIP pixels on the sample's surface with p: pulse separation; h: hatch distance; d_p: pixel size; Λ: DLIP period. The scanning direction is vertical.

2.2. Fabrication of Laser-Induced Periodic Surface Structures

For the manufacturing of LIPSS two laser sources were used, see Table 1. Third harmonics were generated of a pulsed Yb:YAG disk laser source (TruMicro 5050 of Trumpf GmbH, Ditzingen, Germany) emitting a linearly polarized laser beam with a wavelength of 1030 nm, diameter of ≈5 mm, a maximum pulse repetition rate of 400 kHz, pulse energies up to 125 μJ and a fixed pulse duration of 6.7 ps. In other experiments, a frequency-tripled Nd:VAN laser source (Fuego of Time-Bandwidth Products AG, Zurich, Switzerland) emitting a linearly polarized laser beam with a wavelength of 1064 nm, a maximum pulse repetition rate of 8 MHz, pulse energies up to 200 μJ and a fixed pulse duration of 10 ps were used. To obtain homogeneous areas of LIPSS, the laser beam was scanned over the substrate using galvanometer scanners (intelliSCAN14 of ScanLab GmbH, Puchheim, Germany). The laser beam was focused on the surface of the samples, using a telecentric Fθ lens (Ronar of Linos GmbH, Göttingen, Germany) with a focal length of 103 mm. For obtaining large geometrical pulse-to-pulse overlap values in both $x-$ and $y-$directions at a scan speed of 1 m/s, the laser spot diameter on the sample was increased either by decreasing the laser beam diameter to ≈1 mm before focusing using a beam reducing telescope (TRE13 of Optogama, Vilnius, Lithuania), or by defocused laser processing. It was shown in an earlier publication, that the LIPSS morphologies and dimensions do not significantly differ when processing a defocused laser beam compared to processing with the focal spot [52]. The geometrical pulse-to-pulse overlap is given by $OL = (1 - v/(d \times f_F)) \times 100$, with v being the laser scan speed, d the beam spot diameter and f_F the laser pulse repetition rate. The processing parameters for the manufacturing of LIPSS are also summarized in Table 1. The meandering area scanning procedure could be repeated several times, denoted as the number of overscans (N_{OS}). Schematic representations of the laser setups and the scanning trajectory of the laser spot are shown in Figure 2.

The laser power at the sample surface was measured using a photodiode power sensor (S130VC of ThorLabs GmbH, Dachau, Germany) with a measurement uncertainty of ±5%, connected to a readout unit (PM100A of ThorLabs GmbH, Dachau, Germany). Along with the pulse repetition rate, this allowed to determine the energy E per individual laser pulse. The Gaussian (TEM_{00}) focal spot diameter $d = 174 \pm 1.6\ \mu m$ (e^{-2}) was measured in the sample processing plane using a laser beam characterization device (MicroSpotMonitor of Primes GmbH, Pfungstadt, Germany). From both information, the peak laser fluence F_0 in front of the sample surface was calculated according to $F_0 = \frac{2E}{\pi d/2^2}$ [48].

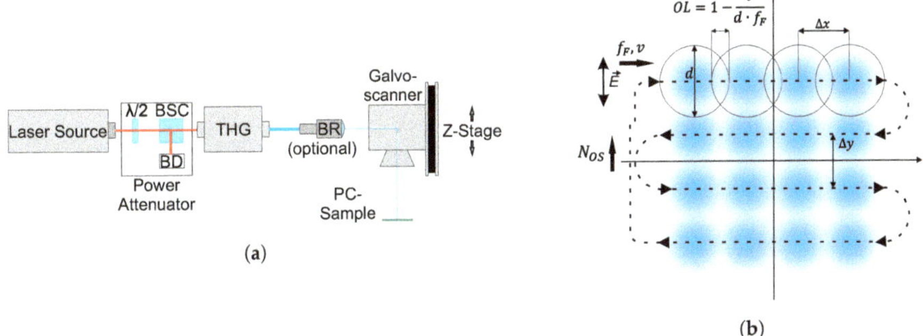

Figure 2. (a) Schematic representation of the laser setup; $\lambda/2$: half-wave plate; BSC: polarizing beam splitter cube; BD: beam dump; THG: third harmonic generator, BR: beam reducer. **(b)** Scanning trajectory of the laser spot; the double-headed arrow indicates the direction of the laser polarization $\vec{E}$; f_F: laser pulse frequency; v: scan velocity; d: laser spot diameter; OL: geometrical pulse-to-pulse overlap; N_{OS}: number of overscans; Δx: geometrical pitch between subsequent laser pulses in x-direction, Δy: Line pitch in y-direction.

2.3. Morphological Characterization

The morphology and dimensions of the processed surface structures were analyzed by a Scanning Electron Microscope (SEM JSM-7200F of JEOL, Tokio, Japan) and an Atomic Force Microscope (AFM NX10, Park Systems Corp., Suwon, Korea) in true non-contact mode using a non-contact cantilever (PPP-NCHR, $125 \times 30 \times 4\ \mu m^3$, nominal tip radius < 10 nm, Park Systems Corp., Suwon, Korea). Prior to SEM characterization, the samples were sputter coated with gold (JFC-1300 coater from JEOL, Tokio, Japan), resulting in a ≈10 nm thick, electrically conductive layer.

From SEM micrographs, the spatial frequencies of LIPSS were analyzed with the help of the 2D fast Fourier transform (FFT) algorithm using a MATLAB script [53]. Details of this script are reported in our earlier work [28]. From cross-sections of AFM micrographs, the amplitude of LIPSS were determined using another MATLAB script, also reported in [52].

2.4. FDTD Simulations

A commercially available photonic *Finite-difference time-domain* (FDTD) simulation software (Lumerical FDTD of Lumerical Inc., Vancouver, Canada) was used to analyze numerically the time-averaged optical intensity distribution of one laser pulse duration induced by a 6.7 ps laser pulse with a wavelength of 343 nm and with the laser beam polarization perpendicular and parallel to 1.5 μm ridge-like DLIP structures on polycarbonate (DLIP-type 1, see Figure 3a). The surface of the DLIP structure was modeled using the period, depth and full width at half maximum (FWHM) of the DLIP-type 1 structure obtained by AFM measurements. The period was found to be 1.5 μm, the depth of was found to be 400 nm and the FWHM of the ridges was found to be about 1 μm. The mesh settings of the two-dimensional computations were set to an automated mesh accuracy of 7 with a minimal step size of 0.25 nm, the time step size was set to 1.62×10^{-17} s, and the periodic boundaries were periodical on *x*-axis were set to *Periodic* and on the *y*-axis to *PML*. The optical properties were taken from Ref. [54].

2.5. Chemical Characterization

In order to analyze the surface chemistry of the polycarbonate substrate before and after laser irradiation of the sample, two different spectroscopy techniques were employed to record the IR spectra, i.e., *micro-Raman spectroscopy* (µ-RS) and *microscopy based Fourier-transform infrared spectroscopy* (FTIR).

µ-RS was performed on the laser-irradiated sample and on a reference position (Alpha 300R, WiTEC, Ulm, Germany). A ruled 600 grooves/mm grating was chosen in the optical spectrometer (UHTS 300, WiTEC, Ulm, Germany), which was equipped with a Peltier-cooled CCD camera (iDus DV401A, Andor Technology Ltd, Belfast, Ireland) operated at a temperature of 210 K. The resulting wavenumber resolution is < 2 cm^{-1}. The ps-laser irradiated surface regions were excited at a power level of 0.6 mW using the 532 nm emission line of a continuous wave laser (Excelsior, Spectra Physics, Santa Clara, USA). The Raman-laser radiation was focused on the sample surface by a microscope objective (EC Epiplan 20× NA 0.4, Carl Zeiss AG, Oberkochen, Germany) probing a circular spot of about 4 µm in diameter. All spectra are presented without background correction.

FTIR spectra were recorded in attenuated total reflection (ATR) mode (Vertex 70 with a Hyperion 3000 microscope, Bruker Optik, Ettlingen, Germany). The ATR microscope objective is equipped with a Ge-crystal tip ensured surface sensitivity through evanescent field coupling. ATR FT-IR spectra were taken at numerous arbitrary positions on laser processed and unprocessed sample areas with a measurement area of 80×80 µm^2. The FTIR spectra were background corrected (see details below) and absorption peaks resulting from ambient air containing CO_2 water vapor were removed.

3. Results and Discussion

3.1. Direct Laser Interference Patterning on Polycarbonate

Four different types of DLIP structures were processed in a first set of experiments. DLIP-type 1 and 2 are ridge- and pillar-like structures, respectively, with a period of 1.5 µm and a height of approximately 400 nm (Figure 3a,b). Otherwise, DLIP-type 3 and 4 types are ridge- and pillar-like structures, respectively, with a period of 10 µm and an approximated height of about 15 µm (Figure 3c,d). The laser processing parameters are listed in Table 1.

3.2. Laser-Induced Periodic Surface Structures on Polycarbonate

In a second processing approach, an area of 5×5 mm^2 was processed with the Trumpf TruMicro 5050 laser system on a pristine polycarbonate sample by scanning the laser spot perpendicular to the laser polarization and using an overlap of $OL = 93\%$ at a pulse frequency of $f = 100$ kHz, number of overscans $N_{OS} = 1000$ and a peak fluence level of $F_0 = 4.42$ mJ/cm^2. These parameters permitted low-spatial frequency LIPSS parallel to the laser polarization (type II [25]) with a very homogeneous morphology , see Figure 4. The period of the LSFL-II was found to be $\Lambda_{LIPSS} = 265 \pm 75$ nm and their amplitude (modulation depth) A was 11 ± 8 nm. On the basis of these laser processing parameters, hierarchical micro-/nanostructures were produced by processing LSFL-II on top of the four different types of DLIP structures (see next section).

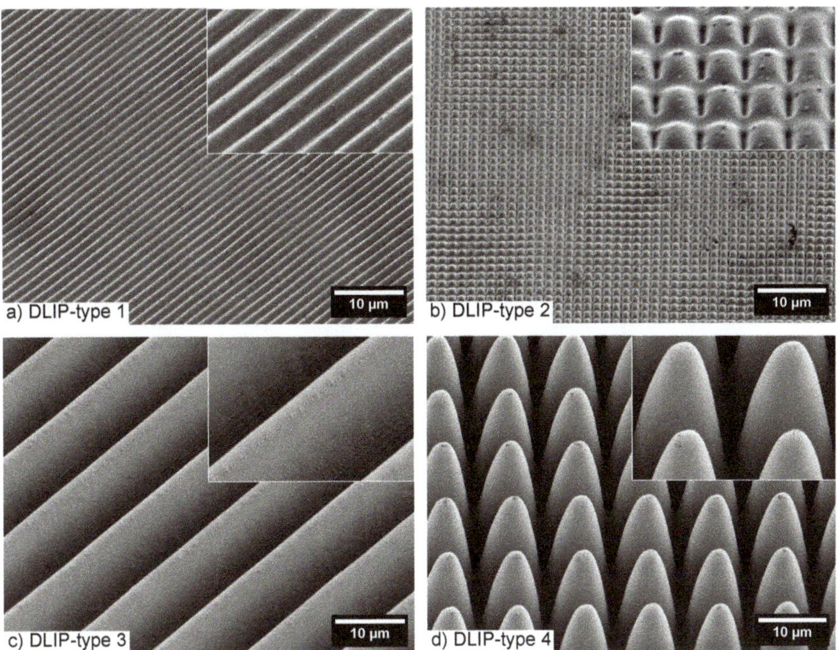

Figure 3. (**a**) ridge-like DLIP structure with a period of 1.5 µm and an depth of 400 nm (DLIP-type 1); (**b**) pillar-like DLIP structure with a period of 1.5 µm and an depth of 400 nm (DLIP-type 2); (**c**) ridge-like DLIP structure with a period of 10 µm and an depth of 15 µm (DLIP-type 3); (**d**) pillar-like DLIP structure with a period of 10 µm and an depth of 15 µm (DLIP-type 4) obtained with laser parameters listed in Table 1.

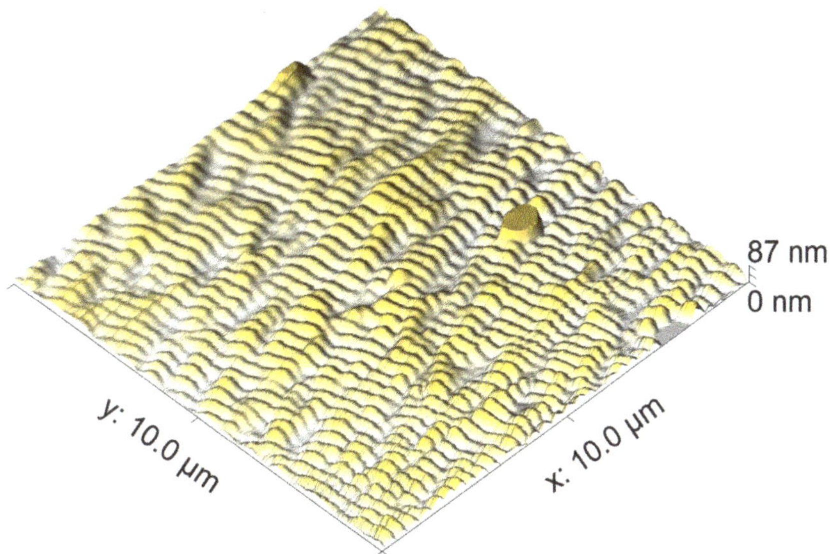

Figure 4. AFM topography of LSFL processed on untreated polycarbonate with the above mentioned parameters.

3.3. Hierarchical Structures on Polycarbonate

In a further processing phase, the DLIP-treated PC samples were irradiated with ultrashort UV radiation with the aim to create a two-level hierarchical microtexture. Figures 5–10 show SEM micrographs of different DLIP structures processed in a second laser processing step in order to produce LIPSS with orthogonal scanning (laser polarization) directions regarding the direction of the DLIP ridges and at various laser fluence levels, indicated as $\vec{E}$ and $\vec{v}$ in sub-figures (b), respectively.

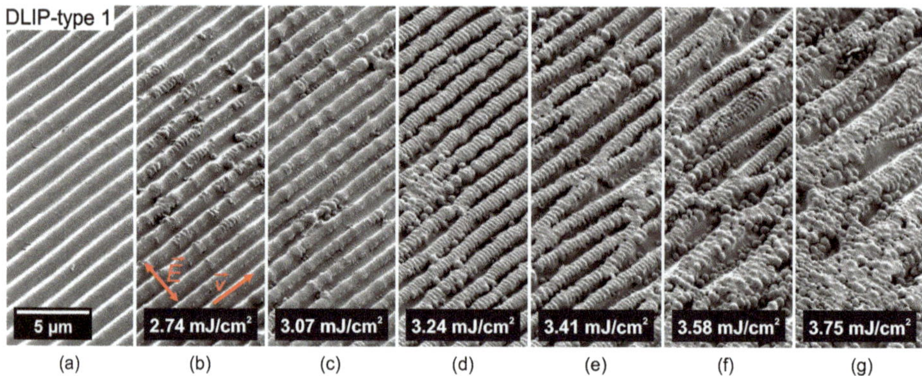

Figure 5. SEM micrographs of (**a**) ridge-like DLIP structure with a period of 1.5 µm (DLIP-type 1). (**b**–**g**) Evolution of surface morphology on top of the DLIP structure upon additional scan-processing with increasing peak fluence levels and a laser beam polarization perpendicular to the DLIP ridges processed with the Trumpf TruMicro 5050 laser system.

It can be observed in Figure 5, that LSFL-II parallel to the laser polarization and perpendicular to the 1.5 µm DLIP ridges start to develop on the latter at a fluence level of 2.74 mJ/cm² and higher. Also, nano-droplets start to appear in the valleys between the DLIP ridges (see Figure 5b). At a slightly higher laser fluence level of 3.07 mJ/cm², LSFL-II also start to develop in the valleys between the DLIP ridges in form of periodic chains of nano-droplets (see Figure 5c). At a fluence of 3.24 mJ/cm², all DLIP ridges are homogeneously covered with LSFL-II. Additionally, it can be observed that the nano-droplets between the DLIP ridges in Figure 5d–f are larger in diameter then at lower fluence levels in Figure 5b,c. The seeding of the periodic nano-droplet chains in Figure 5c can be linked to optical scattering effects, leading to periodic laser-induced defects of the PC with increased absorptivity [26]. The growth of the nano-droplets can be related to thermocapillary forces, pushing molten material from the ridges down in the DLIP valleys towards the nano-droplets [55,56]. At even higher laser fluence levels, nano-droplet appearance increases and the DLIP ridges become thinner due to more molten material that is transferred into nano-droplets until the ridges merge into each other and severe ablation takes place, see Figure 5e–g.

Figure 6 shows SEM micrographs of the evolution of surface morphology on top of the DLIP-type 1 structure with increasing peak fluence levels and a laser polarization parallel to the DLIP ridges. Here, the polarization of the laser beam is rotated by 90° compared to Figure 5, i.e., parallel to the DLIP ridges. It can be observed, that also in this case nano-droplets appear at the valleys of the DLIP ridges at a fluence level of 2.74 mJ/cm², see Figure 6b. At a somewhat higher fluence level of 2.90 mJ/cm², the DLIP ridges start to separate into chains of larger micro-droplets. Similar results were found when processing LSFL on top of LSFL on polyethersulfone, when the sample was rotated by 90° [57] and when processing LIPSS on a chromium thin film [58]. This phenomenon can be related to the Plateau-Rayleigh instability [58]. In brief, the surface energy in a stationary fluid in cylindrical form is larger than the effect of gravity and, hence, changes the shape of the cylinder into droplets in order to reduce the total surface energy. As for the case with the laser polarization perpendicular to the

DLIP pattern (see Figure 5), the amount of nano-droplet increases with increasing laser fluence levels, see Figure 6d–g. LSFL-II appear inhomogeneously and randomly on top of DLIP ridges at fluence levels of 3.41 mJ/cm^2 and 3.58 mJ/cm^2. Similar results were found for the pillar-like DLIP structure with a period of 1.5 µm (DLIP-type 2), as shown in Figure 7. Here, nano-droplets start first to appear between the ridges and droplet appearance and growth dominates. LSFL-II were found sporadically on top of the droplet roughened surface at laser fluence levels of 2.57 to 3.07 mJ/cm^2, see Figure 7a–g.

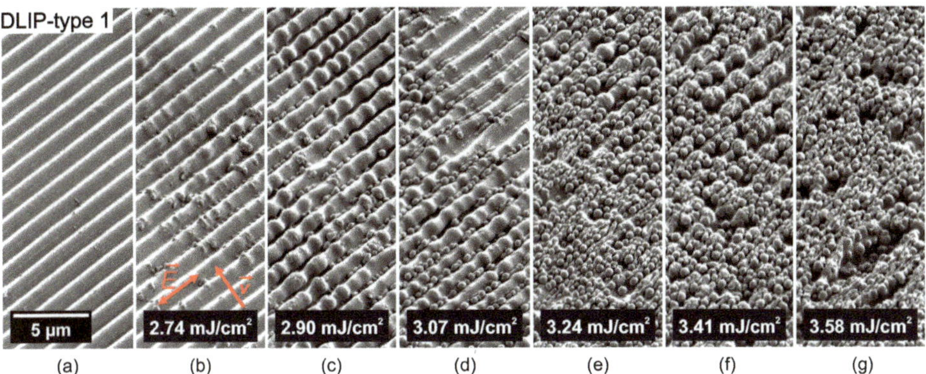

Figure 6. SEM micrographs of (**a**) ridge-like DLIP structure with a period of 1.5 µm (DLIP-type 1). (**b–g**) Evolution of surface morphology on top of the DLIP-type 1 structure with increasing peak fluence levels and a laser polarization parallel to the DLIP ridges processed with the Trumpf TruMicro 5050 laser system.

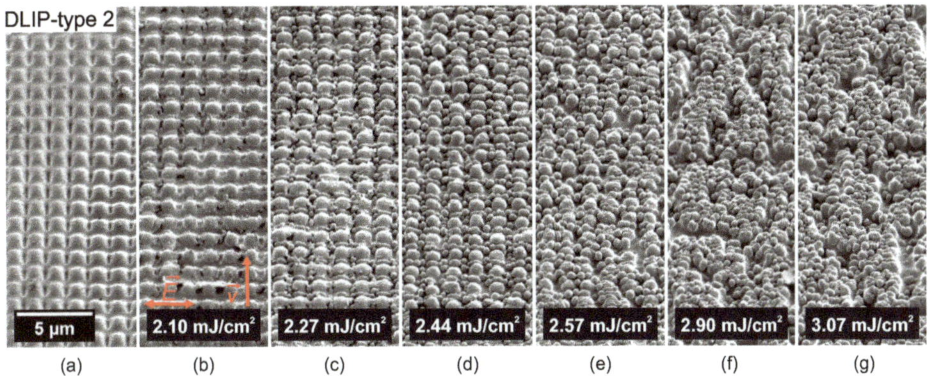

Figure 7. SEM micrographs of (**a**) pillar-like DLIP structure with a period of 1.5 µm (DLIP-type 2). (**b–g**) Evolution of surface morphology on top of DLIP-type 2 structure with increasing peak fluence levels processed with the Trumpf TruMicro 5050 laser system.

The formation of LIPSS on top of polymeric photoresist film microstructures were recently reported by Ehrhardt et al. [42]. The authors studied the formation of LSFL-II on top of pillar-like dot array microstructures with pillar widths of 2 × 2 µm^2 and 5 × 5 µm^2 and a pillar height of about 2.2 µm, as well as on top of ridge-like microstructures with ridge widths of 1 µm and 3 µm and a ridge height of about 1 µm using 100 to 1500 pulses of a nanosecond laser source with a wavelength of 248 nm at a pulse repetition rate of 100 Hz. The authors reported that no laser parameter regime was found to obtain LSFL on top of 1 × 1 µm^2 dot pillar arrays. This is in agreement with our experiments for the pillar-like DLIP-type 2 structure with a period of 1.5 µm reported in this study.

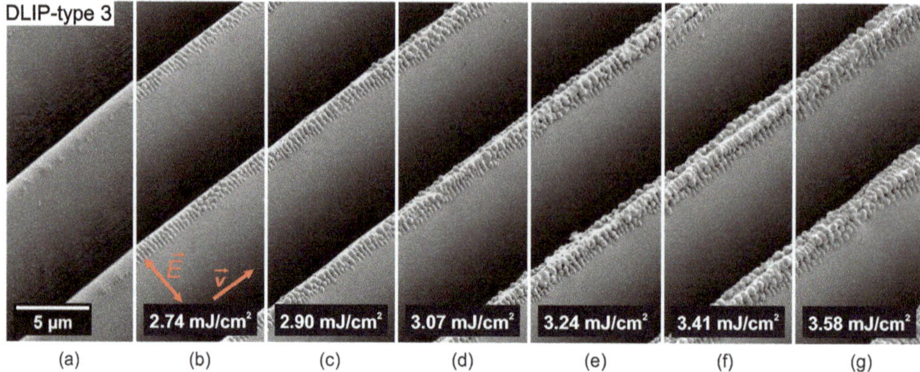

Figure 8. SEM micrographs of (**a**) ridge-like DLIP structure with a period of 10 μm (DLIP-type 3). (**b–g**) Evolution of surface morphology on top of the DLIP structure with increasing peak fluence levels and a laser polarization perpendicular to the DLIP ridges processed with the Trumpf TruMicro 5050 laser system.

Figure 8 shows the evolution of nanostructures on top of ridge-like DLIP structures with a period of 10 μm (DLIP-type 3) when irradiated with UV picosecond pulses linearly polarized perpendicular to the ridges of the DLIP structure. At a laser fluence level of 2.74 mJ/cm^2, LSFL-II and nano-droplets appear on top of the DLIP ridges, see Figure 8b. With increasing laser fluence, the nano-droplet growth on top of the ridges is reinforced, see Figure 8b–g. When irradiating the ridge-like DLIP structures having a period of 10 μm with the laser polarization parallel to the DLIP ridges, LSFL-II only appear on top of the DLIP ridges at laser fluence levels exceeding the melting threshold of the material. As a consequence, the "sharp" DLIP ridges collapse, forming small valleys as shown in Figure 9.

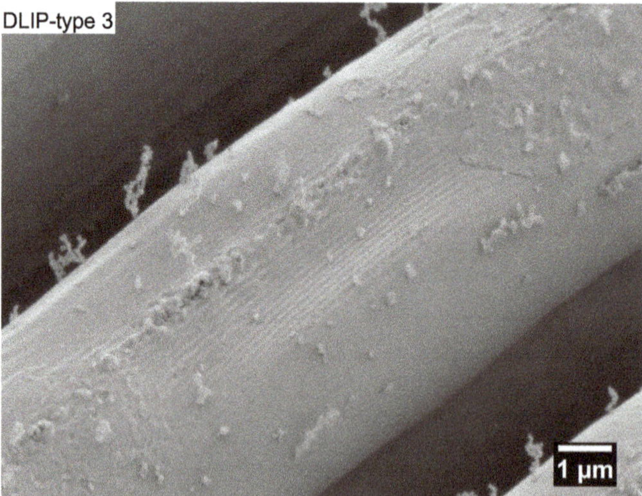

Figure 9. SEM micrograph of LSFL-II on top of the ridge of ridge-like DLIP structure with a period of 10 μm (DLIP-type 3) processed with the Time-Bandwidth Fuego laser system. This sample was contaminated.

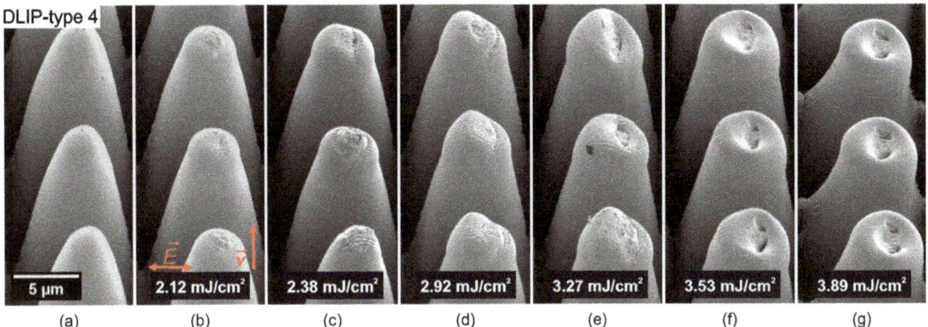

Figure 10. SEM micrographs of (a) pillar-like DLIP structure with a period of 10 µm (DLIP-type 4). (**b–g**) Evolution of surface morphology on top of the DLIP structure with increasing peak fluence levels processed with the Time-Bandwidth Fuego laser system.

The formation of LSFL-II and deformation due to melting of pillar-like DLIP cones due to UV picosecond laser irradiation can be observed in Figure 10. It can be seen that the DLIP cones tip flattens due to laser-induced melting effects. Moreover, LSFL-II develop on top of the flattened parts, see Figure 10a–f. The melting also leads to a collapse of the DLIP cones and leaves some holes in the central region, see Figure 10e–g. The holes may be induced due to the sub-surface release of gaseous photo-thermal reaction products. Similar results were found in the recent publication by Ehrhardt et al. [42] on pre-patterned polymer films.

The appearance of LSFL-II on ridge-like photoresist film microstructures "moved" from the top to the side walls of the ridge-like structures with increasing laser fluence levels, as reported by Ehrhardt et al. [42]. Additionally, these authors reported that melting is responsible for the disappearance of the LSFL-II on top of the ridge-like microstructures [42]. The appearance of the LSFL-II only on top of the DLIP ridges and pillars in this study and ridge-like microstructures reported in [42] at certain fluence levels can both be explained with the decrease of the local laser fluence at the tilted slopes of the DLIP ridge topography. That is, due to the geometrical enlargement of the laser spot on the irradiated surface area at slopes for non-normal incident radiation, the laser fluence level decreases below the LSFL-II threshold.

In order to evaluate the periodicity and amplitude of the LSFL-II on top of the different DLIP structures, the topography of DLIP structures at which the ridges are covered homogeneously with LSFL-II are analyzed using AFM. Figure 11a,c,d show the AFM micrographs of Figures 5d, 8b and 10d, respectively. Note, that the DLIP-type 3 and 4 are too deep for the AFM tip to reach the bottom of the DLIP ridges. For these two cases, the AFM micrograph is cut at the depth at which the AFM measurement lost its signal. The average periods of the LSFL-II are $\Lambda_{\text{LIPSS}} = 254 \pm 9$ nm and the average amplitudes are 33±12 nm for all analyzed hierarchical structures. It is known that LSFL-II are seeded and formed in a sub-surface layer [26,27]. Figure 11b shows cross-sections obtained from AFM measurements of ridge-like DLIP structures with a period of 1.5 µm (DLIP-type 3)—with and without LSFL-II, as it can be seen in Figure 11a. The Figure also shows, that the overall depth of the of the DLIP structure is reduced when LSFL-II are generated. This is an indication, that the LSFL-II indeed are formed below the surface.

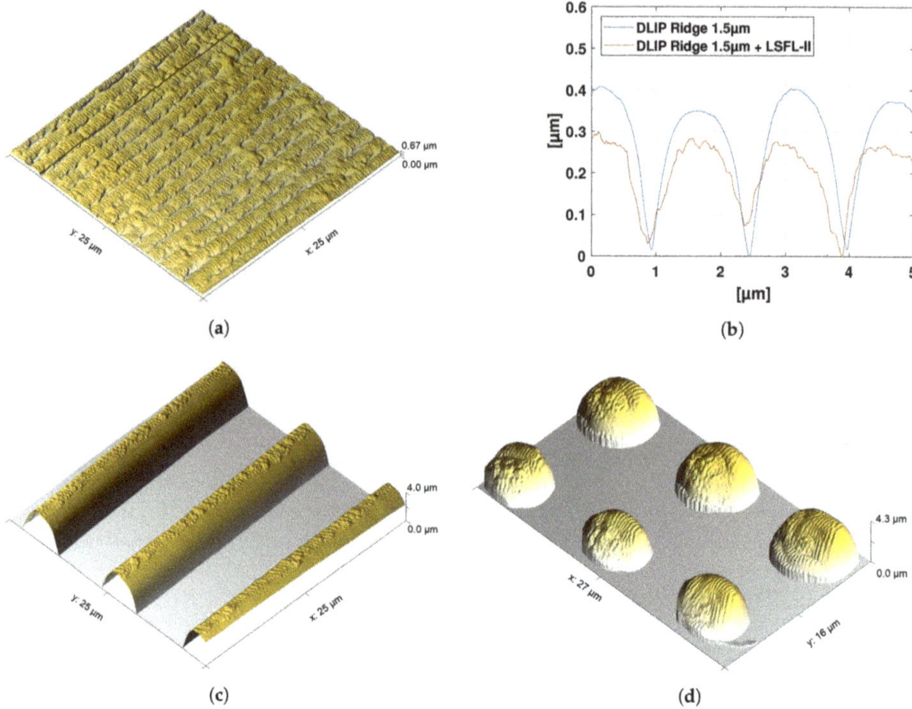

(a)

(b)

(c)

(d)

Figure 11. (**a**) AFM micrograph of LSFL-II processed on DLIP-type 1 with polarization perpendicular to the DLIP ridges at a laser peak fluence of $F_0 = 3.24\,\text{mJ}/\text{cm}^2$, (see Figure 5d). (**b**) AFM cross-sections of ridge-like DLIP structures with a period of 1.5 µm (DLIP-type 3)—with (see Figure 11a) and without LSFL-II. (**c**) AFM micrograph of LSFL-II processed on DLIP-type 3 with polarization perpendicular to the DLIP ridges at a laser peak fluence of $F_0 = 2.74\,\text{mJ}/\text{cm}^2$, (see Figure 8b). (**d**) AFM micrograph of LSFL-II processed on DLIP-type 4 at a laser peak fluence of $F_0 = 2.92\,\text{mJ}/\text{cm}^2$, (see Figure 10d).

3.4. FDTD Simulations

Figure 12 shows the time-averaged optical intensity distribution (as calculated by the photonic simulation software, see Section 2.4) induced by one UV picosecond pulse on and in the surface of DLIP-type 1 structure with the parameters described in Section 2.4. It can be concluded from this Figure, that the maximum intensities differ for each case of polarization. That is, if the orientation of the laser polarization is perpendicular to the DLIP ridges, the maximum intensity is found on top of the ridge and another less localized intensity enhancement is observed several hundreds of nanometers below it. The intensity maximum close to the surface can facilitate the seeding of LIPSS on top of the DLIP ridges, see Figure 5b. If the orientation of the laser polarization is parallel to the DLIP ridges, the maximum intensity is found in the bottom of the DLIP ridges, which can explain the dominance of nano-droplet growth at these positions, see Figure 6b.

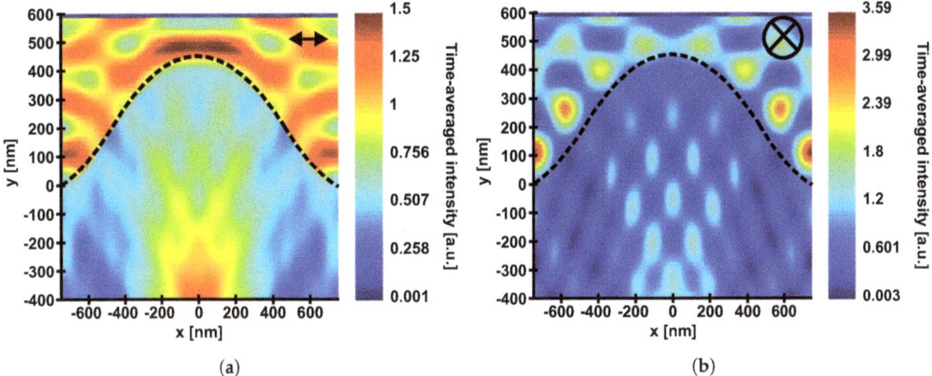

Figure 12. Calculated time-averaged optical intensity distribution of a 6.7 ps laser pulse with a wavelength of 343 nm with a polarization (**a**) perpendicular and (**b**) parallel to the orientation of DLIP-type 1 ridges with a period of 1.5 µm. The arrows in the upper right corners indicate the direction of the laser polarization. Note the different color scales used in (**a**,**b**) for encoding the intensity.

3.5. Chemical Characterization

Figure 13 shows the µ-Raman spectra of non-irradiated and LIPSS-processed PC (LSFL-II as described in Section 3.2). The reference measurement of the non-irradiated material (black curve) shows pronounced characteristic peaks that are typical for this type of bisphenol-A based PC material [59]. The measurement in the LIPSS-covered area (green and red curves) exhibit a very strong and broad background signal that is caused by optical fluorescence. This fluorescence is excited by the Raman laser in the laser-modified PC over the entire depth (Rayleigh-length, $\gtrsim 3$ µm for the given microscope objective) of the probing Raman spot. Moreover, it was noticed that the µ-RS spectra recorded in the LIPSS-covered areas show a characteristic photo-bleaching effect, i.e., the fluorescence level of the spectra drops about 75 percent upon exposure to the Raman laser radiation and then saturates after several tens to hundreds of seconds, see Figure 13. The effect arises from broken bonds in the polymer material that create energetic states within the electronic band gap. These states are capable of being excited by the Raman laser radiation at 532 nm wavelength, causing the strong fluorescence background. Upon continuous Raman laser irradiation, these broken bonds may react with the environment (e.g., via oxidation), hence, reducing the fluorescence again. The red curve of the laser-processed PC was recorded after photo-bleaching the sample. However, at the given signal-to-noise level it is difficult to quantify changes induced upon the UV ps-laser irradiation here. Hence, other DLIP structures were not tested by µ-RS here and a more surface sensitive method (ATR-FTIR) was selected for further material characterizations.

Fourier-transform infrared spectroscopy in attenuated total reflection mode (ATR-FTIR) is capable of selectively examining the near-surface layer of organic films [46]. With this technique, the chemical changes before and after laser irradiation and the resulting degradation of the polycarbonate are analyzed. Figure 14 exemplifies ATR-FTIR spectra of non-irradiated polycarbonate samples (black curves) compared to samples processed homogeneously with LSFL-II (see Figure 14a–c, top row, red curves), DLIP structures with a period of 1.5 µm (see Figure 14d–f, middle row, red and blue curves) and DLIP structures with a period of 10 µm (see Figure 14g–i, bottom row, red and blue curves). For each type of surface structure, three different spectral regions of interest are selected, i.e., left column: 4000 to 2700 cm^{-1}, middle column: 2000 to 1300 cm^{-1}, and right column: 1300 to 600 cm^{-1}, all being representative for the absorption range of specific vibrational excitation modes in the polymeric material. All measured ATR-FTIR spectra were normalized at the peak located at 1014 cm^{-1}, as proposed in Ref. [46]. It can be concluded from the graphs that the different laser processing techniques and irradiation parameters lead to a degradation of characteristic absorption bands and the formation of various absorption bands at specific wavenumbers. The differences of the reference spectra of the unprocessed polycarbonate among the different measurements may arise from inhomogeneities and additives within the polycarbonate samples.

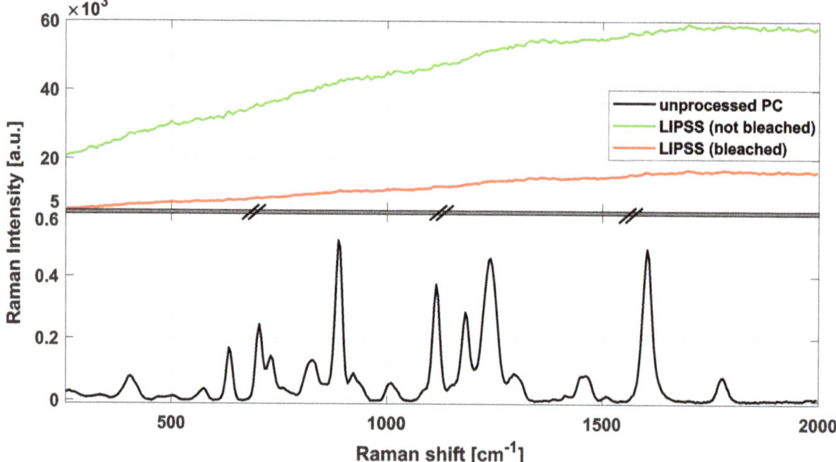

Figure 13. Raman spectra of non-irradiated (unprocessed) PC (black curve) and of UV ps-laser irradiated PC with LIPSS (LSFL-II) as described in Section 3.2 before (green curve) and after (red curve) photo-bleaching (for details see the text). Note that the ordinate is separated and the scaling differs for both separations.

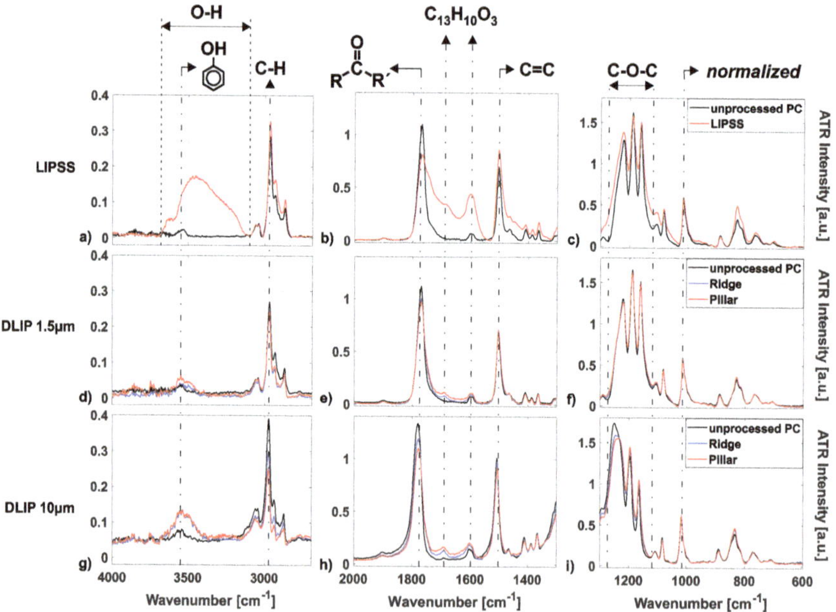

Figure 14. ATR-FTIR spectra of unprocessed samples (black curves) compared to samples processed homogeneously with LIPSS (type LSFL-II, red curves in (**a–c**), ridge-like (blue curves) or pillar-like (red curves) DLIP structures with 1.5 μm period (**d–f**) and ridge-like (blue curves) or pillar-like (red curves) DLIP structures with 10 μm period (**g–i**). Note the different vertical scales in the graphs.

For all processed samples, the laser processing led to a degradation of the C-H vibrational region at a wavenumber of about 3000 cm^{-1} (see Figure 14a,d,f), the carbonyl and C=C vibration peaks at wavenumbers about 1790 and 1500 cm^{-1}, respectively (see Figure 14b,e,h) and the C-O-C vibration features at about 1120 to 1280 cm^{-1} (see Figure 14c,f,i). The degradation of the laser irradiated polycarbonate is also accompanied by the appearance/growth of other bands. While the processing of LIPSS leads to a broad absorption band at about 3400 cm^{-1} (see Figure 14a), which is related to the OH-stretch region [45], the processing of DLIP structures leads to absorption bands at 3550 and 3500 cm^{-1} (see Figure 14d,g), which are attributed to free- and hydrogen-bonded phenolic groups [44]. Additionally, two bands arise at about 1630 and 1690 cm^{-1} (see Figure 14b,e,h), which are attributed to the creation of phenylsalicate and dihydroxybenzophenone (both $C_{13}H_{10}O_3$), respectively [46,47]. The degradation on the polycarbonate upon irradiation with ultrashort pulsed UV radiation is in full accordance with the literature here [44–47].

It can be observed in Figure 14, when comparing the relative changes of the absorption spectra from the untreated PC with the spectra of the processed samples with various micro- and nanostructures, that the relative changes due to laser irradiation are most affected by the production of homogeneous areas of LIPSS (see Figure 14a–c) and least affected by the creation of DLIP structures with a period of 1.5 µm (see Figure 14d–f). Additionally, it can be observed in Figures 14d–i, that the formation of pillar-like DLIP structures affects the degradation of the PC more than the fabrication of ridge-like DLIP structures . The different levels of degradation of the PC due to the processing of different structures can be related to the number of laser pulses irradiating one spot diameter N_{eff} and the corresponding accumulated fluence (F_{acc}) levels, see Table 2. The total amount of pulses impinging the same spot equals $N_{eff} = N_{OS} \times d^2 \times f/(v \times h)$ with h being the hatch distance (line pitch). The accumulated fluence is given by $F_{acc} = N_{eff} \times F_0$. It can be observed in and Table 2, that although the peak fluence to manufacture LIPSS-II is much less than the peak fluence to create the ridge-like and pillar-like DLIP structures, the accumulated fluence is much greater when comparing those parameters. The latter explains the increasing degradation of the polycarbonate when comparing different ATR-FTIR spectra of the different samples in Figure 14.

Table 2. Effective number of pulses and accumulated laser fluence levels due to various laser micro- and nanostructuring techniques.

Structure Type	N_{eff} [-]	F_0 [J/cm^2]	F_{acc} [J/cm^2]
LSFL-II	302760	4.4×10^{-3}	1338
DLIP 1.5 µm Ridge	68	0.25	17
DLIP 1.5 µm Pillar	68 + 68	0.25 + 0.2	27
DLIP 10 µm Ridge	50	1.45	71
DLIP 10 µm Pillar	50 + 50	1.45 + 1.13	111

It must to be noted, that the light penetration depth according to the Lambert-Beer law between the nanosecond and picosecond laser sources differ due to the different wavelength used. The penetration depth will also have an impact on the depth of the changes of the chemical structure of the polymer. Whereas the penetration depth for a wavelength of 266 nm on PC is about 0.6 µm [60], the penetration depths for wavelengths of 343 nm and 355 nm are about 104 and 107 µm, respectively [54]. However, the depth of the chemical changes fall out of the scope of this paper.

4. Conclusions

The evolution of nanometric LIPSS on different types of micrometric DLIP structures with increasing fluence levels was analyzed to achieve hierarchical micro-/nano-structures on a commercial polycarbonate (Makrolon ™). It was found that LIPSS can be formed on top of various forms and sizes of DLIP structures by selecting the laser beam polarization perpendicular to the DLIP ridges. However, the fabrication of LIPSS on a micro-scale structure is limited by the height and width of the pre-processed microscale ridges. FDTD calculations of the time-averaged optical intensity distribution of a picosecond laser pulse with a wavelength of 343 nm at micro-structures with a period of 1.5 µm and a height of 400 nm were conducted with varying laser beam polarization directions. If the latter was set perpendicular to the micro-structures, the time-averaged optical intensity was found to be enhanced on top of the micro-structures, promoting the seeding of LSFL on top of the ridges. However, when the laser beam polarization direction was set parallel to the DLIP ridges, the optical intensity was found to be locally increased at the bottom of the DLIP ridges, enhancing nano-droplet growth at these positions. Moreover, since LSFL appearance is limited to a narrow window of laser fluence levels, the growth of LSFL only on top of the DLIP ridges was was limited by the non-normal angle of incidence of the laser radiation at the side walls of the DLIP structures. The latter decreases the local fluence level below the LSFL threshold. As an important aspect for potential future applications, it was found that a sufficiently broad top of the DLIP-pillars is required to allow LIPSS to be formed there. Additionally, with increasingly accumulated fluence levels, the degradation of the polymer also progresses. This needs to be considered if the original property of the unprocessed polymer is essential for applications.

Author Contributions: Conceptualization: M.M., S.A.; Methodology: M.M., S.A.; Software, M.M., W.A.P.M.H.; Validation: M.M., S.A.; formal analysis: M.M., J.B.; investigation: M.M., S.A., W.A.P.M.H., A.H., A.M.E.; resources: M.M., G.-w.R.B.E.R., J.B., T.K., A.F.L.; data curation: M.M.; writing–original draft preparation: M.M.; writing–review and editing: M.M., S.A., W.A.P.M.H., A.H., J.B., T.K., A.F.L., G.-w.R.B.E.R.; visualization: M.M., S.A., W.A.P.M.H.; supervision: G.-w.R.B.E.R., A.F.L.; project administration: M.M.; funding acquisition: G.-w.R.B.E.R., J.B., A.F.L. All authors have read and agreed to the published version of the manuscript.

Funding: This study was funded partly by the European Union's Horizon 2020 research and innovation programme under the Marie Skłodowska-Curie grant agreement No. 675063 (Laser4Fun project, www. laser4fun.eu) and under the FET Open grant agreement No. 862016 (BioCombs4Nanofibers, www.jku.at/biocombs4nanofibers).

Acknowledgments: We thank Volker Franke and his whole team of Micro Materials Processing at the Fraunhofer Institute for Material and Beam Technologies for their provision of fundamental resources and time.

Conflicts of Interest: The authors declare no conflict of interest.

References

1. Barthlott, W.; Neinhuis, C. Purity of the sacred lotus, or escape from contamination in biological surfaces. *Planta* **1997**, *202*, 1–8. [CrossRef]

2. Hasan, J.; Webb, H.K.; Truong, V.K.; Pogodin, S.; Baulin, V.A.; Watson, G.S.; Watson, J.A.; Crawford, R.J.; Ivanova, E.P. Selective bactericidal activity of nanopatterned superhydrophobic cicada Psaltoda claripennis wing surfaces. *Appl. Microbiol. Biotechnol.* **2013**, *97*, 9257–9262. [CrossRef] [PubMed]

3. Pogodin, S.; Hasan, J.; Baulin, V.A.; Webb, H.K.; Truong, V.K.; Phong Nguyen, T.H.; Boshkovikj, V.; Fluke, C.J.; Watson, G.S.; Watson, J.A.; et al. Biophysical model of bacterial cell interactions with nanopatterned cicada wing surfaces. *Biophys. J.* **2013**, *104*, 835–840. [CrossRef] [PubMed]

4. Bandara, C.D.; Singh, S.; Afara, I.O.; Wolff, A.; Tesfamichael, T.; Ostrikov, K.; Oloyede, A. Bactericidal effects of natural nanotopography of Dragonfly Wing on Escherichia coli. *ACS Appl. Mater. Interfaces* **2017**, *9*, 6746–6760. [CrossRef]

5. Jaggessar, A.; Shahali, H.; Mathew, A.; Yarlagadda, P.K. Bio-mimicking nano and micro-structured surface fabrication for antibacterial properties in medical implants. *J. Nanobiotechnology* **2017**, *15*, 64. [CrossRef]

6. Parker, A.R.; Lawrence, C.R. Water capture by a desert beetle. *Nature* **2001**, *414*, 33–34. [CrossRef]

7. Hermens, U.; Kirner, S.V.; Emonts, C.; Comanns, P.; Skoulas, E.; Mimidis, A.; Mescheder, H.; Winands, K.; Krüger, J.; Stratakis, E.; et al. Mimicking lizard-like surface structures upon ultrashort laser pulse irradiation of inorganic materials. *Appl. Surf. Sci.* **2017**, *418*, 499–507. [CrossRef]

8. Lutey, A.H.; Gemini, L.; Romoli, L.; Lazzini, G.; Fuso, F.; Faucon, M.; Kling, R. Towards laser-textured antibacterial surfaces. *Sci. Rep.* **2018**, *8*, 10112. [CrossRef]

9. van der Poel, S.; Mezera, M.; Römer, G.R.B.E.; de Vries, E.; Matthews, D. Fabricating Laser-Induced Periodic Surface Structures on Medical Grade Cobalt–Chrome–Molybdenum: Tribological, Wetting and Leaching Properties. *Lubricants* **2019**, *7*, 70. [CrossRef]

10. Qin, L.; Wu, H.; Guo, J.; Feng, X.; Dong, G.; Shao, J.; Zeng, Q.; Zhang, Y.; Qin, Y. Fabricating hierarchical micro and nano structures on implantable Co–Cr–Mo alloy for tissue engineering by one-step laser ablation. *Colloids Surfaces B Biointerfaces* **2018**, *161*, 628–635. [CrossRef]

11. Heitz, J.; Plamadeala, C.; Muck, M.; Armbruster, O.; Baumgartner, W.; Weth, A.; Steinwender, C.; Blessberger, H.; Kellermair, J.; Kirner, S.V.; et al. Femtosecond laser-induced microstructures on Ti substrates for reduced cell adhesion. *Appl. Phys. A Mater. Sci. Process.* **2017**, *123*, 734. [CrossRef]

12. Paradisanos, I.; Fotakis, C.; Anastasiadis, S.H.; Stratakis, E. Gradient induced liquid motion on laser structured black Si surfaces. *Appl. Phys. Lett.* **2015**, *107*, 111603. [CrossRef]

13. Stark, T.; Kiedrowski, T.; Marschall, H.; Lasagni, A.F. Avoiding starvation in tribocontact through active lubricant transport in laser textured surfaces. *Lubricants* **2019**, *7*, 54. [CrossRef]

14. Eichstädt, J.; Römer, G.R.B.E.; Huis in't Veld, A.J. Towards friction control using laser-induced periodic Surface Structures. *Phys. Procedia* **2011**, *12*, 7–15.

15. Bonse, J.; Kirner, S.V.; Griepentrog, M.; Spaltmann, D.; Krüger, J. Femtosecond laser texturing of surfaces for tribological applications. *Materials* **2018**, *11*, 801. [CrossRef]

16. Alamri, S.; Aguilar-Morales, A.I.; Lasagni, A.F. Controlling the wettability of polycarbonate substrates by producing hierarchical structures using Direct Laser Interference Patterning. *Eur. Polym. J.* **2018**, *99*, 27–37. [CrossRef]

17. Aguilar-Morales, A.I.; Alamri, S.; Lasagni, A.F. Micro-fabrication of high aspect ratio periodic structures on stainless steel by picosecond direct laser interference patterning. *J. Mater. Process. Technol.* **2018**, *252*, 313–321. [CrossRef]

18. Vorobyev, A.Y.; Guo, C. Colorizing metals with femtosecond laser pulses. *Appl. Phys. Lett.* **2008**, *92*, 041914. [CrossRef]

19. Dusser, B.; Sagan, Z.; Soder, H.; Faure, N.; Colombier, J.; Jourlin, M.; Audouard, E. Controlled nanostructures formation by ultra fast laser pulses for color marking. *Opt. Express* **2010**, *18*, 2913–2924. [CrossRef]

20. Ahsan, M.S.; Ahmed, F.; Kim, Y.G.; Lee, M.S.; Jun, M.B.G. Colorizing stainless steel surface by femtosecond laser induced micro/nano-structures. *Appl. Surf. Sci.* **2011**, *257*, 7771–7777. [CrossRef]

21. Ionin, A.A.; Kudryashov, S.I.; Makarov, S.V.; Seleznev, L.V.; Sinitsyn, D.V.; Golosov, E.V.; Golosova, O.A.; Kolobov, Y.R.; Ligachev, A.E. Femtosecond laser color marking of metal and semiconductor surfaces. *Appl. Phys. A Mater. Sci. Process.* **2012**, *107*, 301–305. [CrossRef]

22. Alamri, S.; Fraggelakis, F.; Kunze, T.; Krupop, B.; Mincuzzi, G.; Kling, R.; Lasagni, A.F. On the interplay of DLIP and LIPSS upon ultra-short laser pulse irradiation. *Materials* **2019**, *12*, 1018. [CrossRef] [PubMed]

23. Fabris, D.; Lasagni, A.F.; Fredel, M.C.; Henriques, B. Direct Laser Interference Patterning of Bioceramics: A Short Review. *Ceramics* **2019**, *2*, 578–586. [CrossRef]

24. Lang, V.; Roch, T.; Lasagni, A.F. High-Speed Surface Structuring of Polycarbonate Using Direct Laser Interference Patterning: Toward $1 m^2 min^{-1}$ Fabrication Speed Barrier. *Adv. Eng. Mater.* **2016**, *18*, 1342–1348. [CrossRef]

25. Bonse, J.; Höhm, S.; Kirner, S.V.; Rosenfeld, A.; Krüger, J. Laser-induced periodic surface structures—A scientific evergreen. *IEEE J. Sel. Top. Quantum Electron.* **2017**, *23*, 9000615. [CrossRef]

26. Rudenko, A.; Colombier, J.P.; Höhm, S.; Rosenfeld, A.; Krüger, J.; Bonse, J.; Itina, T.E. Spontaneous periodic ordering on the surface and in the bulk of dielectrics irradiated by ultrafast laser: A shared electromagnetic origin. *Sci. Rep.* **2017**, *7*, 12306. [CrossRef]

27. Mezera, M.; Bonse, J.; Römer, G.R.B.E. Influence of bulk temperature on laser-induced periodic surface structures on polycarbonate. *Polymers* **2019**, *11*, 1947. [CrossRef]

28. Mezera, M.; Römer, G.R.B.E. Model based optimization of process parameters to produce large homogeneous areas of laser-induced periodic surface structures. *Opt. Express* **2019**, *27*, 6012–6029. [CrossRef]

29. Kirner, S.V.; Hermens, U.; Mimidis, A.; Skoulas, E.; Florian, C.; Hischen, F.; Plamadeala, C.; Baumgartner, W.; Winands, K.; Mescheder, H.; et al. Mimicking bug-like surface structures and their fluid transport produced by ultrashort laser pulse irradiation of steel. *Appl. Phys. A Mater. Sci. Process.* **2017**, *123*, 754. [CrossRef]

30. Dufft, D.; Rosenfeld, A.; Das, S.K.; Grunwald, R.; Bonse, J. Femtosecond laser-induced periodic surface structures revisited: A comparative study on ZnO. *J. Appl. Phys.* **2009**, *105*, 034908. [CrossRef]

31. Bonse, J.; Sturm, H.; Schmidt, D.; Kautek, W. Chemical, morphological and accumulation phenomena in ultrashort-pulse laser ablation of TiN in air. *Appl. Phys. A Mater. Sci. Process.* **2000**, *71*, 657–665. [CrossRef]

32. Yasumaru, N.; Miyazaki, K.; Kiuchi, J. Femtosecond-laser-induced nanostructure formed on hard thin films of TiN and DLC. *Appl. Phys. A Mater. Sci. Process.* **2003**, *76*, 983–985. [CrossRef]

33. Florian, C.; Déziel, J.L.; Kirner, S.V.; Siegel, J.; Bonse, J. The role of the laser-induced oxide layer in the formation of laser-induced periodic surface structures. *Nanomaterials* **2020**, *10*, 147. [CrossRef] [PubMed]

34. Baudach, S.; Bonse, J.; Kautek, W. Ablation experiments on polyimide with femtosecond laser pulses. *Appl. Phys. A: Mater. Sci. Process.* **1999**, *69*, S395–S398. [CrossRef]

35. Rebollar, E.; Pérez, S.; Hernández, J.J.; Martín-Fabiani, I.; Rueda, D.R.; Ezquerra, T.A.; Castillejo, M. Assessment and formation mechanism of laser-induced periodic surface structures on polymer spin-coated films in real and reciprocal space. *Langmuir* **2011**, *27*, 5596–5606. [CrossRef]

36. Castillejo, M.; Ezquerra, T.A.; Martín, M.; Oujja, M.; Pérez, S.; Rebollar, E. Laser nanostructuring of polymers: Ripples and applications. In Proceedings of the AIP Conference Proceedings; American Institute of Physics (AIP), Santa Fe, NM, USA, 30 April–3 May 2012; Volume 1464, pp. 372–380.

37. Mezera, M.; van Drongelen, M.; Römer, G.R.B.E. Laser-Induced Periodic Surface Structures (LIPSS) on polymers processed with picosecond laser pulses. *J. Laser Micro Nanoeng.* **2018**, *13*, 105–116.

38. Ionin, A.A.; Kudryashov, S.I.; Makarov, S.V.; Rudenko, A.A.; Seleznev, L.V.; Sinitsyn, D.V.; Golosov, E.V.; Kolobov, Y.R.; Ligachev, A.E. Beam spatial profile effect on femtosecond laser surface structuring of titanium in scanning regime. *Appl. Surf. Sci.* **2013**, *284*, 634–637. [CrossRef]

39. Kunz, C.; Büttner, T.N.; Naumann, B.; Boehm, A.V.; Gnecco, E.; Bonse, J.; Neumann, C.; Turchanin, A.; Müller, F.A.; Gräf, S. Large-area fabrication of low- and high-spatial-frequency laser-induced periodic surface structures on carbon fibers. *Carbon* **2018**, *133*, 176–185. [CrossRef]

40. Vorobyev, A.Y.; Guo, C. Direct femtosecond laser surface nano/microstructuring and its applications. *Laser Photonics Rev.* **2013**, *7*, 385–407. [CrossRef]

41. Klein-Wiele, J.H.; Blumenstein, A.; Simon, P.; Ihlemann, J. Laser interference ablation by ultrashort UV laser pulses via diffractive beam management. *Adv. Opt. Technol.* **2020**, *9*, 41–52. [CrossRef]

42. Ehrhardt, M.; Lai, S.; Lorenz, P.; Zimmer, K. Guiding of LIPSS formation by excimer laser irradiation of pre-patterned polymer films for tailored hierarchical structures. *Appl. Surf. Sci.* **2020**, *506*, 144785. [CrossRef]

43. Abts, G.; Eckel, T.; Wehrmann, R. *Polycarbonates*, 7th ed.; Wiley-VCH Verlag GmbH: Weinheim, Germany, 2014.

44. Rivaton, A.; Sallet, D.; Lemaire, J. The photo-chemistry of bisphenol-A polycarbonate reconsidered: Part 2—FTIR Analysis of the Solid-state Photo-chemistry in 'Dry' Conditions. *Polym. Degrad. Stab.* **1986**, *14*, 1–22. [CrossRef]

45. Adams, M.R.; Garton, A. Surface modification of bisphenol-A-polycarbonate by far-UV radiation. Part I: In vacuum. *Polym. Degrad. Stab.* **1993**, *41*, 265–273. [CrossRef]

46. Diepens, M.; Gijsman, P. Photodegradation of bisphenol A polycarbonate. *Polym. Degrad. Stab.* **2007**, *92*, 397–406. [CrossRef]

47. Yazdan Mehr, M.; Van Driel, W.D.; Jansen, K.M.; Deeben, P.; Boutelje, M.; Zhang, G.Q. Photodegradation of bisphenol A polycarbonate under blue light radiation and its effect on optical properties. *Opt. Mater.* **2013**, *35*, 504–508. [CrossRef]

48. Liu, J.M. Simple technique for measurements of pulsed Gaussian-beam spot sizes. *Opt. Lett.* **1982**, *7*, 196–198. [CrossRef]

49. Dyer, P.E.; Jenkins, S.D.; Sidhu, J. Development and origin of conical structures on XeCl laser ablated polyimide. *Appl. Phys. Lett.* **1986**, *49*, 453–455. [CrossRef]

50. Lippert, T.; Dickinson, J.T. Chemical and spectroscopic aspects of polymer ablation: Special features and novel directions. *Chem. Rev.* **2003**, *103*, 453–485. [CrossRef]

51. Murthy, N.S.; Prabhu, R.D.; Martin, J.J.; Zhou, L.; Headrick, R.L. Self-assembled and etched cones on laser ablated polymer surfaces. *J. Appl. Phys.* **2006**, *100*, 023538. [CrossRef]

52. Mezera, M.; Römer, G.R.B.E. Upscaling laser-induced periodic surface structures (LIPSS) manufacturing by defocused laser processing. In *SPIE Conference Proceedings*; SPIE: San Francisco, CA, USA, 2019; Volume 10906, p. 109060U.

53. The MathWorks, I. *MATLAB® R2019b*; The MathWorks, Inc.: Natick, MA, USA, 2019.

54. Brissinger, D. Complex refractive index of polycarbonate over the UV-Vis-IR region from 0.2 to 3 µm. *Appl. Opt.* **2019**, *58*, 1341. [CrossRef]

55. Kuchmizhak, A.A.; Vitrik, O.B.; Kulchin, Y.N. Novel hydrodynamic instability of the molten Au/Pd alloy film irradiated by tightly focused femtosecond laser pulses. *Pac. Sci. Rev.* **2014**, *16*, 183–188. [CrossRef]

56. Wang, H.P.; Guan, Y.C.; Zheng, H.Y.; Hong, M.H. Controllable fabrication of metallic micro/nano hybrid structuring surface for antireflection by picosecond laser direct writing. *Appl. Surf. Sci.* **2019**, *471*, 347–354. [CrossRef]

57. Fajstavr, D.; Michaljaničová, I.; Slepička, P.; Neděla, O.; Sajdl, P.; Kolská, Z.; Švorčík, V. Surface instability on polyethersulfone induced by dual laser treatment for husk nanostructure construction. *React. Funct. Polym.* **2018**, *125*, 20–28. [CrossRef]

58. Gedvilas, M.; Račiukaitis, G.; Kučikas, V.; Regelskis, K. Driving forces for self-organization in thin metal films during their partial ablation with a cylindrically focused laser beam. In *AIP Conference Proceedings*; American Institute of Physics (AIP): Santa Fe, NM, USA, 2012; Volume 1464, pp. 229–243.

59. Dybal, J.; Schmidt, P.; Baldrian, J.; Kratochvíl, J. Ordered structures in polycarbonate studied by infrared and Raman spectroscopy, wide-angle X-ray scattering, and differential scanning calorimetry. *Macromolecules* **1998**, *31*, 6611–6619. [CrossRef]

60. Philipp, H.R.; Legrand, D.G.; Cole, H.S.; Liu, Y.S. The Optical Properties of Bisphenol-A Polycarbonate. *Polym. Eng. Sci.* **1987**, *27*, 1148–1155. [CrossRef]

Article

Fabrication of Periodic Nanostructures on Silicon Suboxide Films with Plasmonic Near-Field Ablation Induced by Low-Fluence Femtosecond Laser Pulses

Tatsuyoshi Takaya [1], Godai Miyaji [1,*], Issei Takahashi [1], Lukas Janos Richter [2] and Jürgen Ihlemann [2,*]

[1] Department of Applied Physics, Tokyo University of Agriculture and Technology, 2-24-16 Nakacho, Koganei, Tokyo 184-8588, Japan; s190953t@st.go.tuat.ac.jp (T.T.); s202053s@st.go.tuat.ac.jp (I.T.)

[2] Laser-Laboratorium Göttingen e. V., Hans-Adolf-Krebs-Weg 1, 37077 Göttingen, Germany; lukas.richter@llg-ev.de

* Correspondence: gmiyaji@cc.tuat.ac.jp (G.M.); juergen.ihlemann@llg-ev.de (J.I.)

Received: 30 June 2020; Accepted: 29 July 2020; Published: 30 July 2020

Abstract: Silicon suboxide (SiO_x, $x \approx 1$) is a substoichiometric silicon oxide with a large refractive index and optical absorption coefficient that oxidizes to silica (SiO_2) by annealing in air at ~1000 °C. We demonstrate that nanostructures with a groove period of 200–330 nm can be formed in air on a silicon suboxide film with 800 nm, 100 fs, and 10 Hz laser pulses at a fluence an order of magnitude lower than that needed for glass materials such as fused silica and borosilicate glass. Experimental results show that high-density electrons can be produced with low-fluence femtosecond laser pulses, and plasmonic near-fields are subsequently excited to create nanostructures on the surface because silicon suboxide has a larger optical absorption coefficient than glass. Calculations using a model target reproduce the observed groove periods well and explain the mechanism of the nanostructure formation.

Keywords: femtosecond laser; laser ablation; nanostructure formation; surface plasmon polaritons; near-field; silicon suboxide; glass

---

1. Introduction

Structures smaller than the wavelength of light can induce optical anisotropy and rotatory and resonant scattering [1,2]. Recent developments in material nanoprocessing techniques enabled the regular arrangement of nanostructures on or inside solids to perform many attractive applications such as optical cloaking [3,4], photon trapping [5,6], and structured light generation [7,8]. Because glass is transparent in the visible region, chemically stable, and inexpensive, it is a promising material for use in these applications, and it is popularly used for many kinds of optical elements and optical integrated devices.

Periodic nanostructures with a sub-μm groove period d can be easily formed inside or on glass by superimposed multiple shots of tightly focused femtosecond (fs) laser pulses of a few μm in diameter and at a fluence of a few to 10 J/cm^2 by using a high-numerical-aperture (NA) lens such as a microscope objective [9–11]. Recently, using the birefringence of nanostructures formed in fused silica with fs pulses, Beresna et al. developed a spatial-distributed wave plate that can convert a Gaussian beam to structured light such as a radially polarized beam or an optical vortex [12]. This remarkable optical element is currently commercially available. Since laser nanoprocessing can be rapidly applied to a large area, it has attracted much interest as an industrial nanofabrication technique [13]. However, the focal spots of a few μm and a short focusing depth being required for nanostructuring somewhat restrict the possible applications. Current limitations include slow processing times and short working distances between the target and the focusing lens [14].

Silicon suboxide is a substoichiometric silicon oxide that has a larger refractive index and a larger optical absorption coefficient in the region from ultra-violet (UV) to near-infrared (NIR) than other glass materials such as fused silica, borosilicate, and soda-lime glass [15–18]. In addition, it can be easily oxidized to SiO_2 by thermal treatment (~1000 °C) in air to become transparent in the UV–NIR region [18]. Thermal treatment in an oxygen-free environment leads to the formation of silicon nanocrystals [19]. As a result of these properties, this material has attracted attention for applications such as anti-reflection coating [20], photo luminescence [21], giant Raman scattering [22], and SiO_2 precursors for laser processing [23].

In this paper, we describe the successful formation of a nanostructure with a groove period of $d = 200$–330 nm in air on a silicon suboxide (SiO_x, $x \approx 1$) surface irradiated with 800 nm, 100 fs laser pulses at ~700 mJ/cm^2 using a low NA lens. Preliminary results were presented in [24]. The spot size was 120 μm in diameter and the fluence was an order of magnitude lower than that needed for structuring glass materials. However, thus far, the fs-laser-induced nanostructuring of silicon suboxide has not been well understood, and its successful control has not been achieved. In this work, based on the formation process of a nanostructure observed on silicon suboxide, we discuss its physical mechanism. The groove period calculated for a model target closely matches the observed groove period. These results showed that the formation of a thin layer of high-density electrons and the excitation of surface plasmon polaritons (SPPs) are responsible for the nanostructure formation on a silicon suboxide induced with intense femtosecond laser pulses.

2. Experimental

As a target, we used a silicon suboxide, SiO_x ($x \approx 1$) film of 1.4 μm thickness deposited on a fused silica substrate by thermal evaporation (Leybold UNIVEX 350, Cologne, Germany) [18]. Figure 1a shows the refractive index n and the extinction coefficient κ measured as a function of wavelength $\lambda = 300$–900 nm with a spectroscopic ellipsometer (M-220, JASCO Corporation, Tokyo, Japan). As seen in Figure 1b, the film is colored yellow because κ is especially large in the UV to blue region. The nonlinear absorption coefficient β of the film was not measured in this work, though values of $\beta = 10^{-7}$–10^{-5} cm/W were measured using an 800 nm fs laser, which are much larger than the value for fused silica [16,17]. In addition, because the refractive index of silicon suboxide was measured as $n = 1.8$–2.4 in the UV–NIR region, which is much larger than that of glass such as fused silica as shown in Figure 1a, β of the film applied in this work should also be larger. In a previous report, a silicon suboxide film 700 nm thick was confirmed to be able to oxidize to SiO_2 by annealing in air at ~1000 °C for 48 h, resulting in an increase of the transmittance in the UV–NIR region [18].

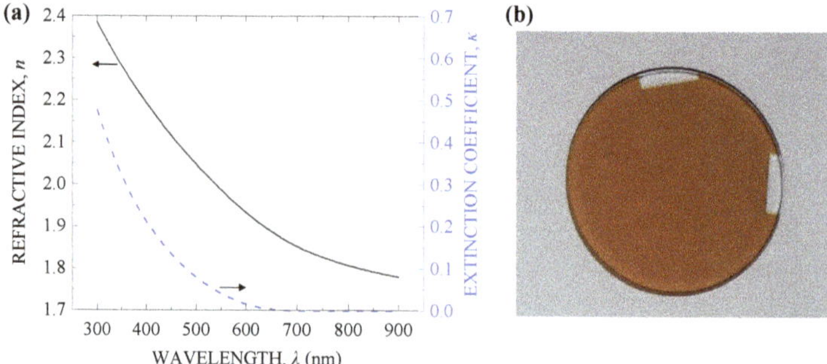

Figure 1. (a) Refractive index n (black solid line) and extinction coefficient κ (blue dashed line) of the silicon suboxide film measured as a function of the wavelength λ. (b) Photograph of the film on a fused silica substrate 25 mm in diameter.

Figure 2a shows a schematic diagram of the optical configuration used in the ablation experiment. We used 800 nm and 100 fs laser pulses delivered from a Ti:sapphire laser system at a repetition rate of 10 Hz. The incident laser pulses were controlled with a mechanical shutter to be either a superimposed number of pulses of N = 250–1000 or a single shot N = 1. The pulse energy was controlled with a pair of a half-wave plate and a polarizer to set the laser fluence to F = 625–750 mJ/cm^2. To measure the energy shot by shot, we acquired the energy of the pulse reflected at a glass surface with a photodiode. The linearly polarized fs laser pulses were focused in air at normal incidence on the silicon suboxide films with a lens with a 250 mm focal length. To monitor the surface irradiated by the pulses, the microscopic image of the target surface was observed with a charge-coupled device (CCD) camera with a He–Ne laser beam used for illumination. For measuring the intensity profile of the focal spot, we tilted the target to observe the fs pulse reflected at the surface with the camera. The result is shown in Figure 2b. The intensity distribution of the focal spot fitted the lowest-order Gaussian profile well. The focal spot size was 120 μm in diameter at $1/e^2$ of the maximum intensity.

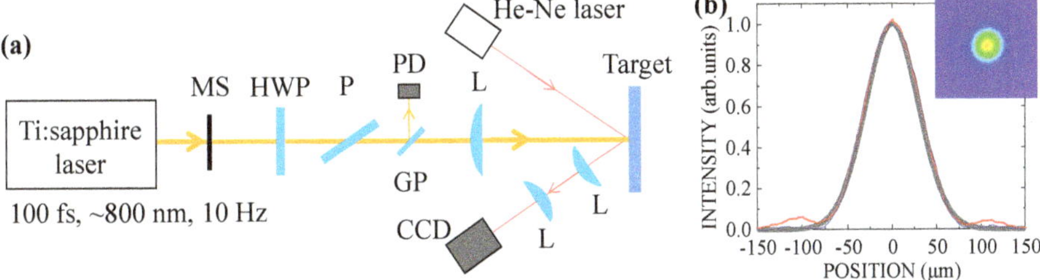

Figure 2. (**a**) Schematic diagram of the optical configuration for the ablation experiment. MS, mechanical shutter; HWP, half-wave plate; P, polarizer; GP, glass plate; PD, photodetector; L, lens. (**b**) Intensity profile of the focal spot of a fs pulse. The red and blue curves represent the horizontal and vertical profiles, respectively, with a Gaussian profile shown in gray for comparison. The inset shows a CCD image of the focal spot.

The surface morphology of the target was observed with an optical microscope (VH-Z500R, Keyence, Osaka, Japan) and a scanning electron microscope (SEM, JSM-6510, JEOL, Tokyo, Japan). A 10 nm thick platinum layer was applied homogeneously onto the irradiated surface with an ion sputter coater to improve the conductivity of the surface, thus improving the clarity of the SEM image. To evaluate the spatial periodicity of the surface structure, a two-dimensional Fourier transform (using the image processing software, SPIP, Image Metrology, Lyngby, Denmark) was performed on the SEM image in a 5 × 5 μm^2 region to acquire the spatial frequency distribution along the polarization direction.

3. Results and Discussion

The nanostructures formed on various kinds of materials irradiated with fs laser pulses at a fluence F lower than the single-shot ablation threshold F_1 [25]. To compare the F_1 of the silicon suboxide film to those of the glass materials [26,27], we measured F_1 from the optical microscope images of the single-shot ablation traces. Figure 3a shows an example of an optical microscope image of the film surface irradiated with a single fs pulse at F = 1200 mJ/cm^2. The light gray area in the center of the image is the ablated trace. Then, while varying F, we measured the area S of the corresponding ablation traces. Figure 3b shows S plotted as a function of F. By extrapolating the fitting line (solid line in Figure 3b), the single-shot ablation threshold of the silicon suboxide was estimated to be F_1 = 1060 (±10) mJ/cm^2, which is much smaller than F_1 = 2 J/cm^2 for fused silica [26], F_1 = 4.1 J/cm^2 for borosilicate glass [27], and F_1 = 3.4 J/cm^2 for soda-lime-silicate glass [27]. The results showed that the small F_1 for silicon suboxide could be attributed to the large optical absorption coefficient [15].

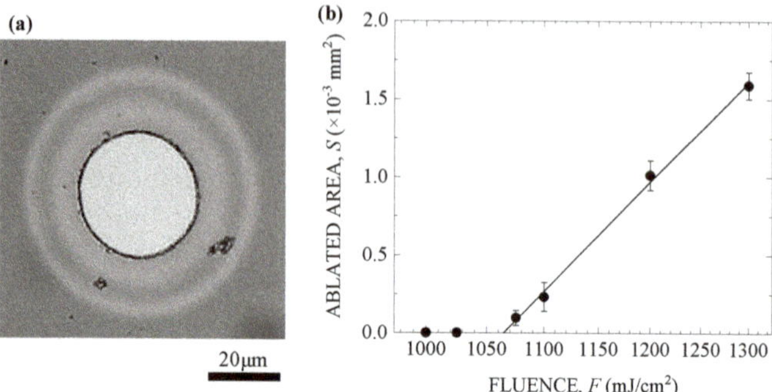

Figure 3. (**a**) Optical microscope image of the silicon suboxide film irradiated with a single fs pulse at $F = 1200$ mJ/cm^2, and (**b**) single-shot fs laser ablated area S plotted as a function of the laser fluence F.

Figure 4 shows an SEM image and the spatial frequency spectrum of the silicon suboxide film irradiated at $F = 675$ mJ/cm^2 for $N = 250$, 500, and 1000 pulses. For $N \leq 200$, no ablation trace was observed. Increasing N to 250, ablation traces were observed on the surface at several places in the focal spot. As shown in Figure 4a, dot-like nanostructures formed over the whole area of the traces, while line-like nanostructures expanding perpendicular to the polarization direction were generated at the center of the traces. With a further increase of N to 500 and 1000, as seen in Figure 4b,c, the area over which the dot-like nanostructures formed extended more widely than that for the line-like nanostructures. In the spatial-frequency spectrum, therefore, the spectral peak indicating the periodic structures in the ablation trace could not be identified.

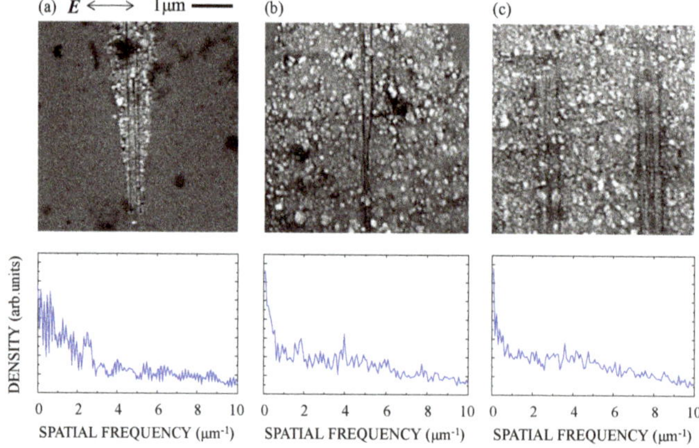

Figure 4. SEM image (top) and Fourier spectrum (bottom) of the silicon suboxide films irradiated with (**a**) $N = 250$, (**b**) $N = 500$, and (**c**) $N = 1000$ fs laser pulses at $F = 675$ mJ/cm^2. E denotes the direction of the polarization.

An increase in F was expected to increase the density of the free electrons produced in the surface layer to a level sufficient to change the surface morphology. To confirm this, we irradiated the film surfaces with fs pulses at $F = 700$ mJ/cm^2. The results are shown in Figure 5. At $N = 250$ (Figure 5a), the line-like nanostructures clearly formed with a period of ~160 nm in a larger area than those generated at $F = 675$ mJ/cm^2. Increasing N to 500 (Figure 5b), the area over which both the dot-like and line-like nanostructures formed widened from the center of the focal spot to the edge. At $N = 1000$, as shown in Figure 5c, the spectral peak of the ablation trace appeared faintly at $d \approx 220$ nm, corresponding to the peak at 4.5 μm^{-1}.

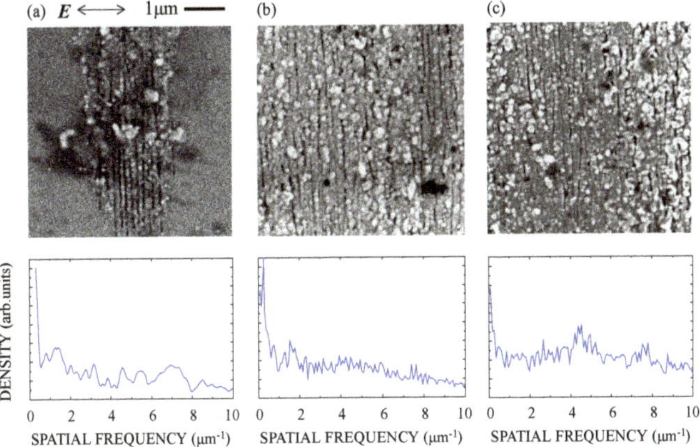

Figure 5. SEM image (top) and Fourier spectrum (bottom) of the silicon suboxide films irradiated with (**a**) $N = 250$, (**b**) $N = 500$, and (**c**) $N = 1000$ fs laser pulses at $F = 700$ mJ/cm^2. E denotes the direction of the polarization.

We increased F to 750 mJ/cm^2, and observed the surface morphology of the films irradiated with the fs pulses. The results are shown in Figure 6. At $N = 250$ (Figure 6a), the line-like nanostructures clearly formed with a spectral peak at $d = 210$–320 nm, where d is defined as the full width at a half maximum of the spectrum with the background signal subtracted (dashed line in Figure 6). With increasing N, the line-like nanostructures were formed with $d = 230$–290 nm at $N = 500$ (Figure 6b) and $d = 200$–330 nm at $N = 1000$ (Figure 6c), where the other spectral peaks at the respective harmonic frequencies were virtual. In this experiment, we could not clearly observe any change in the spectral peak position for increasing N. At $N = 1000$, the nanostructures homogeneously formed in an area of ~20 μm in diameter in the center of the focal spot, which is an order of magnitude larger than that on glass irradiated with tightly focused fs pulses. These results showed that fs laser pulses are strongly absorbed in a silicon suboxide surface to produce high-density electrons in the vicinity of the surface, leading to nanoablation with intense near-fields of SPPs.

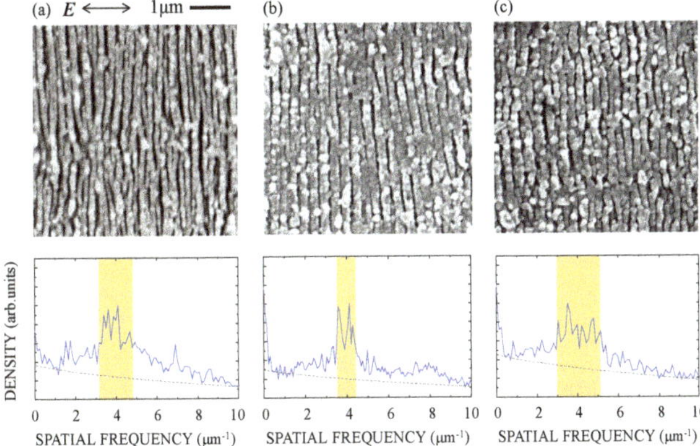

Figure 6. SEM image (upper) and Fourier spectrum (lower) of the silicon suboxide films irradiated with (**a**) $N = 250$, (**b**) $N = 500$, and (**c**) $N = 1000$ fs laser pulses at $F = 750$ mJ/cm^2. E denotes the direction of the polarization. The period d is estimated from the half maximum of the spectrum (orange-hatched area in the spectrum).

Based on the experimental results and the calculation using a model target, we discussed the excitation of SPPs in the silicon suboxide surface with the fs pulses and the subsequent nanostructure formation. The reflectivity of 1.4% at the interface between SiO_x and substrate was calculated at normal incidence. Assuming the constructive interference between the incident and reflected pulses [28], the fluence at the surface could be enhanced by a factor of 1.25, leading to a decrease in the effective F_1 from 1060 to 850 mJ/cm^2. The experimental results for $F = 675$–750 mJ/cm^2, as shown in Figures 4–6, clearly showed that these fluences are lower than the single-shot ablation threshold and higher than the multiple-shot ablation threshold for the pulses with $N \geq 250$. Assuming that the superimposed multiple fs laser pulses at a fluence lower than the single-shot ablation threshold induce surface modifications within a thin surface layer [29–31], subsequent fs pulses could induce a high density of electrons suitable to excite SPPs in this layer [32,33]. As shown in a previous report on pump-probe reflectivity measurements, an intense fs laser pulse can produce electrons with a high density of ~10^{22} cm^{-3} in fused silica [34]. Assuming the production of the high-density electrons in the silicon suboxide, the permittivity of the silicon suboxide can be described using a Drude model [35,36]:

$$\varepsilon_a = \varepsilon_{siox} - \omega_p^2/\left(\omega^2 + i\omega/\tau\right), \tag{1}$$

where $\varepsilon_{siox} = 3.24$ is the permittivity of the silicon suboxide at $\lambda = 800$ nm measured with the ellipsometer, and the second term represents the modulation by free-carrier response at the electron density N_e produced in the silicon suboxide surface. Here, $\omega = 2\pi c/\lambda$ is the laser frequency in vacuum, c is the light speed in vacuum, τ is the Drude damping time, and $\omega_p = (e^2 N_e/(\varepsilon_0 m^* m))^{1/2}$ is the plasma frequency with the permittivity of vacuum ε_0, the electron charge e and mass m, and the optical effective mass of an electron m^*. We ignored other effects modulating the permittivity such as band and state filling [37,38] and band gap renormalization [39–41] because they are very small.

The calculation method to determine the plasmon wavelength λ_{spp} in the surface layer irradiated with the fs pulse is almost the same as that used in previous studies [32,33]. Briefly, $\lambda_{spp} = 2\pi/\text{Re}[k_{spp}]$ was calculated using the following relation between light and SPPs:

$$k_{spp} = k_0 \sqrt{\frac{\varepsilon_a\, \varepsilon_b}{\varepsilon_a + \varepsilon_b}}, \tag{2}$$

where $k_0 = 2\pi/\lambda$ is the laser wavenumber in vacuum. Assuming that the SPPs are excited at the interface between the silicon suboxide and the surface layer with high-density electrons produced with the fs laser pulse, as shown in the inset of Figure 7, we set $\varepsilon_b = \varepsilon_{siox}$. For the excitation of SPPs, that is, for evanescent waves to exist in the vicinity of the surface, the relation $\text{Re}[\varepsilon_a] < 0$ should be satisfied [42]. The excitation of the SPPs is the origin of the periodicity of the fs laser-induced nanostructure formation, and periodic nanoablation is induced by a fine spatial distribution of electromagnetic energy in the surface layer [32,33,43–51]. To form the stationary energy distribution, the following two processes were proposed: the interference between the incident laser beam and the SPPs [43–45], and the counter-propagating SPP interference, i.e., the generation of a standing wave mode of SPPs [32,33,46–51]. The groove periods of the aforementioned types of interference are λ_{SPP} and $\lambda_{SPP}/2$, respectively. Assuming that either these types of interference occur simultaneously or the latter occurs dominantly, the period could become $\lambda_{SPP}/2$.

Figure 7 shows the calculated groove period D plotted as a function of N_e. We reported the damping time and optical effective mass of an electron to be $\tau = 0.1$–2 fs [52,53] and $m^* = 1$, respectively. Here, we show two results at $\tau = 0.5$ fs (thick red curve) and $\tau = 1$ fs (thin blue curve) for $m^* = 1$. Excitation of SPPs is allowed in the region $\text{Re}[\varepsilon_a] < 0$, as shown by the solid curves in Figure 7. For both values of τ, when the high-density electrons are excited in the surface layer and metalize the surface with the fs pulse, the calculated period is $D = 180$–430 nm. At $\tau < 0.5$ fs, D was calculated to approach ~200 nm. If the SPPs are resonantly excited, where $\varepsilon_a + \varepsilon_b$ becomes zero [42,50], D could be 267 nm for $\tau = 1$ fs at $N_e = 1.0 \times 10^{22}$ cm^{-3}. Here, N_e for these values of D is much larger than the critical plasma density of 1.7×10^{21} cm^{-3}. N_e can reach a higher value than this critical density because the electrons are generated in the solid surface [34,54]. These calculated results are in good agreement with the observed period d. The results showed that plasmonic near-fields generated in the silicon suboxide surface induce nanoablation to form a periodic nanostructure.

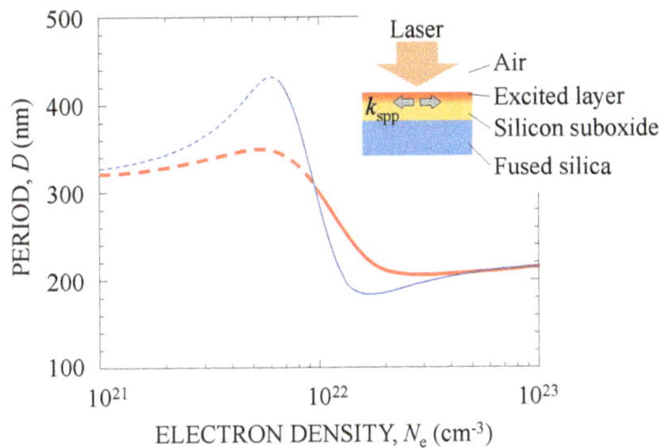

Figure 7. Period D of nanostructures calculated as a function of electron density N_e. The thin blue and thick red curves denote D at the damping times of $\tau = 1$ fs and $\tau = 0.5$ fs, respectively. Excitation of SPPs is allowed in the region Re[ε_a] < 0, where D is drawn with solid lines.

4. Conclusions

We found that clear periodic nanostructures can be formed on a silicon suboxide film with superimposed fs laser pulses at low fluence. The experimental and calculated results obtained showed that the low-fluence fs pulses are absorbed near the surface due to the large absorption coefficient of the film, forming a thin layer having high-density electrons and leading to nanoablation by plasmonic near-fields. The interfering fs pulses can form grooves with a period of $\lambda/(2\sin\theta)$, where θ is the incident angle of the laser pulse [55]. By using a high-NA optical configuration, the period can reach $\lambda/2$. For nanostructuring in this present work, the period $\lambda_{SPP}/2$ was much smaller than $\lambda/2$, as denoted in Equation (2). As silicon suboxide can easily be transformed into transparent glass, the proposed technique should provide a useful approach for rapidly and homogeneously fabricating nanostructures on glass.

Author Contributions: Conceptualization, G.M. and J.I.; methodology, G.M. and J.I.; software, G.M.; validation, G.M.; formal analysis, G.M.; investigation, G.M.; resources, G.M., L.J.R., and J.I.; data curation, T.T. and I.T.; writing—original draft preparation, T.T.; writing—review and editing, G.M. and J.I.; visualization, T.T. and G.M.; supervision, G.M. and J.I.; project administration, G.M. and J.I.; funding acquisition, G.M. and J.I. All authors have read and agreed to the published version of the manuscript.

Funding: This research was partially supported by the Grant-in-Aid for Scientific Research (B) (18H01894) and the Joint Usage/Research Program on Zero-Emission Energy Research, Institute of Advanced Energy, Kyoto University (ZE30C-02, 2018) and (ZE31C-01, 2019).

Acknowledgments: The authors would like to thank Y. Oki for performing the preliminary experiment and K. Sugioka and A. Narazaki for their advice on the physical process of femtosecond laser processing of glass.

Conflicts of Interest: The authors declare no conflicts of interest. The funders had no role in the design of the study; in the collection, analyses, or interpretation of data; in the writing of the manuscript, or in the decision to publish the results.

References

1. Yariv, A.; Yeh, P. *Optical Waves in Crystals: Propagation and Control of Laser Radiation*; Wiley: New York, NY, USA, 2003; pp. 69–154.
2. Bohren, C.F.; Huffman, D.R. *Absorption and Scattering of Light by Small Particles*; Wiley-VCH Verlag: Weinheim, Germany, 1998; pp. 82–129.

3. Cai, W.; Chettiar, U.K.; Kildishev, A.V.; Shalaev, V.M. Optical cloaking with metamaterials. *Nat. Photo.* **2007**, *1*, 224–227. [CrossRef]

4. Valentine, J.; Li, J.; Zentgraf, T.; Bartal, G.; Zhang, X. An optical cloak made of dielectrics. *Nat. Mat.* **2009**, *8*, 568–571. [CrossRef] [PubMed]

5. Yablonovitch, E. Photonic band-gap structures. *J. Opt. Soc. Am. B* **1993**, *10*, 283–295. [CrossRef]

6. Noda, S.; Chutinan, A.; Imada, M. Trapping and emission of photons by a single defect in a photonic bandgap structure. *Nature* **2000**, *407*, 608–610. [CrossRef] [PubMed]

7. Yu, N.; Genevet, P.; Kats, M.A.; Aieta, F.; Tetienne, J.-P.; Capasso, F.; Gaburro, Z. Light Propagation with Phase Discontinuities: Generalized Laws of Reflection and Refraction. *Science* **2011**, *334*, 333–337. [CrossRef]

8. Yu, N.; Capasso, F. Flat optics with designer metasurfaces. *Nat Mater.* **2014**, *13*, 139–150. [CrossRef]

9. Shimotsuma, Y.; Kazansky, P.G.; Qiu, J.; Hirao, K. Self-Organized Nanogratings in Glass Irradiated by Ultrashort Light Pulses. *Phys. Rev. Lett.* **2003**, *91*, 247405. [CrossRef]

10. Bhardwaj, V.R.; Simova, E.; Rajeev, P.P.; Hnatovsky, C.; Taylor, R.S.; Rayner, D.M.; Corkum, P.B. Optically Produced Arrays of Planar Nanostructures inside Fused Silica. *Phys. Rev. Lett.* **2006**, *96*, 057404. [CrossRef]

11. Wagner, R.; Gottmann, J.; Horn, A.; Kreutz, E.W. Subwavelength ripple formation induced by tightly focused femtosecond laser radiation. *Appl. Surf. Sci.* **2006**, *252*, 8576–8579. [CrossRef]

12. Beresna, M.; Gecevičius, M.; Kazansky, P.G.; Gertus, T. Radially polarized optical vortex converter created by femtosecond laser nanostructuring of glass. *Appl. Phys. Lett.* **2011**, *98*, 201101. [CrossRef]

13. Ion, J.C. *Laser Processing of Engineering Materials*; Elsevier Butterworth-Heinemann: Oxford, UK, 2005; pp. 104–138.

14. Kannatey-Asibu, E., Jr. *Principles of Laser Materials Processing*; John Wiley & Sons: Hoboken, NJ, USA, 2008; pp. 409–430.

15. Phillip, H.R. Optical properties of non-crystalline Si, SiO, SiO_x and SiO_2. *J. Phys. Chem. Solids* **1971**, *32*, 1935–1945. [CrossRef]

16. Hernández, S.; Pellegrino, P.; Martínez, A.; Lebour, Y.; Garrido, B.; Spano, R.; Cazzanelli, M.; Daldosso, N.; Pavesi, L.; Jordana, E.; et al. Linear and nonlinear optical properties of Si nanocrystals in SiO_2 deposited by plasma-enhanced chemical-vapor deposition. *J. Appl. Phys.* **2008**, *103*, 064309. [CrossRef]

17. Minissale, S.; Yerci, S.; Dal Negro, L. Nonlinear optical properties of low temperature annealed silicon-rich oxide and silicon-rich nitride materials for silicon photonics. *Appl. Phys. Lett.* **2012**, *100*, 021109. [CrossRef]

18. Fricke-Begemann, T.; Meinertz, J.; Weichenhain-Schriever, R.; Ihlemann, J. Silicon suboxide (SiO_x): Laser processing and applications. *Appl. Phys. A Mater. Sci. Process.* **2014**, *117*, 13–18. [CrossRef]

19. Kahler, U.; Hofmeister, H. Size evolution and photoluminescence of silicon nanocrystallites in evaporated SiOx thin films upon thermal processing. *Appl. Phys. A* **2002**, *74*, 13–17. [CrossRef]

20. Friz, M.; Waibel, F. Coating Material. In *Optical Interference Coatings*; Kaiser, N., Pulker, H.K., Eds.; Springer-Verlag: Berlin, Germany, 2003; pp. 105–130.

21. Roschuk, T.; Li, J.; Wojcik, J.; Mascher, P.; Calder, I.D. Lighting Applications of Rare Earth-Doped Silicon Oxides. In *Silicon Nanocrystals: Fundamentals, Synthesis and Applications*; Pavesi, L., Turan, R., Eds.; Wiley-VCH Verlag: Weinheim, Germany, 2010; pp. 487–506.

22. Sirleto, L.; Ferrara, M.A.; Nikitin, T.; Novikov, S.; Khriachtchev, L. Giant Raman gain in silicon nanocrystals. *Nat. Commun.* **2012**, *3*, 1220. [CrossRef]

23. Schulz-Ruhtenberg, M.; Ihlemann, J.; Heber, J. Laser patterning of SiO_x-layers for the fabrication of UV diffractive phase elements. *Appl. Surf. Sci.* **2005**, *248*, 190–195. [CrossRef]

24. Takaya, T.; Miyaji, G.; Richter, L.J. Ihlemann, Nanostructure Formation on Silicon Suboxide with Plasmonic Near-Field Ablation Induced by Femtosecond Laser Pulses. In Proceedings of the 2019 Conference on Lasers and Electro-Optics Europe & European Quantum Electronics Conference (CLEO/Europe-EQEC), Munich, Germany, 23–27 June 2019.

25. Miyazaki, K.; Miyaji, G. Mechanism and control of periodic surface nanostructure formation with femtosecond laser pulses. *Appl. Phys. A* **2014**, *114*, 177–185. [CrossRef]

26. Gamaly, E.G.; Juodkazis, S.; Nishimura, K.; Misawa, H.; Luther-Davies, B.; Hallo, L.; Nicolai, P.; Tikhonchuk, V.T. Laser-matter interaction in the bulk of a transparent solid: Confined microexplosion and void formation. *Phys. Rev. B-Condens. Matter Mater. Phys.* **2006**, *73*, 1–15. [CrossRef]

27. Gräf, S.; Kunz, C.; Müller, F.A. Formation and Properties of Laser-Induced Periodic Surface Structures on 254 Different Glasses. *Materials* **2017**, *10*, 933. [CrossRef]

28. Juodkazis, S.; Nishi, Y.; Misawa, H.; Mizeikis, V.; Schecker, O.; Waitz, R.; Leiderer, P.; Scheer, E. Optical transmission and laser structuring of silicon membranes. *Opt. Express* **2009**, 15308–15317. [CrossRef] [PubMed]

29. Bonse, J.; Rudolph, P.; Krüger, J.; Baudach, S.; Kautek, W. Femtosecond pulse laser processing of TiN on silicon. *Appl. Surf. Sci.* **2000**, *154*, 659–663. [CrossRef]

30. Yasumaru, N.; Miyazaki, K.; Kiuchi, J. Glassy carbon layer formed in diamond-like carbon films with femtosecond laser pulses. *Appl. Phys. A* **2004**, *79*, 425–427. [CrossRef]

31. Žemaitis, A.; Gaidys, M.; Brikas, M.; Gečys, P.; Račiukaitis, G.; Gedvilas, M. Advanced laser scanning for highly-efficient ablation and ultrafast surface structuring: Experiment and model. *Sci. Rep.* **2018**, *8*, 1–14. [CrossRef]

32. Miyaji, G.; Miyazaki, K. Origin of periodicity in nanostructuring on thin film surfaces ablated with femtosecond laser pulses. *Opt. Express* **2008**, *16*, 16265–16271. [CrossRef] [PubMed]

33. Miyaji, G.; Miyazaki, K.; Zhang, K.; Yoshifuji, T.; Fujita, J. Mechanism of femtosecond-laser-induced periodic nanostructure formation on crystalline silicon surface immersed in water. *Opt. Express* **2012**, *20*, 14848–14856. [CrossRef] [PubMed]

34. Puerto, D.; Gawelda, W.; Siegel, J.; Bonse, J.; Bachelier, G.; Solis, J. Transient reflectivity and transmission changes during plasma formation and ablation in fused silica induced by femtosecond laser pulses. *Appl. Phys. A* **2008**, *92*, 803–808. [CrossRef]

35. Ibach, H.; Lüth, H. *Solid-State Physics, An Introduction to Principles of Materials Science*, 4th ed.; Springer: Berlin/Heidelberg, Germany, 2009; pp. 371–403.

36. Wang, S. *Solid-State Electronics*; McGraw-Hill: New York, NY, USA, 1966; pp. 262–308.

37. Shah, J.; Leheny, R.F.; Lin, C. Dynamic Burstein shift in GaAs. *Solid State Commun.* **1976**, *18*, 1035–1037. [CrossRef]

38. Oudar, J.L.; Hulin, D.; Migus, A.; Antonetti, A.; Alexandre, F. Subpicosecond spectral hole burning due to nonthermalized photoexcited carriers in GaAs. *Phys. Rev. Lett.* **1985**, *55*, 2074–2077. [CrossRef]

39. Wolff, P.A. Theory of the Band Structure of Very Degenerate Semiconductors. *Phys. Rev.* **1962**, *126*, 405–412. [CrossRef]

40. Berggren, K.-F.; Sernelius, B.E. Band-gap narrowing in heavily doped many-valley semiconductors. *Phys. Rev. B* **1981**, *24*, 1971–1986. [CrossRef]

41. Abram, R.A.; Childs, G.N.; Saunderson, P.A. Band gap narrowing due to many-body effects in silicon and gallium arsenide. *J. Phys. C* **1984**, *17*, 6105–6125. [CrossRef]

42. Raether, H. *Surface Plasmons on Smooth and Rough Surfaces and on Gratings*; Springer-Verlag: Heidelberg, Germany, 1988; pp. 4–7.

43. Sipe, J.E.; Young, J.F.; Preston, J.S.; Van Driel, H.M. Laser-induced periodic surface structure. I. Theory. *Phys. Rev. B* **1983**, *27*, 1141–1154. [CrossRef]

44. Bonse, J.; Rosenfeld, A.; Krüger, J. On the role of surface plasmon polaritons in the formation of laser-induced periodic surface structures upon irradiation of silicon by femtosecond-laser pulses. *J. Appl. Phys.* **2009**, *106*, 104910. [CrossRef]

45. Garrelie, F.; Colombier, J.-P.; Pigeon, F.; Tonchev, S.; Faure, N.; Bounhalli, M.; Reynaud, S.; Parriaux, O. Evidence of surface plasmon resonance in ultrafast laser-induced ripples. *Opt. Express* **2011**, *19*, 9035–9043. [CrossRef]

46. Novotny, L.; Hecht, B.; Pohl, D.W. Interference of locally excited surface plasmons. *J. Appl. Phys.* **1997**, *81*, 1798–1806. [CrossRef]

47. Liu, Z.; Wei, Q.; Zhang, X. Surface Plasmon Interference Nanolithography. *Nano Lett.* **2005**, *5*, 957–961. [CrossRef]

48. Vorobyev, A.Y.; Guo, C. Direct femtosecond laser surface nano/microstructuring and its applications. *Laser Photonics Rev.* **2013**, *7*, 385–407. [CrossRef]

49. Buividas, R.; Mikutis, M.; Juodkazis, S. Surface and bulk structuring of materials by ripples with long and short laser pulses: Recent advances. *Prog. Quantum Electron.* **2014**, *38*, 119–156. [CrossRef]

50. Stankevič, V.; Račiukaitis, G.; Bragheri, F.; Wang, X.; Gamaly, E.G.; Osellame, R.; Juodkazis, S. Laser printed nano-gratings: Orientation and period peculiarities. *Sci. Rep.* **2017**, *7*, 1–8. [CrossRef]

51. Wang, L.; Chen, Q.-D.; Cao, X.-W.; Buividas, R.; Wang, X.; Juodkazis, S.; Sun, H.-B. Plasmonic nano-printing: Large-area nanoscale energy deposition for efficient surface texturing. *Light Sci. Appl.* **2017**, *6*, e17112. [CrossRef] [PubMed]

52. Arnold, D.; Cartier, E.; Dimaria, D.J. Acoustic-phonon runaway and impact ionization by hot electrons in silicon dioxide. *Phys. Rev. B* **1992**, *45*, 1477–1480. [CrossRef]

53. Sun, Q.; Jiang, H.; Liu, Y.; Wu, Z.; Yang, H.; Gong, Q. Measurement of the collision time of dense electronic plasma induced by a femtosecond laser in fused silica. *Opt. Lett.* **2005**, *30*, 320. [CrossRef]

54. Shank, C.V.; Yen, R.; Hirlimann, C. Time-resolved reflectivity measurements of femtosecond-optical-pulse-induced phase transitions in silicon. *Phys. Rev. Lett.* **1983**, 454–457. [CrossRef]

55. Voisiat, B.; Gedvilas, M.; Indrišiunas, S.; Raciukaitis, G. Flexible microstructuring of thin films using multi-beam interference: Ablation with ultrashort lasers. *J. Laser Micro Nanoeng.* **2011**, *6*, 185–190. [CrossRef]

 *nanomaterials*

Article

Formation of Periodic Nanoridge Patterns by Ultrashort Single Pulse UV Laser Irradiation of Gold

Andreas Blumenstein [1,2,*], **Martin E. Garcia** [2], **Baerbel Rethfeld** [3], **Peter Simon** [1], **Jürgen Ihlemann** [1,*] and **Dmitry S. Ivanov** [2,3,4,*]

1   Laser-Laboratorium Göttingen e.V., Hans-Adolf-Krebs-Weg 1, 37077 Göttingen, Germany; peter.simon@llg-ev.de
2   Physics Department, University of Kassel, Heinrich-Plett-Str. 40, 34132 Kassel, Germany; garcia@physik.uni-kassel.de
3   Department of Physics and OPTIMAS Research Center, Technical University of Kaiserslautern, Erwin-Schrödinger-Str. 46, 67663 Kaiserslautern, Germany; rethfeld@physik.uni-kl.de
4   Quantum Electronics Division, Lebedev Physical Institute, 119991 Moscow, Russia
*   Correspondence: andreas.blumenstein@llg-ev.de (A.B.); juergen.ihlemann@llg-ev.de (J.I.); ivanov@uni-kassel.de (D.S.I.)

Received: 4 September 2020; Accepted: 3 October 2020; Published: 10 October 2020

Abstract: A direct comparison of simulation and experimental results of UV laser-induced surface nanostructuring of gold is presented. Theoretical simulations and experiments are performed on an identical spatial scale. The experimental results have been obtained by using a laser wavelength of 248 nm and a pulse length of 1.6 ps. A mask projection setup is applied to generate a spatially periodic intensity profile on a gold surface with a sinusoidal shape and periods of 270 nm, 350 nm, and 500 nm. The formation of structures at the surface upon single pulse irradiation is analyzed by scanning and transmission electron microscopy (SEM and TEM). For the simulations, a hybrid atomistic-continuum model capable of capturing the essential mechanisms responsible for the nanostructuring process is used to model the interaction of the laser pulse with the gold target and the subsequent time evolution of the system. The formation of narrow ridges composed of two colliding side walls is found in the simulation as well as in the experiment and the structures generated as a result of the material processing are categorized depending on the range of applied fluencies and periodicities.

Keywords: periodic nanostructures; molecular dynamics; two-temperature model; laser material processing; ultrashort laser pulses; laser ablation; laser interference ablation

---

1. Introduction

Since the beginning of ablative laser materials processing, the fabrication of periodic surface patterns has always been a subject of interest. Besides the spontaneous development of periodic ripple structures (LIPSS) [1–3] and surface plasmon-polariton (SPP)-induced structures under the controlled conditions [4–6], the beam interference concepts have been applied to create deterministic patterns with a predefined period and control of the surface profile [7,8]. For moderate period sizes in the range of a few µm to tens of µm, mostly nanosecond pulsed lasers have been applied [9–12]. If shorter periods in the µm or sub-µm range are desired, especially in case of metals and semiconductors, ultrashort pulses of picosecond or femtosecond duration are used [13–15]. The latter have the advantage that thermal diffusion of the absorbed energy and thus the extent of the heat affected zone can be minimized enabling precise patterning [16,17]. However, in many cases it has been observed that the local ablation depth does not correlate with the local laser fluence in a simple way as it could be expected from large-area ablation experiments, where a monotonously increasing ablation depth with increasing fluence is observed. Instead, the formation of bumps, voids, ridges, and droplets is observed, indicating that on this scale, other contributions like lateral material movement have to be considered [18]. Irradiation with a line pattern leads to grooves with protruding side walls [19]. Thin films disintegrate and form nanowires and nanodots [20,21] or nanospikes [22]. Thermocapillary and Marangoni effects as well as Rayleigh–Plateau instability have been identified to influence the structure formation and its disintegration process [23,24]. Later on, however, the relaxation of the laser-induced stresses, generated in the vicinity of the target's surface as a result of the laser heating, has been determined as a main driving mechanism responsible for the nanostructures growth [25–27]. The specifics of the final nanostructures shape depend on the material as well as on the laser parameters, the dynamical changes of the reflectivity of the material [28,29], and the geometric parameters of the irradiation pattern.

In this study, we investigate the formation of structure on gold irradiated by an UV ultrashort pulse at a 248-nm wavelength with a periodic line pattern with periods of 270–500 nm. Experimentally obtained results are directly compared to simulations based on a hybrid atomistic-continuum model, where the atomic motion is described within a molecular dynamics (MD) approach, the effect of free carriers is considered with a diffusion differential equation describing the temperature dynamics of electrons, and the energy exchange between electrons and the lattice is treated as in the Two Temperature Model (TTM) [30]. With such combined MD-TTM approach [31], the mechanisms of ultrashort laser pulse melting, spallation, and ablation of mono- and poly-crystalline metal targets [32], nanostructuring in air [33] and under water confinement [34,35], as well as laser-induced generation of nanoparticles in different media [36] have been successfully studied. The results of simulations are analyzed and the target's evolution is characterized via dynamic evolution of thermophysical dynamics quantities such as temperature, pressure, density, and the lift-off velocity. The strength and accuracy of the derived mechanisms are verified with the experimental measurements and observations. Such a methodology allows for extracting the mechanisms of the nanostructures formation and manipulating with them targeting the problem of generation of periodic patterns with predesigned morphology. The obtained structures are classified into four different fluence regimes resulting in (1) surface swelling, (2) internal void formation including structures growth, (3) surface walls formation, and (4) broad melting. The first two lower fluence regimes around the material's modification threshold, corresponding to the onset of surface melting, sub surface voids generation and subsequent formation of periodic lines were described in [37]. The higher fluence regimes, around and above the ablation threshold of 210 mJ/cm^2 [38], resulting in the formation of steep ridges and broad melting, including the process of generation of a subsurface layer (here 100 nm depth), with polycrystalline structures are studied in the present paper. The nanostructuring processes are well described by the simulations that match the experimental results for different irradiation conditions with great detail.

2. Experiment

The surface structuring experiments are performed by single pulse exposure using a combination of interference and mask projection to form a large area of sinusoidal intensity distribution on the sample surface (Figure 1). The used laser system consists of a frequency tripled Ti:Sa system (Coherent Libra) seeding a KrF excimer amplifier (LLG TwinAmp) producing pulse energies of up to 40 mJ at a wavelength of 248 nm and a repetition rate of 10 Hz. The pulse length of 1.6 ps is measured by using a frequency resolved optical gating (FROG) trace obtained with a self build device described in Ref. [39]. Single pulses are selected by a shutter. The used pulse energy is set by a variable attenuator. The flat top central part of the laser beam illuminates the mask comprising a square aperture of 1 mm edge length in contact with a Cr-on-quartz grating of lines and spaces with 25 μm period and duty cycle 0.5. The demagnification of the Schwarzschild objective used for mask imaging is adjusted to be 25x. Consequently, the edge length of the overall projected area on the workpiece is 40 μm. Since all diffracted orders emerging from the transmission grating except the two first orders are blocked, the image is formed via two-beam interference, resulting in a sinusoidal intensity distribution across the 40 μm area. The spatial frequency of this modulation is doubled by blocking the zeroth order diffraction, thus resulting in a period of 500 nm on the workpiece (instead of a pitch size of 1 μm based on the nominal demagnification). In order to obtain other periods, the mask period and objective demagnification have been changed accordingly. For the 350-nm structure a 36x objective and a 25-μm grating, and for the 270-nm period a 74x objective and a 40-μm grating were used. A schematic view of the intensity distribution is shown in Figure 1b. The purpose of the lens with f = 1000 mm in the setup is to reduce the size of the beam within the Schwarzschild objective. All experiments are conducted at the Laser-Laboratorium Göttingen e.V. (Göttingen, Germany).

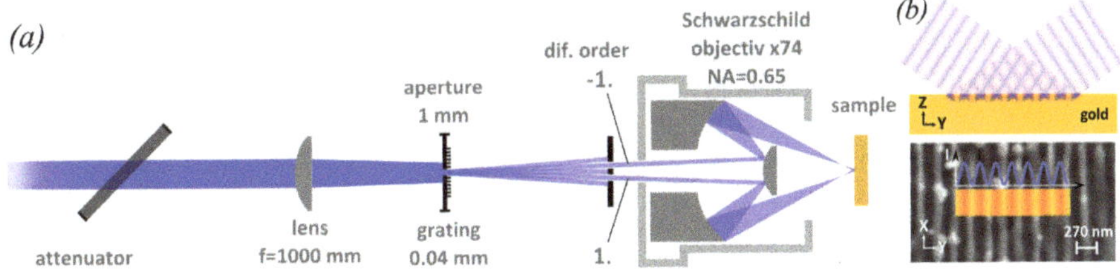

Figure 1. (**a**) Mask projection setup for periodic surface structuring. A sinusoidal interference pattern is obtained by using only the ± 1st diffraction orders for mask imaging. (**b**) Shape of the intensity distribution generated by two beam interference.

3. Theory

The essential concepts and applications of the combined atomistic-continuum model MD-TTM to study the evolution of metallic solids excited with an ultrashort laser pulse are explained in Ref. [31]. The model consists of two parts: the atomic motion is described within the MD approach, whereas a diffusion differential equation for electrons is accounting for the effect of free carriers via their temperature dynamics. Thus, the combined model provides a detailed atomic-level description of the kinetics of fast non-equilibrium processes of laser melting at the atomic resolution and, at the same time, in continuum ensures for the adequate description of the laser light absorption by the conduction band electrons, the energy transfer to the lattice due to the electron–phonon coupling, and the fast electron heat conduction. Based on the atomistic-continuum approach, the MD-TTM model was further developed for simulation of the nanostructuring processes on metal targets in multiprocessing algorithm (running in parallel) to be applied in large scale modeling tasks. The model applied here and its parametrization for the investigation of UV ultrashort pulse generation of periodic pattern on metal surfaces in a high fluence regime is described in detail in Ref. [37].

The initial structures represented by supercells of (40 x 270 x 200) nm^3 and (40 x 500 x 200) nm^3 in X, Y, and Z directions correspondingly were prepared and consisted of 125,000,000 and 230,000,000 atoms, respectively. The MD supercells were equilibrated at room temperature before exposure to a single 1.6-ps laser pulse of wavelength equal to 248 nm with a spatially sinusoidal intensity profile with periods of 270 nm and 500 nm for the smaller and the larger sample, correspondingly. Then, the prepared computational cells were run across 256 and 480 processor cores, respectively, to simulate the nanostructuring process up to 1.5 ns of the experimental time.

The geometry and boundary conditions of the computational supercell were the same as those reported in Ref. [37], whereas, depending on the investigated feature, various techniques of the visualization and numerical analysis of the results from the point of thermo-dynamical properties were used. Thus, monitoring of the pressure and lift-off dynamics at the same time allows for extracting the mechanism of the nanostructures' formation, whereas the simultaneous observation of the electronic and lattice temperatures explains their fast solidification [25]. Additionally, depending on the particular conditions of laser-induced phase transition processes (ultrafast melting and resolidification), local order parameter (LOP) [31] or central symmetry parameter (CSP) [40] have been used to identify the local crystal structure. The former parameter, due to its high sensitivity is frequently utilized for tracking down the ultrafast laser-induced melting process under conditions of strong superheating and high pressure and temperature gradients, whereas the latter is independent on crystal orientation and therefore allows for convenient identification of the forming nanocrystals during the solidification process. Such an approach has been in particular successfully used during the procedure of direct comparison of the simulation results with the obtained nanostructures in the experiments. Thus, the local crystal structures are readily highlighted by LOP or CSP values during the visual analysis of the atomic configurations.

4. Results and Discussions

Figure 2 displays scanning electron micrographs (SEM) taken from a gold surface irradiated with a 350-nm-periodic intensity profile (single pulse at 248 nm, 1.6 ps). At low fluence (a) some slight elevations at the lines of maximum fluence are visible (surface swelling). Increasing the fluence (b), these elevations become more pronounced and open up at some points forming holes (void formation). At further increased fluence (c), the elevations turn into grooves by opening up over the whole length and forming side walls. In some regions the side walls of neighboring grooves are well separated, in other regions the walls of neighboring grooves merge to form one nanoridge with some droplets on top (wall formation). Taking an even higher fluence (d), the line pattern blurs giving the impression of broad melting.

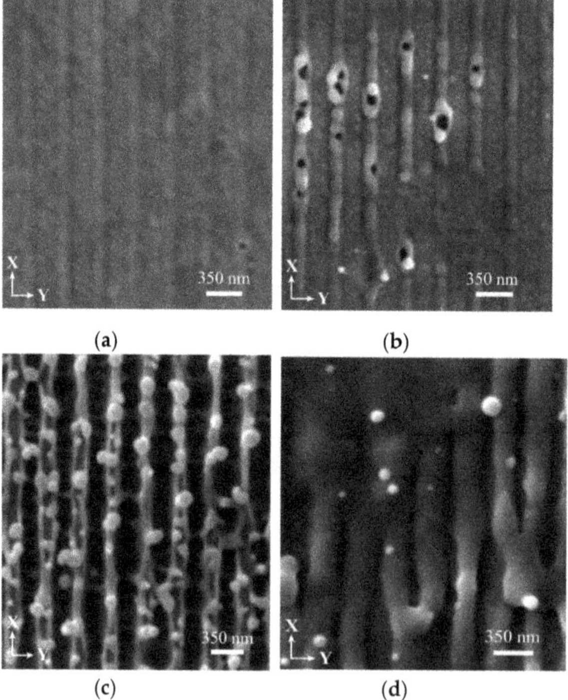

(a)   (b)

(c)   (d)

Figure 2. Surface topography obtained on gold after single pulse irradiation with a 350-nm period. (a) Surface swelling at 100 mJ/cm^2, (b) void formation at 150 mJ/cm^2, (c) wall formation at 200 mJ/cm^2, (d) broad melting at 350 mJ/cm^2.

In the case of a larger structure period of 500 nm, the neighboring features at the high fluence lines are a bit more separated. Figure 3a displays the 3D simulated atomic snapshots taken 1 ns after the laser pulse absorption for three different fluences: at 130 mJ/cm^2 slight surface elevations are visible. At 160 mJ/cm^2 these voids are much larger, and at 250 mJ/cm^2 the voids open leading to a rather shallow groove with steep side walls. In Figure 3b this simulation result is directly compared to the corresponding experimental data. There is a wide agreement; even details like droplets on the walls and spikes on the ground are found in both representations. Only the thickness and the height of the walls seem to deviate in the simulation from the experimental observation. While further dynamics of molten material is not excluded, a better agreement may also be reached applying alternative models for material parameters. For example, the electron-phonon coupling parameter is heavily under discussion [41–43]. Here, we have applied the result of Lin et al. [44], while recent calculations have yielded a lower value and showed a strong dependence on density [45]. In addition, the model for the electron heat conductivity has a strong impact on the resulting laser energy deposition profile. In the present work it was described as a function of electron and lattice temperatures [46], whereas an alternative approach accounting for the density of matter can be utilized [47]. Thus, as a result of different descriptions of the above quantities, the effective penetration depth of the laser-deposited energy [48] (was identified here as 150 nm) can deviate as well as the profile and magnitude of the established temperature and pressure gradients [16]. The relaxation of the laser-induced stresses, depending on the temperature distribution and the peak position of tensile forces, results in nucleation of the internal voids (onset of the spallation process) at a certain depth, therefore, determining the thickness of the spalled layer as a result [49].

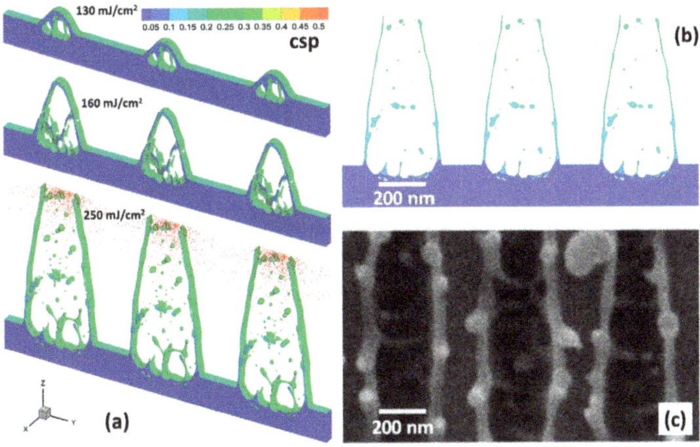

Figure 3. (**a**) Simulated 3D atomic snapshots 1 ns after 1.6 ps single pulse irradiation with a structure period of 500 nm. Fluence: 130 mJ/cm² (top), 160 mJ/cm² (middle), 250 mJ/cm² (bottom). The atoms are colored by central symmetry parameter (CSP) for distinguishing a local crystal structure as follows: crystal < 0.08 < defects (dislocations) < 0.11 < liquid < 0.25 < surfaces < 0.50 < vapor (free atoms). (**b**) Direct comparison of the simulation (cross section for the case of a 500-nm period) with the (**c**) experimental result (SEM, top view) on the same length scale at the incident fluence of 250 mJ/cm².

Eventually, the high linear momentum of the upward motion of the solidifying structure results in a propagation of cracks along the X direction of the forming nanostructure. This is indicated by arrows in the Figure 4a at the last atomic snapshot at 1000 ps. The development of the crack leads to a complete detachment of the nanostructure tip resulting in spallation, leaving an opened hole as it is experimentally observed, see Figure 2b. The tip detachment is obvious in the analysis of the velocity lift-off field where a clear jump in its value is observed at the crack position. Interestingly, the spalled tip in Figure 4a is already in the solid phase, which is supported by the CSP analysis (CSP < 0.11) and the tip temperature has a value of 950 K, whereas the equilibrium melting point of the modeled Au material is 1337 K.

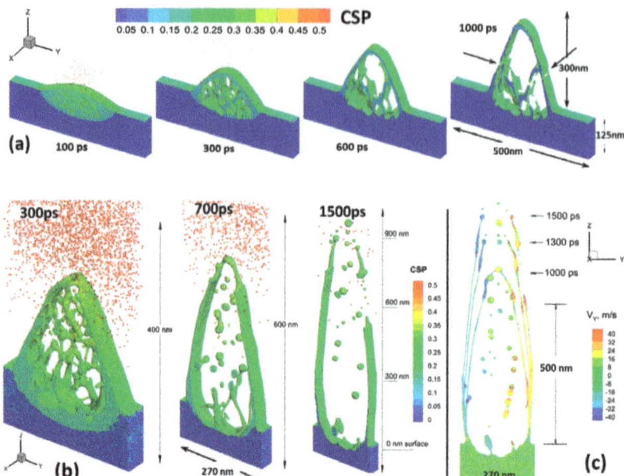

Figure 4. (**a**) Sequence of atomistic snapshot configurations (100 ps, 300 ps, 600 ps, and 1000 ps) of a 500-nm period structures obtained at the incident fluence of 160 mJ/cm^2. The arrows on the 1000 ps snapshot indicates the formation of a crack along the nanostructure direction X. (**b**) Calculated shapes for 300 ps, 700 ps, 1.5 ns after the pulse for a 270-nm period structure obtained at an incident fluence of 250 mJ/cm^2. The atoms are colored by CSP for distinguishing a local crystal structure as follows: crystal < 0.08 < defects (dislocations) < 0.11 < liquid < 0.25 < surfaces < 0.50 < vapor (free atoms). (**c**) The tangential (perpendicular to the incident pulse) Y component of the velocity field of the growing structure in (**b**) is shown by color for the set of times: 1000 ps, 1300 ps, and 1500 ps.

It can be clearly seen that in the non-damaged parts of the surface (more precisely: in the areas where the fluence is below the threshold) the original morphology remains intact even at the highest fluence for the case of a 500-nm period (Figure 4a). However, as the incident fluence increases (Figure 3a), or at a smaller structure period of 270 nm, this non-influenced zone becomes very narrow. This is observed in Figure 4b, where the high incident fluence of 250 mJ/cm^2 results in an active formation of subsurface voids that develop fast and open to the surface by the time of 700 ps. While the non-influenced zone is getting narrow, the walls of neighboring grooves will collide and form one merged wall (nanoridge), as it experimentally can be seen in Figure 2c. This can also be concluded from Figure 4c, where the tangential component of the local velocities of the material in Y direction is plotted as a color field. Specifically, at the moment of the separation of the left and right wings of the growing structure, the surface tension forces add up to the tangential momentum for their further motion and the left wing has a negative Y component (−20 to −40 m/s), whereas the right wing has a positive one (+20 to +40 m/s). This means that the walls are moving outwards approaching their merging point in between the grooves. In fact, already upon the beginning of the spallation process, the side walls of the forming structure already have the tangential component of the velocity due to outward motion of the melt driven by the unloading pressure wave. At 1.5 ns after the pulse, the walls of liquid material are still moving outwards without any resistance and they will collide afterwards. Note that the smaller the period, the lower the required fluence at which merging of side walls occurs. This is supported by the experimental measurements (see Figure 5) where at a period of 270 nm, the incident fluence of 300 mJ/cm^2 is sufficient to lead to single walls formation between the grooves, whereas it does not happen yet below the incident fluence of 400 mJ/cm^2 at a 500-nm period. In Figure 5a–d on top of the nanoridges, droplets are forming at some irregular positions, caused by instabilities, they are also visible forming already in the MD simulation in Figure 4c after 1500 ps on the tip of the filament. The formation of droplets occurs due to Rayleigh–Plateau instability and minimizes the surface tension of the elevating and elongating wall-like structures.

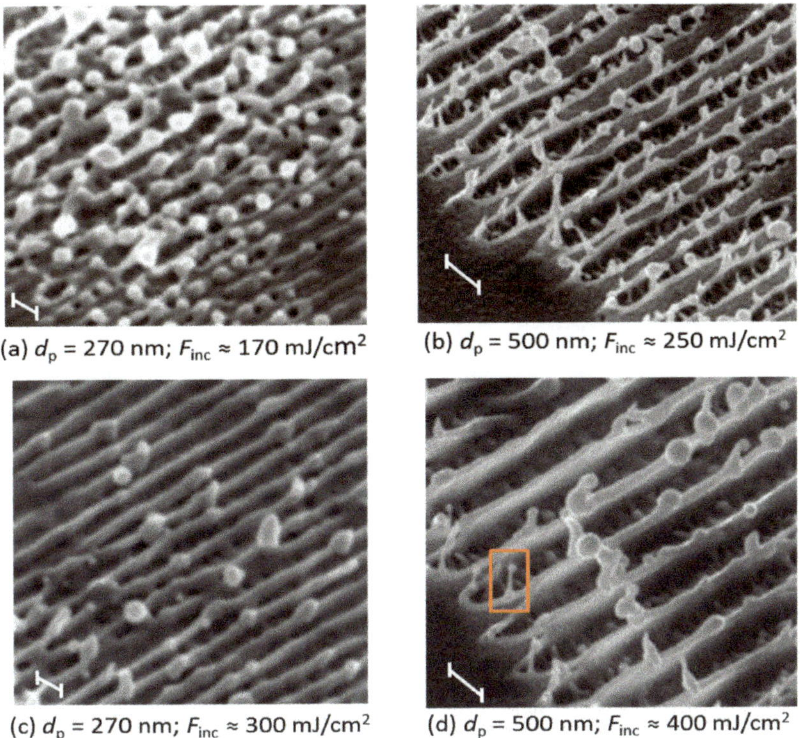

(a) d_p = 270 nm; F_{inc} ≈ 170 mJ/cm^2 (b) d_p = 500 nm; F_{inc} ≈ 250 mJ/cm^2

(c) d_p = 270 nm; F_{inc} ≈ 300 mJ/cm^2 (d) d_p = 500 nm; F_{inc} ≈ 400 mJ/cm^2

Figure 5. Periodic patterns obtained with periods of 270 nm in (**a**) and (**c**) and 500 nm in (**b**) and (**d**) at various fluences, at a tilt angle 45°, the period is indicated by the scale bars. For larger periods, a higher fluence is required to obtain nanoridges composed of two collided walls. The squared structure in (**d**) is experimentally identified and obtained in simulations (Figure 6).

Figure 6 shows the direct comparison of the modeling results with the experimental observations for the case of a 500-nm period structures formation at an incident fluence of 250 mJ/cm^2. One can see that the final structures are opened to the surface (a) and the simulation is reproducing this process (c). Amazingly, not only the formation of the wall-like structures is observed in both simulations and experiments (green lines relate the experimental observation with corresponding cross sections in the simulated structure), but even the tiny surface and subsurface nanograins, nanovoids, nanowalls, nanocolumns, and polycrystalline structures at the base of the modeled structures reveal a perfect match between the simulation and experiments. Namely, the process of spallation creates an excessively rough surface with nanostructures and nanovoids, captured inside the solidified material (b). To make it clear, the corresponding nanocolumn (orange rectangle) and nanovoid structures (red rectangle) are identified and boxed in both the TEM (a) and FIB (b) pictures and related to the atomic snapshot (c). The formation of such structures perfectly agrees with recent theoretical findings presented in [54,55].

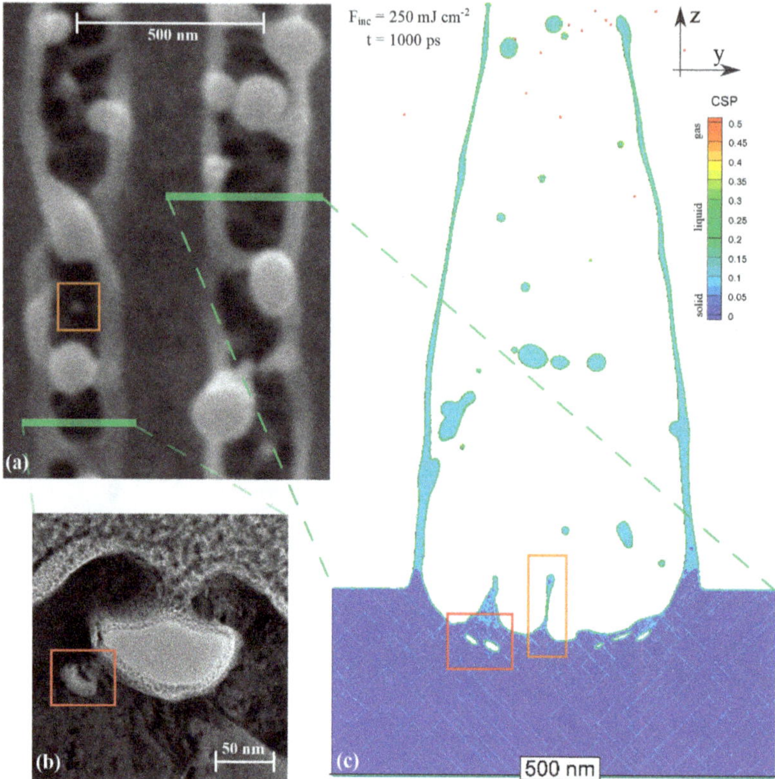

Figure 6. Direct comparison of SEM top view in (**a**) and TEM cross-section in (**b**) with the simulation results of the atomic configuration 40 x 500 x 200 nm^3 of a 500-nm period at 1000 ps in (**c**). The atoms are colored by CSP parameter. The process of spallation results in an opening of the generated structure and formation of a number of droplets and pillar-like structures on the remaining surface (orange rectangle). The spallation process is accompanied with the generation of subsurface voids (red rectangle) registered in both experiment and simulation and is magnified in the squared window in Figure 7d below.

Apart of the generation of a great number of dislocation planes, highlighted by a light blue color in Figure 6c, the results of the simulation also predict the polycrystalline character of the solidified structures as it was emphasized in [56]. As it was found in [25], the fast cooling process of the developing structure is provided by the strong electronic heat conduction mechanism. Since the classical solidification is governed by the diffusion process, which takes time, the liquid can undergo a significant cooling below the equilibrium melting point by more than 20% [57]. Thus, at such strong undercooling conditions, the classical solidification process governed by the heterogeneous mechanism (propagation of the well-defined solid–liquid interface) cannot keep up due to a too strong cooling rate and the solidification process turns into a massive nucleation of the solid phase inside the liquid due to the homogeneous nucleation mechanism. This results in the formation of nanograins of the solid phase that have random orientation forming nanograin boundaries constituting the polycrystalline character of the final nanostructures. Moreover, the stronger the cooling rate, the smaller the resulting nanograin size. Since the cooling rate depends on the established temperature gradient, it therefore depends on both the physical size of the excited volume and the applied incident fluence, see Figure 7.

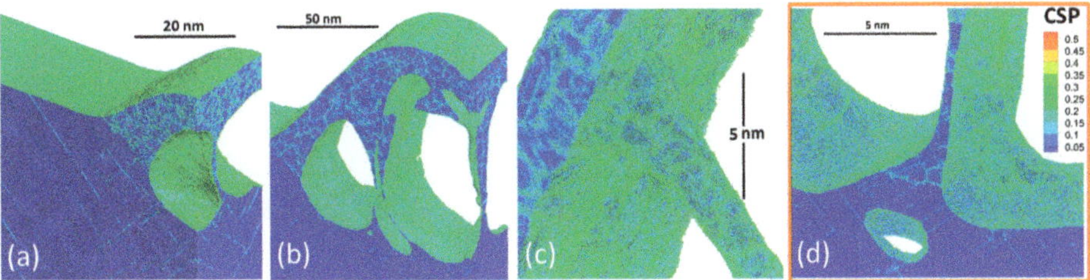

Figure 7. The nanostructures generated in the simulations on periodic nanostructuring are magnified for (**a**) the period of 270 nm at the incident fluence of 120 mJ/cm^2 and (**b**) 145 mJ/cm^2. (**c**) Structures obtained for the case of a 500-nm period at the incident fluence of 160 mJ/cm^2 and (**d**) 250 mJ/cm^2. The atoms are colored by CSP parameter.

The analysis of the generated crystal structures, presented in Figure 7a–d, indicates their polycrystalline character, where mainly small nanocrystals are constituting the nanostructures generated at a smaller period and at a smaller incident fluence. The last figure in (d) allows for a direct comparison of the simulated bottom and subsurface structure generation, shown in an orange rectangle region in Figure 6a–c. The noticeably larger size of the nanocrystals is explained by a relatively weaker cooling rate limited to the nearly 1D case due to the elongated geometry of the pillar-, column- or wall-like structures. The detailed analysis of the experimental picture (Figure 6b) indicates the polycrystalline character of the subsurface region via darker lines and spots due to different optical properties of the nanograin boundaries and the relative defects. While the experimentally observed and numerically simulated nanograin structures are in agreement with recent theoretical works [49,58,59], it is worth mentioning that the crystal structures obtained during the nanostructuring process due to extremely fast laser-induced cooling mechanism can be of unique character and never met in nature [58]. Thus, generation of such structures can be of great interest for potential technological applications.

Figure 8 illustrates that by increasing the incident fluence to a higher value of 500 mJ/cm^2 (well above the ablation threshold ~210 mJ/cm^2), the melting extends over the whole irradiated area (40×40) μm^2, and the contrast of the resulting surface pattern strongly diminishes (500-nm period) or totally vanishes (270-nm period). We refer to this observation as broad melting. In order to follow the development of an ultrafast melting process, induced in the material at such strong heating, we used here the LOP parameter that is very sensitive to the crystal structure disordering and allows for a reliable identification of the liquid phase [31,33]. Thus, in both Figure 8a,c, a well distinguished melting front due to propagation of a solid–liquid interface referred to as heterogeneous melting is not observed. Instead, due to significant overheating the phase transition proceeds by a massive nucleation of liquid phase inside the crystal referred to as homogeneous mechanism of the phase transition. Additionally, a number of dislocation planes, generated due to rapid expansion of the most heated material volume, is less resistant to the phase transition. This provides an additional input to the nucleation sites for the liquid phase, visible as red lines inside the solid phase shown in yellow/orange.

The experimental as well as the simulation results show that in addition to the laser induced upward motion of material, a lateral displacement of material plays an important role in the structure formation process. This observation is in agreement with a recent experimental and theoretical work on redistribution of the processed material volume during the femtosecond laser processing [51]. In contrast to material irradiation by a single focused beam to a small spot where the material can expand freely in lateral directions, in the case of a periodic lateral pattern, the material expulsions from neighboring grooves at high fluence affect each other. This leads to the observed specific features like steep ridges at a moderate incident fluence and a flatter pattern at higher energy deposition values.

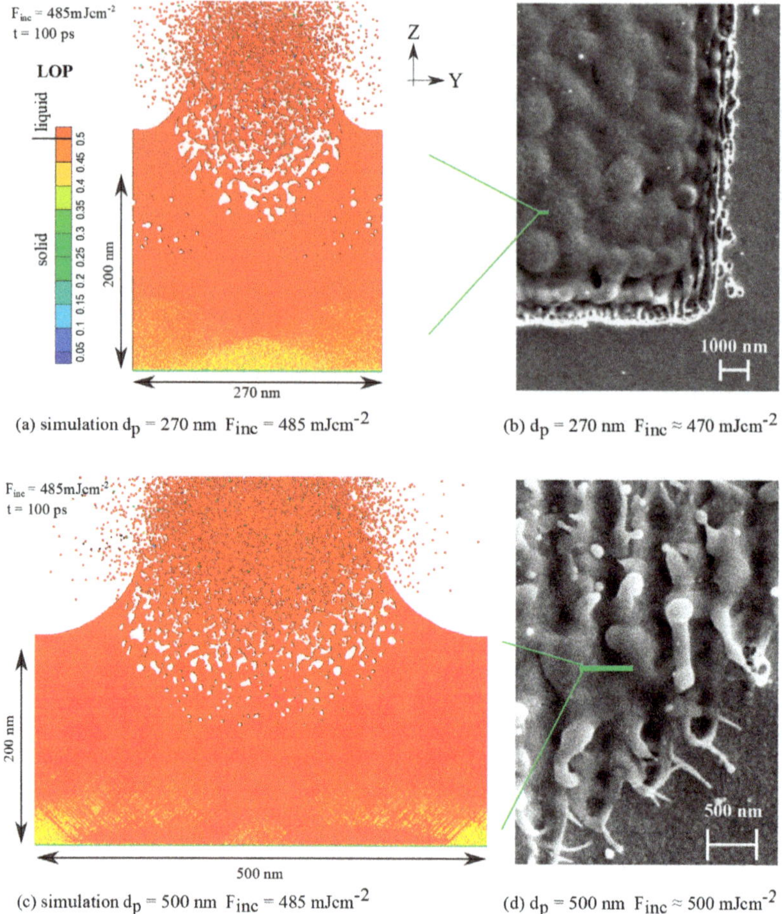

(a) simulation d_p = 270 nm F_{inc} = 485 mJcm^{-2}

(b) d_p = 270 nm F_{inc} ≈ 470 mJcm^{-2}

(c) simulation d_p = 500 nm F_{inc} = 485 mJcm^{-2}

(d) d_p = 500 nm F_{inc} ≈ 500 mJcm^{-2}

Figure 8. (**a**) Direct comparison of the simulation results of 270-nm period nanostructuring at a high incident fluence of 485 mJ/cm^2 with (**b**) the experimental data obtained for 470 mJ/cm^2. (**c**) Direct comparison of the simulation results of 500-nm period nanostructuring at a high incident fluence of 485 mJ/cm^2 with (**d**) the experimental data obtained for 500 mJ/cm^2. The atoms are colored by local order parameter (LOP) used here for identification of the extremely fast melting process. The liquid phase is characterized by a rapid jump of LOP above cut-off value 0.5.

Finally, the good agreement between the modeling results and the experimental data justifies the proposed approach as a powerful tool revealing the physics behind the nanostructuring process at a gold surface and providing a microscopic insight into the dynamics of the structuring processes of metals in general. Moreover, the results discussed above are applicable to other metals as well, whenever fast and localized laser energy deposition occurs. In particular, transition metals such as Al and Ni, due to a shorter characteristic distance (75–125nm as compared to Au 150–200nm) of the laser-deposited heat penetration [48] (heat affected zone), could be better candidates for the achievement of cleaner structures at a shorter structure period. Reducing the structure period below this distance, however, can deteriorate the resulting surface quality due to laser-deposited dissipated heat (governed by the fast electron heat conduction process) and therefore smoothing the resulting pattern.

5. Conclusions

The performed experiments of periodic patterning with a single UV ultrashort laser pulse and corresponding simulations with the combined atomistic-continuum model allowed us to draw the following conclusions about the mechanism responsible for the nanostructures formation: the outward hydrodynamic motion of the expelled melted surface layer and its subsequent opening result in a tangential component of the velocity and lead to merging of neighboring walls of the forming structures and the formation of nanoridges on the target surface. The complete solidification of the forming structures at the fluencies well above the ablation threshold as well as the solidification of the generated droplets takes much longer (tens of ns) time scale. The strong cooling rate of the forming structures is responsible for the formation of the nanofeatures with a polycrystalline character constituting the nanosize features of various shapes. The broad melting observed at high fluencies was identified as a reason for diminished contrast of the generated structures as a result of the molten material redistribution. The applied numerical model has been identified as a powerful numerical tool for the investigation of the ultrashort laser pulse nanostructuring process on metals. The results of the performed study can have a strong impact on potential industrial applications for the generation of material surfaces with predesigned properties and topography.

Author Contributions: P.S., J.I., B.R., M.E.G. conceived and designed the study. The Experiments were conducted by A.B. and supervised by P.S. and J.I. The MD-TTM Simulation was developed and run by D.S.I. The analyzation, comparison and discussion of the experimental and simulated data were done by A.B., D.S.I., B.R. The text was written by A.B., D.S.I. and J.I. All authors have read and agreed to the published version of the manuscript.

Funding: The presented work is completed under financial support by Deutsche Forschungsgemeinschaft (DFG) projects RE 1141/14-2, IV 122/4-1, IH 17/18-1,2 and GA 465/18-1. The MD simulations were performed at Lichtenberg Super Computer Facility TU-Darmstadt (Germany) and analyzed at the ITS computer center of the University of Kassel (Germany).

Acknowledgments: The TEM measurements were made by V. Roddatis at the University of Göttingen.

Conflicts of Interest: The authors declare no conflict of interest.

References

1. Bonse, J.; Höhm, S.; Kirner, S.V.; Rosenfeld, A.; Krüger, J. Laser-Induced Periodic Surface Structures—A Scientific Evergreen. *IEEE J. Selec. Top. Quantum Electron.* **2017**, *23*, 9000615. [CrossRef]
2. Reif, J.; Varlamova, O.; Varlamov, S.; Bestehorn, M. The role of asymmetric excitation in self-organized nanostructure formation upon femtosecond laser ablation. *Appl. Phys. A* **2011**, *104*, 969–973. [CrossRef]
3. Drogowska-Horna, K.A.; Mirza, I.; Rodriguez, A.; Kovaříček, P.; Sládek, J.; Derrien, T.J.Y.; Gedvilas, M.; Račiukaitis, G.; Frank, O.; Bulgakova, N.M. Periodic surface functional group density on graphene via laser-induced substrate patterning at Si/SiO$_2$ interface. *Nano Res.* **2020**, *13*, 2332–2339. [CrossRef]
4. Miyaji, G.; Miyazaki, K. Fabrication of 50-nm period gratings on GaN in air through plasmonic near-field ablation induced by ultraviolet femtosecond laser pulses. *Opt. Express* **2016**, *24*, 4648–4653. [CrossRef]
5. Murphy, R.D.; Torralva, B.; Adams, D.P.; Yalisove, S.M. Polarization dependent formation of femtosecond laser-induced periodic surface structures near stepped features. *Appl. Phys. Lett.* **2014**, *104*, 231117. [CrossRef]
6. Terekhin, P.N.; Benhayoun, O.; Weber, S.T.; Ivanov, D.S.; Garcia, M.E.; Rethfeld, B. Influence of Surface Plasmon Polaritons on Laser Energy Absorption and Structuring of Surfaces. *Appl. Surf. Sci.* **2020**, *512*, 144420. [CrossRef]
7. Ilcisin, K.J.; Fedosejevs, R. Direct production of gratings on plastic substrates using 248-nm KrF laser radiation. *Appl. Opt.* **1987**, *26*, 396. [CrossRef]
8. Phillips, H.M.; Sauerbrey, R.A. Excimer-laser-produced nanostructures in polymers. *Opt. Eng.* **1993**, *32*, 2424. [CrossRef]
9. Lasagni, A.F.; Acevedo, D.F.; Barbero, C.A.; Mücklich, F. One-Step Production of Organized Surface Architectures on Polymeric Materials by Direct Laser Interference Patterning. *Adv. Eng. Mat.* **2007**, *9*, 99. [CrossRef]

10. Wang, D.; Wang, Z.; Zhang, Z.; Yue, Y.; Li, D.; Qiu, R.; Maple, C. Both antireflection and superhydrophobicity structures achieved by direct laser interference nanomanufacturing. *J. Appl. Phys.* **2014**, *115*, 233101. [CrossRef]

11. Beckemper, S.; Huang, J.; Gillner, A.; Wang, K. Generation of periodic micro- and nano-structures by parameter-controlled three-beam laser interference technique. *J. Laser Micro Nanoen.* **2011**, *6*, 49. [CrossRef]

12. Tsunetomo, K.; Koyama, T. Direct formation of a surface-relief grating on glass by ultraviolet–visible laser irradiation. *Opt. Lett.* **1997**, *4*, 411–413. [CrossRef] [PubMed]

13. Simon, P.; Ihlemann, J. Machining of submicron structures on metals and semiconductors by ultrashort UV-laser pulses. *Appl. Phys. A* **1996**, *63*, 505–508. [CrossRef]

14. Kaakkunen, J.J.J.; Paivasaari, K.; Vahimaa, P. Fabrication of large-area hole arrays using high-efficiency two-grating interference system and femtosecond laser ablation. *Appl. Phys. A* **2011**, *103*, 267–270. [CrossRef]

15. Nakata, Y.; Momoo, K.; Miyanaga, N.; Hiromoto, T.; Tsuchida, K. Organized metamaterials comprised of gold nanoneedles in a lattice generated on silicon (100) wafer substrates by interfering femtosecond laser processing. *Appl. Phys. A* **2013**, *112*, 173–177. [CrossRef]

16. Ivanov, D.S.; Lipp, V.P.; Veiko, V.P.; Jakovlev, E.; Rethfeld, B.; Garcia, M.E. Molecular Dynamics Study of the Short Laser Pulse Ablation: Quality and Efficiency in Production. *Appl. Phys. A* **2014**, *117*, 2133–2141. [CrossRef]

17. Rethfeld, B.; Ivanov, D.S.; Garcia, M.E.; Anisimov, S.I. Modelling Ultrafast Laser Ablation. *J. Phys. D* **2017**, *50*, 193001. [CrossRef]

18. Inogamov, N.A.; Zhakhovsky, V.V.; Khokhlov, V.A.; Petrov, Y.V.; Migdal, K.P. Solitary Nanostructures Produced by Ultrashort Laser Pulse. *Nanoscale Res. Lett.* **2016**, *11*, 177. [CrossRef]

19. Apel, O.; Beinhorn, F.; Ihlemann, J.; Klein-Wiele, J.-H.; Marowsky, G.; Simon, P. Periodic nanostructures. *Z. Phys. Chem.* **2000**, *214*, 1233. [CrossRef]

20. Molotokaitė, E.; Gedvilas, M.; Račiukaitis, G.; Girdauskas, V. Picosecond laser beam interference ablation of thin metal film on glass substrate. *J. Laser Micro Nanoen.* **2010**, *5*, 74–79. [CrossRef]

21. Höche, T.; Böhme, R.; Gerlach, J.W.; Frost, F.; Zimmer, K.; Rauschenbach, B. Semiconductor Nanowires Prepared by Diffraction-Mask-Projection Excimer-Laser Patterning. *Nano Lett.* **2004**, *4*, 895–897. [CrossRef]

22. Nakata, Y.; Matsuba, Y.; Miyanaga, N. Sub-micron period metal lattices fabricated by interfering ultraviolet femtosecond laser processing. *Appl. Phys. A* **2016**, *122*, 532. [CrossRef]

23. Peláez, R.J.; Rebollar, E.; Serna, R.; Acosta-Zepeda, C.; Saavedra, P.; Bonse, J.; Haro-Poniatowski, E. Nanosecond laser-induced interference grating formation on silicon. *J. Phys. D Appl. Phys.* **2019**, *52*, 225302. [CrossRef]

24. Gedvilas, M.; Voisiat, B.; Regelskis, K.; Raciukaitis, G. Instability-triggered transformations in thin chromium film on glass under laser irradiation. *Appl. Surf. Sci.* **2013**, *278*, 26–32. [CrossRef]

25. Ivanov, D.S.; Rethfeld, B.C.; O'Connor, G.M.; Glynn, T.J.; Volkov, A.N.; Zhigilei, L.V. The Mechanism of Nanobump Formation in Femtosecond Pulse Laser Nanostructuring of Thin Metal Films. *Appl. Phys. A* **2008**, *92*, 791–796. [CrossRef]

26. Ashitkov, S.I.; Komarov, P.S.; Ovchinnikov, A.V.; Struleva, E.V.; Zhakhovsky, V.V.; Inogamov, N.A.; Agranat, M.B. Ablation and nanostructuring of metals by femtosecond laser pulses. *Quantum Electron.* **2014**, *44*, 535–539. [CrossRef]

27. Anisimov, S.I.; Zhakhovsky, V.V.; Inogamov, N.A.; Murzov, S.A.; Khokhlov, V.A. Formation and crystallisation of a liquid jet in a film exposed to a tightly focused laser beam. *Quantum Electron.* **2017**, *47*, 509–521. [CrossRef]

28. Fourment, C.; Deneuville, F.; Descamps, D.; Dorchies, F.; Petit, S.; Peyrusse, O.; Holst, B.; Recoules, V. Experimental Determination of Temperature-Dependent Electron-Electron Collision Frequency in Isochorically Heated Warm Dense Gold. *Phys. Rev. B* **2014**, *89*, 161110. [CrossRef]

29. Blumenstein, A.; Zijlstra, E.S.; Ivanov, D.S.; Weber, S.T.; Zier, T.; Kleinwort, F.; Rethfeld, B.; Ihlemann, J.; Simon, P.; Garcia, M.E. Transient Optics of Gold during Laser Irradiation: From First Principles to Experiment. *Phys. Rev. B* **2020**, *101*, 165140. [CrossRef]

30. Anisimov, S.I.; Kapeliovich, B.L.; Perel'man, T.L. Electron emission from metal surfaces exposed to ultrashort laser pulses. *Soviet Phys. JETP* **1974**, *39*, 375–377.

31. Ivanov, D.S.; Zhigilei, L.V. Combined atomistic-continuum modeling of short-pulse laser melting and disintegration of metal films. *Phys. Rev. B* **2003**, *68*, 064114. [CrossRef]

32. Ivanov, D.S.; Lipp, V.P.; Rethfeld, B.; Garcia, M.E. Molecular-dynamics study of the mechanism of short-pulse laser ablation of single-crystal and polycrystalline metallic targets. *J. Opt. Technol.* **2014**, *81*, 250–253. [CrossRef]

33. Ivanov, D.S.; Kuznetsov, A.I.; Lipp, V.P.; Rethfeld, B.; Chichkov, B.N.; Garcia, M.E.; Schulz, W. Short Laser Pulse Surface Nanostructuring on Thin Metal Films: Direct Comparison of Molecular Dynamics Modeling and Experiment. *Appl. Phys. A* **2013**, *111*, 675–687. [CrossRef]

34. Ivanov, D.S.; Blumenstein, A.; Ihlemann, J.; Simon, P.; Garcia, M.E.; Rethfeld, B. Molecular Dynamics Modeling of Periodic Nanostructuring of Metals with a Short UV Laser Pulse under Spatial Confinement by a Water Layer. *Appl. Phys. A* **2017**, *123*, 744. [CrossRef]

35. Shih, C.Y.; Gnilitskyi, I.; Shugaev, M.V.; Skoulas, E.; Stratakis, E.; Zhigilei, L.V. Effect of a liquid environment on single-pulse generation of laser induced periodic surface structures and nanoparticles. *Nanoscale* **2020**, *12*, 7674–7687. [CrossRef]

36. Iqbal, M.; Khan, S.A.; Ivanov, D.S.; Ganeev, R.A.; Kim, V.V.; Boltaev, G.S.; Abbasi, N.A.; Shaju, S.; Garcia, M.E.; Rethfeld, B.; et al. The mechanism of laser assisted generation of aluminum nanoparticles, their wettability and nonlinearity properties. *Appl. Surf. Sci.* **2020**, *527*, 146702. [CrossRef]

37. Ivanov, D.S.; Lipp, V.P.; Blumenstein, A.; Kleinwort, F.; Veiko, V.P.; Yakovlev, E.; Roddatis, V.; Garcia, M.E.; Rethfeld, B.; Ihlemann, J.; et al. Experimental and Theoretical Investigation of Periodic Nanostructuring of Au with Ultrashort UV Laser Pulses near the Damage Threshold. *Phys. Rev. Appl.* **2015**, *4*, 064006. [CrossRef]

38. Preuss, S.; Demchuk, A.; Stuke, M. Sub-picosecond UV laser ablation of metals. *Appl. Phys. A* **1995**, *61*, 33–37. [CrossRef]

39. Nagy, T.; Simon, P. Single-shot TG FROG for the characterization of ultrashort DUV pulses. *Opt. Express* **2009**, *17*, 8144–8151. [CrossRef]

40. Gilvarry, J.J. The Lindemann and Gruneisen laws. *Phys. Rev.* **1956**, *102*, 308. [CrossRef]

41. Waldecker, L.; Bertoni, R.; Ernstorfer, R.; Vorberger, J. Electron-Phonon Coupling and Energy Flow in a Simple Metal beyond the Two-Temperature Approximation. *Phys. Rev. X* **2016**, *6*, 021003. [CrossRef]

42. Mo, M.Z.; Chen, Z.; Li, R.K.; Dunning, M.; Witte, B.B.L.; Baldwin, J.K.; Fletcher, L.B.; Kim, J.B.; Ng, A.; Redmer, R.; et al. Heterogeneous to homogeneous melting transition visualized with ultrafast electron diffraction. *Science* **2018**, *360*, 1451–1455. [CrossRef] [PubMed]

43. Smirnov, N.A. Copper, gold, and platinum under femtosecond irradiation: Results of first-principles calculations. *Phys. Rev. B* **2020**, *101*, 094103. [CrossRef]

44. Lin, Z.; Zhigilei, L.V.; Celli, V. Electron-phonon coupling and electron heat capacity of metals under conditions of strong electron-phonon nonequilibrium. *Phys. Rev. B* **2008**, *77*, 075133. [CrossRef]

45. Medvedev, N.; Milov, I. Electron-phonon coupling in metals at high electronic temperatures. *Phys. Rev. B* **2020**, *102*, 064302. [CrossRef]

46. Anisimov, S.I.; Rethfeld, B. On the theory of ultrashort laser pulse interaction with a metal. *SPIE Proc.* **1997**, *3093*, 192.

47. Petrov, Y.V.; Inogamov, N.A.; Anisimov, S.I.; Migdal, K.P.; Khokhlov, V.A.; Khishchenko, K.V. Thermal conductivity of condensed gold in states with the strongly excited electron subsystem. *J. Phys. Conf. Ser.* **2015**, *653*, 012087. [CrossRef]

48. Wellershoff, S.-S.; Hohlfeld, J.; Güdde, J.; Matthias, E. The role of electron–phonon coupling in femtosecond laser damage of metals. *Appl. Phys. A* **1999**, *69*, S99–S107.

49. Wu, C.; Christensen, M.S.; Savolainen, J.-M.; Balling, P.; Zhigilei, L.V. Generation of sub-surface voids and a nanocrystalline surface layer in femtosecond laser irradiation of a single crystal Ag target. *Phys. Rev. B* **2015**, *91*, 035413. [CrossRef]

50. Nakata, Y.; Miyanaga, N.; Okada, T. Effect of pulse width and fluence of femtosecond laser on the size of nanobump array. *Appl. Surf. Sci.* **2007**, *253*, 6555–6557. [CrossRef]

51. Danilov, P.A.; Zayarny, D.A.; Ionin, A.A.; Kudryashov, S.I.; Rudenko, A.A.; Kuchmizhak, A.A.; Vitrik, O.B.; Yu, N.; Kulchin, V.V.; Zhakhovsky, N.A. Redistribution of a material at femtosecond laser ablation of a thin silver film. *JETP Lett.* **2016**, *104*, 759–765. [CrossRef]

52. Wu, C.; Zhigilei, L.V. Microscopic mechanisms of laser spallation and ablation of metal targets from large-scale molecular dynamics simulations. *Appl. Phys. A* **2014**, *114*, 11–32. [CrossRef]

53. Kuchmizhak, A.; Vitrik, O.; Kulchin, Y.; Storozhenko, D.; Mayor, A.; Mirochnik, A.; Makarov, S.; Milichko, V.; Kudryashov, S.; Zhakhovsky, V.; et al. Laser printing of resonant plasmonic nanovoids. *Nanoscale* **2016**, *8*, 12352–12361. [CrossRef] [PubMed]

54. Shugaev, M.V.; Gnilitskyi, I.; Bulgakova, N.M.; Zhigilei, L.V. Mechanism of single-pulse ablative generation of laser-induced periodic surface structures. *Phys. Rev. B* **2017**, *96*, 205429. [CrossRef]

55. Abou-Saleh, A.; Karim, E.T.; Maurice, C.; Reynaud, S.; Pigeon, F.; Garrelie, F.; Zhigilei, L.V.; Colombier, J.P. Spallation-induced roughness promoting high spatial frequency nanostructure formation on Cr. *Appl. Phys. A* **2018**, *124*, 308. [CrossRef]

56. Ivanov, D.S.; Rethfeld, B.C.; O'Connor, G.M.; Glynn, T.J.; Lin, Z.; Zhigilei, L.V. Nanocrystalline Structure of Nanobump Generated by Localized Photoexcitation of Metal Film. *J. Appl. Phys.* **2010**, *107*, 013519. [CrossRef]

57. Ivanov, D.S.; Zhigilei, L.V. Kinetic Limit of Heterogeneous Melting in Metals. *Phys. Rev. Lett.* **2007**, *98*, 195701. [CrossRef]

58. Wu, C.; Zhigilei, L.V. Nanocrystalline and polyicosahedral structure of a nanospike generated on metal surface irradiated by a single femtosecond laser pulse. *J. Phys. Chem. C* **2016**, *120*, 4438–4447. [CrossRef]

59. Shugaev, M.V.; Wu, C.; Armbruster, O.; Naghilou, A.; Brouwer, N.; Ivanov, D.S.; Derrien, J.-Y.; Bulgakova, N.M.; Kautek, W.; Rethfeld, B.; et al. Fundamentals of Ultrafast Laser–Material Interaction. *MRS Bull.* **2016**, *41*, 960–968. [CrossRef]

 *nanomaterials*

Article

Formation of the Submicron Oxidative LIPSS on Thin Titanium Films During Nanosecond Laser Recording

Dmitry A. Sinev [1,2,*], Daria S. Yuzhakova [1], Mikhail K. Moskvin [1] and Vadim P. Veiko [1,2]

[1] Faculty of Laser Photonics and Optoelectronics, ITMO University, 49 Kronverksky Pr.,
    bldg. A, 197101 St. Petersburg, Russia; dsyuzhakova@itmo.ru (D.S.Y.); mkmoskvin@itmo.ru (M.K.M.);
    vadim.veiko@mail.ru (V.P.V.)
[2] Institute of Automation and Electrometry of the Siberian Branch of the Russian Academy of Sciences
    (IA&E SB RAS), 1 Academician Koptyug ave., 630090 Novosibirsk, Russia
* Correspondence: sinev@itmo.ru

Received: 1 October 2020; Accepted: 27 October 2020; Published: 29 October 2020

Abstract: Laser-induced periodic surface structures (LIPSSs) spontaneously appearing on the laser-treated (melted or evaporated) surfaces of bulk solid materials seem to be a well-studied phenomenon. Peculiarities of oxidative mechanisms of LIPSS formation on thin films though are far less clear. In this work, the appearance of oxidative LIPSSs on thin titanium films was demonstrated under the action of commercially available nanosecond-pulsed Yb-fiber laser. The temperature and energy regimes favoring their formation were revealed, and their geometric characteristics were determined. The period of these LIPSSs was found to be about 0.7 λ, while the modulation depth varied between 70 and 110 nm, with high stability and reproducibility. It was shown that LIPSS orientation is rather easily manageable in the regimes of our interest, which could provide a way of controlling their properties.

Keywords: LIPSS; LSFL; laser thermochemical recording; titanium films; laser-induced oxidation

---

1. Introduction

The formation of laser-induced periodic surface structures (LIPSSs), which was observed on a wide range of materials under the action of laser radiation, is a well-known phenomenon that attracts both fundamental and applied attention in modern photonics. LIPSS submicron reliefs are used to control optical [1–4], mechanical [5], chemical [6–8], and other properties of functional surfaces. A significant number of studies have been devoted to the study of LIPSSs appearing under pico- and femtosecond laser exposure [4,9–11], often localized in small areas of the irradiated region. LIPSS uniformity and long-range order are also usually disturbed [11] by the simultaneous initiation from multiple independent seed locations [12]. In general, there are two types of LIPSSs: structures with a period comparable to the radiation wavelength λ (low-spatial-frequency LIPSS (LSFLs)) and small-scale structures, the period of which is much less than λ (high-spatial-frequency LIPSSs (HSFLs)) [13]. In [14], LSFLs were obtained on a Cr film under femtosecond exposure to Yb:KGW laser radiation (wavelength 1026 nm, pulse duration 232 fs) in a nondestructive regime. Those structures, however, significantly protruded above the surface of the initial film, which according to the authors indicated the significant oxidation of the irradiated metal in the areas of the protrusions.

Despite the fact that the mechanisms, controllability, and reproducibility of the LIPSS appearance, presumably, should not significantly depend on the duration of laser exposure, studies of LIPSS formation under nanosecond pulses [15–17] are uncommon. This is due to the narrow range of recording parameters and the sensitivity of LIPSS characteristics to the physical mechanisms of their formation, which are easier to identify under a short exposure. The physical mechanisms of LIPSS formation have been discussed in many works, e.g., [13,18–20], mainly, however, for bulk metals irradiated by ultrashort pulses with energy densities above the ablation threshold. The mechanisms of LIPSS formation on thin metal films in nondestructive regimes have been studied less thoroughly, although these films are widely used for different applications—from diffraction optical components recording to color laser marking [21]—which increases the interest in the phenomena (oxidation, melting, ablation, etc.) of their laser treatment. The formation of oxidative LIPSSs in [12,14,22] can be represented as the result of the interference of the incident laser radiation, with that scattered by the initial film roughness. Interference leads to the formation of a periodic absorption profile and appropriate modulation of the temperature distribution along the surface. At the maxima of the temperature profile, oxidation occurs more rapidly due to the nonlinear nature of the oxidation process [23], which completes a positive feedback. In addition, the protruding oxide lines and dots can serve as cylindrical or spherical convex nano- or microlenses that redistribute and concentrate the energy of the next laser pulses closer to the lower oxide–metal interface, thereby exacerbating temperature modulation and promoting the formation of oxidative LIPSSs.

In [12], it was suggested that the usage of ultrashort pulses is necessary for LIPSS formation, in order to ensure that heat diffusion does not smear out the nanometer-scale localization of the deposited laser energy. In this work, we present experimental results showing that regular oxidative LIPSSs, which have not been previously observed in the range of low-temperature (below the melting point) processes on thin titanium films, can be obtained under the action of nanosecond laser pulses. Recording, in this case, was conducted using the method of direct laser thermochemical writing on thin metal films (for instance, on Ti [21,23–27], Sn [28], V [29], Zr [29], etc.), in which the image is created by direct oxidation of the original film under local laser heating until the formation of a contrasting transparent oxide layer (for further information about this method, see [23]). The formation of LIPSSs accompanies the oxidation of the initial film and creates an additional relief over the recorded image.

Since the main field of application of the technology of laser thermochemical recording on thin metal films is recording diffractive optical elements [23,30], self-organization of LIPSSs in the irradiated region can be regarded either as a defect (leading to distortion of the wavefront of radiation interacting with the diffractive element) or, on the contrary, as an advantage (as it makes it easier to form structures with a period on the order of the wavelength), with the prospects of being used for direct writing of photonic crystals, specific diffractive elements, etc.

2. Materials and Methods

To obtain LIPSSs, in the present work, we used titanium films with the initial transmission in the visible range of about 10% (thickness on the order of 10 nm) on a BK7 glass substrate with a thickness of 2.85 mm. Yb-fiber laser ($\lambda \approx 1.07$ μm) with random (partial) polarization and Gaussian spatial distribution of radiation was used as a radiation source, as part of the MiniMarker laser processing complex (Laser Center Ltd., St. Petersburg, Russia) (Figure 1a,b, for technical parameters see [31]).

For determining the regimes suitable for obtaining LIPSSs, the following parameters were varied: pulse duration τ (from 4 to 50 ns), average power P_{av} (from 20 to 400 mW), and pulse repetition rate f (from 10 to 100 kHz). Ti films were irradiated with a scanning focused laser beam; the scanning speed V remained constant and equal to 0.1 mm/s. With an increase in the scanning speed, LIPSS formation was not registered, possibly due to a decrease in the number of pulses per unit area, down to values at which the oxide did not have enough time to form. Laser radiation was initially partially polarized, and in some experiments a Glan–Taylor prism was introduced into the optical scheme, changing the polarization to mostly linear. The diameter of a Gaussian laser beam on the surface of the film d was about 80 μm, and the irradiation of the film was carried out perpendicularly to the surface. Note that under the studied conditions, the pulse overlap coefficient calculated by formula [32]

$$k = \left(1 - \frac{V}{fd}\right) \times 100\%$$ (1)

exceeded 99%, and the distance between the centers of neighboring irradiated zones under the action of successive pulses was on the order of 1–2 nm, which, as will be seen below, does not coincide with the period of emerging LIPSSs. Thus, the LSFLs obtained in this work were indeed formed by the mechanism of LIPSS formation and do not represent the edges of the heat-affected zone (HAZ) from neighboring pulses or other similar structures. The obtained structures were examined using the Carl Zeiss Axio Imager A1.m optical microscope (Carl Zeiss Microscopy GmbH, Munich, Germany) and the AFM NT-MDT Nanoeducator (LLC "NT-MDT", Moscow, Russia). 2D-FFT analysis of optical micrographs was carried out using a homemade image processing program written in Python. The FLIR Titanium 520M (FLIR Systems, Inc., Wilsonville, OR, USA) thermal imaging camera was used to estimate the temperature on the film surface during irradiation, and the power meter Gentec-EO SOLO2 (Gentec Electro-Optics, Inc., Quebec, Canada) was used to measure the average laser power.

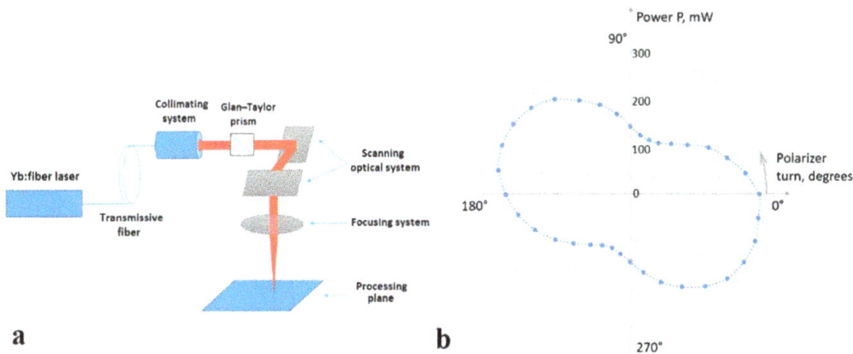

Figure 1. (**a**) Experimental setup diagram; (**b**) results of measuring the power of laser radiation after the optical system depending on the rotation angle of the polarizer (Glan–Taylor prism).

3. Results and Discussion

Figure 2 shows optical and AFM micrographs of structures recorded at pulse durations τ from 4 to 8 ns and fluences ε from 20 to 44 mJ/cm^2, as well as the corresponding measured temperature distributions. There was no dependence of the LIPSS period on the laser energy parameters to be found; in all the cases studied, the LIPSS period was about 0.72 ± 0.02 µm, which corresponds to the values acquired for the films of similar thickness in [12]. The AFM results (Figure 2g–i) show that the LIPSS amplitude also averages at 70–110 nm without clear dependence on the exposure parameters, which indicates the reproducibility and stability of such structures.

Based on the data obtained, it was determined that the optimal value of the average energy density for the appearance of LIPSSs in the oxidative mode lies in the range from 20 to 60 mJ/cm^2, while the experimentally determined average temperature on the surface does not exceed 350 °C, based on thermal imaging pictures (Figure 2j–l). The peak temperature value reached in the center of the irradiated region on the film at the end of a single pulse was estimated by the following formula [32]:

$$T_1 = \frac{qA_1\tau}{\rho_1 c_1 h\left(1 + \frac{\sqrt{\pi}}{2\psi}\right)} + T_{in}, \tag{2}$$

where q is the intensity of laser radiation; A is the absorbance coefficient; τ is the pulse duration; ρ is the density; c is the thermal capacity; a is the thermal diffusivity of the materials; $\psi = \frac{\rho_1 c_1 h}{\rho_2 c_2 \sqrt{a_2 t}}$ is the coefficient, defining the amount of heat transfer to the substrate; index "1" is related to the metal film; index "2" is related to the oxide layer; and T_{in} is the initial temperature (≈20 °C). The values of the physical and optical parameters of the materials used in the calculations are given in Table 1.

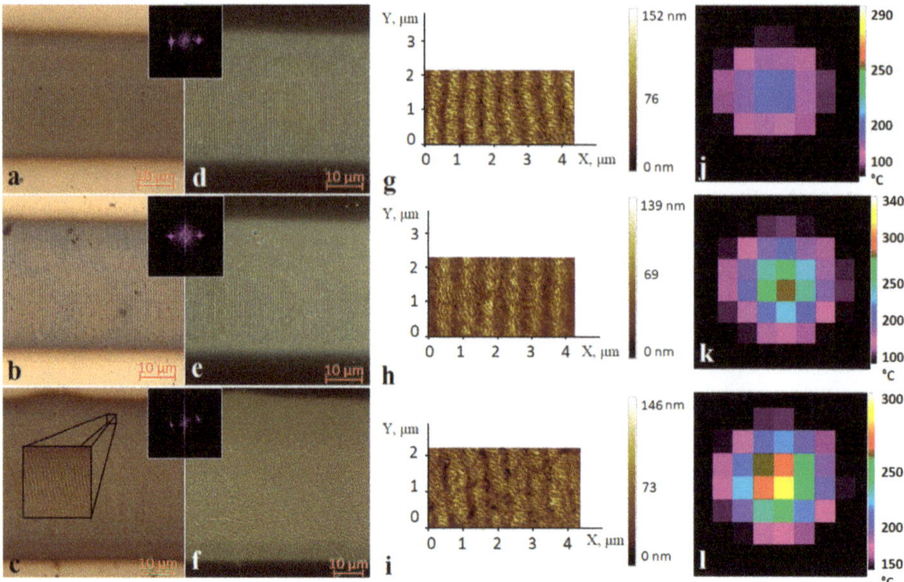

Figure 2. Optical micrographs of the recorded oxide microstructures in reflected (**a–c**) and transmitted (**d–f**) light, AFM images of surface fragments containing laser-induced periodic surface structures (LIPSSs) (**g–i**), and the corresponding thermal images (**j–l**). Insets show 2D-FFT spectra of fragments of the corresponding images (spatial frequencies up to 2 μm^{-1}). The scanning direction by the laser beam in the images (**a–f**) is from left to right. All recordings were conducted with initial partially polarized laser beam. Recording regimes: (**a,d,g,j**) τ = 4 ns, P_{av} = 90 mW, f = 50 kHz, ε = 20 mJ/cm^2; (**b,e,h,k**) τ = 4 ns, P_{av} = 140 mW, f = 100 kHz, ε = 24 mJ/cm^2; (**c,f,i,l**) τ = 8 ns, P_{av} = 160 mW, f = 50 kHz, ε = 44 mJ/cm^2.

Table 1. Physical and optical characteristics of titanium and BK7 glass.

Titanium [33,34]	
Density, p_1	4.5×10^3 kg/m^3
Thermal capacity, c_1	530.8 J/(kg·K)
Absorbance coefficient A_1 at λ = 1.07 µm	0.42
BK7 glass [35]	
Density, p_1	2.5×10^3 kg/m^3
Thermal capacity, c_2	720 J/(kg·K)
Thermal diffusivity, a_2	0.6×10^{-6} m^2/s

Since titanium oxides are substantially transparent in the visible and near-IR ranges, the estimates obtained by formula (2) using the absorbance of the initial metal film A_1 show the upper limit of possible temperature values. Heat accumulation effects under the studied conditions are negligible due to heat dissipation into the substrate during the time between pulses. Model estimations of peak temperature values (as well as experimental average evaluations (Figure 2i–l)) show numbers significantly lower than titanium melting point (1660 °C [33]) and, to be precise, range from 470 to 1065 °C for pulse durations of 4–14 ns within the experimentally determined fluence regimes of oxidation with the appearance of clearly defined LIPSSs. Interference modulation during LIPSS formation may have increased the peak temperature values in the maxima of the interference pattern, but microscopy results (Figure 2a–i) show no evidence of melting, or other signs of thermally-induced aberrations, which makes plausible the nonablative, oxidative nature of the LIPSS formation process.

Although direct comparison of our results with those from other sources is difficult due to differences in selected materials or exposure durations, LIPSSs with similar quality were formed under laser irradiation of the same wavelength, and average powers of the same order, in the benchmark article [12]. However, the laser spot coverage area size in our case is 7 times larger (80 μm vs. 12 μm), and scanning speed is far higher (100 μm/s vs. 2–8 μm/s), which results in much more efficient LIPSS recording, all without violation of long-range order. In addition, the usage of a commercially available nanosecond-pulsed Yb-fiber laser system instead of a femtosecond laser system is economically beneficial.

The well-known result, according to which the orientation axis of the LSFLs is defined by the polarization axis of laser radiation [13], was experimentally confirmed using a Glan–Taylor prism. The original partially polarized laser beam (Figure 3a) was converted using a prism to a linearly polarized one, and the angle of the LIPSS orientation corresponded to the rotation angle of the prism polarization axis. Based on the results of 2D-FFT analysis for structures formed under the action of linearly polarized radiation (Figure 3b–e), the period was found to be the same, i.e., 0.72 ± 0.02 μm. Separately, it was noted that the scanning direction (left-to-right, top-down) of the linearly polarized laser beam does not affect the rotation angle of the LIPSSs.

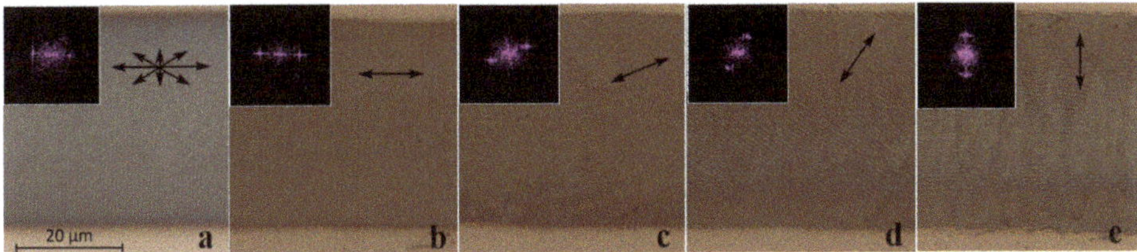

Figure 3. Optical micrographs of LIPSSs recorded in the same energy regime: τ = 4 ns, P_{av} = 90 mW, f = 50 kHz, ε = 20 mJ/cm². (**a**) Original partially polarized laser beam. (**b–e**) Linearly polarized radiation, where double-sided arrows indicate the direction of the polarization plane (at 0, 30, 60, and 90° to the scanning direction respectively). Insets show 2D-FFT spectra of fragments of the corresponding images (spatial frequencies up to 2 μm⁻¹).

The experimentally defined regimes favorable for the formation the oxidative LIPSSs on the thin metallic titanium films are shown in Figure 4a. With an increase in the duration or fluence of the pulses, a shift of the LIPSSs from the center of the irradiated region to the zone with a lower radiation intensity was observed (see Figures 2c and 4c), until their complete disappearance. The optimal repetition rate of nanosecond pulses lies in the range from 40 to 70 kHz (Figure 4d); however, this range expands with a reduction in the duration of a single pulse. Thus, contrast, ordered, and regular LIPSSs were observed when exposed to pulses with duration of 4 ns in a wide range of their repetition rates from 30 to 100 kHz. Observed restrictions on the pulse repetition rates are most likely to be explained by their connection with laser fluence. A decrease in the pulse repetition rate while maintaining the average power leads to an increase in the fluence and, consequently, an increase in the maximum peak value of the film temperature. This is followed by an increase of thermal-induced stresses, and, finally, the experimentally observed cracking of the film material and the disappearance of the regular LIPSSs (Figure 4b). An increase in the pulse repetition rate beyond the optimal range most likely leads to a decrease in the average fluence below the threshold values required for the LIPSS excitation. Reproduction of the LIPSS orientation as a result of positive feedback was also demonstrated by recording several tracks with a slight overlap (about 8%) (Figure 5). It can be noted that the direction, rotation angle, and location of the LIPSSs were reproduced on different tracks recorded sequentially and independently of each other. This reproducibility can be useful for functional structuring of the film surface under the considered conditions.

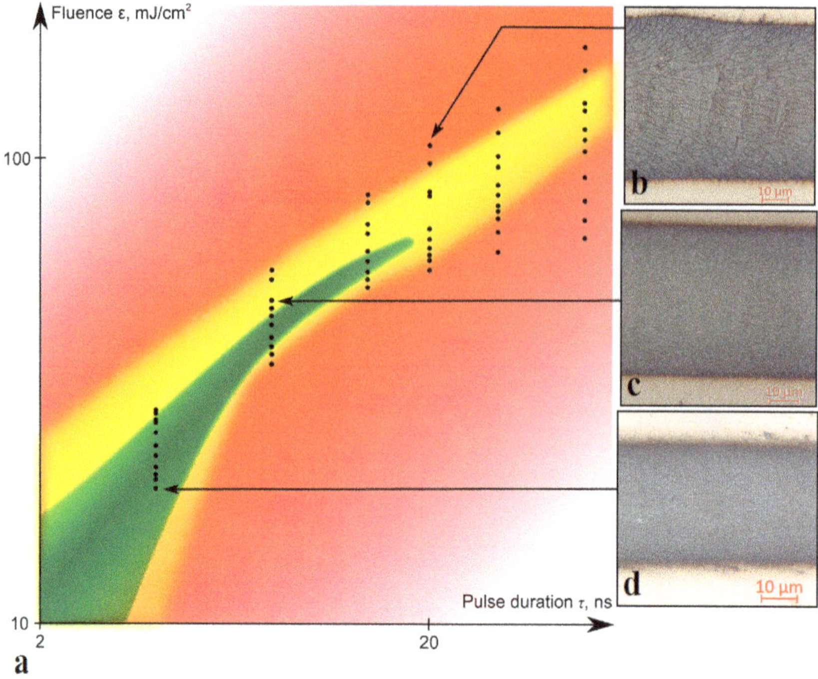

Figure 4. (**a**) Experimentally determined recording regimes for obtaining oxidative LIPSSs on titanium films of the selected thickness: the green area is homogeneous, with contrast structures over the entire track area; the yellow area has structures predominantly along the edges of the tracks; the red area has no structures or film cracking over the entire track area. Black dots mark the experimentally obtained results. Recording regimes (with the corresponding points marked in (**a**)): (**b**) $\tau = 20$ ns, $P_{av} = 100$ mW, $f = 15$ kHz, $\varepsilon = 100$ mJ/cm^2; (**c**) $\tau = 8$ ns, $P_{av} = 100$ mW, $f = 30$ kHz, $\varepsilon = 48$ mJ/cm^2; (**d**) $\tau = 4$ ns, $P_{av} = 100$ mW, $f = 60$ kHz, $\varepsilon = 23$ mJ/cm^2.

Figure 5. Fragment of a structure consisting of sequentially recorded tracks with LIPSSs using an initial partially polarized laser beam. Recording regime: $\tau = 4$ ns, $P_{av} = 90$ mW, $f = 50$ kHz, $\varepsilon = 20$ mJ/cm^2 (for a single track see Figure 2a). Sample photo is shown in the inset.

4. Conclusions

In the presented work, we demonstrated the formation of oxidative LIPSSs on titanium films under the action of nanosecond pulses. The mechanism of LIPSS formation under the considered conditions confirms the model proposed earlier in [12,14,22], which is based on the interference of the initial laser radiation, with that scattered by the initial film roughness. The results of AFM, optical microscopy, and subsequent 2D-FFT analysis show that the depth of the oxidative relief formed on the titanium film is stable, with an average of 90 nm, and its period is about 0.7 λ. It is significant that the formation of LIPSSs occurs in the preablative oxidative temperature regime at temperatures below the melting point, due to which the structures have high reproducibility and order (deviations in the value of the period do not exceed 0.02 μm). The LIPSS reproducibility is also facilitated by the recording method, in which the formation of the relief occurs under the influence of 24,000–80,000 nanosecond pulses (for 0.8 s) that successively irradiate the affected area. The action of subsequent pulses consolidates and significantly enhances the relief formed as a result of the scattering of the first pulses on the initial roughness of the film. This positive feedback contributes to the appearance of LIPSSs (under the influence of partially polarized radiation) with the predominant formation direction. The experimentally defined optimal regimes for the LIPSS formation were pulse durations in the range of 4–14 ns and fluences in the range of 10–70 mJ/cm^2, while the range of the latter is limited by fluence values sufficient for the excitation of LIPSSs but not exceeding the films' cracking thresholds. It was shown that since the pulse repetition rate is a technological factor that affects the energy density, it can be used to control the occurrence of LIPSSs (for example, under studied conditions, the frequency values suitable for the formation of LIPSS were 30–70 kHz at τ = 8 ns ceteris paribus). Yb-laser as part of the MiniMarker laser processing complex is a commercially-available, economically competitive system that allows obtaining high-quality uniform structures far more efficiently than by using a femtosecond laser system [12]. Controlling the direction of formation of the oxidative relief under the studied conditions can be achieved relatively easily, simply by changing the scanning direction of a partially polarized laser beam or by rotating the plane of polarization of radiation, which opens up new possibilities for applied surface functionalization.

Author Contributions: Methodology, D.A.S.; resources, D.A.S; writing—review and editing, D.A.S., M.K.M. and V.P.V.; visualization, D.A.S. and D.S.Y.; investigation, D.S.Y. and M.K.M.; software, D.S.Y.; validation, D.S.Y; writing—original draft, D.S.Y; conceptualization, V.P.V.; project administration, D.A.S.; funding acquisition, D.A.S. and V.P.V.; supervision, V.P.V. All authors have read and agreed to the published version of the manuscript.

Funding: This work was supported by the Russian Science Foundation grant #17-19-01721-P.

Conflicts of Interest: The authors declare no conflict of interest.

References

1. Veiko, V.; Karlagina, Y.; Moskvin, M.; Mikhailovskii, V.Y.; Одинцова, Г.; Olshin, P.; Pankin, D.; Romanov, V.; Yatsuk, R. Metal surface coloration by oxide periodic structures formed with nanosecond laser pulses. *Opt. Lasers Eng.* **2017**, *96*, 63–67. [CrossRef]

2. Ageev, E.I.; Veiko, V.P.; Vlasova, E.A.; Karlagina, Y.Y.; Krivonosov, A.; Moskvin, M.K.; Odintsova, G.V.; Pshenichnov, V.E.; Romanov, V.V.; Yatsuk, R.M. Controlled nanostructures formation on stainless steel by short laser pulses for products protection against falsification. *Opt. Express* **2018**, *26*, 2117–2122. [CrossRef] [PubMed]

3. Kunz, C.; Engel, S.; Müller, F.A.; Gräf, S. Large-Area Fabrication of Laser-Induced Periodic Surface Structures on Fused Silica Using Thin Gold Layers. *Nanomaterials* **2020**, *10*, 1187. [CrossRef] [PubMed]

4. San-Blas, A.; Martinez-Calderon, M.; Buencuerpo, J.; Sanchez-Brea, L.; Del Hoyo, J.; Gómez-Aranzadi, M.; Rodríguez, A.; Olaizola, S. Femtosecond laser fabrication of LIPSS-based waveplates on metallic surfaces. *Appl. Surf. Sci.* **2020**, *520*, 146328. [CrossRef]

5. Vorobyev, A.; Guo, C. Femtosecond laser structuring of titanium implants. *Appl. Surf. Sci.* **2007**, *253*, 7272–7280. [CrossRef]

6. Klos, A.; Sedao, X.; Itina, T.E.; Helfenstein-Didier, C.; Donnet, C.; Peyroche, S.; Vico, L.; Guignandon, A.; Dumas, V. Ultrafast Laser Processing of Nanostructured Patterns for the Control of Cell Adhesion and Migration on Titanium Alloy. *Nanomaterials* **2020**, *10*, 864. [CrossRef]

7. Martínez-Calderon, M.; Rodríguez, A.; Dias-Ponte, A.; Morant-Miñana, M.; Gómez-Aranzadi, M.; Olaizola, S. Femtosecond laser fabrication of highly hydrophobic stainless steel surface with hierarchical structures fabricated by combining ordered microstructures and LIPSS. *Appl. Surf. Sci.* **2016**, *374*, 81–89. [CrossRef]

8. Long, J.; Fan, P.; Zhong, M.; Zhang, H.; Xie, Y.; Lin, C. Superhydrophobic and colorful copper surfaces fabricated by picosecond laser induced periodic nanostructures. *Appl. Surf. Sci.* **2014**, *311*, 461–467. [CrossRef]

9. Nathala, C.S.; Ajami, A.; Ionin, A.A.; Kudryashov, S.I.; Makarov, S.V.; Ganz, T.; Assion, A.; Husinsky, W. Experimental study of fs-laser induced sub-100-nm periodic surface structures on titanium. *Opt. Express* **2015**, *23*, 5915–5929. [CrossRef]

10. Han, W.; Jiang, L.; Li, X.; Liu, P.; Xu, L.; Guo, L. Continuous modulations of femtosecond laser-induced periodic surface structures and scanned line-widths on silicon by polarization changes. *Opt. Express* **2013**, *21*, 15505–15513. [CrossRef]

11. Bonse, J.; Höhm, S.; Rosenfeld, A.; Kruger, J. Sub-100-nm laser-induced periodic surface structures upon irradiation of titanium by Ti:sapphire femtosecond laser pulses in air. *Appl. Phys. A* **2012**, *110*, 547–551. [CrossRef]

12. Öktem, B.; Pavlov, I.; Ilday, S.; Kalaycıoğlu, H.; Rybak, A.; Yavaş, S.; Erdoğan, M.; Ilday, F. Ömer Nonlinear laser lithography for indefinitely large-area nanostructuring with femtosecond pulses. *Nat. Photon.* **2013**, *7*, 897–901. [CrossRef]

13. Bonse, J.; Hohm, S.; Kirner, S.V.; Rosenfeld, A.; Kruger, J. Laser-Induced Periodic Surface Structures— A Scientific Evergreen. *IEEE J. Sel. Top. Quantum Electron.* **2016**, *23*, 1–15. [CrossRef]

14. Dostovalov, A.V.; Derrien, T.J.-Y.; Lizunov, S.A.; Přeučil, F.; Okotrub, K.A.; Mocek, T.; Korolkov, V.P.; Babin, S.A.; Bulgakova, N.M. LIPSS on thin metallic films: New insights from multiplicity of laser-excited electromagnetic modes and efficiency of metal oxidation. *Appl. Surf. Sci.* **2019**, *491*, 650–658. [CrossRef]

15. Одинцова, Г.; Andreeva, Y.M.; Salminen, A.; Roozbahani, H.; Van Cuong, L.; Yatsuk, R.; Golubeva, V.; Romanov, V.; Veiko, V.; Luong, V.C. Investigation of production related impact on the optical properties of color laser marking. *J. Mater. Process. Technol.* **2019**, *274*, 116263. [CrossRef]

16. Kang, M.J.; Park, T.S.; Kim, M.; Hwang, E.S.; Kim, S.H.; Shin, S.T.; Cheong, B.-H. Periodic surface texturing of amorphous-Si thin film irradiated by UV nanosecond laser. *Opt. Mater. Express* **2019**, *9*, 4247–4255. [CrossRef]

17. Reyes-Contreras, A.; Camacho-López, M.A.; Camacho-López, S.; Olea-Mejía, O.; Esparza-García, A.; Bañuelos-Muñetón, J.G. Laser-induced periodic surface structures on bismuth thin films with ns laser pulses below ablation threshold. *Opt. Mater. Express* **2017**, *7*, 1777. [CrossRef]

18. Dusser, B.; Sagan, S.; Soder, H.; Faure, N.; Colombier, J.-P.; Jourlin, M.; Audouard, E. Controlled nanostructrures formation by ultra fast laser pulses for color marking. *Opt. Express* **2010**, *18*, 2913–2924. [CrossRef]

19. Makin, V.S.; Logacheva, E.I.; Makin, R.S. Localized surface plasmon polaritons and nonlinear overcoming of the diffraction optical limit. *Opt. Spectrosc.* **2016**, *120*, 610–614. [CrossRef]

20. Ehrhardt, M.; Han, B.; Frost, F.; Lorenz, P.; Zimmer, K. Generation of laser-induced periodic surface structures (LIPSS) in fused silica by single NIR nanosecond laser pulse irradiation in confinement. *Appl. Surf. Sci.* **2019**, *470*, 56–62. [CrossRef]

21. Andreeva, Y.M.; Luong, V.C.; Lutoshina, D.S.; Medvedev, O.S.; Mikhailovskii, V.Y.; Moskvin, M.K.; Одинцова, Г.; Romanov, V.V.; Shchedrina, N.N.; Veiko, V.P. Laser coloration of metals in visual art and design. *Opt. Mater. Express* **2019**, *9*, 1310–1319. [CrossRef]

22. Florian, C.; Déziel, J.-L.; Kirner, S.V.; Siegel, J.; Bonse, J. The Role of the Laser-Induced Oxide Layer in the Formation of Laser-Induced Periodic Surface Structures. *Nanomaterials* **2020**, *10*, 147. [CrossRef] [PubMed]

23. Veiko, V.P.; Shakhno, E.A.; Sinev, D.A. Laser thermochemical writing: Pursuing the resolution. *Opt. Quantum Electron.* **2016**, *48*, 322. [CrossRef]

24. Wang, Y.; Wang, R.; Guo, C.F.; Miao, J.; Tian, Y.; Ren, T.; Liu, Q. Path-directed and maskless fabrication of ordered TiO2 nanoribbons. *Nanoscale* **2012**, *4*, 1545–1548. [CrossRef]

25. Veiko, V.P.; Zakoldaev, R.A.; Shakhno, E.A.; Sinev, D.A.; Nguyen, Z.K.; Baranov, A.V.; Bogdanov, K.V.; Gedvilas, M.; Raciukaitis, G.; Vishnevskaya, L.V.; et al. Thermochemical writing with high spatial resolution on Ti films utilising picosecond laser. *Opt. Mater. Express* **2019**, *9*, 2729–2737. [CrossRef]

26. Veiko, V.P.; Nguyen, Q.D.; Shakhno, E.A.; Sinev, D.A.; Lebedeva, E.V. Physical similarity of the processes of laser thermochemical recording on thin metal films and modeling the recording of submicron structures. *Opt. Quantum Electron.* **2019**, *51*, 348. [CrossRef]

27. Shakhno, E.A.; Sinev, D.A.; Kulazhkin, A.M. Features of laser oxidation of thin films of titanium. *J. Opt. Technol.* **2014**, *81*, 298. [CrossRef]

28. Guo, C.F.; Cao, S.; Jiang, P.; Fang, Y.; Zhang, J.; Fan, Y.; Wang, Y.; Xu, W.; Zhao, Z.; Liu, Q. Grayscale photomask fabricated by laser direct writing in metallic nano-films. *Opt. Express* **2009**, *17*, 19981–19987. [CrossRef]

29. Korolkov, V.P.; Sedukhin, A.G.; Mikerin, S.L. Technological and optical methods for increasing the spatial resolution of thermochemical laser writing on thin metal films. *Opt. Quantum Electron.* **2019**, *51*, 389. [CrossRef]

30. Veiko, V.P.; Korolkov, V.P.; Poleshchuk, A.G.; Sinev, D.A.; Shakhno, E.A. Laser technologies in micro-optics. Part 1. Fabrication of diffractive optical elements and photomasks with amplitude transmission. *Optoelectron. Instrum. Data Process.* **2017**, *53*, 474–483. [CrossRef]

31. Laser Center. Available online: https://www.newlaser.ru/laser/lc/minimarker2.php (accessed on 27 September 2020).

32. Metev, S.M.; Veiko, V.P. *Laser-Assisted Microtechnology*; Springer: Berlin, Germany, 1998; p. 180.

33. Kikoin, I.K. *Tables of Physical Quantities*; Atomizdat: Moscow, Russia, 1976.

34. Ordal, M.A.; Bell, R.J.; Alexander, R.W.; Newquist, L.A.; Querry, M.R. Optical properties of Al, Fe, Ti, Ta, W, and Mo at submillimeter wavelengths. *Appl. Opt.* **1988**, *27*, 1203–1209. [CrossRef]

35. Gedvilas, M.; Voisiat, B.; Indrisiunas, S.; Raciukaitis, G.; Veiko, V.; Zakoldaev, R.A.; Sinev, D.; Shakhno, E. Thermo-chemical microstructuring of thin metal films using multi-beam interference by short (nano- & picosecond) laser pulses. *Thin Solid Films* **2017**, *634*, 134–140. [CrossRef]

Publisher's Note: MDPI stays neutral with regard to jurisdictional claims in published maps and institutional affiliations.

Article

Surface Superconductivity Changes of Niobium Sheets by Femtosecond Laser-Induced Periodic Nanostructures

Álvaro Cubero [1], Elena Martínez [1,*], Luis A. Angurel [1], Germán F. de la Fuente [1], Rafael Navarro [1], Herbert Legall [2], Jörg Krüger [2] and Jörn Bonse [2]

[1] Instituto de Nanociencia y Materiales de Aragón (INMA), CSIC-Universidad de Zaragoza, 50009 Zaragoza, Spain; acubero@unizar.es (Á.C.); angurel@unizar.es (L.A.A.); german.delafuente.leis@csic.es (G.F.d.l.F.)

[2] Bundesanstalt für Materialforschung und -prüfung (BAM), Unter den Eichen 87, 12205 Berlin, Germany; herbert.legall@bam.de (H.L.); joerg.krueger@bam.de (J.K.); joern.bonse@bam.de (J.B.)

* Correspondence: elenamar@unizar.es

Received: 23 November 2020; Accepted: 14 December 2020; Published: 16 December 2020

Abstract: Irradiation with ultra-short (femtosecond) laser beams enables the generation of sub-wavelength laser-induced periodic surface structures (LIPSS) over large areas with controlled spatial periodicity, orientation, and depths affecting only a material layer on the sub-micrometer scale. This study reports on how fs-laser irradiation of commercially available Nb foil samples affects their superconducting behavior. DC magnetization and AC susceptibility measurements at cryogenic temperatures and with magnetic fields of different amplitude and orientation are thus analyzed and reported. This study pays special attention to the surface superconducting layer that persists above the upper critical magnetic field strength H_{c2}, and disappears at a higher nucleation field strength H_{c3}. Characteristic changes were distinguished between the surface properties of the laser-irradiated samples, as compared to the corresponding reference samples (non-irradiated). Clear correlations have been observed between the surface nanostructures and the nucleation field H_{c3}, which depends on the relative orientation of the magnetic field and the surface patterns developed by the laser irradiation.

Keywords: niobium; surface superconductivity; laser-induced periodic surface structures (LIPSS); femtosecond n-IR laser

---

1. Introduction

It is well established for type II superconductors that both a lack of material's extended lattice periodicity (grain boundaries, stacking faults, etc.) and local crystallographic defects (vacancies, substitutions) interact with magnetic vortices and act as effective pinning centers of the magnetic flux lines. The vortex pinning forces generated by these structural imperfections cause flux density gradients that contribute to the irreversible behavior of the magnetization [1] and to low frequency AC losses. Vortex pinning is a phenomenon of great relevance for practical conductors since it allows the superconductor to carry resistance-less current.

Type II superconductors with negligible bulk pinning may also present hysteresis effects, due to Bean–Livingston surface barriers [2] or due to geometrical edge barriers (specimen-shape dependent) [3]. Both types have been observed in low- and high-temperature superconductors (LTS and HTS, respectively) [2–8]. The former type of barriers is generally observed in clean single crystals, whereas the latter is more pronounced for thin films of constant thickness in perpendicular magnetic fields. In both cases, the magnetic irreversibility is caused by the asymmetry between the magnetic flux entry and exit. In type II superconductors, vortex cores overlap at the upper critical field strength, H_{c2}, and superconductivity becomes extinguished from the bulk. However, it is worth noting that a surface superconducting layer can persist above H_{c2}, up to the surface critical field strength, H_{c3}. For flat surfaces, Saint James and De Gennes [9] predicted a superconducting layer of a thickness approximately equal to the superconducting coherence length, up to a field $H_{c3} \approx 1.69 \times H_{c2}$ when the field is applied parallel to the superconductor's surface. These surface current vortices can be pinned, resulting in a surface critical current (i_c), which depends on the surface characteristics, such as roughness and morphology. These can be changed, for example, by different polishing procedures [10–12], or by low-energy Ar^+ ion irradiation of the surface [13].

Within the class of type II LTS, pure niobium has been widely studied in the literature by its intrinsic properties and by its application in superconducting radio frequency (SRF) cavities, where surface control has the greatest relevance. In SRF applications, very clean surfaces are required to achieve a high quality factor, Q_0, which is inversely proportional to the surface resistance [11]. Surface treatment procedures including thermal etching, electropolishing and buffered chemical polishing are usually used to achieve the former requirement [10,12]. Grassallino et al. [14] also found that the annealing in a partial pressure of nitrogen or argon gas, followed by the electropolishing of the niobium cavity, yields very low values of the microwave surface resistance and, therefore, more efficient accelerating structures.

Nowadays, focused laser pulsed sources allow surface control on the near micrometer and submicrometer scales, enabling surface specialized functionalities in a wide variety of materials and with increasing processing speeds. Different processing techniques, such as laser direct writing, laser interference patterning or laser-induced self-organization, enable the control and modification of the laser-processed surface morphologies [15–21]. Particularly here, a laser-induced self-ordering process [22] has been chosen that enables the generation of laser-induced periodic surface structures (LIPSS) in a robust single-step approach. In general, such surface structures may exhibit the shape of grating-like ripples, grooves, spikes, pillars, cones, etc., featuring spatial periods ranging from a few tens of micrometers to a few tens of nanometers far beyond the wavelength diffraction limit [17,18,21,22]. Furthermore, femtosecond (fs) laser pulses facilitate the attainment of LIPSS with minimum thermal heating effects on the irradiated target.

Ultra-short laser processing, thus, opens the possibility of changing the surface morphology through periodic surface structures on large areas and in continuous fabrication processes featuring currently maximal areal processing rates at the m^2/min level for both the laser interference patterning and the self-organization approaches [23]. Moreover, the spatial period of the LIPSS can be controlled via the laser irradiation wavelength, the laser pulse fluence or by the effective number of incident pulses [21]. Particularly, close to the ablation threshold of fs-laser irradiated metals, a significant variation of the ripple period can be realized [24–26]. Hence, the localized generation and control of surface defects, such as ripples, is a promising approach to affect the surface superconducting properties.

In this work, two different near-infrared (n-IR) fs-lasers, which differ in wavelength and pulse duration, have been used for laser structuring of Nb foil samples. Ultra-short laser irradiation produces distinctive quasi-periodic nanostructures that vary with the laser pulse fluence or with the effective number of incident pulses. Here, these values have been chosen to produce elongated quasi-parallel morphologies of ripples that are not isotropic. The microstructural changes following irradiation have been analyzed and compared with non-irradiated samples. The effects of applying different atmospheres (argon, nitrogen and air) during the laser treatment process have also been analyzed. In a previous work, it was reported that the laser-generated structures on Nb foils produce some irreversible changes in the magnetic behavior of this superconductor [18]. Nevertheless, discriminating between bulk and surface effects frequently presents important challenges. The aim of the present work is to study the effect of femtosecond n-IR laser structuring of Nb foil samples on their surface superconductivity characteristics. These properties were characterized using DC magnetization and AC susceptibility measurements with the magnetic field applied parallel to the Nb foil surface in order to maximize H_{c3} values and to minimize the geometric specimen-shape factor. The effect of the LIPSS anisotropy on the surface superconductivity behavior has been studied and is reported here.

2. Materials and Methods

Commercial Nb foil samples (rolled, 25 μm thickness, 99.8% purity and typical roughness of 0.30–0.35 μm) (Sigma-Aldrich, Darmstadt, Germany) have been irradiated with two different n-IR femtosecond lasers located in different laboratories, i.e., at the Institute of Nanoscience and Materials of Aragón (INMA) in Zaragoza (Spain), and at the German Federal Institute for Materials Research and Testing (BAM) in Berlin: (i) L1 at INMA: n-IR Yb:YAG laser (Light Conversion, Vilnius, Lithuania), (Carbide CB3-40W), center wavelength $\lambda = 1030$ nm, pulse duration $\tau_p = 280$ fs. The focusing of the laser beam was realized by means of a cylindrical lens system (focal length of 150 mm) leading to an elliptical beam with $1/e^2$ diameters of $2a_b = 1500$ μm and $2b_b = 26$ μm. (ii) L2 at BAM: n-IR Ti:Sapphire laser (Femtolasers, Vienna, Austria) (Compact Pro), center wavelength $\lambda = 790$ nm, pulse duration $\tau_p = 30$ fs. For properly handling such extremely short femtosecond laser pulses, the focusing of the laser beam was realized by a spherical dielectric mirror (Layertec GmbH, Mellingen, Germany) (focal length of 500 mm) resulting in a circular beam of $1/e^2$ diameter $D_b = 2\,r_b = 130$ μm. For both laser systems, the experiments were performed with a pulse repetition frequency $f_{rep} = 1$ kHz and the focused $1/e^2$ beam diameters were determined in the sample processing plane using the D^2-method proposed by Liu [27,28]. At the given conditions, the areal processing rates were approximately 1.0 mm^2/s for L1 and 0.12 mm^2/s for L2 here. For irradiation with L1, the sample was placed inside a chamber that allows laser processing in different gaseous atmospheres, such as air, Ar or N_2; and the line-wise laser scanning was performed in the direction perpendicular to the major axis of the elliptical beam. In all experiments, the laser scanning direction coincides with that of the linear laser beam polarization.

Table 1 collects the laser processing conditions of the irradiated Nb samples. With laser L1, two samples were processed with the same conditions but changing the atmosphere (Ar and N_2). In both cases, the laser polarization was parallel to the rolling direction. The two samples that were processed with laser L2 were treated in air. The difference between samples "FS_air1" and "FS_air2" is that they were irradiated with orthogonal orientation with respect to the Nb foil rolling orientation. Sample FS_air2 sample was oriented in order to have laser beam polarization perpendicular to the rolling direction. A reference non-irradiated sample, named "REF", was also studied for comparison.

Table 1. Laser processing parameters used in 25 μm thick Nb foil samples irradiated with fs-lasers.

Sample	Laser	Atm.	P (W)	f_{rep} (kHz)	v_L (mm/s)	d_s (μm)	F_p (J/cm^2)	I_p (GW/cm^2)	F_{2D} (J/cm^2)
FS_Ar	L1	Ar	0.18	1	1	1000	0.61	2671	17.1
FS_N	L1	N_2	0.18	1	1	1000	0.61	2671	17.1
FS_air1	L2	Air	0.02	1	6	20	0.15	5023	16.6
FS_air2	L2	Air	0.02	1	6	20	0.15	5023	16.6

The main parameters of laser processing were described in detail in [18] but are summarized here as follows. P is the nominal average laser power and f_{rep} the pulse repetition frequency. Each laser pulse is characterized by the peak fluence (F_p) and the peak irradiance (I_p). For a given value of f_{rep}, the laser beam scanning velocity (v_L) controls the distance between two laser pulses in a line, $d_p = v_L/f_{rep}$. Taking into account the Gaussian beam profile in the processing plane, a uniform fluence distribution is obtained in the full 2D scanned area if line-to-line overlap is above a critical threshold value (d_s/r_b for L2 or d_s/a_b for L1 is below 0.9, where d_s is the distance between adjacent lines). This is fulfilled for the treatments performed with laser L2, where d_s is 20 μm while the characteristic beam radius is r_b = 65 μm. In this situation, the total accumulated fluence can be calculated as

$$F_{2D} = \frac{\pi\, r_b^2}{d_p\, d_s}\, F_p \tag{1}$$

Laser parameters were selected in order to have similar F_{2D} values with both lasers. A constant value of F_{2D} = 16.6 J/cm^2 was obtained for samples treated with laser L2. In the case of laser L1, the distance between two lines (d_s = 1 mm) is larger than the characteristic line size (a_b = 0.75 mm) leading to a non-uniform fluence distribution on the surface, with an average value of 17.1 J/cm^2.

The optical penetration depth $1/\alpha = \lambda/(4\pi k)$, with α being the linear absorption coefficient and k the imaginary part of the complex valued refractive index, accounts to ~16 nm at both laser irradiation wavelengths.

The surface microstructural characterization was performed in a MERLIN field-emission scanning electron microscope (FE-SEM) (Carl Zeiss, Jena, Germany) equipped with an energy dispersive X-ray spectroscopy (EDX) system (Oxford Instruments, Abingdon, UK) operated at 5 kV. Surface topographic cross-sections were analyzed by Scanning Transmission Electron Microscopy (STEM) using a Tecnai F30 microscope (FEI Company, Hillsboro, OR, USA), also equipped with a high-angle annular dark field (HAADF) detector. Sample preparation was performed with a Focused Ion Beam (FIB) in a Dual Beam Helios 650 (FEI Company, Hillsboro, OR, USA) apparatus, using 30 kV Ga$^+$ ions for the initial steps and 5 kV for final thinning. Prior to the preparation of a FIB lamella, a protective Pt cap layer was deposited at the region of interest. X-ray photoelectron spectroscopy (XPS) was applied to characterize the changes in the chemical state of the surface using an AXIS Supra spectrometer (Kratos, Manchester, UK). The photoemission was excited with monochromatic Al K$_\alpha$ X-ray radiation at 1486 eV over a spot size of 700 × 300 μm^2, resulting in an XPS information depth between 5 and 10 nm. The carbon C1s peak at 284.8 eV served as the reference signal for energy calibration.

Measurements of DC magnetization (M) and complex AC susceptibility, χ_{ac} (with in-phase, χ', and out-of-phase, χ'', components) were carried out in a SQUID-based MPMS-5T system (Quantum Design, San Diego, CA, USA). The Reciprocating Sample Option (RSO) of the system was used for DC measurements. For χ_{ac} measurements, the AC drive magnetic field (sine wave of amplitude $\mu_0 h_0$ and frequency f) is superimposed on the constant DC magnetic field, $\mu_0 H$. In this work, AC and DC components of the applied magnetic field have the same direction. Values of f = 10 Hz and $\mu_0 h_0$ = 10 and 100 μT have been used. The magnetic field was applied parallel to x- and y- axes of the foil's surface plane (see Figure 1), in order to analyze if laser polarization and rolling orientations have any effect on their magnetic and superconducting behavior. For magnetic measurements, the size of the measured sample area is in the range of (3–4) mm × (3–4) mm and both surfaces of the samples were irradiated. The critical temperature (T_c) was determined as the onset of diamagnetism from $\chi'(T)$, being T the temperature, obtaining T_c values ranging between 9.30 and 9.35 K in all samples. The heat capacity was measured in a PPMS-9T (Quantum Design, San Diego, CA, USA) apparatus.

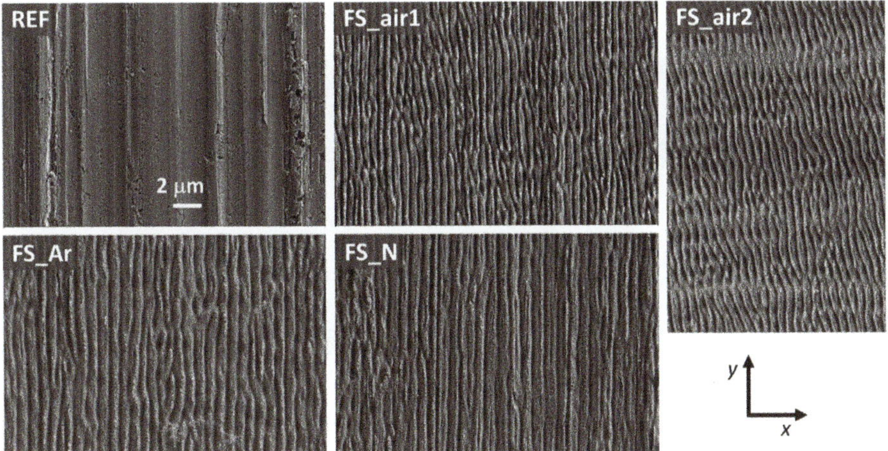

Figure 1. FE-SEM micrographs (secondary electrons) of the surfaces of all analyzed representative samples (same magnification for all images). The rolling direction is parallel to the *y*-axis in all samples, except for FS_air2, where it is parallel to *x*. The linear laser beam polarization is parallel to *x* in all cases.

3. Results and Discussion

3.1. Microstructural Characterization

Figure 1 shows SEM micrographs from the surface of each of the five analyzed samples. The reference sample clearly shows the microstructural characteristics arising from rolling, with defects parallel to the rolling direction (*y*-axis in the figure). These are also visible, although with less clarity, after laser irradiation. This rolling direction coincides with the *y*-axis in all the samples, except in the case of sample FS_air2, which was rotated by 90°. Surface irradiation of the samples with both fs-lasers produces, in all cases, elongated quasi-periodic surface structures aligned perpendicular to the laser polarization (direction *x* in the figure). This type of near-wavelength sized ripples is known as low spatial frequency LIPSS (LSFL) [17,22]. These structures are caused by the excitation of surface plasmon polaritons (SPPs) at the rough metallic surface and their interference with the incident laser radiation. The intra-pulse interference modulates the spatial pattern of optical energy absorbed by the electronic system of the solid and leads—after electron-phonon energy relaxation—to spatially modulated ablation [22]. This formation mechanism is supported by the values of the dielectric permittivity of the material Nb at the laser irradiation wavelengths that account to $\varepsilon = -10.1 + i \times 15.6$ at 790 nm and $\varepsilon = -24.4 + i \times 16.8$ at 1030 nm, respectively [29]. Hence, at both irradiation wavelengths, the condition $\Re(\varepsilon) < -1$ is fulfilled here—a prerequisite for the excitation of SPPs [30,31]. The relevance of specific hydrodynamically driven supra-wavelength sized ripples that were observed upon irradiation of thin metal films on dielectric substrates with high aspect ratio elliptical ns-laser beams parallel to the direction of scanning [32] can be ruled out here, since, in our case, the ripples (LSFL) are formed perpendicular to the direction of beam scanning—always perpendicular to the laser beam polarization and they exhibit near wavelength sized periods (see Figure 1).

The EDX analyses of the sample surfaces revealed an increase in the % of O elemental composition in the samples processed in air (6.2 wt%, 27.7 at%) compared to the reference sample (4.1 wt%, 19.9 at%), and a reduction for samples processed in N_2 or Ar (3.4 wt%, 17.0 at%). The standard deviations of the wt% values range between 0.2 and 0.4. A small amount of nitrogen (0.6 wt%, 3.0 at%, 0.1 wt% sigma) was detected in sample FS_N. Note that the absolute percentage values must be taken with care here due to the large EDX information depth and due to surface corrugations being present in the laser processed areas.

Quasi-periodic submicrometer-structures induced by both lasers can be better analyzed using higher magnification, as it is demonstrated in Figure 2 for samples FS_Ar and FS_air2. The upper row images in the figure correspond to cross-sectional views obtained by STEM for a lamella extracted from the samples (cut parallel to *x*-axis). The lower row images correspond to surface views by FE-SEM using an in-lens secondary detector. LSFL were formed in both samples and have similar modulation depths (peak-to-valley distances of about 200 nm). These structures are more homogeneous for sample FS_air2, as it is clearly seen in the cross-sectional view. The spatial period of the LSFL is 775 nm for samples irradiated with laser L1 and 570 nm for those irradiated with laser L2. These are mean values that have been calculated by analyzing statistically representative areas observed on samples by FE-SEM, and have standard deviations of 68 and 35 nm, respectively. The differences in the observed periods are assigned to the different laser wavelengths emitted by L1 and L2. It must be pointed out that the geometrical characteristics of the LSFL are in good agreement with the atomic force microscopy measurements published for fs-laser irradiated niobium in [18,33]. High spatial frequency LIPSS (HSFL) with periods of 50 to 80 nm, which form between LSFL structures, being perpendicular to them, are also observed in both samples (see Figure 2c,d), in line with the observations reported in [34].

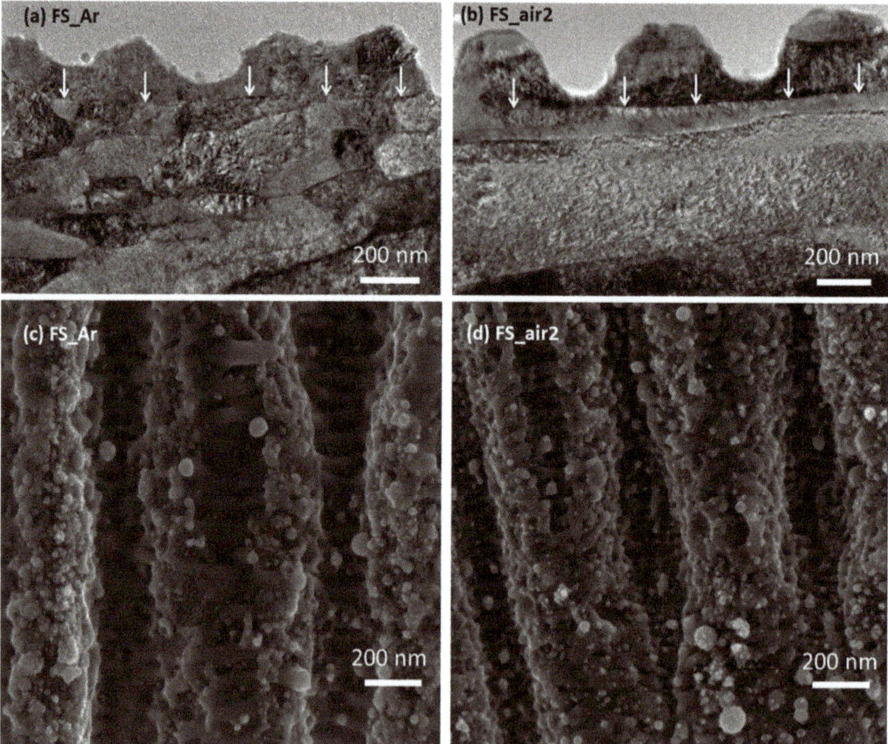

Figure 2. (**a**,**b**) STEM micrographs of the cross-sectional images near the surface. (**c**,**d**) FE-SEM top-view images (in-lens detector) of the surface of the same samples. (Left column) FS_Ar, (Right column) FS_air2. The white arrows in (**a**,**b**) point at the boundary between laser-affected and non-affected regions.

It is important to note here that, due to the rolling process involved in the manufacturing of the Nb foils, the grains are elongated in the bulk of the material, with the longest grain axis oriented parallel to the rolling direction. This is also parallel to the ripple orientation in sample FS_Ar, but perpendicular in sample FS_air2. This is the reason for the different grain shapes observed in Figure 2a,b, with a clear texture of grains oriented perpendicular to the nanostructure observed in sample FS_air2. Interestingly, a boundary can be observed in both samples that is separating the laser-affected region from the non-affected bulk material underneath (marked by white arrows in Figure 2a,b. Supposedly, the Nb surface was melted up to this depth during the fs-laser scan processing, resulting in a re-solidified layer of 40 to 300 nm here (depending on the position across the LIPSS). This thickness is larger than the optical penetration depth of the laser radiation in Nb, which can be explained by the multi-pulse laser treatment upon scan-processing.

Figure 3 shows HAADF-STEM images of the cross-section of samples FS_N and FS_air2. Corresponding STEM-EDX analyses confirm the presence of O and Nb elements in the darker areas observed near the sample surface (data not shown here). These zones are more abundant in the sample processed in air (Figure 3b) than in the samples processed in nitrogen (Figure 3a) or argon (not shown here). The oxide layer can be associated with the dark interfacial zones and exhibits a thickness of a few nanometers for the sample FS_N ($\approx$5 nm in the zone shown in the inset of Figure 3a). For sample FS_air2, however, this layer exhibits an increased thickness (reaching values up to 20 nm) and is much less uniform. Note that the thickness of the oxide layer formed upon fs-laser processing in nitrogen here is very similar to that found for the natural Nb passivation layers that are characterized by a thickness of about 6–8 nm and Nb_2O_5 as the outermost layer [33].

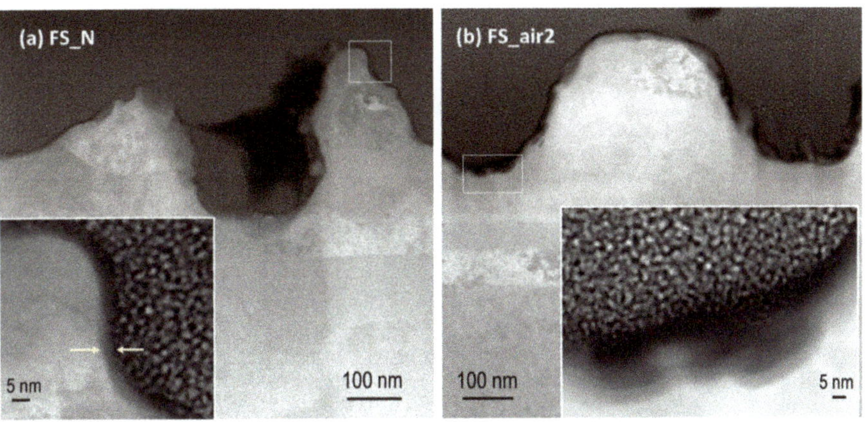

Figure 3. HAADF-STEM images of a cross-section of sample (**a**) FS_N and (**b**) FS_air2, near the surface. The insets show the areas highlighted by a white rectangle in the corresponding main images, with higher magnification.

The high resolution XPS spectra of niobium Nb 3d, plotted in Figure 4, show very similar behavior for all samples, with the presence of the Nb 3d doublet at binding energies (BEs) of 209.80 and 207.05 eV, corresponding to Nb_2O_5, as the main chemical compound at the surface [35–37]. The peaks corresponding to metallic Nb, Nb^0, are also observed in all samples except in the one processed in air, FS_air2. The latter indicates a thickness of the laser-induced oxide layer exceeding the XPS information depth here—fully in line with Figure 3b and with previous observations made for fs-irradiation of Ti in the air environment [38]. The sample irradiated in nitrogen atmosphere exhibits a very weak signal in the N 1s spectrum with a BE of approximately 400.0 eV (see Figure 5), thus discarding the presence of niobium nitrides, NbN_x, which are expected to appear at lower binding energies (396.5 eV [39], 397.5 eV [40]). This is very similar to the results reported for nitrogen doped niobium [37], where the peak at 399.84 eV in the N 1s spectrum was attributed to the formation of CH_3CN.

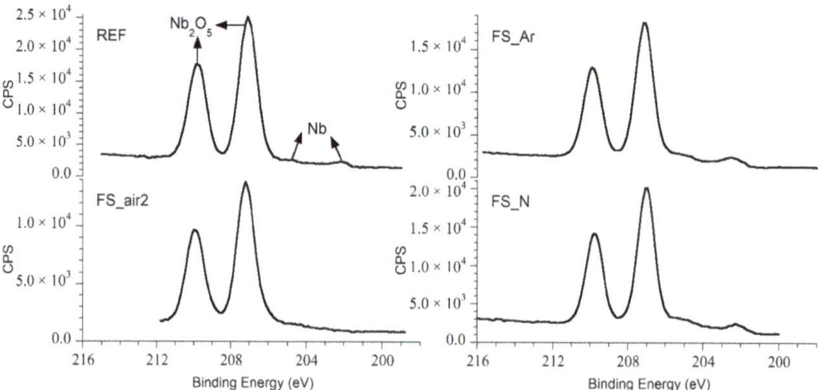

Figure 4. High resolution Nb 3d XPS spectra for the analyzed samples discussed in the text. CPS: counts per second.

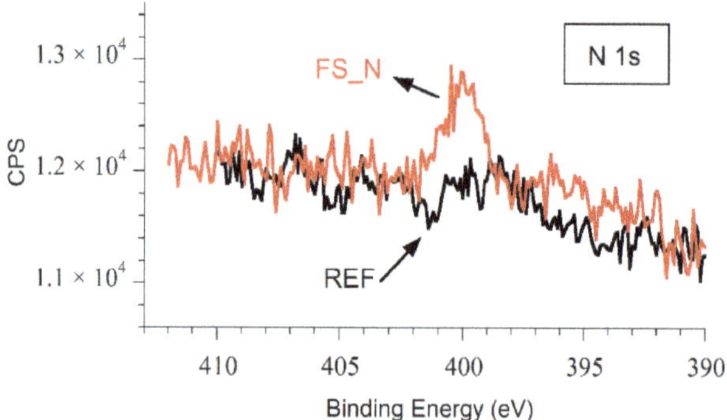

Figure 5. High resolution XPS spectra of the N 1s line of samples REF (non-irradiated, black line) and FS_N (laser irradiated in nitrogen atmosphere, red line). CPS: counts per second.

3.2. Irreversible Magnetization and Upper Critical Field, H_{c2}

Figure 6a shows the width of the magnetic hysteresis loop, ΔM, as a function of the magnetic field for the reference non-irradiated sample (REF) and the irradiated sample (FS_N), at the temperature $T = 5$ K, and with the magnetic field applied parallel to the x- and y-axes. ΔM was obtained for each H value as $\Delta M = M_\downarrow - M_\uparrow$, where $M_\uparrow$ and $M_\downarrow$ are the corresponding values measured for increasing and decreasing fields, respectively.

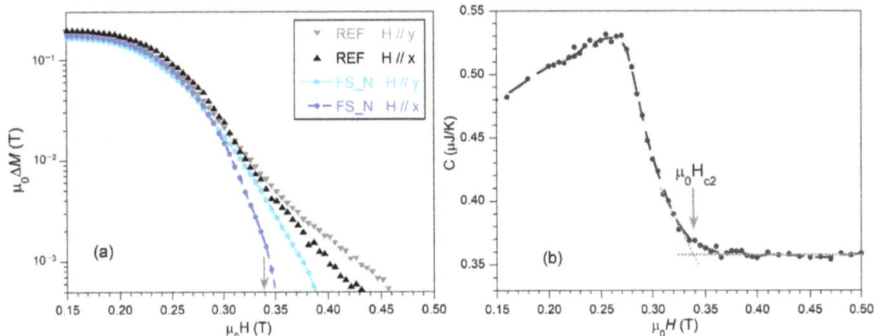

Figure 6. (**a**) Field dependence of the magnetic hysteresis loop width, $\mu_0\Delta M(H)$ at $T = 5$ K of the REF and FS_N samples with the external DC field applied in the x- and y-directions, both parallel to the surface as indicated in Figure 1; (**b**) heat capacity of the non-irradiated reference (REF) sample recorded at 5 K. The derived upper critical field $\mu_0 H_{c2}$ is marked by an arrow. Both arrows point to the same magnetic field, for better comparison.

In these measurements, the sharp increase of the magnetic irreversibility, which is related to ΔM, is frequently associated to the onset of bulk superconductivity and, therefore, to the upper critical field $\mu_0 H_{c2}$ [10]. Nevertheless, this value is not always easy to derive from magnetization curves, due to the appearance of a tail at high magnetic fields in some conditions, as in this case [41]. With this aim, heat capacity, which is essentially a bulk property, could better allow the determination of $\mu_0 H_{c2}$, as marked by the arrow in Figure 6b [12,13]. It must be noted that the $C(H)$ curve for the FS_N sample (not shown here for clarity reasons) is very similar to that of the reference sample, indicative of similar bulk properties (and $\mu_0 H_{c2}$) between both samples.

The results in Figure 6a thus clearly show irreversible magnetization values above $\mu_0 H_{c2}$, with differences between samples and orientations. The existence of non-zero ΔM values above $\mu_0 H_{c2}$ is indicative of the presence of surface critical currents, i_c. As it is observed, fs-laser irradiation produces a significant decrease of ΔM above H_{c2}, particularly when the magnetic field is applied in the direction perpendicular to the nanostructures (x-direction, as indicated in Figure 1). This effect is observed for all analyzed irradiated samples, independently of the processing atmosphere. It must be noted that the reference sample also exhibits some anisotropy in ΔM, with higher values above H_{c2} for the field applied parallel to the rolling direction (y-axis). This is indicative of an influence of the anisotropy of the microscopic grain structure induced by rolling, as visualized in Figures 2a,b and 3, in agreement with previous studies [41].

Changes in the ΔM (or i_c) at fields above H_{c2} for different surface treatments have been reported by several groups. For example, Scola et al. [13] observed an increase of i_c values after irradiating the Nb surface with low-energy Ar$^+$ ions; Aburas et al. [12] observed changes by polishing the surface using different processes (sandpaper, diamond, colloidal silica and chemical polishing), with lower i_c values for smoother surfaces; Casalbuoni et al. [10] also reported differences in i_c values of Nb cylinders subjected to buffered chemical polishing or to electropolishing; and van Gurp [41] observed higher i_c values for cold rolled Nb foils compared to electrolytic Nb foils.

The behavior observed here would thus suggest that surface critical currents present a marked anisotropy, with higher values for magnetic fields applied along the ripples in the irradiated samples. Further analysis of the surface superconducting characteristics and the effects of the different laser irradiation conditions can be better performed from $\chi_{ac}(H)$ measurements because of their higher sensitivity, as discussed in the following section.

3.3. Surface Superconductivity Characterization

H_{c3} values can be estimated from $\chi_{ac}(H)$ following a procedure similar to that described in [10,11,42]. The complex $\chi_{ac}(H)$ curve was measured in descending DC magnetic fields applied parallel to the Nb sheet surface, starting from normal-state conditions and then approaching the superconducting transition by decreasing the DC magnetic field. Low frequency (f = 10 Hz) and low amplitude ($\mu_0 h_0$ = 10 μT) of the AC alternating magnetic field were used for the present measurements, similarly to those reported in references [10,11], for ease of comparison. Figure 7 shows $\chi'(H)$ and $\chi''(H)$ curves measured at 5 K with decreasing fields from 1.5 T down to 0 T for different samples and for two orientations of the external magnetic field. Note that the value of the initial DC magnetic field ($\mu_0 H$ = 1.5 T) was chosen considerably larger than $\mu_0 H_{c2}$ to ensure that the entire sample, including its surface, was in the normal state at the beginning of each measurement run [11]. All values in the graphs have been scaled by χ'_{-1}, which is the value of χ' measured at zero field, and it is close to the expected value for perfect diamagnetism for each sample, i.e., χ'_{-1}(emu) $\approx V$(cm^3)/(4π), where V is the volume of the sample. H_{c3} at each temperature can be defined as the first deviation point from the values at normal state of χ'' or as the onset of χ' (above noise level). In the figure, arrows indicate the position of the H_{c3} values of two of the samples (REF and FS_N), for clarity purposes.

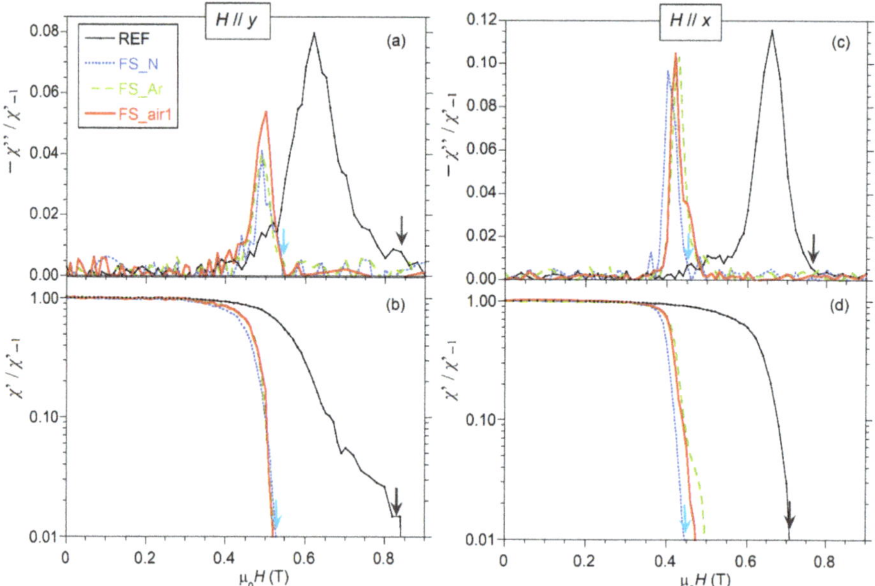

Figure 7. Components of the χ_{ac}: χ'' (**a,c**) and χ' (**b,d**), scaled by χ'_{-1}, as a function of the DC magnetic field, $\mu_0 H$, applied parallel to the y-axis (**a,b**) or the x-axis (**c,d**), at T = 5 K, f = 10 Hz and $\mu_0 h_0$ = 10 μT measured in descending fields from the initial field 1.5 T. The arrows mark $\mu_0 H_{c3}$ values for samples REF (black arrow) and FS_N (blue arrow), derived as explained in the text.

From Figure 7, it is worth pointing out that fs-laser processing produces a pronounced shift of the transition towards smaller magnetic fields together with much narrower transitional widths, as compared to the reference sample. Two non-irradiated samples were measured, showing very similar curves (see Figure S1 presented in Supplementary Material), thus confirming this laser-modified behavior. The onset of the normal-to-superconducting transition of fs-laser processed samples occurs at higher fields when H is applied parallel to the nanostructures produced by the laser (y-axis) than when it is applied perpendicular to them (x-axis). The effect of the used atmosphere during laser processing is not significantly relevant, although some differences are observed, especially for fields applied along the *x*-axis (Figure 7c,d). The non-irradiated sample REF also shows some anisotropy, assigned to the elongated defects/grains produced by the rolling during its manufacture. Note that the maximum of $\chi''(H)$ curves is lower (thus indicating lower AC losses, which represent the energy dissipation during a cycle), for the fs-laser processed samples, especially for those irradiated in Ar and N_2 atmosphere and for magnetic fields applied parallel to the induced nanostructures (*y*-axis). This could suggest lower surface resistance in these samples and conditions [43].

$\chi''(H)$ curves also allow the estimation of H_{c2}, which is defined as the point where $\chi''(H)$ decreases to zero after the transition when ramping down the field [10,11]. As seen in Figure 7a, the increase of $\chi''(H)$ around H_{c2} is smooth at these conditions, so that we can only have an estimate of this value in the range between 0.35 T and 0.40 T in all samples, as the low-intensity signal makes the noise relevant, thus preventing more precise estimation of H_{c2} here. Nevertheless, it is possible to derive H_{c2} values using higher amplitude of the AC field, since the directly measured signal by the magnetometer is proportional to $h_0 \cdot \chi_{ac}(H, f, h_0)$, so that the signal-to-noise ratio increases with h_0. Figure 8 compares the same samples and conditions as in Figure 7a but for $\mu_0 h_0 = 100$ µT. H_{c2} values defined this way would be about 0.34 T for all the samples (dotted arrow in the figure). This is in good agreement with the result obtained previously from the heat capacity measurement shown in Figure 6b. It must be pointed out that, as expected, H_{c2} does not depend on the field orientation, since similar H_{c2} values are derived when the field is oriented parallel to the *x*-axis (see Figure S2 in the Supplementary Material).

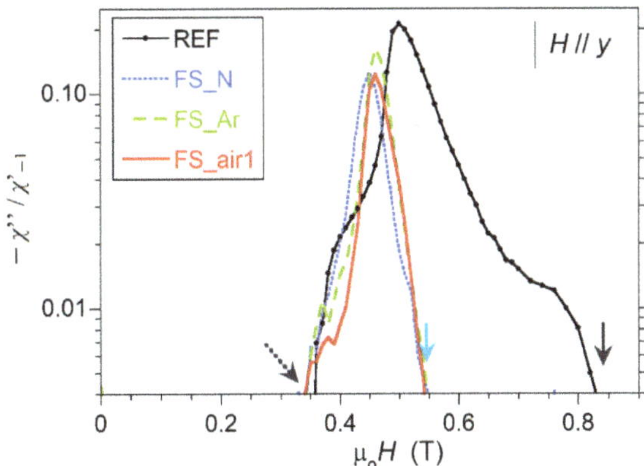

Figure 8. $\chi''(H)$ for the same samples and under the same conditions as in Figure 7a, but in this case $\mu_0 h_0 = 100$ µT. χ'' is plotted in a log scale. The dotted arrow indicates the estimated $\mu_0 H_{c2}$ value for these samples. Note that the two continuous arrows point to the same DC field values as in Figure 7a, to allow for a direct comparison between the two sets of data.

The choice of small $\mu_0 h_0$ values in these measurements aims at improving the surface sensitivity. It can be observed that the onset of the normal-to-superconductor transition does not vary significantly by increasing $\mu_0 h_0$ from 10 to 100 µT, for laser processed samples, but it shifts at slightly lower external DC fields for the reference sample. This indicates that both AC field amplitudes would provide good sensitivity to analyze near-surface regions. It should be underlined that, upon increasing $\mu_0 h_0$, the AC field will sense deeper into the sample so that the peak of the $\chi''(H)$ curve shifts towards lower fields [5]. More specifically, for the fs-laser treated samples, the peak shifts from 0.49–0.50 T for the $\mu_0 h_0 = 10$ µT AC field down to the 0.45–0.46 T for the 100 µT AC field, when *H* is parallel to the *y*-axis (Figures 7 and 8). The behavior is similar in all laser treated samples, although the peak of the sample processed in argon appears at slightly higher fields. In the case of the non-irradiated sample, the observed shift is larger, as the $\chi''(H)$ peak moves from 0.60 T (at 10 µT AC field) down to 0.50 T (at 100 µT AC field). In all samples, these fields are well above the $\mu_0 H_{c2}$ field, thus clearly indicative of surface phenomena.

The surface critical fields, H_{c3}, derived from $\chi''(H)$ measurements at 5 K and AC fields of 10 Hz and amplitude 10 µT are plotted in Figure 9 for all analyzed samples presented in this study and for both DC field orientations. Note that the values H_{c3} obtained from $\chi'(H)$ curves, which are not displayed here for clarity purposes, show the same trends, but are slightly shifted towards lower magnetic fields than those corresponding to $\chi''(H)$. The estimated $\mu_0 H_{c3}$ values for external magnetic fields applied parallel to the surface nanostructures (y-axis) decrease from 0.84 T for the REF sample to about 0.54–0.58 T. That is, the r_{32} factor ($=H_{c3}/H_{c2}$) decreases from $\approx$2.4 for the as-rolled reference sample to 1.6–1.7 for fs-laser irradiated samples. Moreover, H_{c3} values are smaller when applying the magnetic field perpendicular to the nanostructures (x-axis). For this field orientation, we observe some changes depending on the processing atmosphere, with samples irradiated in argon and nitrogen having the smallest and the largest anisotropy, respectively. More precisely, for this field orientation, the transition of FS_N shifts slightly towards lower fields compared to the other irradiated samples (see Figure 7d), whereas the transitions of FS_air1 and FS_Ar are very similar for $\chi'/\chi'_{-1} > 0.06$, but differ on the onset of diamagnetism, which occurs at higher DC fields for FS_Ar, thus exhibiting a larger H_{c3} value. The reasons for the observed differences among the irradiated samples are still unknown. They are, in any case, small compared with those found between fs-laser irradiated and non-irradiated samples. It seems clear, however, that irradiation with fs-lasers, as used in this work, is sufficiently short in time to limit the chemical reaction between niobium and oxygen or nitrogen, thus resulting in very similar critical surface fields, independently of the atmosphere used during the irradiation process. Hence, this similarity points toward a topographic effect of the LIPSS here. Moreover, it was observed that the laser polarization orientation with respect to the rolling direction of the sample does not show a very important effect on the superconducting properties of the laser processed surface as the laser treatment diminishes rolling defects and both FS_air1 and FS_air2 samples exhibit a similar nanostructure and magnetic behavior.

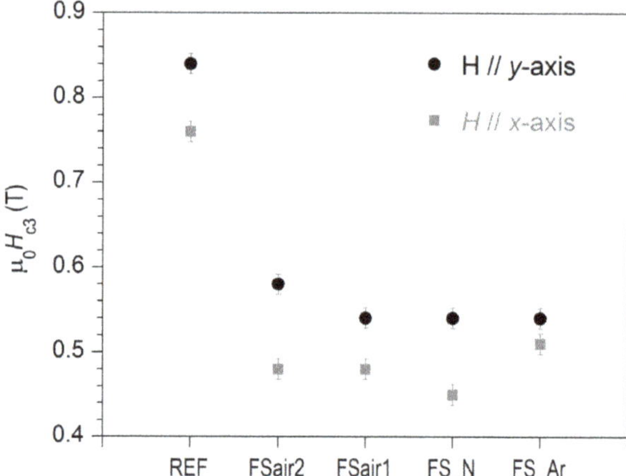

Figure 9. Critical surface fields, $\mu_0 H_{c3}$, of all analyzed samples derived from $\chi''(H)$ measurements at $T = 5\,\text{K}$, $f = 10\,\text{Hz}$ and $\mu_0 h_0 = 10\,\text{µT}$, with the external magnetic field applied parallel to the x- and y-axes.

Note that the effect of fs-laser processing on H_{c3} is similar to the effect of electropolishing and buffered chemical polishing of Nb wire reported by Sung et al. [11]. In that study, the authors observed a decrease of H_{c3} in all polished samples as compared with the as-drawn wires. For example, $\mu_0 H_{c3}$ at 5 K was reduced from $\approx$0.84 T (as-drawn) to $\approx$0.68 T after electropolishing for 3 h. They also found some differences in H_{c3} and r_{32} values by annealing (at 800 °C) or by baking (at 120 °C) the niobium samples. Following these results, it is thus worth further exploring the effect of different LIPSS topographies, together with annealing and/or baking of the Nb samples, on the surface superconducting properties of niobium.

4. Conclusions

Laser-induced periodic surface structures (LIPSS) were generated on oppositely aligned (3–4) mm × (3–4) mm areas on the top and bottom surfaces of as-rolled commercial niobium foils using two different fs-laser systems under different atmospheres (air, argon, and nitrogen). After laser treatment, the surface roughness is very similar in all cases. The formed low spatial period LIPSS (LSFL) have similar modulation depths (peak-to-valley distance of about 200 nm) and are aligned perpendicular to the laser beam polarization. They are characterized by a spatial frequency that depends on the laser wavelength: 775 nm (std 68 nm) and 570 nm (std 35 nm), i.e., approximately 73–74% of the irradiation wavelengths of λ = 1030 and 790 nm, respectively. High spatial frequency LIPSS (HSFL) with periods between 50 and 80 nm, are also formed between the LSFL structures, in the perpendicular direction.

Chemical analyses by EDX and XPS indicated some laser-induced oxidation effects, with Nb_2O_5 being the dominating type of oxide. Upon fs-laser processing in ambient air, a thin surface oxide layer of a few tens of nanometers extent was detected through cross-sectional STEM/EDX analyses. When laser processing takes place under inert gases, however, the laser-induced oxide layer thickness of ~5 nm is similar to that found for the native oxide on the non-irradiated reference sample. A small nitrogen signal is observed by EDX and XPS on the surface of the laser irradiated sample under nitrogen, not being consistent with the definite presence of niobium nitrides.

The surface critical field, H_{c3}, was derived from AC susceptibility measurements as a function of the externally applied DC magnetic field, which was applied parallel to the surface and in both orientations with respect to the generated LIPSS. Clear effects of laser irradiation on the surface superconducting properties of the niobium foil samples have been shown. In particular, H_{c3} decreases for laser irradiated samples as compared to the non-irradiated reference sample. Laser irradiation also results in significantly narrower normal-superconductor transitions and lower $\chi''(H)$ peaks, whereas the upper critical field, H_{c2}, remains unaffected. The microstructural anisotropy of the fs-laser-generated LIPSS is clearly reflected on the surface superconducting properties of the samples, with higher H_{c3} when the external magnetic field is parallel to the LIPSS (LSFL). Nevertheless, the differences in the spatial periodicity of LIPSS between these samples do not affect significantly the H_{c3} values. Moreover, it seems clear that irradiation with ultra-short pulsed lasers, as used in this work (with pulse durations of 280 fs and 30 fs), is of sufficiently short interaction time to limit the chemical reaction between niobium and oxygen or nitrogen, thus resulting in similar critical surface fields independently of the irradiation atmosphere.

The observed behavior indicates that fs-laser processing is useful to control the surface superconducting properties of niobium and could be an alternative to some well-stablished procedures, such as electropolishing or buffered chemical polishing. This work demonstrates that surface modification associated with ultra-short pulse laser processing strongly affects surface superconductivity. In order to understand completely the interaction between the laser generated surface nanostructures and superconductivity, additional studies using different laser sources (wavelengths or longer pulse durations) should be necessary.

Supplementary Materials: The following are available online at http://www.mdpi.com/2079-4991/10/12/2525/s1, Figure S1: Comparison between $\chi'(H)$ and $\chi''(H)$ curves of two different non-irradiated samples, REF and REF(2), nominally identical. Magnetic fields were applied parallel to the y-axis and the measurements were performed at 5 K, 10 Hz and $\mu_0 h_0$ = 10 μT with descending fields from the initial field of 1.5 T. Figure S2: $\chi''(H)$ for different samples at 5 K, 10 Hz and $\mu_0 h_0$ = 100 μT, measured in descending DC fields from the initial magnetic field 1.5 T. AC and DC magnetic fields were applied parallel to the x-axis.

Author Contributions: Conceptualization, L.A.A., G.F.d.l.F., and E.M.; methodology, E.M., G.F.d.l.F., L.A.A., and J.B.; investigation, Á.C., E.M., G.F.d.l.F., and L.A.A.; formal analysis, Á.C., R.N., and E.M.; resources, G.F.d.l.F., L.A.A., H.L., J.K., and J.B.; writing—original draft preparation, E.M. and Á.C.; writing—review and editing, R.N.,

L.A.A., G.F.d.l.F., J.B., and J.K.; software, J.B.; project administration, L.A.A., E.M., G.F.d.l.F., J.K., and J.B. All authors have read and agreed to the published version of the manuscript.

Funding: This work was funded by project ENE2017-83669-C4-1-R (MCIU/AEI/FEDER, EU) and by the Gobierno de Aragón "Construyendo Europa desde Aragón" (research group T54_20R).

Acknowledgments: This paper is dedicated to the memory of Rafael Navarro. The authors acknowledge the use of Servicio General de Apoyo a la Investigación-SAI and Laboratorio de Microscopías Avanzadas, University of Zaragoza. The authors are extremely grateful to Agustín González-Elipe for very valuable discussions regarding the XPS measurements.

Conflicts of Interest: The authors declare no conflict of interest.

References

1. Dew-Hughes, D. The metallurgical enhancement of type II superconductors. *Rep. Prog. Phys.* **1971**, *34*, 821–873. [CrossRef]

2. Bean, C.P.; Livingston, J.D. Surface barrier in type-II superconductors. *Phys. Rev. Lett.* **1964**, *12*, 14–16. [CrossRef]

3. Zeldov, E.; Larkin, A.I.; Geshkenbein, V.B.; Konczykowski, M.; Majer, D.; Khaykovich, B.; Vinokur, V.M.; Shtrikman, H. Geometrical Barriers in high-temperature superconductors. *Phys. Rev. Lett.* **1994**, *73*, 1428–1431. [CrossRef] [PubMed]

4. van der Klein, C.A.M.; Kes, P.H.; van Beelen, H.; de Klerk, D. The effect of neutron irradiation damage on the magnetic behavior of superconducting niobium in stationary fields. *J. Low Temp. Phys.* **1974**, *16*, 169–191. [CrossRef]

5. Hopkins, J.R.; Finnemore, D.K. Surface superconductivity in niobium and niobium-tantalum alloys. *Phys. Rev. B* **1974**, *9*, 108–114. [CrossRef]

6. Mishra, P.K.; Ravikumar, G.; Sahni, V.C.; Koblischka, M.R.; Grover, A.K. Surface pinning in niobium and a high-T_c superconductor. *Physica C* **1996**, *269*, 71–75. [CrossRef]

7. Chen, D.-X.; Cross, R.W.; Sanchez, A. Effects of critical current density, equilibrium magnetization and surface barrier on magnetization of high temperature superconductors. *Cryogenics* **1993**, *33*, 695–702. [CrossRef]

8. Konczykowski, M.; Burlachkov, L.I.; Yeshurun, Y.; Holtzberg, F. Evidence of surface barriers and their effect on irreversibility and lower critical field measurements in Y-Ba-Cu-O crystals. *Phys. Rev. B* **1991**, *43*, 13707–13710. [CrossRef]

9. Saint-James, D.; de Gennes, P.G. Onset of superconductivity in decreasing fields. *Phys. Lett.* **1963**, *7*, 306–308. [CrossRef]

10. Casalbuoni, S.; Knabbe, E.A.; Kötzler, J.; Lilje, L.; von Sawilski, L.; Schmüser, P.; Steffen, B. Surface superconductivity in niobium for superconducting RF cavities. *Nucl. Instrum. Methods Phys. Res. Sect. A* **2005**, *538*, 45–64. [CrossRef]

11. Sung, Z.-H.; Dzyuba, A.; Lee, P.J.; Larbalestier, D.C.; Cooley, L.D. Evidence of incomplete annealing at 800 °C and the effects of 120 °C baking on the crystal orientation and the surface superconducting properties of cold-worked and chemically polished Nb. *Supercond. Sci. Technol.* **2015**, *28*, 075003. [CrossRef]

12. Aburas, M.; Pautrat, A.; Bellido, N. Change of surface critical current in the surface superconductivity and mixed states of superconducting niobium. *Supercond. Sci. Technol.* **2017**, *30*, 015009. [CrossRef]

13. Scola, J.; Pautrat, A.; Goupil, C.; Méchin, L.; Hardy, V.; Simon, C. Voltage noise and surface current fluctuation in the superconducting surface sheath. *Phys. Rev. B* **2005**, *72*, 012507. [CrossRef]

14. Grassallino, A.; Romanenko, A.; Sergatskov, D.; Melnychuk, O.; Trenikhina, Y.; Crawford, A.; Rowe, A.; Wong, M.; Khabiboulline, T.; Barkov, F. Nitrogen and argon doping of niobium for superconducting radio frequency cavities: A pathway to highly efficient accelerating structures. *Supercond. Sci. Technol.* **2013**, *26*, 102001. [CrossRef]

15. Lasagni, A.; Benke, D.; Kunze, T.; Bieda, M.; Eckhardt, S.; Roch, T.; Langheinrich, D.; Nerger, J. Bringing the Direct Laser Interference Patterning Method to Industry: A One Tool-Complete Solution for Surface Functionalization. *J. Laser Micro/Nanoeng.* **2015**, *10*, 340–344. [CrossRef]

16. Pronko, P.P.; Dutta, S.K.; Squier, J.; Rudd, J.V.; Du, D.; Mourou, G. Machining of sub-micron holes using a femtosecond laser at 800 nm. *Opt. Commun.* **1995**, *114*, 106–110. [CrossRef]

17. Bonse, J.; Höhm, S.; Kirner, S.V.; Rosenfeld, A.; Krüger, J. Laser-Induced Periodic Surface Structures—A Scientific Evergreen. *IEEE J. Sel. Top. Quantum Electron.* **2017**, *23*, 9000615. [CrossRef]

18. Cubero, A.; Martínez, E.; Angurel, L.A.; de la Fuente, G.F.; Navarro, R.; Legall, H.; Krüger, J.; Bonse, J. Effects of laser-induced periodic surface structures on the superconducting properties of Niobium. *Appl. Surf. Sci.* **2020**, *508*, 145140. [CrossRef]

19. Voisiat, B.; Gedvilas, M.; Indrisiunas, S.; Raciukaitis, G. Flexible microstructuring of thin films using multi-beam interference. *J. Laser Micro/Nanoeng.* **2011**, *6*, 185–190. [CrossRef]

20. Gedvilas, M.; Indrišiūnas, S.; Voisiat, B.; Stankevičius, E.; Selskis, A.; Račiukaitis, G. Nanoscale thermal diffusion during the laser interference ablation using femto-, pico-, and nanosecond pulses in silicon. *Phys. Chem. Chem. Phys.* **2018**, *20*, 12166–12174. [CrossRef]

21. Bonse, J.; Kirner, S.V.; Krüger, J. Laser-Induced Periodic Surface Structures (LIPSS). In *Handbook of Laser Micro and Nano-Engineering*; Sugioka, K., Ed.; Springer: Cham, Switzerland, 2020. [CrossRef]

22. Bonse, J.; Gräf, S. Maxwell Meets Marangoni—A Review of Theories on Laser-Induced Periodic Surface Structures. *Laser Photonics Rev.* **2020**, *14*, 2000215. [CrossRef]

23. Bonse, J. Quo Vadis LIPSS?—Recent and Future Trends on Laser-Induced Periodic Surface Structures. *Nanomaterials* **2020**, *10*, 1950. [CrossRef] [PubMed]

24. Okamuro, K.; Hashida, M.; Miyasaka, Y.; Ikuta, Y.; Tokita, S.; Sakabe, S. Laser fluence dependence of periodic grating structures formed on metal surfaces under femtosecond laser pulse irradiation. *Phys. Rev. B* **2010**, *82*, 165417. [CrossRef]

25. Hou, S.; Huo, Y.; Xiong, P.; Zhang, Y.; Zhang, S.; Jia, T.; Sun, Z.; Qiu, J.; Xu, Z. Formation of long-and short-periodic nanoripples on stainless steel irradiated by femtosecond laser pulses. *J. Phys. D* **2011**, *44*, 505401. [CrossRef]

26. Gecys, P.; Vinciunas, A.; Gedvilas, M.; Kasparaitis, A.; Lazdinas, R.; Raciukaitis, G. Ripple formation by femtosecond laser pulses for enhanced absorptance of stainless steel. *J. Laser Micro/Nanoeng.* **2015**, *10*, 129–133. [CrossRef]

27. Liu, J.M. Simple technique for measurements of pulsed Gaussian-beam spot sizes. *Opt. Lett.* **1982**, *7*, 196. [CrossRef] [PubMed]

28. Zemaitis, A.; Gaidys, M.; Brikas, M.; Gecys, P.; Raciukaitis, G.; Gedvilas, M. Advanced laser scanning for highly efficient ablation and ultrafast surface structuring: Experiment and model. *Sci. Rep.* **2018**, *8*, 17376. [CrossRef]

29. Golovashkin, A.I.; Leksina, I.E.; Motulevich, G.P.; Shubin, A.A. The optical properties of niobium. *Sov. Phys. JETP* **1969**, *29*, 27–34.

30. Raether, H. *Surface Plasmons on Smooth and Rough Surfaces and on Gratings*; Springer: Berlin/Heidelberg, Germany, 1988.

31. Derrien, T.J.Y.; Krüger, J.; Bonse, J. Properties of surface plasmon polaritons on lossy materials: Lifetimes, periods and excitation conditions. *J. Opt.* **2016**, *18*, 115007. [CrossRef]

32. Gedvilas, M.; Raciukaitis, G.; Regelskis, K. Self-organization in a chromium thin film under laser irradiation. *Appl. Phys. A* **2008**, *93*, 203–208. [CrossRef]

33. Kunz, C.; Bonse, J.; Spaltmann, D.; Neumann, C.; Turchanin, A.; Bartolomé, J.F.; Müller, F.A.; Gräf, S. Tribological performance of metal-reinforced ceramic composites selectively structured with femtosecond laser-induced periodic surface structures. *Appl. Surf. Sci.* **2020**, *499*, 143917. [CrossRef]

34. Pan, A.; Dias, A.; Gomez-Aranzadi, M.; Olaizola, S.M.; Rodriguez, A. Formation of laser-induced periodic surface structures on niobium by femtosecond laser irradiation. *J. Appl. Phys.* **2014**, *115*, 173101. [CrossRef]

35. Choudhury, T.; Saied, S.O.; Sullivan, J.L.; Abbot, A.M. Reduction of oxides of iron, cobalt, titanium and niobium by low-energy ion bombardment. *J. Phys. D Appl. Phys.* **1989**, *22*, 1185–1195. [CrossRef]

36. Dacca, A.; Gemme, G.; Mattera, L.; Parodi, R. XPS analysis of the surface composition of niobium for superconducting RF cavities. *Appl. Surf. Sci.* **1998**, *126*, 219–230. [CrossRef]

37. Yang, Z.; Lu, X.; Tan, W.; Zhao, J.; Yang, D.; Yang, Y. XPS studies of nitrogen doping niobium used for accelerator applications. *Appl. Surf. Sci.* **2018**, *439*, 1119–1126. [CrossRef]

38. Kirner, S.V.; Wirth, T.; Sturm, H.; Krüger, J.; Bonse, J. Nanometer-resolved chemical analyses of femtosecond laser-induced periodic surface structures on titanium. *J. Appl. Phys.* **2017**, *122*, 104901. [CrossRef]

39. Badrinarayanan, S.; Sinha, S. X-ray photoelectron spectroscopy studies of the reaction of N^+_2-ion beams with niobium and tantalum metals. *J. Appl. Phys.* **1991**, *69*, 1141–1146. [CrossRef]

40. Prieto, P.; Galán, L.; Sanz, J.M. An XPS study of NbN$_x$ prepared by ion implantation and the near-surface effects induced by Ar$^+$ bombardment. *Surf. Sci.* **1991**, *251–252*, 701–705. [CrossRef]

41. van Gurp, G.J. The effect of structure on the superconducting properties of vanadium and niobium foils. *Philips Res. Rep.* **1967**, *22*, 10–35.

42. Finnemore, D.K.; Stromberg, T.F.; Swenson, C.A. Superconducting properties of high-purity niobium. *Phys. Rev.* **1966**, *149*, 231–243. [CrossRef]

43. Castel, X.; Guilloux-Viry, M.; Perrin, A.; Le Paven-Thivet, C.; Debuigne, J. Correlation between microwave surface resistance, AC susceptibility and in-plane ordering in YBa$_2$Cu$_3$O$_7$ thin films epitaxially grown on (100) MgO substrates. *Physica C* **1995**, *255*, 281–292. [CrossRef]

Article

Response of NIH 3T3 Fibroblast Cells on Laser-Induced Periodic Surface Structures on a 15×(Ti/Zr)/Si Multilayer System

Suzana Petrović [1,*], Davor Peruško [1], Alexandros Mimidis [2], Paraskeva Kavatzikidou [2], Janez Kovač [3], Anthi Ranella [2], Mirjana Novaković [1], Maja Popović [1] and Emmanuel Stratakis [2]

[1] Department of Atomic Physics, "VINČA" Institute of Nuclear Sciences—National Institute of the Republic of Serbia, University of Belgrade, P.O. Box 522, 11001 Belgrade, Serbia; dperusko@vin.bg.ac.rs (D.P.); mnovakov@vin.bg.ac.rs (M.N.); majap@vin.bg.ac.rs (M.P.)

[2] Institute of Electronic Structure and Laser (IESL), Foundation for Research and Technology (FORTH), N. Plastira 100, Vassilika Vouton, 70013 Heraklion, Greece; amimidis@iesl.forth.gr (A.M.); ekavatzi@iesl.forth.gr (P.K.); ranthi@iesl.forth.gr (A.R.); stratak@iesl.forth.gr (E.S.)

[3] Jožef Stefan Institute, Jamova 39, 1000 Ljubljana, Slovenia; janez.kovac@ijs.si

* Correspondence: spetro@vin.bg.ac.rs; Tel.: +381-11-3408-560

Received: 9 November 2020; Accepted: 11 December 2020; Published: 16 December 2020

Abstract: Ultrafast laser processing with the formation of periodic surface nanostructures on the 15×(Ti/Zr)/Si multilayers is studied in order to the improve cell response. A novel nanocomposite structure in the form of 15×(Ti/Zr)/Si multilayer thin films, with satisfying mechanical properties and moderate biocompatibility, was deposited by ion sputtering on an Si substrate. The multilayer 15×(Ti/Zr)/Si thin films were modified by femtosecond laser pulses in air to induce the following modifications: (i) mixing of components inside of the multilayer structures, (ii) the formation of an ultrathin oxide layer at the surfaces, and (iii) surface nano-texturing with the creation of laser-induced periodic surface structure (LIPSS). The focus of this study was an examination of the novel Ti/Zr multilayer thin films in order to create a surface texture with suitable composition and structure for cell integration. Using the SEM and confocal microscopies of the laser-modified Ti/Zr surfaces with seeded cell culture (NIH 3T3 fibroblasts), it was found that cell adhesion and growth depend on the surface composition and morphological patterns. These results indicated a good proliferation of cells after two and four days with some tendency of the cell orientation along the LIPSSs.

Keywords: multilayer thin film; ultrafast laser processing; laser-induced periodic surface structure; cell response

---

1. Introduction

The role of biomaterials as implants based on natural or synthetic materials in clinical treatment is the replacement of damaged, non-functional organs, and tissues. The selection and manufacturing of biomaterials are predominantly based on imitating the chemical and physical properties of the natural tissue, causing a minimal response of foreign body [1,2]. For years, metallic biomaterials have been widely used in biomedical devices due to an excellent combination of mechanical properties and durability. However, the metallic biomaterials still lack satisfactory biofunctionalities for certain applications, such as blood compatibility for blood-contacting devices and bone conductivity in orthopedic applications [3]. Titanium-based materials are nowadays well integrated into the body, attributing to their constructive properties, such as high specific strength, relatively low Young's modulus values, excellent corrosion resistance, and good biocompatibility [4,5]. Pure titanium and titanium alloys are ideal as biomaterials due to the superior biocompatibility based on an outer oxide layer with a negative charge at physiological pH and protects dissolving metallic components into biological fluids [6,7]. New biomedical Ti-based alloys have been developed with a high concentration of β-stabilizer elements (β phase of titanium) as potential solutions to the mismatch between the Young's modulus of the implant and the surrounding hard tissues. The most common alloying elements added to these new alloys are niobium, tantalum, zirconium, and molybdenum, as they do not exhibit any cytotoxic reaction in contact with cells [8–10].

The surface properties of the biomaterial play critical roles in the interactions between the biological environment and the implant. Surface functionalization is one of particular interest and a requirement to improve the surface bioactivity and other biofunctionalities, in terms of the adsorption of proteins on the material surface, which is determinant for the subsequent processes of cell growth, differentiation, and extracellular matrix formation [11]. Due to the limitation of biocompatibility on the surface properties, thin films and coatings have been considered as a useful alternative. The surface properties of thin films including dissolution and corrosion, fouling resistance, and hydrophilicity/hydrophobicity are modulated to improve the materials response in biological environments [12,13]. The appropriate way for the preparation of Ti-based alloy coatings is the deposition of a multilayer structure with an alternate distribution of Ti and other (Zr, Ta, Mo, and Nb) components in thin films [14–16]. These multilayer systems exhibit desirable mechanical properties, because the alloying Ti with mentioned non-cytotoxic metals can be achieved with the further lowering of Young's modulus [17–20].

The surface modification of the biomaterial aims to create specific chemical and physical properties that offer a favorable cellular response. In cases where tissue integration is desired, the physical environment should include macro, micro, and even nanoscale features that allow for cells to adhere, proliferate, and migrate [21–23]. Femtosecond laser texturing holds promise for the surface modification of biomaterials due to a wide application to all materials; the possibility of getting a wide variety of structures with micro- and nano-scaled features; and a fast, repeatable, and contactless process. The femtosecond laser pulses exhibit some advantages such as less debris contamination, reproducibility, precision, and a minimal heat-affected zone, making it a very promising technique for the surface modification of biomaterials [24]. With laser fluence close to the damage threshold for given materials, the nanoscale periodic ripple structures usually are formed due to interference between the incident laser beam and the surface scattered wave [25–29]. The interaction with intense laser pulses involves a lot of physical processes simultaneously (heating, melting, ejection of species, vaporization, shock wave propagation, and expansion) inducing the formation of rougher surface texture-like grooves and spikes. Laser surface modification is a unique method that allows the production of a bioactive surface with formation of the desired oxide and alloy, the creation of nano/micro textures, and a change in the wettability of the surface. The functionalization of these bioactive surfaces is reflected in the adhesion, spreading, and proliferation of the different cells and wherein to inhibit the development of bacterial infections. Surfaces with micro/nano morphological characteristics support cell growth. Furthermore, bacteria adhere preferentially to polished surfaces and to surfaces with cavities larger than the bacteria size, but the development of bacterial infections on the nano-textured surfaces is reduced [30–34].

The orientation of cells is a crucial issue for nerve regeneration and muscle reparation. The cell growth is determined by the structure of the extracellular matrix, where the tri-dimensional patterns can induce the spread of cells in the desired direction. At the surface with roughness determined by randomly distributed features, cells are oriented in all directions. When muscle tissue made up from elongated cells (fibers) are cultured on well-defined patterns with laser-etched micro and nanometer lines, the cells will tend to localize in the grooves between the lines, favoring oriented growth. In the healthy tendons, fibroblasts are oriented in the direction of stretching; however, the fibroblasts tend to orient themselves randomly during the tendon healing. By incorporating an extracellular matrix with suitable patterns, the direction of fibroblast growth and better tendon healing can be influenced [35,36].

Surface modification in terms of the oxidation and texturing of Ti-based materials shows cell repellence explained by limitation of the cell flexibility as well as a limited contact area, which considerably weakens the cell adhesion. On the other hand, morphological features (laser-induced periodic surface structure (LIPSS), spike etc.) reduce adhesion of the trans-membrane proteins in the cell membrane, which are in contact with their ridges, leading also to weaker adsorption. By studying the cell response, it was found that most cells are attached to the tops of morphological objects and do not reach the valleys. In addition, it has been observed that the certain number of cells is oriented to follow the direction of the LIPSS [37,38]. In the future, special attention should be focused on researching the possibility of deposition of Ti-based multilayer thin films and their laser processing, which would achieve an appropriate distribution of composition (alloy and oxide) in a small thickness and the creation of bioactive surface structures.

In this study, the main focus was directed to the examination of a novel titanium–zirconium (Ti/Zr) bimetallic nanocomposite in order to create a bioactive surface with adjusted composition and morphology for cell responses with a high degree of proliferation, as well as cell growth along laser-induced morphological features. Following the previous sub-ablative femtosecond laser studies for the fabrication of a micro/nano laser-induced periodic surface structure (LIPSS) on the multilayer thin films, the formation of an LIPSS was explored for the possible creation of alloys in the multilayer Ti/Zr system and a very thin oxide layer on its surface. The micro/nano patterns distribution with changed composition and wettability are evaluated with the aim of determining their influence in cell proliferation and cell morphology with the possibility of cell orientation along the laser-induced periodic structure.

2. Materials and Methods

2.1. Deposition of Ti/Zr Multilayers

The multilayer structures composed of consecutively distributed titanium and zirconium layers were deposited in a Balzers Sputtron II system, using 1.3 keV argon ions and 99.9% pure Ti and Zr targets. Before deposition, the chamber was evacuated to the base pressure of 1×10^{-6} mbar, while the Ar partial pressure during deposition was 1×10^{-3} mbar. For the substrate, we selected a silicon wafer Si (100), which was cleaned by etching in hydrofluoric acid (HF) and immersion in deionized water before mounting in the chamber. The deposition of multilayers was performed in a single vacuum run at a deposition rate of 0.17 nm s^{-1} for both Ti and Zr components, without heating of the substrates. The total thickness of the complete multilayer structure, which consisted of fifteen (Ti/Zr) bilayers, was 500 nm, where thickness of individual Ti and Zr layers was about 17 nm.

2.2. Laser Processing of Ti/Zr Multilayers

Laser processing of the multilayer 15×(Ti/Zr) thin films was performed with the Yb:KGW laser source Pharos SP from Light Conversion. The surface of thin films was irradiated by focused linearly p-polarized pulses with the following characteristics: repetition rate of 1 kHz, pulse duration equal to 160 fs, central wavelength of 1030 nm, and 43 μm Gaussian spot diameter. Samples were laser processed in an open air ambient environment and mounted on a motorized, computer controlled, X-Y-Z translation stage, at normal incidence to the laser beam. For higher precision, the irradiations were conducted at identical conditions covering a surface of 5 × 5 mm at a pulse energy of 2.5 μJ and scan velocity of 3 mm s^{-1} with constant distance between lines of 15 μm. In each line, the energy per pulse was assumed to be constant, since the pulse energy deviation was less than 1%.

2.3. Characterization of Ti/Zr Multilayers

The depth profile in the unmodified and laser-modified areas of the 15×(Ti/Zr)/Si system was analyzed by Auger electron spectroscopy (AES) in a PHI SAM 545 spectrometer. During depth profiling, the sample was sputtered by inclined Ar ion beams of 1 keV at an ion incidence angle of 47° with respect to the surface normal. Surface compositional analysis of the 15×(Ti/Zr)/Si system was done by X-ray photoelectron spectroscopy (XPS) using the PHI-TFA XPS spectrometer. The relative sensitivity factors were used for the calculation of surface concentrations [39], and they were provided by an instrument producer. Detailed surface morphology after irradiation was examined firstly by optical microscopy and then by scanning electron microscopy (JEOL JSM-7500F) equipped with energy-dispersive X-ray spectroscopy (EDS) (Oxford Instruments INCA) and by atomic force microscopy (AFM, Solver PRO 47) in the contact mode with a standard preamplifier. Structural characterization of the samples was done by cross-sectional transmission electron microscopy (TEM) in conventional and high-resolution modes, using an FEI Talos F200X microscope (Thermo Fisher Scientific, 168 Third Avenue, Waltham, MA USA) which was operated at 200 keV voltages. The samples for TEM examination were prepared by a conventional procedure including the cutting and mounting of the samples onto a copper slot, followed by polishing and Ar ions milling. In addition, the samples were also analyzed in scanning transmission (STEM) mode with energy-dispersive spectrometry (EDS) elemental profiling and element color mapping.

2.4. Cell Study

Cell integration was studied with the NIH 3T3 established adherent mouse fibroblast cell line (obtained by Mr. G. Vrentzos, IMBB, FORTH). Fibroblast NIH 3T3 cells were grown in cell culture flasks using Dulbecco's modified Eagle's medium DMEM (Invitrogen, Grand Island, NY, USA) supplemented with 10% fetal bovine serum (Biosera, Sussex, UK) in a 5% CO_2 incubator (Thermo Scientific) at 37 °C. Laser-processed samples were autoclaved and transferred into sterile wells of 24-well plates (Sarstedt; Numbrecht, Germany). Culture medium with 3×104 cells were seeded onto the samples, where they were cultured in different time periods, ranging from two to four days in order to estimate the cell orientation, adhesion, and proliferation. The cultured samples with the NIH 3T3 fibroblast were washed twice with 0.1 M sodium cacodylate buffer (SCB) and fixed with 2% glutaraldehyde (GDA) and 2% paraformaldehyde (PFA) in 0.1 M SCB for 30 min. Subsequently, samples were washed twice with 0.1 M SCB and dehydrated in increasing concentrations (from 30 to 100%) of ethanol. Finally, before the determination of NIH 3T3 fibroblasts cell morphology, samples were dried in a critical point drier (Baltec CPD 030), sputter-coated with thin (10 nm) gold/palladium layer (Baltec SCD 050) and observed by Scanning Electron Microscope (JEOL JSM-6390 LV).

The cytoskeleton, focal adhesion points, and nucleus of NIH 3T3 fibroblasts were stained for actin filaments, vinculin, and 4′,6-Diamidino-2-Phenylindole (DAPI). Specifically, after 2 and 4 days of cell culture, the samples were fixed with 4% paraformaldehyde (PFA) for 15 min and permeabilized with 0.1% Triton ×-100 in phosphate-buffered saline (PBS) for 5 min. The non-specific binding sites were blocked with 2% Bovine Serum Albumen (BSA) in PBS for 30 min. The samples were incubated with α-vinculin primary antibody (FAK 100, Actin Cytoskeleton/Focal Adhesion Staining Kit, Sigma Aldrich) (1:500 in PBS–BSA 1%), and then, the next day, double labeling for 2 h at room temperature with Alexa Flour 48 2nd Antibody, mouse (Thermo Scientific) for focal adhesion points staining, and Phalloidin–Tetramethylrhodamine B isothiocyanate (TRITC-conjugated) phalloidin (FAK 100, Actin Cytoskeleton/Focal Adhesion Staining Kit, Sigma Aldrich) (1:500 in PBS–BSA 1%) for actin filaments in the cytoskeleton staining. Finally, the samples were washed with PBS and put on coverslips with 4′,6-Diamidino-2-Phenylindole (DAPI) (Molecular Probes by Life Technologies, Carlsbad, CA, USA) for nuclei staining. Cell imaging was performed using a Leica SP8 inverted scanning confocal microscope with ×40 objective. Changes in the directional orientation of cell cytoskeleton, "Local gradient orientation" were performed using the Fiji Image-J plug-in "Directionality" [40]. In this way, the data of the structures, presented in the input image (Actin image on the Confocal), in a given direction was extracted and plotted as a polar plot. The amount of cells was normalized with the maximum value and expressed as a normalized amount. For good statistics, ten images of each time point and each area were recorded.

3. Results and Discussion

The laser processing of materials with multiple, linearly polarized ultrafast laser pulses is used to create the arrayed surface structure, such as a laser-induced periodic surface structures (LIPSS). An optimal combination of several laser parameters (pulse energy, number of pulses, repetition rate, and scanning rate) has achieved conditions for the generation of LIPSS morphology on the surface of a 15×(Ti/Zr)/Si multilayer system (Figure 1a). The observed surface structure is defined as low spatial frequency LIPSS (LSFL), originating from an interference of the incident laser beam with a surface electromagnetic wave excited during the laser treatment [41]. Polarization of the laser beam was chosen to be normal on the scanning direction, in order to obtain LIPSSs for as long as possible. The relatively good uniformity of the LIPSS can be characterized by their length in the range of 4 to 6 μm, where some LIPSSs are ended, while others appeared or bifurcation points can be recognized. These values of LIPSS length are coincided with the mean free path of excited surface plasmon polaritons (SPP) for Ti component irradiated with laser pulses at a wavelength of 1030 nm [42]. It has been found that titanium has a small SPP mean free path, which supports a good coherence between excited SPP and incident laser radiation, favoring the formation of high regularity LIPSS [42]. The formation of LIPSSs is followed by the ablation of a multilayer Ti/Zr system without visible hydrodynamic features, but ripples are somewhere covered with a nanoparticles dimension up to 100 nm (Figure 1b). The significant ablation of a multilayer Ti/Zr structure, especially Ti as the top layer, is expected due to the applied fluence of 0.662 Jcm^{-2} being higher than the ablation threshold of 0.228 Jcm^{-2} for the Ti component [42]. By comparing the EDS spectra before and after the laser irradiation of a 15×(Ti/Zr)/Si multilayer system (Figure 1c), it has been found that approximately half of the multilayer thin film was removed. The concentration of both components Ti and Zr were reduced by about 50%, while the concentration of Si was increased as a contribution of the substrate. At the same time, the content of oxygen was increased, which indicates that the surface oxidation of Ti and Zr has occurred during the laser processing.

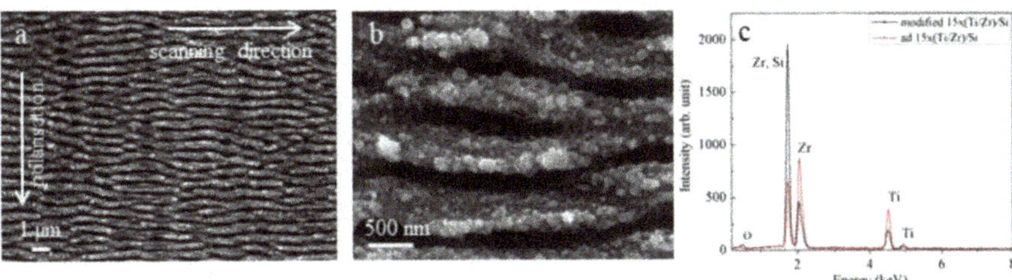

Figure 1. SEM images (**a,b**) of laser-modified 15×(Ti/Zr)/Si multilayer thin films with a fluence of 0.662 Jcm^{-2}, and energy-dispersive X-ray spectroscopy (EDS) spectra (**c**) recorded from un-modified and laser-treated surface.

The morphological features of the created LIPSS at the surface of the 15×(Ti/Zr)/Si multilayer thin film are specified in more detail by atomic force microscopy (Figure 2). The whole laser-processed surface (5 × 5 mm) was covered by uniform distributed LIPSSs (Figure 2a), which are almost regular and identical with lengths in the range from 4 to 7 μm. The consequence of the ripple structure is a higher roughness of the modified surfaces (Ra = 65 nm), causing the higher total surface. The average periodicity between LIPSSs was about 880 nm (Figure 2b), which would correspond to the formation of low spatial frequency LIPSS (LSFL). The depth between ripples reaches a value up to 200 nm (Figure 2b), which would mean that it almost touches the Si substrate at the bottom of the valley between two ripples if half of the multilayer structure was removed.

The information about the surface composition and chemical state of constituents (Ti and Zr) before and after laser processing is obtained by an analysis of binding energies in the corresponding XPS spectra. At the surface of the as-deposited 15×(Ti/Zr)/Si multilayer system, where titanium was the top layer (Figure 3a), the signal of the Zr component did not appear (Figure 3b). Titanium on the surface appeared in the oxidation state of +4 with a binding energy of 458.5 eV, corresponding to titanium dioxide (TiO$_2$). However, a small amount (approximately 3%) of Ti exists in a metallic state at the binding energy of 454.2 eV. Together with the laser-assisted ablation of the thin film material, laser-induced surface oxidation is also achieved, since both components Ti and Zr exist as oxide phases at a binding energy of 458.5 eV for Ti (Figure 3c) and 182.4 eV for Zr (Figure 3d). The interaction between the laser beam and 15×(Ti/Zr)/Si multilayer thin film generates a dense plasma with lot of energetic species such as Ti^+, Ti^{3+}, and TiO^+, making the irradiated surface very chemically reactive [43]. The titanium and zirconium react with the gaseous surrounding, forming a surface ultrathin oxide layer rich with carbon due to the decomposition of CO_2 [44]. On the laser-treated surface, the TiO_2 constitutes about 58%, while ZrO_2 covers the surface with 42%, which would mean that slightly more Zr components are removed by laser treatment. The interesting finding was that there was no appearance of silicon at the contact surface, which is very important for cell study: only the biocompatible phases of Ti-oxide and Zr-oxide exist on the surface.

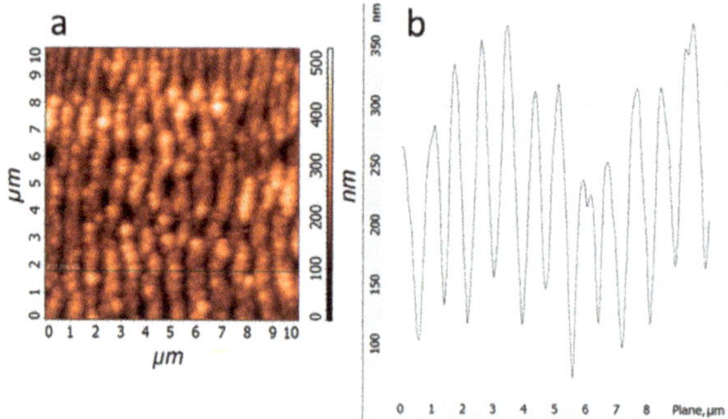

Figure 2. Atomic Force Micrpscopy (AFM) image (**a**) and surface profile (**b**) for laser-treated 15×(Ti/Zr)/Si multilayer thin film at fluence of 0.662 Jcm^{-2}.

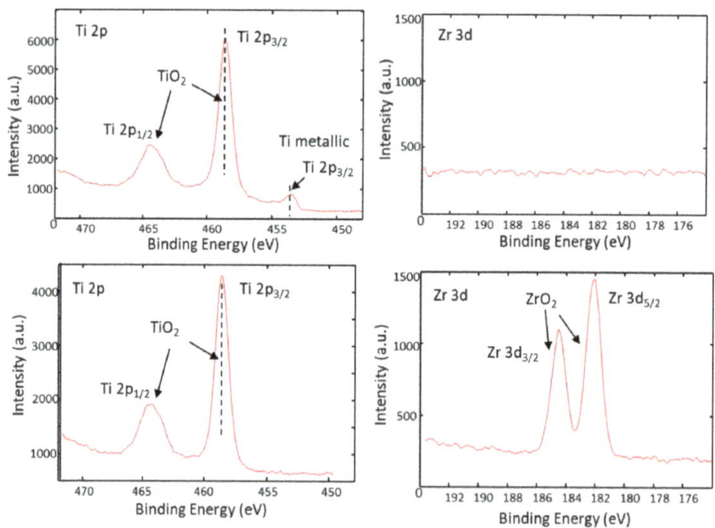

Figure 3. XPS spectra of (**a**) Ti 2p and (**b**) Zr 3d for the surface of as-deposited 15×(Ti/Zr)/Si multilayer thin films and XPS spectra (**c**) Ti 2p and (**d**) Zr 3d recorded after laser modification.

The concentration depth profiles through the 15×(Ti/Zr)/Si multilayer system were recorded with AES spectrometry (Figure 4). The spectrum for as-deposited multilayer thin films (Figure 4, Top) is shown to have very well separated Ti and Zr layers with approximately the same thicknesses as the individual layers. Initially well-separated layers are intermixed after the laser irradiation (Figure 4, Bottom), which could be attributed to the alloying between Ti and Zr components. However, a weak periodicity has been retained on curves corresponding to Ti and Zr, indicating that the layers are not completely intermixed. At the surface and near-surface region, the concentration of oxygen was relatively high; this is a consequence of the laser-assisted surface oxidation and the certain penetration of oxygen inside the multilayer structure. On the other hand, silicon atoms were diffused in the multilayer structure and almost reached the surface.

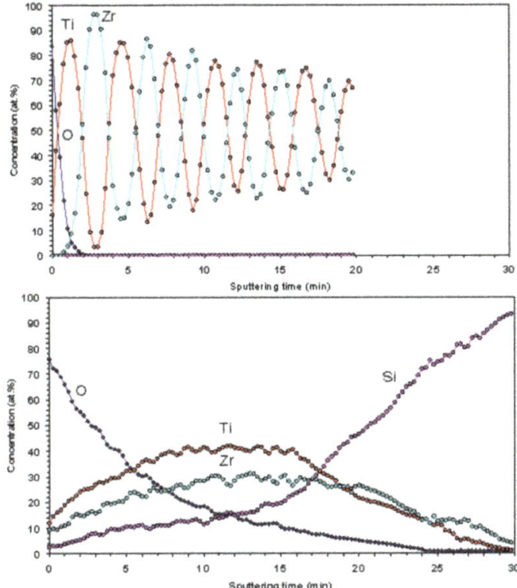

Figure 4. Auger electron spectroscopy (AES) spectra before (**Top**) and after (**Bottom**) laser modification of the 15×(Ti/Zr)/Si multilayer thin film.

A cross-section view of the 15×(Ti/Zr)/Si multilayer system was obtained with TEM microscopy, which illustrated the volume changes made inside of the thin films after laser processing. A very well-defined multilayer structure composed of a total 30 layers was confirmed for the as-deposited 15×(Ti/Zr)/Si sample (Figure 5a). The thicknesses of individual Ti and Zr layers are identical with values of about 17 nm (Figure 5b). The TEM cross-section view after laser processing (Figure 5c) confirms previous assumptions obtained by the SEM-EDS method that almost half of the multilayer structure was removed, because thirteen layers remained from the deposited thirty. The TEM microphotograph (Figure 5c) has shown that the part of the thin films close to the surface has mixed layers (Figure 5d), but the layers that are close to the Si substrate remain quite clearly separated. A small part of the sub-surface region, including a few layers (no more than three), has very fine grain structures, where Ti and Zr are totally intermixed with a high possibility to form alloy.

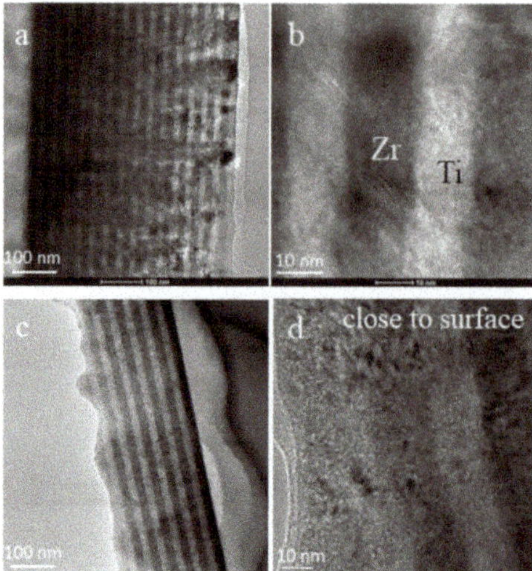

Figure 5. TEM images of a cross-section view for (**a**) a whole unmodified 15×(Ti/Zr)/Si multilayer thin film, (**b**) part of an unmodified structure at high magnification, (**c**) a whole laser-modified 15×(Ti/Zr)/Si multilayer thin film, and (**d**) part of a modified structure close to the surface.

The EDS/TEM elemental mapping images are shown in Figure 6 to identify the spatial distributions of Ti, Zr, and Si components in the multilayer Ti/Zr system before and after laser processing. In elemental maps for the unmodified sample (Figure 6a), it can be found that Si atoms appeared only in a narrow area attributed to the substrate (Figure 6a1), while Zr (Figure 6a2) and Ti (Figure 6a3) atoms were grouped in periodically repeated areas associated with their individual layers. In the case where all three elemental maps were overlapped (Figure 6a4), the edges between the Ti and Zr layers are very sharp, pointing to their complete separation. On the contrary, in the laser-processed sample (Figure 6b), the spatial distribution of Si atoms (Figure 6b1) shows the penetration of Si atoms into the multilayer structure. The spatial distribution of Si atoms reflects that the Si atoms on their tracing through the multilayer structure are more collected in the Zr layer than in the Ti layer. Generally, it is known that Si is the mainly dominant diffusion species, and Si atoms as smaller atoms regarding Ti and Zr are easily diffused in their crystal lattice, locating in interstitially positions [45]. Elemental maps for Zr (Figure 6b2) and Ti (Figure 6b3) reveal that these atoms are intermixed, indicating the final result to the formation of Ti-Zr alloy due to their unlimited dissolving [46,47]. From the image obtained by the mutual overlap of all three elemental maps (Figure 6b4), it can be concluded that the degree of Ti and Zr mixing is reduced with the distance from the surface, whereby they keep a layered structure inside of the thin films.

The biocompatibility of this laser-patterned thin films is closely related to the cell viability and proliferation with attachment, adhesion, and spreading in the early phase of the cell–material interaction, and at the same time, it demonstrates the role of surface topography in influencing the cell behavior [48]. A morphological examination of NIH 3T3 fibroblast cell proliferation on as-deposited and laser-modified 15×(Ti/Zr)/Si multilayer surfaces was estimated by SEM analysis after two- and four-day cultivation (Figure 7). On the flat area of the as-deposited 15×(Ti/Zr)/Si sample, there is an arbitrary cell growth occurring in all directions (Figure 7a), especially after four days when the number of cells is increased (Figure 7b). After two-day cultivation, the laser-modified surface of 15×(Ti/Zr)/Si with LIPSS structures is partially covered by cells, forming elongated groups (Figure 7c). The fibroblast cells adhered and proliferated on the LIPSSs in 2 days, with tendency for a growth along the LIPSSs orientation. Almost the whole laser-modified surface is covered by cells, with the significant longer groups of cells in the orientation of LIPSSs after 4 days of cultivation (Figure 7d). In addition to the evident directed growth of fibroblast cells on a laser-created surface topography, it can be observed that the surface topography induced faster cell proliferation due to the significantly greater coverage of the surface with higher and longer cell groups. On the other hand, fibroblast cells are oriented along periodic structures, especially after 4 days of cultivation (Figure 8).

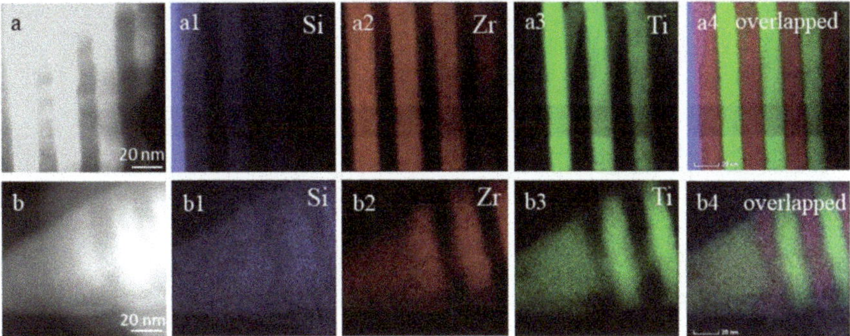

Figure 6. Elemental mapping (Si, Zr, and Ti) and their overlapping for (**a**) an as-deposited, (**a1**) elemental map for Si, (**a2**) elemental map for Zr, (**a3**) elemental map for Ti, (**a4**) overlapped all elemental map and (**b**) laser-modified 15×(Ti/Zr)/Si multilayer thin film, (**b1**) elemental map for Si, (**b2**) elemental map for Zr, (**b3**) elemental map for Ti, (**b4**) overlapped all elemental map

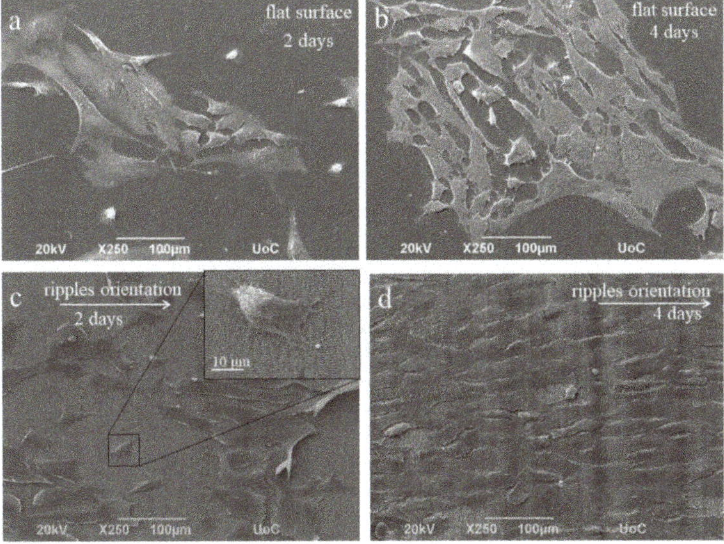

Figure 7. SEM images of NIH 3T3 fibroblasts cultivated on the as-deposited (**a,b**), and laser-processed 15×(Ti/Zr)/Si multilayer thin film (**c,d**), for 2 and 4 days, respectively.

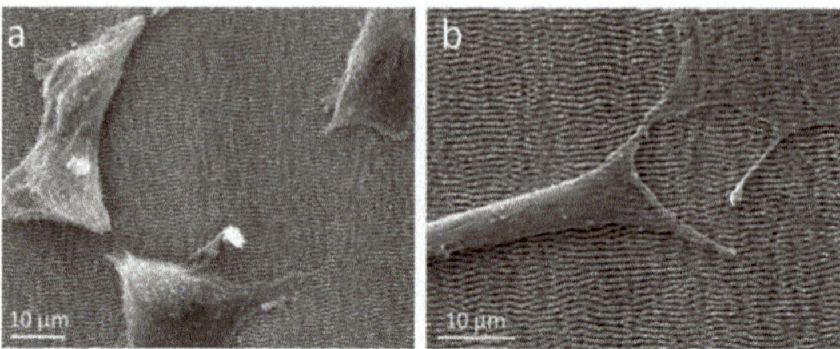

Figure 8. SEM images of NIH 3T3 fibroblasts cultivated on laser-modified 15×(Ti/Zr)/Si multilayer after (**a**) two and (**b**) four days.

A relatively high degree of cell proliferation is caused by good cell adhesion, which is directly related to the surface characteristics such as composition and morphology. Increasing the surface roughness of the multilayer 15×(Ti/Zr)/Si system, which is reflected in the creation of parallel periodic structures, has a positive effect on cell adhesion. Surface roughness at order of microns increased the effective surface area of cell adhesion. It could be concluded that cells on the laser-modified surface are immobilized, and the cell deposition/adhesion rates increase as a consequence of cells anchored in morphology features especially in sub-micro- and micro-channels [49–51].

This study of the laser-modified 15×(Ti/Zr)/Si multilayer demonstrated that in sub-micro features, the ridge width is commonly larger than or equal to the size of a single cell, which is permissive for cell attachment and migration, as well as cell alignment following the geometrical guidance. In contrast, nanometer features are similar to the ECM (extracellular matrix) architectures and typically much smaller than a single cell, consequently inducing cell alignment in a more fundamental way such as mimicking or signaling the cell membrane receptors [52].

The Directionality Polar Plot demonstrates the effect of the flat and ripples area of the 15×(Ti/Zr)/Si multilayer on the fibroblast cytoskeleton in terms of orientation at both time points (Figure 9). It was observed that there was a random orientation of cells on the flat area at both time points. In particular, the normalized amount of the flat area (black line) after 2 days showed a broad distribution at a range of ±90°, and this was also observed after 4 days (blue line). After 2 days, the fibroblasts oriented on the LIPSS area (red line) with the narrowest distribution ranging from ±15° corresponding to an orientation profile following the LIPSSs direction. The orientation of fibroblasts after 4 days was a little disturbed (pink line), in which the distribution ranged ±30° due to the greater covering of the surface with cells.

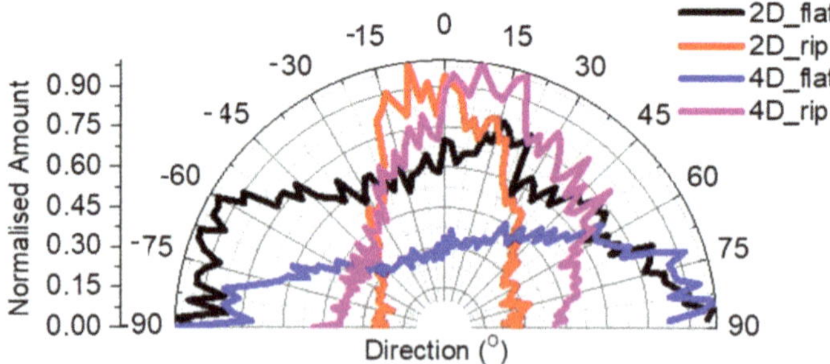

Figure 9. Directional Polar Plot of NIH 3T3 fibroblasts' cytoskeleton on the flat and ripples area of sample 2 for 2 and 4 days; the black and blue lines represent the flat area at 2 days and 4 days respectively, while the red and magenta lines represent the ripples area at 2 days and 4 days, respectively.

4. Conclusions

This study included laser processing of the nanocomposite 15×(Ti/Zr)/Si structure in order to achieve surface bioactivation by adjusting the morphology and composition. An optimal combination of laser parameters, a pulse energy of 2.5 µJ and scan velocity of 3 mm s^{-1}, achieved conditions for the formation of a high regularity laser-induced periodic surface structure (LIPSS) on a relatively large 5×5 mm surface. The formation of an LIPSS was accompanied by an intense ablation of almost half of the Ti/Zr multilayer structure. Laser modification of the nanocomposite 15×(Ti/Zr)/Si samples induced the desired composition changes to achieve optimum biocompatibility. Primarily, the formation of an ultrathin oxide layer on a surface composed of Ti and Zr oxide, and then alloying Ti and Zr components in a sub-surface layer, was reached.

The biocompatibility of the laser-processed nanocomposite 15×(Ti/Zr)/Si thin films is confirmed by a seed NIH 3T3 established adherent mouse fibroblast cell line. The fibroblast cells adhered and proliferated on the laser-induced morphology in 2 days with tendency for a growth along the ripples (LIPSS) orientation. Almost the whole laser-modified surface of 15×(Ti/Zr)/Si system is covered by cells, with the significant longer groups of cells in the orientation of ripples after 4 days of cultivation. The bioactivation of this specific 15×(Ti/Zr)/Si multilayer system with laser surface texturing and an adjusting of surface composition could be potentially useful for tissue engineering and the application of this material as an implant.

The presence of titanium and zirconium oxides at the contact surface with a laser-induced ripple structure enabled physical and molecular interaction between the observed surface and the cell membrane, which contributed to a relatively large number of cells after 2 days and especially 4 days after cultivation. It can be suggested that the newly developed multilayer Ti/Zr thin films achieved a high degree of proliferation and oriented cell growth, and thus, they are suitable implants.

Author Contributions: Conceptualization, S.P. and E.S.; data curation, S.P., D.P. and P.K.; methodology, S.P. and D.P.; formal analysis, A.M., P.K., J.K., A.R., M.N. and M.P.; investigation, S.P. and D.P.; resources, S.P., M.N. and E.S.; writing—original draft preparation, S.P.; writing—review and editing, S.P.; visualization, P.K. and M.N.; supervision, E.S.; project administration, E.S.; funding acquisition, S.P. and E.S. All authors have read and agreed to the published version of the manuscript.

Funding: This project has received funding from the EU-H2020 research and innovation programme under grant agreement No 654360 having benefitting from the access provide by Institute of Electronic Structure and Laser (IESL), Foundation for Research and Technology (FORTH) in access site Heraklion, Crete, Greece with in the frame work of the NFFA/Europe Transnational Access Activity. The research was funded by the Ministry of Education, Science and Technological Development of the Republic of Serbia.

References

1.   Othman, Z.; Pastor, B.C.; van Rijt, S.; Habibovic, P. Understanding interactions between biomaterials and biological systems using proteomic. *Biomaterial* **2018**, *167*, 191–204. [CrossRef] [PubMed]

2.   Su, Y.; Luo, C.; Zhang, Z.; Hermawan, H.; Zhu, D.; Huang, J.; Liang, Y.; Li, G.; Ren, L. Bioinspired surface functionalization of metallic biomaterials. *J. Mech. Behav. Biomed. Mater.* **2018**, *77*, 90–105. [CrossRef] [PubMed]

3.   Bose, S.; Robertson, S.F.; Bandyopadhyay, A. Surface modification of biomaterials and biomedical devices using additive manufacturing. *Acta Biomater.* **2018**, *66*, 6–22. [CrossRef] [PubMed]

4.   Correa, D.R.; Kuroda, P.A.; Lourenço, M.L.; Fernandes, C.J.; Buzalaf, M.A.; Zambuzzi, W.F.; Grandini, C.R. Development of Ti-15Zr-Mo alloys for applying as implantable biomedical devices. *J. Alloy. Compd.* **2018**, *749*, 163–171. [CrossRef]

5.   Rack, H.J.; Qazi, J.I. Titanium alloys for biomedical applications. *Mater. Sci. Eng. C* **2006**, *26*, 1269–1277. [CrossRef]

6.   Zhao, D.; Chen, C.; Yao, K.; Shi, X.; Wang, Z.; Hahn, H.; Gleiter, H.; Chen, N. Designing biocompatible Ti-based amorphous thin films with no toxic element. *J. Alloy. Compd.* **2017**, *707*, 142–147. [CrossRef]

7.	Asri, R.I.M.; Harun, W.S.W.; Samykano, M.; Lahc, N.A.C.; Ghani, S.A.C.; Tarlochand, F.; Raza, M.R. Corrosion and surface modification on biocompatible metals: A review. *Mater. Sci. Eng. C* **2017**, *77*, 1261–1274. [CrossRef]

8.	Wang, W.; Zhang, X.; Sun, J. Phase stability and tensile behavior of metastable β Ti-V-Fe and Ti-V-Fe-Al alloys. *Mater. Charact.* **2018**, *142*, 398–405. [CrossRef]

9.	Jung, I.S.; Jang, H.S.; Oh, M.H.; Lee, J.H.; Wee, D.M. Microstructure control of TiAl alloys containing β stabilizers by directional solidification. *Mater. Sci. Eng.* **2002**, *A329–A331*, 13–18. [CrossRef]

10.	Donato, T.A.G.; De Almeida, L.H.; Nogueira, R.A.; Niemeyer, T.C.; Grandini, C.R.; Caram, R.; Schneider, S.G.; Santos, A.R., Jr. Cytotoxicity study of some Ti alloys used as biomaterial. *Mater. Sci. Eng. C* **2009**, *29*, 1365–1369. [CrossRef]

11.	Chen, W.C.; Chen, Y.S.; Ko, C.L.; Lin, I.; Kuo, T.H.; Kuo, H.N. Interaction of progenitor bone cells with different surface modifications of titanium implant. *Mater. Sci. Eng. C* **2014**, *37*, 305–313. [CrossRef] [PubMed]

12.	Mozetič, M.; Vesel, A.; Primc, G.; Eisenmenger-Sittner, C.; Bauer, J.; Eder, A.; Schmid, G.H.; Ruzic, D.N.; Ahmed, Z.; Barker, D.; et al. Recent developments in surface science and engineering, thin films, nanoscience, biomaterials, plasma science, and vacuum technology. *Thin Solid Films* **2018**, *660*, 120–160.

13.	Mahapatro, A. Bio-functional nano-coatings on metallic biomaterials. *Mater. Sci. Eng. C* **2015**, *55*, 227–251. [CrossRef] [PubMed]

14.	Chui, P.; Jing, R.; Zhang, F.; Li, J.; Feng, F. Mechanical properties and corrosion behavior of b-type Ti-Zr-Nb-Mo alloys for biomedical application. *J. Alloy. Compd.* **2020**, *842*, 155693. [CrossRef]

15.	Shea, H.; Hu, J.; Li, P.C.; Huang, G.; Zhang, J.; Zhang, J.; Mao, Y.; Xiao, H.; Zhou, X.; Zu, X.; et al. Compositional dependence of hydrogenation performance of Ti-Zr-Hf-Mo-Nb high-entropy alloys for hydrogen/tritium storage. *J. Mater. Sci. Technol.* **2020**, *55*, 116–125.

16.	Sevostyanov, M.A.; Kolmakov, A.G.; Sergiyenko, K.V.; Kaplan, M.A.; Baikin, S.A.; Gudkov, S.V. Mechanical, physical–chemical and biological properties of the new Ti–30Nb–13Ta–5Zr alloy. *J. Mater. Sci.* **2020**, *55*, 14516–14529. [CrossRef]

17.	Tallarico, D.A.; Gobbi, A.L.; Paulin Filho, P.I.; Maia da Costa, M.E.H.; Nascente, P.A.P. Growth and surface characterization of TiNbZr thin films deposited by magnetron sputtering for biomedical applications. *Mater. Sci. Eng. C* **2014**, *43*, 45–49. [CrossRef]

18.	Ke, J.L.; Huang, C.H.; Chen, Y.H.; Tsai, W.Y.; Wei, T.Y.; Huang, J.C. In vitro biocompatibility response of Ti–Zr–Si thin film metallic glasses. *Appl. Surf. Sci.* **2014**, *322*, 41–46. [CrossRef]

19.	Busuioc, C.; Voicu, G.; Zuzu, I.D.; Miu, D.; Sima, C.; Iordache, F.; Jinga, S.I. Vitroceramic coatings deposited by laser ablation on Ti-Zr substrates for implantable medical applications with improved biocompatibility. *Ceram. Int.* **2017**, *43*, 5498–5504. [CrossRef]

20.	Soon, G.; Pingguan-Murphy, B.; Lai, K.W.; Akbar, S.A. Review of zirconia-based bioceramic: Surface modification and cellular response. *Ceram. Int.* **2016**, *42*, 12543–12555. [CrossRef]

21.	Jenko, M.; Gorensek, M.; Godec, M.; Hodnik, M.; Setina-Batic, B.; Donik, C.; Grant, J.T.; Dolinar, D. Surface chemistry and microstructure of metallic biomaterials for hip and knee endoprostheses. *Appl. Surf. Sci.* **2018**, *427*, 584–593. [CrossRef]

22.	Wu, W.; Cheng, R.; Neves, J.; Tang, J.; Xiao, J.; Ni, Q.; Liu, X.; Pan, G.; Li, D.; Cui, W.; et al. Advances in biomaterials for preventing tissue adhesion. *J. Control. Release* **2017**, *261*, 318–336. [CrossRef] [PubMed]

23.	Simitzi, C.; Ranella, A.; Stratakis, E. Controlling the morphology and outgrowth of nerve and neuroglial cells: The effect of surface topography. *Acta Biomater.* **2017**, *51*, 21–52. [CrossRef] [PubMed]

24.	Vilar, R. *Laser Surface Modification of Biomaterials: Techniques and Applications*; Woodhead Publishing Elsevier Science: Cambridge, UK, 2016.

25.	Martinez-Calderon, M.; Manso-Silvan, M.; Rodriguez, A.; Gomez-Aranzadi, M.; Garcia-Ruiz, J.P.; Olaizola, S.M.; Martin-Palma, R.J. Surface micro- and nano-texturing of stainless steel by femtosecond laser for the control of cell migration. *Sci. Rep.* **2016**, *6*, 36296. [CrossRef] [PubMed]

26.	Voisey, K.T.; Scotchford, C.A.; Martin, L.; Gill, H.S. Effect of Q-switched laser surface texturing of titanium on osteoblast cell response. *Phys. Procedia* **2014**, *56*, 1126–1135. [CrossRef]

27.	Lawrence, J.; Hao, L.; Chew, H.R. On the correlation between Nd:YAG laser-induced wettability characteristics modification and osteoblast cell bioactivity on a titanium alloy. *Surf. Coat. Technol.* **2006**, *200*, 5581–5589. [CrossRef]

28. Kovačević, A.; Petrović, S.; Bokić, B.; Gaković, B.; Bokorov, M.; Vasić, B.; Gajić, R.; Trtica, M.; Jelenković, B. Surface nanopatterning of Al/Ti multilayer thin films and Al single layer by a low-fluence UV femtosecond laser beam. *Appl. Surf. Sci.* **2015**, *326*, 91–98. [CrossRef]

29. Raimbault, O.; Benayoun, S.; Anselme, K.; Mauclair, C.; Bourgade, T.; Kietzig, A.M.; Girard-Lauriault, P.L.; Valette, S.; Donnet, C. The effects of femtosecond laser-textured Ti-6Al-4V on wettability and cell response. *Mater. Sci. Eng. C* **2016**, *69*, 311–320. [CrossRef]

30. Peruško, D.; Čizmović, M.; Petrović, S.; Siketić, Z.; Mitrić, M.; Pelicon, P.; Dražić, G.; Kovač, J.; Milinović, V.; Milosavljević, M. Laser irradiation of nano-metric Al/Ti multilayers. *Laser Phys.* **2013**, *23*, 036005. [CrossRef]

31. Simitzi, C.; Efstathopoulos, P.; Kourgiantaki, A.; Ranella, A.; Charalampopoulos, I.; Fotakis, C.; Athanassakis, I.; Stratakis, E.; Gravanis, A. Laser fabricated discontinuous anisotropic microconical substrates as a new model scaffold to control the directionality of neuronal network outgrowth. *Biomaterials* **2015**, *67*, 115–128. [CrossRef]

32. Stanciuc, A.M.; Flamant, Q.; Sprecher, C.M.; Alini, M.; Anglada, M.; Peroglio, M. Femtosecond laser multi-patterning of zirconia for screening of cell-surface interactions. *J. Eur. Ceram. Soc.* **2018**, *38*, 939–948. [CrossRef]

33. Stašić, J.; Gaković, B.; Perrie, W.; Watkins, K.; Petrović, S.; Trtica, M. Surface texturing of the carbon steel AISI 1045 using femtosecond laser in single pulse and scanning regime. *Appl. Surf. Sci.* **2011**, *258*, 290–296. [CrossRef]

34. Xu, J.; Chen, X.; Zhang, C.; Liu, Y.; Wang, J.; Deng, F. Improved bioactivity of selective laser melting titanium: Surface modification with micro-/nano-textured hierarchical topography and bone regeneration performance evaluation. *Mater. Sci. Eng. C* **2016**, *68*, 229–240. [CrossRef] [PubMed]

35. Fan, J.; Zhang, Y.; Liu, Y.; Wang, Y.; Cao, F.; Yang, O.; Tian, F. Explanation of the cell orientation in a nanofiber membrane by the geometric potential theory. *Results Phys.* **2019**, *15*, 102537. [CrossRef]

36. Wang, J.; Dai, J.; Ya, Y. Nanofibers with tailored degree of directional orientation regulate cell movement. *Mater. Today Commun.* **2020**, *25*, 101496. [CrossRef]

37. Heitz, J.; Plamadeala, C.; Muck, M.; Armbruster, O.; Baumgartner, W.; Weth, A.; Steinwender, C.; Blessberger, H.; Kellermair, J.; Kirner, S.V.; et al. Femtosecond laser-induced microstructures on Ti substrates for reduced cell adhesion. *Appl. Phys. A* **2017**, *123*, 734. [CrossRef]

38. Fosodeder, P.; Baumgartner, W.; Steinwender, C.; Hassel, A.W.; Florian, C.; Bonse, J.; Heitz, J. Repellent rings at titanium cylinders against overgrowth by fibroblasts. *Adv. Opt. Technol.* **2020**, *9*, 113–120. [CrossRef]

39. Moulder, J.F.; Stickle, W.F.; Sobol, P.E.; Bomben, K.D. *Handbook of X-ray Photoelectron Spectroscopy*; Physical Electronics Inc.: Eden Prairie, MN, USA, 1995.

40. Schindelin, J.; Arganda-Carreras, I.; Frise, E.; Kaynig, V.; Longair, M.; Pietzsch, T.; Preibisch, S.; Rueden, C.; Saalfeld, S.; Schmid, B.; et al. Fiji: An open-source platform for biological-image analysis. *Nat. Methods* **2012**, *9*, 676–682. [CrossRef]

41. Bonse, J.; Koter, R.; Hartelt, M.; Spaltmann, D.; Pentzien, S.; Höhm, S.; Rosenfeld, A.; Krüger, J. Tribological performance of femtosecond laser-induced periodicsurface structures on titanium and a high toughness bearing steel. *Appl. Surf. Sci.* **2015**, *336*, 21–27. [CrossRef]

42. Gnilitskyi, I.; Derrien, T.J.-Y.; Levy, Y.; Bulgakova, N.M.; Mocek, T.; Orazi, L. High-speed manufacturing of highly regular femtosecond laser-induced periodic surface structures: Physical origin of regularity. *Sci. Rep.* **2017**, *7*, 1–11. [CrossRef]

43. Ye, M.; Grigoropoulos, C.P. Time-of-flight and emission spectroscopy study of femtosecond laser ablation of titanium. *J. Appl. Phys.* **2001**, *89*, 5183–5390. [CrossRef]

44. De Giacomo, A.; Dell'Aglio, M.; Santagata, A.; Teghil, R. Early stage emission spectroscopy study of metallic titanium plasma induced in air by femtosecond and nanosecond- laser pulses. *Spectrochim. Acta B* **2005**, *60*, 935–947. [CrossRef]

45. Yan, L.; Yuan, Y.; Ouyang, L.; Li, H.; Mirzasadeghi, A.; Li, L. Improved mechanical properties of the new Ti-15Ta-xZr alloys fabricated by selective laser melting for biomedical application. *J. Alloy. Compd.* **2016**, *688*, 156–162. [CrossRef]

46. Mukashev, K.M.; Umarov, F.F. Physicochemical and Radiation Modification of Titanium Alloys Structure. In *Titanium Alloys: Advances in Properties Control*; InTech: Rijeka, Croatia, 2013. [CrossRef]

47. Ivanova, A.A.; Surmeneva, M.A.; Shugurov, V.V.; Koval, N.N.; Shulepov, I.A.; Surmenev, R.A. Physico-mechanical properties of Ti-Zr coatings fabricated via ionassisted arc-plasma deposition. *Vacuum* **2018**, *149*, 129–133. [CrossRef]

48. Anselme, K. Osteoblast adhesion on biomaterials. *Biomaterials* **2000**, *21*, 667–681. [CrossRef]

49. Dai, S.; Yu, T.; Zhang, J.; Lu, H.; Dou, D.; Zhang, M.; Dong, M.; Di, J.; Zhao, J. Real-time and wide-field mapping of cell-substrate adhesion gap and its evolution via surface plasmon resonance holographic microscopy. *Biosens. Bioelectron.* **2020**, 112826. [CrossRef]

50. Saxena, K.K.; Das, R.; Calius, E.P. Three Decades of Auxetics Research Materials with Negative Poisson's Ratio: A Review. *Adv. Eng. Mater.* **2016**, *18*, 1847–1870. [CrossRef]

51. Boyd, J.D.; Stromberg, A.J.; Miller, C.S.; Grady, M.E. Biofilm and cell adhesion strength on dental implant surfaces via the laser spallation technique. *Dent. Mater.* **2020**. [CrossRef]

52. Babaliari, E.; Kavatzikidou, P.; Angelaki, D.; Chaniotaki, L.; Manousaki, A.; Siakouli-Galanopoulou, A.; Ranella, A.; Stratakis, E. Engineering Cell Adhesion and Orientation via Ultrafast Laser Fabricated Microstructured Substrates. *Int. J. Mol. Sci.* **2018**, *19*, 2053. [CrossRef]

Publisher's Note: MDPI stays neutral with regard to jurisdictional claims in published maps and institutional affiliations.

 *nanomaterials*

Article

Femtosecond Laser-Induced Periodic Surface Structures on Different Tilted Metal Surfaces

Yi-Hsien Liu, Kong-Kai Kuo and Chung-Wei Cheng *

Department of Mechanical Engineering, National Chiao Tung University, No. 1001 Ta Hsueh Road, Hsinchu 300, Taiwan; kevin820624@gmail.com (Y.-H.L.); kevinkuo031@gmail.com (K.-K.K.)
* Correspondence: weicheng@nctu.edu.tw

Received: 19 November 2020; Accepted: 16 December 2020; Published: 17 December 2020

Abstract: Laser-induced periodic surface structures (LIPSS) are used for the precision surface treatment of 3D components. However, with LIPSS, the non-normal incident angle between the irradiated laser beam and the specimen surface occurs. This study investigated LIPSS on four different metals (SUS 304, Ti, Al, and Cu), processed on a tilted surface by an s-polarized femtosecond fiber laser. A rotated low spatial frequency LIPSS (LSFL) was obtained on SUS 304 and Ti materials by the line scanning process. However, LSFL on Cu and Al materials was still perpendicular to the laser polarization. The reason for the rotated and un-rotated LSFL on tilted metal surfaces was presented. The electron-phonon coupling factor and thermal conductivity properties might induce rotational LSFL on tilted SUS 304 and Ti surfaces. When fabricating LSFL on an inclined plane, a calibration model between the LSFL orientation and inclined plane angle must be established. Hence, the laser polarization direction must be controlled to obtain suitable LSFL characteristics on a 3D surface.

Keywords: femtosecond laser; laser-induced periodic surface structures; LIPSS; LSFL

1. Introduction

LIPSS (Laser-induced periodic surface structures) were first discovered by Birnbaum in 1965 [1] and gradually extended from basic research to application [2–4]. They can be quickly generated on metal, semiconductor, and dielectrics by a single laser beam [5]. In the past, most papers focused on generating LIPSS on plant surfaces by different scanning strategies or polarization [6–9]. Recent studies recently investigated LIPSS by the p-polarized laser beam with oblique incidence on the sample surface [10–13]. These studies found that the polarization portion irradiated on the hole and groove sidewall changes with the incidence angle. The low spatial frequency LIPSS (LSFL) on the sidewall changes direction, for instance, not being perpendicular to the polarization direction. However, few studies investigated LIPSS by the s-polarized laser beam with oblique incidence on the sample surface. The studies showed that in this condition, the laser polarization portion on the specimen surface will not change.

Recently, Schwarz et al. used a scanning s-polarized femtosecond laser to irradiate materials (e.g., fused silica) with a tilted angle [14]. The study found that the LSFL orientation was not perpendicular to the laser polarization, and the rotation angle increased as the specimen tilted angle increased. In another study, Liu et al. [15] investigated the LIPSS generated by linear polarization (s-polarized and p-polarized) femtosecond laser beam on a tilted SUS 304 surface. The experiments found that LSFL was rotated by the s-polarized beam and was not rotated by the p-polarized beam. Hence, the mechanism of rotational LSFL on the tilted specimen by s-polarized laser beam was deduced. However, although [14,15] have published the experimental results for rotational LIPSS on different materials by s-polarized femtosecond laser beam, there is no study to validate whether the LSFL rotates on the tilted surface for more metal materials.

In this study, LIPSS was generated on four metals (SUS 304, Ti, Al, and Cu) by an s-polarized femtosecond laser beam (1030 nm) with specimen tilted angles (0°, 30°, and 45°) and validated whether the LSFL rotated or not. These metal materials were chosen because they have different optical and thermophysical properties and are suitable to investigate the mechanism of rotational LSFL. The results reveal that only specific metals at the processing wavelength can exhibit rotational LSFL structures on the tilted surface.

2. Materials and Methods

The mechanically polished materials (SUS 304, Ti, Al, and Cu) were irradiated using a femtosecond fiber laser (KASMORO, mRadian Inc, Hsinchu, Taiwan) with a wavelength of 1030 nm, a pulse width of 420 fs, a repetition rate of 100 kHz, and a maximum laser power of 2 W. The laser beam was linearly polarized and was passed through a liquid lens (EL-10-42-OF, Optotune Inc., Dietikon, Switzerland) before entering a galvanometric scanner that had an F-theta lens with a focal length of 100 mm. The liquid lens was used to control the focus position on the tilted surface. The focused spot size was determined to be 60 µm. It is important to note that the experiment was mainly based on the wavelength of 1030 nm, but for the copper material, to explore the influence of the laser wavelength, the laser beam was passed through the BBO crystal to convert the laser beam with wavelength 515 nm. The laser beam was entering a galvanometric scanner that had an F-theta lens with a focal length of 165 mm. The focused spot size was determined to be 50 µm.

The experiment process is shown in Figure 1. The line scanning with the s-polarized beam parallel to the scanning direction was used to investigate LSFL on four metals (SUS 304, Ti, Al, and Cu) with different sample tilted angles (θ). For the femtosecond laser processing, all the specimens were cleaned in an ultrasonic bath (in water) machine for 10 min, and the experiment was conducted in an airy atmosphere. After the experiment, scanning electron microscopy (SEM, Hitachi SU-8010, Tokyo, Japan) was used to measure the structure profile. The orientation and the period of structures were determined using a fast Fourier transformation (FFT) analysis of the SEM images.

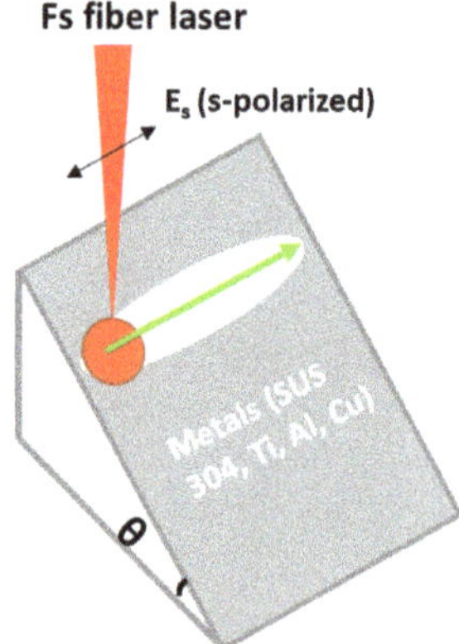

Figure 1. Schematic of fabricating the low spatial frequency LIPSS (LSFL) on different metals.

3. Results and Discussion

Figure 2 shows a line pattern with LSFL fabricated by the femtosecond laser of wavelength 1030 nm and different θ (0°, 30°, and 45°). The scanning direction is from left to right (see Figure 1). The laser power (P) and scanning speed (V) were shown on the top side of the figures. Note that these parameters were experimentally determined so that LSFL can be uniformly fabricated on different metals. Table 1 summarizes LSFL experimental results and material related optical and thermal parameters. In the θ = 0° case, as shown in Table 1, the period of LSFL was near the wavelength, that is, in the range of 907~976 nm by laser wavelength 1030 nm.

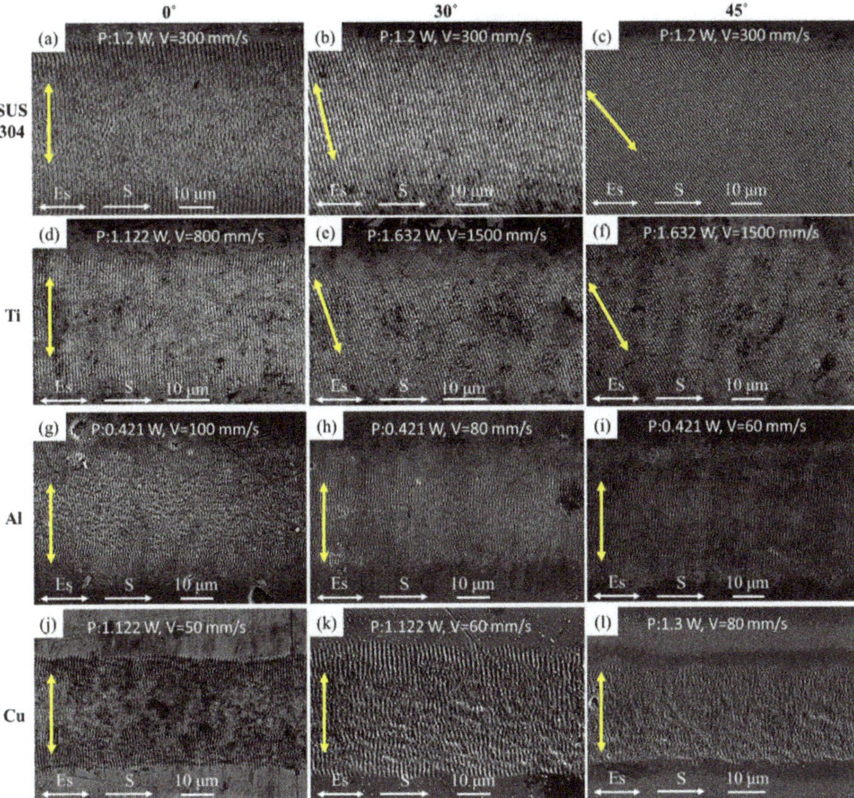

Figure 2. SEM images of the LSFL generated by the line scanning process for the four different metals irradiated by 1030 nm laser with different θ (0°, 30°, and 45°): (**a–c**) SUS 304, (**d–f**) Ti, (**g–i**) Al, and (**j–l**) Cu.

Table 1. Summary of LSFL experimental results and material related optical and thermal parameters.

Material	SUS 304	Ti	Al	Cu	Cu
Wavelength λ (nm)	1030	1030	1030	1030	515
LSFL period (nm) at θ = 0°	959	976	907	919	420
LSFL period (nm) at θ = 30°	893	917	906	932	438
LSFL period (nm) at θ = 45°	788	864	814	1019	496
Optical parameters					
Re (ϵ) at 300 K [16]	−6.6275	−4.2656	−95.117	−45.761	−5.3328
Im (ϵ) at 300 K [16]	23.004	27.277	27.653	4.5744	6.1794
SPP propagation length L_{spp} (µm)	4.02	4.52	57.32	73.35	0.78
Thermal parameters					
Electron-phonon coupling factor G at T_e = 10,000 K (Wm^{-3}K^{-1} × 10^{17}) [17]	30.04	36.74	3.62	5.29	5.29
Thermal conductivity at 300 K (W/m-K)	16.2	22	235	385	385
Melting Temperature (K)	1723	1930	933	1356	1356

In Figure 2a,d,g,j, the orientation of LSFL (yellow double arrows) was nearly perpendicular to the laser polarization. The mechanism of generating LSFL can be explained by the interference between the incident laser beam and the surface plasmon polariton (SPP) [3]. Interestingly, at θ = 30° and 45°, as shown in Figure 2b,c,e,f, LSFL presented a different orientation for SUS 304 and Ti. When θ was increased, the LSFL rotation angle is increased. However, for Al and Cu, the LSFL is always perpendicular to the laser polarization rather than rotation, as shown in Figure 2h,i,k,l.

Figure 3 shows the SEM images of the line scanning start region for four materials. An interesting phenomenon was observed in the scanning start region for SUS 304 and Ti, as shown in Figure 3a,b. The LSFL was perpendicular to the polarization at the center of the starting point (pink arrow) before gradually rotating along the scanning direction. It then remained at a constant rotated angle (red arrow). The LSFL rotation angle was around 14° and 19° for SUS 304 and Ti, respectively. However, in the Al and Cu material, as shown in Figure 3c,d, the LSFL was not rotated in the scanning start region. Their orientation remained perpendicular to the laser polarization.

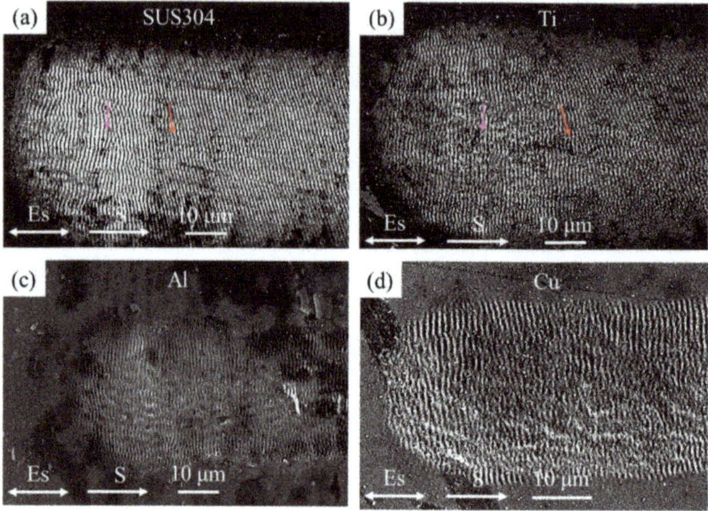

Figure 3. SEM images of the line scanning start regions for four materials with θ = 30°; (a) SUS304, (b) Ti, (c) Al, and (d) Cu.

Why is it that not all tilted metal materials can produce LSFL rotation? The possible reasons need to be further discussed. Bonse and Gräf show that the formation of LSFL may be caused by electromagnetic effects or matter reorganization [3]. This study will examine the two issues and discuss why only SUS 304 and Ti produce LSFL rotation among the four experimental materials.

Assume that SPP orientation on the tilted sample differs from the flat surface and induces the LSFL rotation. The experiment results show that the rotational LSFL only exists in SUS304 and Ti. We can deduce that the SPP generated on the tilted sample is different among the four materials. The LSFL related to SPP propagation length (Lspp) is presented in [18,19]. Recently, studies have further proposed that Lspp is related to the uniformity of LSFL. The shorter the Lspp length, the better the LSFL uniformity in metals [20]. Metals (Ti, Steel, Mo) can produce highly regular LSFL by femtosecond laser irradiation with wavelength 1030 nm. However, the LSFL were not regular for Cu and Al. The Lspp can be calculated by:

$$L_{spp} = \frac{1}{2\mathrm{Im}(\beta)} \quad \beta = \frac{2\pi}{\lambda}\sqrt{\frac{\varepsilon}{1+\varepsilon}} \tag{1}$$

where λ is the laser wavelength, ε is the relative permittivity, and β is the SPP wave number.

The calculated Lspp at room temperature is shown in Table 1. It was found that the SUS 304 and Ti materials with shorter Lspp (<5 µm) and the Cu and Al materials with Lspp larger than 50 µm at laser wavelength 1030 nm. It will be necessary to conduct experiments with different wavelengths to validate the speculation that Lspp is a possible factor to induce LSFL rotation on a tilted surface. As shown in Table 1, the Lspp is shorter in wavelength of 515 nm (<1 µm). Therefore, the LSFL generated by a 515 nm laser was inferred to be rotated. However, as shown in Figure 4, the LSFL remains perpendicular to the laser polarization. The different wavelengths (1030 nm and 515 nm) irradiation on the Cu demonstrate that Lspp is not a critical factor to induce LSFL rotation. However, the findings validate that the uniformity of LSFL generated by 515 nm laser (Figure 4c) is better than that generated by 1030 nm laser (see Figure 3d).

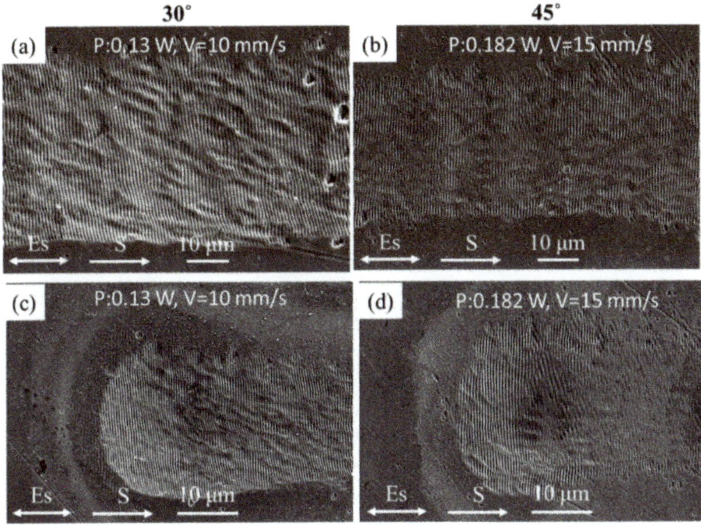

Figure 4. SEM images of the LSFL generated by the line scanning process for the Cu irradiated by 515 nm laser with (**a,c**) θ = 30° and (**b,d**) 45°; (**c,d**) scanning start regions.

When the femtosecond laser was irradiated on the metal sample, the surface reflectivity was varied due to the dynamic electron temperature. We then further investigated the dynamic permittivity and Lspp to know whether they induced the rotated LSFL or not. Winter et al. used pump-probe ellipsometry to measure the dynamic relative permittivity ε when the Cu was irradiated by 528 nm laser and 0.4 J/cm^2 [21]. The dynamic Lspp during femtosecond laser irradiation can be calculated by Equation (1) with the experimental data (dynamic ε) from [21]. As shown in Figure 5, Lspp was increased during laser irradiation. The maximum Lspp was 1.03 (at 0.14 ps) and was slightly higher than 0.78 at room temperature. However, the LSFL was still perpendicular to the laser polarization. Thus, this study speculates that Lspp was not a factor that affected LSFL rotation on the tilted sample.

We tried to compare whether the matter reorganization by material properties could influence rotational LSFL. Gurevich et al. proposed the three steps for the generation of LIPSS [22]: (1) modulation of the electron temperature (T_e) by the interference of the incident laser beam and surface plasmon wave; (2) modulation of lattice temperature (T_l); (3) hydrodynamic instabilities and reorganization of the molten material on the surface.

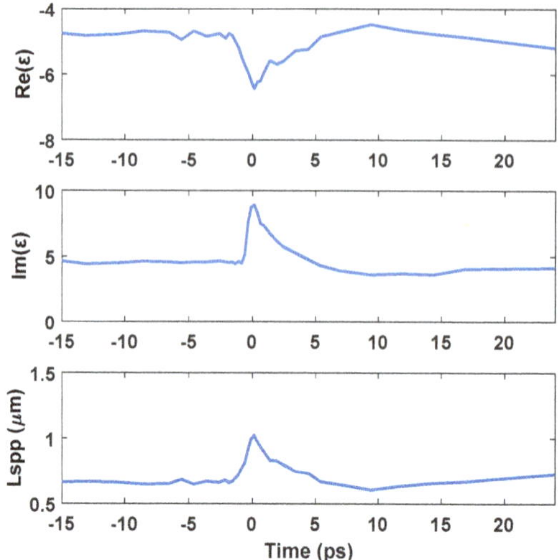

Figure 5. Dynamic permittivity ε and Lspp.

In this study, the electron-phonon coupling factor (G) and the thermal conductivity of the four materials were compared (see Table 1) in an attempt to deduce the reason for rotational LSFL. The two-temperature model (TTM) can be used to describe the electron-lattice coupling effect $(G(T_e - T_l))$ in the ultrafast laser process. If the G is larger and the electron thermal diffusivity is smaller, the energy that can be absorbed in the irradiated area is faster and more concentrated, providing a positive feedback loop for the amplification of the lattice temperature modulation [22,23]. The G values of the four materials at T_e = 10,000 K are shown in Table 1. SUS 304 and Ti were about 6–10 times the ratio of Al and Cu. Therefore, for SUS 304 and Ti, the electron temperature coupled to the lattice temperature faster, causing a larger temperature gradient in the molten pool. In addition, SUS 304 and Ti had higher melting temperatures (1723 K and 1930 K) and lower thermal conductivity (16.2 and 22 W/m-K at 300 K), which could also make the melting pool temperature gradient larger. When the melting pool produces a larger temperature gradient, it can create a larger Marangoni force and push the liquid material to a lower temperature area [24].

The influence of the thermophysical parameters of SUS 304 and Ti when processing metals on an inclined surface can easily cause local instability on the melting pool. Accordingly, the LSFL was symmetrically rotated at both sides of the irradiation area, as shown in Figure 3a,b. According to the experimental results, it was found that when fabricating LSFL on the inclined plane, a calibration model between the LSFL orientation and inclined plane angle must be established, and the laser polarization direction must be controlled to obtain a suitable LSFL orientation.

4. Conclusions

This study investigated s-polarized femtosecond laser beams irradiated on four metals (SUS 304, Ti, Al, and Cu) with different sample tilted angles (θ). The experimental results of the line scanning demonstrated that: (1) only SUS304 and Ti material could induce the rotation of LSFL on the tilted sample, and the LSFL rotation angle increased as θ increased; and (2) for Al and Cu material, the LSFL orientation remained perpendicular to the laser polarization as θ increased. We deduced that the electron-phonon coupling factor and thermal conductivity were quite different on the SUS 304 and Ti and could influence the LSFL rotation on the tilted surface. The higher electron-phonon coupling factor and lower thermal conductivity of materials would cause the instability molten pool and induce the rotational LSFL on the tilted surface. Thus, for 3D surface processing, precise, controlled polarization should be conducted to generate the uniform LSFL for specific materials.

Author Contributions: Conceptualization, C.-W.C. supervision; C.-W.C., K.-K.K., Y.-H.L., proposed the concept and designed the experiment; K.-K.K. and Y.-H.L. contributed to the measurement results; Y.-H.L. and C.-W.C. wrote the original draft. Y.-H.L. and C.-W.C. contributed to the simulation results. C.-W.C. funding acquisition. All authors have read and agreed to the published version of the manuscript.

Funding: This work was supported by the Ministry of Science and Technology of Republic of China, under Contract MOST 107-2221-E-009-061-MY3.

Acknowledgments: The femtosecond laser support from mRadian Co., Ltd.

Conflicts of Interest: The authors declare no conflict of interest.

References

1. Birnbaum, M. Semiconductor Surface Damage Produced by Ruby Lasers. *J. Appl. Phys.* **1965**, *36*, 3688–3689. [CrossRef]

2. Bonse, J. Quo Vadis LIPSS?—Recent and Future Trends on Laser-Induced Periodic Surface Structures. *Nanomaterials* **2020**, *10*, 1950. [CrossRef] [PubMed]

3. Bonse, J.; Gräf, S. Maxwell meets Marangoni—A review of theories on laser-induced periodic surface structures. *Laser Photonics Rev.* **2020**, *14*, 2000215. [CrossRef]

4. Florian, C.; Kirner, S.V.; Krüger, J.; Bonse, J. Surface functionalization by laser-induced periodic surface structures. *J. Laser Appl.* **2020**, *32*, 022063. [CrossRef]

5. Buividas, R.; Mikutis, M.; Juodkazis, S. Surface and bulk structuring of materials by ripples with long and short laser pulses: Recent advances. *Prog. Quantum Electron.* **2014**, *38*, 119–156. [CrossRef]

6. Gečys, P. Ripple Formation by Femtosecond Laser Pulses for Enhanced Absorptance of Stainless Steel. *J. Laser Micro/Nanoeng.* **2015**, *10*, 129–133. [CrossRef]

7. Lazzini, G.; Romoli, L.; Tantussi, F.; Fuso, F. Nanostructure patterns on stainless-steel upon ultrafast laser ablation with circular polarization. *Opt. Laser Technol.* **2018**, *107*, 435–442. [CrossRef]

8. Razi, S.; Varlamova, O.; Reif, J.; Bestehorn, M.; Varlamov, S.; Mollabashi, M.; Madanipour, K.; Ratzke, M. Birth of periodic Micro/Nano structures on 316L stainless steel surface following femtosecond laser irradiation; single and multi scanning study. *Opt. Laser Technol.* **2018**, *104*, 8–16. [CrossRef]

9. Ding, K.; Li, M.; Wang, C.; Lin, N.; Wang, H.; Luo, Z.; Duan, J. Sequential Evolution of Colored Copper Surface Irradiated by Defocused Femtosecond Laser. *Adv. Eng. Mater.* **2020**, *22*, 1901310. [CrossRef]

10. Pham, K.X.; Tanabe, R.; Ito, Y. Laser-induced periodic surface structures formed on the sidewalls of microholes trepanned by a femtosecond laser. *Appl. Phys. A* **2012**, *112*, 485–493. [CrossRef]

11. Austin, D.R.; Kafka, K.R.P.; Trendafilov, S.; Shvets, G.; Li, H.; Yi, A.Y.; Szafruga, U.B.; Wang, Z.; Lai, Y.H.; Blaga, C.I.; et al. Laser induced periodic surface structure formation in germanium by strong field mid IR laser solid interaction at oblique incidence. *Opt. Express* **2015**, *23*, 19522–19534. [CrossRef]

12. Garcell, E.M.; Lam, B.; Guo, C. Femtosecond laser-induced herringbone patterns. *Appl. Phys. A* **2008**, *124*, 405. [CrossRef] [PubMed]

13. Zheng, X.; Cong, C.; Lei, Y.; Yang, J.; Guo, C. Formation of Slantwise Surface Ripples by Femtosecond Laser Irradiation. *Nanomaterials* **2018**, *8*, 458. [CrossRef]

14. Schwarz, S.; Rung, S.; Hellmann, R. One-dimensional low spatial frequency LIPSS with rotating orientation on fused silica. *Appl. Surf. Sci.* **2017**, *411*, 113–116. [CrossRef]

15. Liu, Y.H.; Tseng, Y.K.; Cheng, C.W. Direct fabrication of rotational femtosecond laser-induced periodic surface structure on a tilted stainless steel surface. *Opt. Laser Technol.* **2021**, *134*, 106648. [CrossRef]

16. R. Info. Available online: https://refractiveindex.info/?shelf=main&book=Cu&page=Johnson (accessed on 14 August 2020).

17. Lin, Z.; Zhigilei, L.V.; Celli, V. Electron-phonon coupling and electron heat capacity of metals under conditions of strong electron-phonon nonequilibrium. *Phys. Rev. B* **2008**, *77*, 075133. [CrossRef]

18. Huang, M.; Zhao, F.; Cheng, Y.; Xu, N.; Xu, Z. Origin of Laser-Induced Near-Subwavelength Ripples: Interference between Surface Plasmons and Incident Laser. *ACS Nano* **2009**, *3*, 4062–4070. [CrossRef] [PubMed]

19. Terekhin, P.; Benhayoun, O.; Weber, S.; Ivanov, D.; Garcia, M.; Rethfeld, B. Influence of surface plasmon polaritons on laser energy absorption and structuring of surfaces. *Appl. Surf. Sci.* **2020**, *512*, 144420. [CrossRef]

20. Gnilitskyi, I.; Derrien, T.J.-Y.; Levy, Y.; Bulgakova, N.M.; Mocek, T.; Orazi, L. High-speed manufacturing of highly regular femtosecond laser-induced periodic surface structures: Physical origin of regularity. *Sci. Rep.* **2017**, *7*, 1–11. [CrossRef]

21. Winter, J.; Rapp, S.; Schmidt, M.; Huber, H.P. Ultrafast laser energy deposition in copper revealed by simulation and experimental determination of optical properties with pump-probe ellipsometry. In *Laser Applications in Microelectronic and Optoelectronic Manufacturing (LAMOM) XXII*; International Society for Optics and Photonics: San Francisco, CA, USA, 2017; Volume 10091, p. 100910.

22. Gurevich, E.L.; Levy, Y.; Bulgakova, N.M. Three-Step Description of Single-Pulse Formation of Laser-Induced Periodic Surface Structures on Metals. *Nanomaterials* **2020**, *10*, 1836. [CrossRef]

23. Wang, J.; Guo, C. Numerical study of ultrafast dynamics of femtosecond laser-induced periodic surface structure formation on noble metals. *J. Appl. Phys.* **2007**, *102*, 53522. [CrossRef]

24. Hwang, T.Y.; Vorobyev, A.Y.; Guo, C. Ultrafast dynamics of femtosecond laser-induced nanostructure formation on metals. *Appl. Phys. Lett.* **2009**, *95*, 123111. [CrossRef]

nanomaterials

Article

Laser-Induced Transfer of Noble Metal Nanodots with Femtosecond Laser-Interference Processing

Yoshiki Nakata [1,*], Koji Tsubakimoto [1], Noriaki Miyanaga [2], Aiko Narazaki [3], Tatsuya Shoji [4] and Yasuyuki Tsuboi [5]

[1] Institute of Laser Engineering, Osaka University, 2-6 Yamadaoka, Suita, Osaka 565-0871, Japan; tsubaki@ile.osaka-u.ac.jp

[2] Institute for Laser Technology, 1-8-4 Utsubo-honmachi, Nishi-ku, Osaka 550-0004, Japan; miyanaga@ilt.or.jp

[3] National Institute of Advanced Industrial Science and Technology, Central 2, Umezono 1-1-1, Tsukuba, Ibaraki 305-8568, Japan; narazaki-aiko@aist.go.jp

[4] Faculty of Science, Kanagawa University, 2946, Tsuchiya, Hiratsuka, Kanagawa 259-1293, Japan; t-shoji@kanagawa-u.ac.jp

[5] Graduate School of Science, Osaka City University, 3-3-138 Sugimoto Sumiyoshi-ku, Osaka 558-8585, Japan; twoboys@sci.osaka-cu.ac.jp

* Correspondence: nakata-y@ile.osaka-u.ac.jp; Tel.: +81-6-6879-8729

Abstract: Noble metal nanodots have been applied to plasmonic devices, catalysts, and highly sensitive detection in bioinstruments. We have been studying the fabrications of them through a laser-induced dot transfer (LIDT) technique, a type of laser-induced forward transfer (LIFT), in which nanodots several hundred nm in diameter are produced via a solid–liquid–solid (SLS) mechanism. In the previous study, an interference laser processing technique was applied to LIDT, and aligned Au nanodots were successfully deposited onto an acceptor substrate in a single shot of femtosecond laser irradiation. In the present experiment, Pt thin film was applied to this technique, and the deposited nanodots were measured by scanning electron microscopy (SEM) and compared with the Au nanodots. A typical nanodot had a roundness $f_r = 0.98$ and circularity $f_{circ} = 0.90$. Compared to the previous experiment using Au thin film, the size distribution was more diffuse, and it was difficult to see the periodic alignment of the nanodots in the parameter range of this experiment. This method is promising as a method for producing large quantities of Pt particles with diameters of several hundred nm.

Keywords: interference laser processing; laser-induced dot transfer; laser-induced forward transfer; nanodot; nanoparticle; array; femtosecond laser; solid–liquid–solid mechanism; Pt

Citation: Nakata, Y.; Tsubakimoto, K.; Miyanaga, N.; Narazaki, A.; Shoji, T.; Tsuboi, Y. Laser-Induced Transfer of Noble Metal Nanodots with Femtosecond Laser-Interference Processing. *Nanomaterials* **2021**, *11*, 305. https://doi.org/10.3390/nano11020305

Academic Editor: Jörn Bonse

Received: 31 December 2020

Accepted: 21 January 2021

Published: 25 January 2021

1. Introduction

The first laser-induced forward transfer (LIFT) technique was examined in 1970 [1]. It was called as "laser typewriter", in which an ink ribbon was used as a source target. In this technique the irradiated region is transferred to the receiver substrate, as shown in Figure 1a. The target thin films were extended to various materials such as metals [2], dielectrics [3], biomaterials [4], and fluorophore [5], and microscopic process measurement techniques using image laser spectroscopy have been applied to investigate the LIFT process [6,7]. The laser-induced dot transfer (LIDT) technique, a type of LIFT, was developed for depositing nanodots which are smaller than laser wavelengths [8–11]. The mechanism involves the liquid behavior of solute metals and their freezing, called the solid–liquid–solid (SLS) mechanism [12]. Here, SLS enables the fabrication of a variety of nanostructures other than nanodots using LIDT, such as the nanobump [13], nanodrop [13,14], nanowhisker [12,15], nanocrown [15,16], etc. Furthermore, we have investigated interference laser processing for two decades. In these studies, the aligned nanostructures mentioned above have been successfully fabricated in a single shot of laser irradiation. We applied this technique to

LIDT as shown in Figure 1b, and aligned Au nanodots with $\Lambda = 3.6$ μm square matrices were successfully deposited [17].

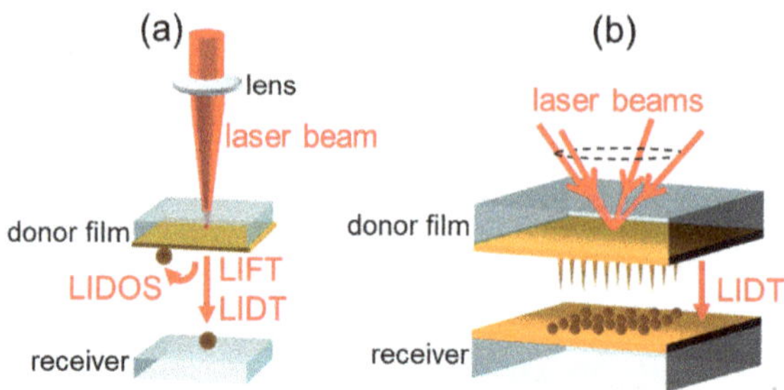

Figure 1. (**a**) Schematic illustration of laser-induced forward transfer (LIFT), laser-induced dot transfer (LIDT), and the laser-induced dot caught on source target (LIDOS) [18] process. (**b**) LIDT with interference pattern [17].

In this study, we used Pt thin film as a source target for LIDT combined with the interference laser processing technique, and the nanodots deposited on the receiver substrate were measured by scanning electron microscopy (SEM). The size distributions were analyzed on parameters such as pulse energy. The results that we obtained are supported by a heuristic model.

2. Materials and Methods

The experimental setup is shown in Figure 2. A laser operated at a 785 nm center wavelength with a 240 fs pulse width was used. The pointing was stabilized with a piezo-actuated mirror by <10 μrad h^{-1}. The beam was split by a diffractive optical element (DOE) into four 1st order diffracted beams. They were correlated and formed an interference pattern on the backside of the donor Pt thin film through a de-magnification system consisting of two achromatic convex lenses ($f_1 = 200$ nm, $f_2 = 50$ nm). The two lenses and the zero-order beam were placed coaxially with the DOE perpendicular to the axis, and at the same time they were properly spaced. The verticality was set by aligning the direction of the return light to the light source. The donor substrate was placed with mechanical precision so that it was perpendicular to the axis. In this way, an interference pattern was formed on the donor substrate. The zero-order beam was dumped between the lenses. The right inset in Figure 2 is a simulation result of the interference pattern [19], and the period was $\Lambda = 3.6$ μm. The 50 nm thick Pt film deposited onto a 1mm thick silica glass substrate that was used as a donor film target. Here, in a quote from a previous paper, the thickness of the donor Au film was 40 nm. An Au receiver film 100 nm thick deposited onto a 1mm thick silica glass substrate which was placed in contact facing the donor film. The LIDT experiment was performed in a vacuum chamber ($P < 1.3$ kPa). The off-line image analysis explained in Section 3.2 was performed using ImageJ (NIH).

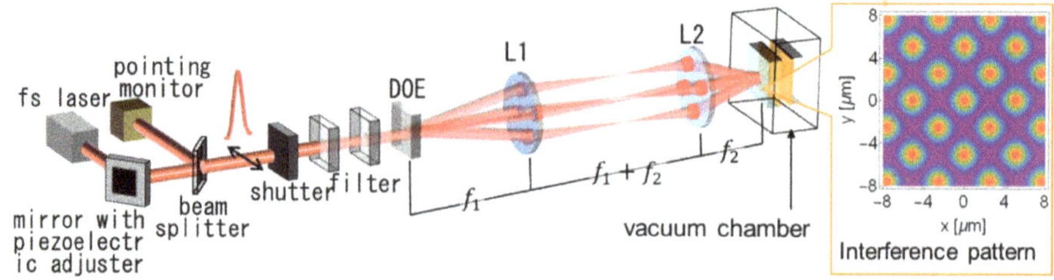

Figure 2. Experimental setup. Right inset shows the simulation result of the interference pattern on the irradiation plane. fs laser: femtosecond laser; DOE: diffractive optical element; L1 and L2: achromatic lenses.

3. Results and Discussion

3.1. SEM Images of the Pt Nanodots

Typical SEM images of the Pt and Au nanodots [17] are shown in Figure 3. The surface morphology was measured by a scanning electron microscope (JSM-7400FS, JEOL, Tokyo, Japan). When F_{peak} was the peak fluence, the area with a fluence higher than F_{peak}/e^2 was 0.073 mm². The pulse energy for the Pt and Au target was 69 and 97 µJ, and the average fluence $(1/e^2)$ in the area was 94.2 and 133 mJ/cm², respectively. It is apparent that the Pt and Au nanodots were transferred successfully. On the other hand, the Pt nanodots are deposited in a dispersed position, though the Au nanodots are in array corresponding to the interference pattern which is shown in the right inset in Figure 2. This tendency did not change with the fluence, and small nanodots are found between the large nanodots in the case of Pt.

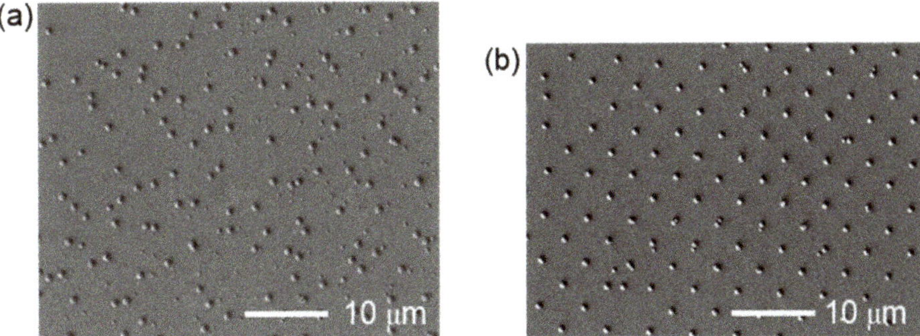

Figure 3. Scanning electron microscopy (SEM) images of (**a**) Pt nanodots and (**b**) Au nanodots deposited by LIDT with the interference pattern of a fs laser. The original image of (**b**) is the same as in [17] Nakata et al. 2020. Int. J. Exrem. Manuf. 2, 025101.1-5. CC-BY-3.0.

The SEM images of Pt nanodots as a function of the averaged fluence are summarized in Figure 4. The measurement range is located approximately at the center of the processed area. At a lower fluence of 75 mJ/cm², the number of nanodots is small. It seems that the size of the nanodots appears to be classified into two types at a higher fluence, the statistical analysis of which is explained in the following Section 3.3.

In the previous experiment using Au donor film, a variety of unit structures such as solo, adjoining, and stacking nanodots were observed [17]. We found similar unit structures in this experiment, as shown in Figure 5. Here, all images are of structures fabricated under the fluence of 106 mJ/cm², but similar structures were seen in the other fluences of 94 or 84 mJ/cm². As mentioned below, the size of the nanodots are classified into two types. Figure 5a is a representative large nanodot, which has circular shape with a diameter of $D = 627$ nm. Here, the circularity f_{circ} and roundness f_r of the nanodot are defined by the following equations:

$$f_{circ} = 4\pi S/P^2 \tag{1}$$

$$f_r = 4S/(2a)^2, \tag{2}$$

where S is the surface area of the nanodot, P is the perimeter, and $2a$ is the length of the major axis, assuming the shape to be an ellipsoid. For this nanodot, we obtained $f_{circ} = 0.90$ and $f_r = 0.98$, so it is a fairly round circle. The three-dimensional shape is considered to be a collapsed sphere, as in previous experiments using Au thin film [17]. On the other hand, a representative small nanodot is shown in Figure 5b. The diameter is $D = 125$ nm. Such fine nanodots were seen in the entire measurement area at all fluences.

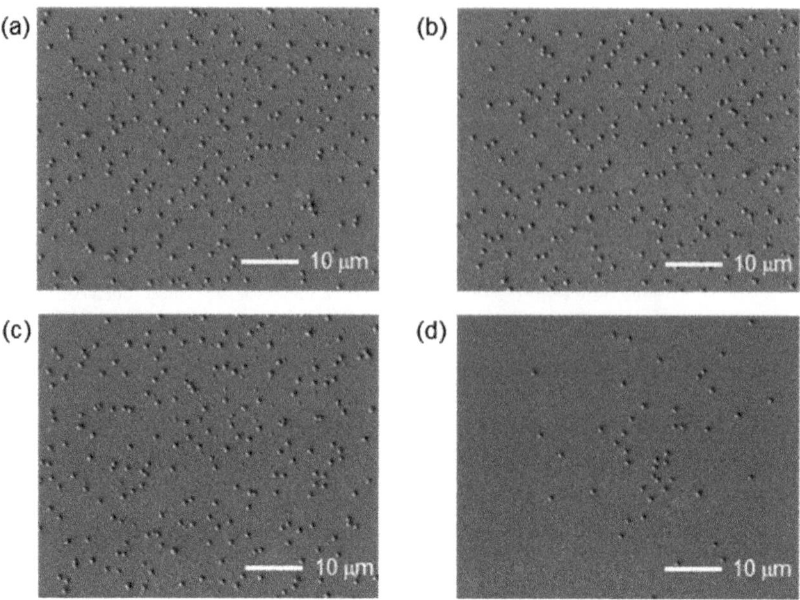

Figure 4. SEM images of Pt nanodots with different averaged fluences. (**a**) 106 mJ/cm^2, (**b**) 94 mJ/cm^2, (**c**) 84 mJ/cm^2, (**d**) 75 mJ/cm^2. (**b**) shows the same area as in Figure 3a, but is slightly wider.

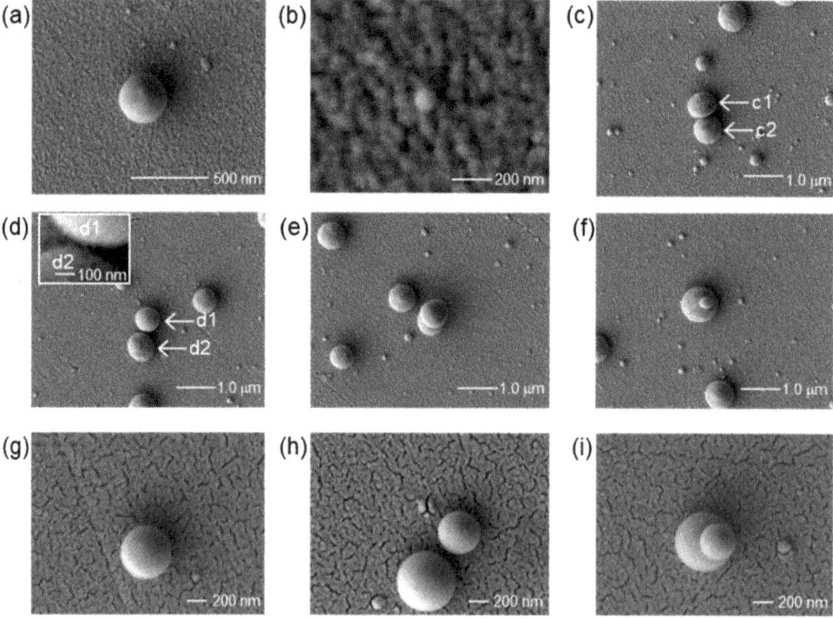

Figure 5. SEM images of the nanodots. (**a,b**) Pt solo nanodots, (**c**) Pt adjoining nanodots, (**d**) Pt nanodots with gap, and (**e,f**) Pt stacking nanodots. (**g,h**) and (**i**) are solo, adjoining and stacking Au nanodots, respectively. The pixel size of the image is (**b**) 2.36 nm/pixel, (**g–i**) 1.88 nm/pixel, and 4.7 nm/pixel otherwise. Therefore, the first digit of the size in the text is for reference.

Similar to the case of Au donor film, we found solo, adjoining, and stacking nanodots, as shown in Figure 5c–f. Here, Figure 5g–i are images of Au nanodots [17]. In the case of the adjoining nanodots that are in contact as shown in Figure 5c, the shape of the c1 nanodot is squashed. It is an ellipse with a minor axis of 600 nm and a major axis of 746 nm. This implies that at the timing of deposition near c2, c1 was at high temperature and soft. The diameter of the c2 nanodot is 707 nm. Whereas, Figure 5d shows nanodots with a nanogap. Both d1 and d2 are circular with diameters of $D = 621$ and 702 nm, respectively. The gap length is $\Delta l = 46$ nm.

In contrast to the case with Au donor film, the number of stacking nanodots, which are shown in Figure 5e,f, seemed to be quite small. The mechanism will be discussed in Section 3.3.

3.2. Statistical Analysis

In a past LIDT experiment, Dr. Narazaki reported that single or multi nanodot ejection occurs from a focal spot of a laser [8]. The same phenomena was observed in the case of LIDT using an interference pattern with an Au film target [17]. In this subsection, nanodot size distribution is analyzed from the images in Figure 4. Here, adjoining and stacking nanodots are excluded from the calculation. The minimum size measurement limit due to the resolution of the image is approximately $D = 280$ nm. Figure 6 shows the nanodot size distribution as a function of the averaged fluence. In all conditions, the distribution is split into two parts: small dot and large dot groups, as divided by the red arrows. The total number of each group is summarized in Table 1, and plotted in Figure 7a. Here, the number of the spots in the corresponding area shown in the images in Figure 4 is 223. So, it is interesting that the number of nanodots is always larger than the number of ablation spots in the interference pattern. This directly proves that multiple nanodots can be generated from a single spot.

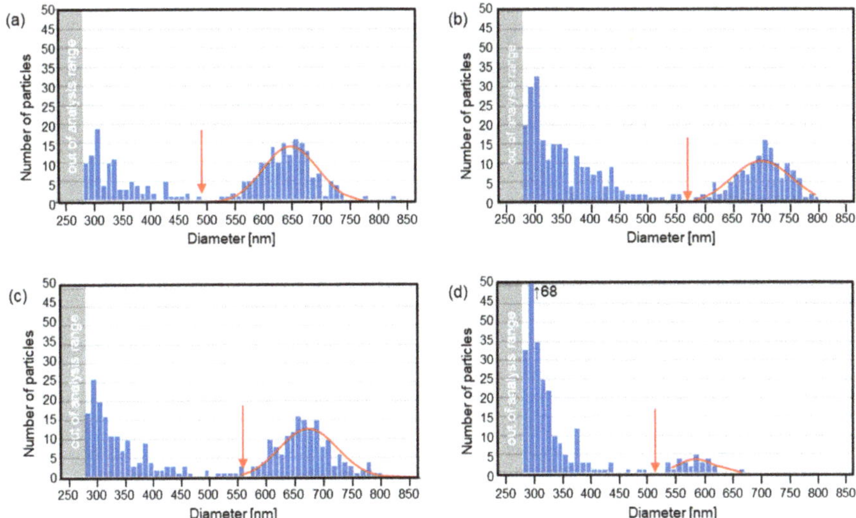

Figure 6. Nanodot size distributions. The averaged fluence was (**a**) 106, (**b**) 94, (**c**) 84, and (**d**) 75 mJ/cm^2, respectively. Alphabetical numerals correspond to those in Figure 4. The red curve on each graph corresponds to a normal distribution fit on the data set starting from the red arrows to the right in each case.

Table 1. Results of the analysis of particle size distribution: number, mean size, standard deviation, and values for each group divided by size (indicated by red arrow in Figure 6). Alphabetical numerals correspond to those in Figures 4 and 7.

Averaged Fluence (mJ/cm^2)	(a) 106	(b) 94	(c) 84	(d) 75
(a-1) total number	263	375	322	264
(a-2) number of small size group	100	233	162	236
(a-3) number of large size group	163	142	160	28
(b-1) average particle size	526	480	506	347
(b-2) average size of small size group	339	349	345	319
(b-3) average size of large size group	641	698	668	581
(c-1) standard deviation (SD)	151	176	171	88
(c-2) SD of small size group	48	58	62	36
(c-3) SD of large size group	46	52	50	30

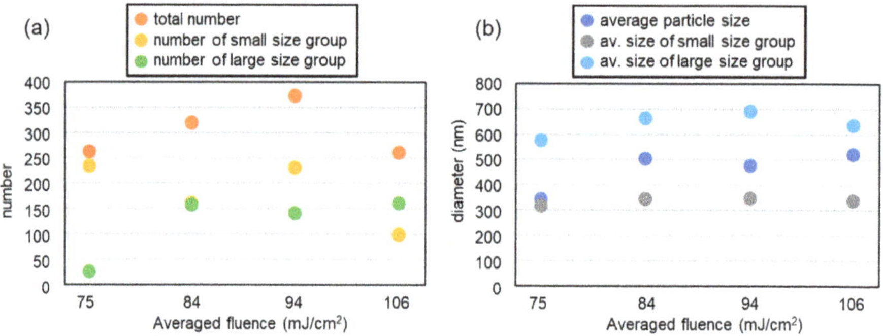

Figure 7. (**a**) nanodot number and (**b**) size distributions. The average fluence was 106, 94, 84, and 75 mJ/cm^2, respectively.

At the lowest fluence, the number of large dots is low, though that of small dots is high. On the other hand, many large nanodots are formed at higher fluences. The total number in of nanodots in the large group does not change much above 84 mJ/cm^2. The red curve on each graph corresponds to a normal distribution fit on the data set starting from the red arrows to the right in each case. The averaged diameter of the large nanodot group is largest at around 84 to 94 mJ/cm^2, and decreased at the highest fluence, as shown in Figure 7b. In addition, the number of the small nanodot group decreased at the highest fluence. These results suggest that the vaporized metal dissipated or was deposited as a thin film on the receiver substrate without condensation at the highest fluence. Here, it should be noted that there is a possibility of the nanodots' shapes being squashed, depending on the temperature at which the particles are deposited, which is likely to vary with the average fluence and the thermal properties of the receiver substrate.

Here, the shape and diameter of the nanodots fabricated by various methods is compared. The circularity and roundness are excellent in most of the LIDT experiments including this paper. On the other hand, the size distribution was one group in the previous experiment, but it is divided into two groups in this experiment. The large nanodot group has an average size of about 580 to 690 nm, which is comparable to the previous interference LIDT experiment using Au thin film, and other group's single LIDT experimental results [8,20]. On the other hand, the nanodots fabricated using LIDOS with an Au thin film, in which nanodots are deposited on a source target placed in air as seen in Figure 1a, have a smaller average particle size of less than 200 nm. This is probably due to the fact that the interference pattern period is 1.93 μμm, which is about half that of the present study [18]. It is important to note that for precise comparison, it is necessary to prepare the same experimental systems. The smallest standard deviation in LIDOS was 3 nm [18], which is far smaller than the values shown in Table 1. With even larger spot sizes, higher pulse energies, and thicker film targets, the deposit becomes a micron-sized dot or larger diameter film structure, and this process is called LIFT [1–5,21]. In the case of pulsed-laser deposition (PLD), micron-sized droplets are ejected from the ablation spot [22]. In this technique, some tens of nm-sized particles are formed due to the condensation at high atmospheric gas pressure conditions [23,24]. Here, chemosynthesis is useful to fabricate a number of nm-sized nanodots and nanorods. In summary, utilizing LIDT with an interference pattern is a good alternative method to PLD. In addition, there is still a possibility that the array structure and uniform size distribution of Pt nanodots can be achieved by optimizing the parameters. We will discuss this in the next section.

3.3. Heuristic Model of LIDT

In this subsection, the formation mechanism of nanodots is discussed based on the above observations and previous experiments. Figure 8 provides a schematic explanation of the mechanisms of LIDT via SLS, which has been discussed based on the experimental results of the interference laser processing. The upper left figure illustrates the interference laser processing. Here, we focus on a spot in the interference pattern. In the past experiment with Au and Ag thin film targets, we have selectively fabricated metal nanowhiskers [12,15] and nanocrowns [15,16], as shown in the inserted pictures in Figure 8b,c, respectively. Usually, nanowhiskers and nanodrops [13,25] are formed as shown in Figure 8b, but nanocrowns are formed as shown in Figure 8c when the film thickness is relatively thin or the interference pattern has a wider period, i.e., when the spot size is relatively large [15,16]. In the former case, a nanodot is formed by surface tension and deposited onto the receiver substrate, which was shown in the previous paper using Au donor film [17]. In this case, the position of the nanodots and the interference pattern are in good agreement. In addition, as seen in the water droplet behavior [26–28], the formation of a second and subsequent nanodots may occur, which are deposited as the neighbor of the first nanodots [8,17]. Thus, the film thickness and spot size have been considered to be the key parameters for determining either of the processes explained in Figure 8b,c, up to this point.

On the other hand, in this experiment, only by changing the thin film material to Pt did we find a dispersion in the deposition position and the generation of a greater number of nanodots than the number of spots. Here, three general mechanisms of nanodot formation can be considered: formation at the spot center by surface tension of the liquid metal, formation at the edge of the nanocrown, and condensation of the metal vapor. Here, the last mechanism does not occur in this experiment because it only occurs under high pressure. As shown in Figure 4, the nanodots are not aligned according to the interference pattern. In addition, the stacking nanodots in this experiment, which are shown in Figure 5e,f, were not frequently found which may be due to the fact that major nanodots are ejected in a splashy manner, not sequentially from the same spot [26–28]. These suggest the ejection of nanodots from the nanocrowns. On the other hand, the sizes are divided into two groups, as shown in Figure 6. The average size of the large sized group is similar to the previous results of the LIDT using Au, where an ablation and a Au nanodot corresponded one-to-one. Additionally, their number is smaller than that of the number of ablation spots. Thus, it is possible that the nanodots in the larger size group were formed by the process explained in Figure 8b, and the smaller group nanodots were formed by the process explained in Figure 8c. In addition, the most reliable way to confirm the type of LIDT mechanism is to measure the donor thin film after LIDT with SEM. Recovery of the donor thin film substrate after the process is possible by improving the method of fixing the donor thin film target, which will be done in the future. Furthermore, imaging measurement of the LIDT process is also an alternative. In past experiments, the resolution of the microscopic imaging laser-induced fluorescence (2D-LIF) [6] or scattering imaging technique [7] were suited to measure the behavior of the micron-sized ejecta in LIFT. On the other hand, the very small size of the nanodots in this LIDT setup makes it difficult to measure them with these techniques.

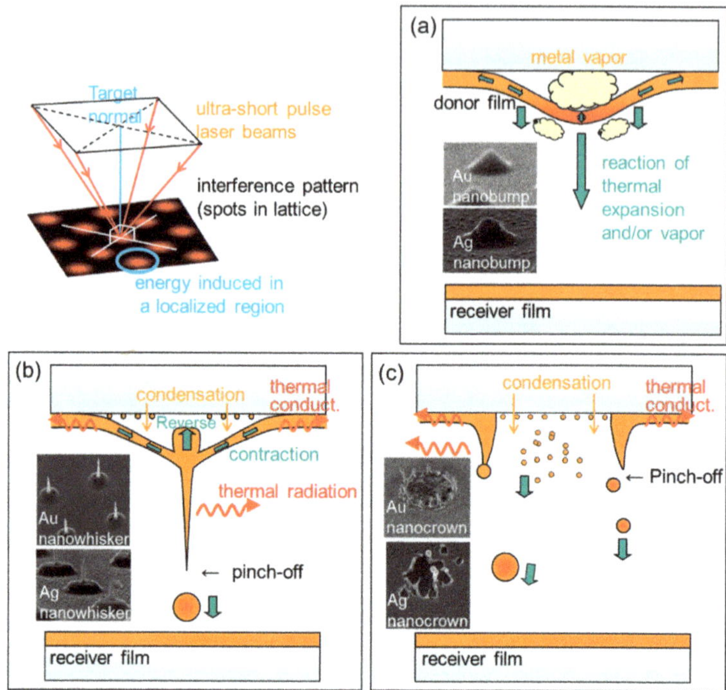

Figure 8. Schematic explanation of solid–liquid–solid (SLS) process in LIDT. The upper left sketch is a schematic of four-beam interference, in which the bright spots are in lattice. (**a**) induction of energy into a localized region results in a partial liquid motion of solute metal thin film by the reaction due to the thermal expansion of the film or vapor pressure. (**b**) partial liquid motion results in the simultaneous formation of a nanodot and a nanowhisker. (**c**) nanodots are ejected in a splashy manner. Insets in each illustration are copied from our past experiments [12,13,15,16].

Here, we discuss on the cause of the difference in results by comparing the experimental conditions with the previous experiment where Au donor thin film was used, and the period of the interference pattern was the same, 3.6 μm. The film thickness and averaged fluence agree within a difference of 20 to 30%. In addition, the melting points of Au and Pt are 1064 and 1768 °C, respectively. The surface tensions at those temperatures are 907 and 1784 dyn/cm respectively [29], and the latter is almost twice as large. The viscosity is between 3.7 and 4.9 cP at 1333–1640 °C in the case of liquid Au [30], and is 6.74 cP at 1768 °C in the case of liquid Pt [31]. At present, it is not clear which physical properties are the key parameters to govern the process. Here, molecular dynamic simulation will be a powerful tool to investigate the underlying mechanisms and key parameters [32]. There is still the possibility of achieving an array structure and uniform size distribution of the Pt nanodots by optimizing the parameters, and further experiments are needed.

4. Conclusions

The interference pattern of a femtosecond laser was applied to the LIDT technique, and a number of nanodots with diameters of some hundreds of nm with good roundness and circularity were fabricated. A typical nanodot had a circularity $f_{circ} = 0.90$ and roundness $f_r = 0.98$. A comparison of the number of nanodots produced and the number of processed spots in the interference pattern revealed that multiple nanodots were formed from a single spot. The size distribution was divided into two groups, and the ratio was affected by the average fluence. The smallest average diameter of the large nanodot group was 581 nm with a smallest standard deviation of 30 nm at 75 mJ/cm^2.

In terms of applications, Pt is useful as a catalyst. Its size is relatively larger than that of a typical catalyst [33] and so the combined use of plasmonic interactions is promising. Compared with bottom-up methods such as chemosynthesis, SLS can fabricate pure nanodots because no catalyst or chemosynthetic solutions are required. It requires no further processes, such as rinsing or annealing. The size distribution of Pt nanodots is reasonably uniform, but the processing parameters for fabricating aligned nanodots needs to be further explored. These advantages will broaden the range of applications in nano-3D printing and catalyst and plasmonic applications.

Author Contributions: Conceptualization, Y.N.; Methodology, Y.N.; Validation, Y.N.; Formal Analysis, Y.N.; Investigation, Y.N.; Resources, Y.N., K.T., N.M.; Data Curation, Y.N.; Writing—Original Draft Preparation, Y.N.; Writing—Review & Editing, Y.N.; Visualization, Y.N.; Supervision, Y.N.; Project Administration, Y.N.; Funding Acquisition, Y.N., A.N., T.S., Y.T. All authors have read and agreed to the published version of the manuscript.

Funding: This research was funded by the Japan Society for the Promotion of Science (JSPS) (16H038850).

Conflicts of Interest: The authors declare no conflict of interest.

References

1. Levene, M.L.; Scott, R.D.; Siryj, B.W. Material Transfer Recording. *Appl. Opt.* **1970**, *9*, 2260–2265. [CrossRef] [PubMed]
2. Bohandy, J. Metal deposition from a supported metal film using an excimer laser. *J. Appl. Phys.* **1986**, *60*, 10–12. [CrossRef]
3. Piqué, A.; Chrisey, D.B.; Auyeung, R.C.Y.; Fitz-Gerald, J.; Wu, H.D.; McGill, R.A.; Lakeou, S.; Wu, P.K.; Nguyen, V.; Duignan, M. A novel laser transfer process for direct writing of electronic and sensor materials. *Appl. Phys. A Mater. Sci. Process.* **1999**, *69*, 279–284. [CrossRef]
4. Colina, M.; Serra, P.; Fernández-Pradas, J.M.; Sevilla, L.; Morenza, J.L. DNA deposition through laser induced forward transfer. *Biosens. Bioelectron.* **2005**, *20*, 1638–1642. [CrossRef] [PubMed]
5. Nakata, Y.; Okada, T.; Maeda, M. Transfer of Laser Dye by Laser-Induced Forward Transfer. *Jpn. J. Appl. Phys.* **2002**, *41*, L839–L841. [CrossRef]
6. Nakata, Y.; Okada, T. Time-resolved microscopic imaging of the laser-induced forward transfer process. *Appl. Phys. A Mater. Sci. Process.* **1999**, *69*, S275–S278. [CrossRef]
7. Nakata, Y.; Okada, T.; Maeda, M. Application and observation of laser-induced forward transfer process. *Proc. SPIE* **2002**, *4637*, 435–444. [CrossRef]
8. Narazaki, A.; Sato, T.; Kurosaki, R.; Kawaguchi, Y.; Niino, H. Nano- and microdot array formation of FeSi$_2$ by nanosecond excimer laser-induced forward transfer. *Appl. Phys. Express* **2008**, *1*, 0570011–0570013. [CrossRef]
9. Narazaki, A.; Sato, T.; Kurosaki, R.; Kawaguchi, Y.; Niino, H. Nano- and microdot array formation by laser-induced dot transfer. *Appl. Surf. Sci.* **2009**, *255*, 9703–9706. [CrossRef]
10. Kuznetsov, A.I.; Evlyukhin, A.B.; Reinhardt, C.; Seidel, A.; Kiyan, R.; Cheng, W.; Ovsianikov, A.; Chichkov, B.N. Laser-induced transfer of metallic nanodroplets for plasmonics and metamaterial applications. *J. Opt. Soc. Am. B* **2009**, *26*, B130. [CrossRef]
11. Willis, D.A.; Grosu, V. Microdroplet deposition by laser-induced forward transfer. *Appl. Phys. Lett.* **2005**, *86*, 244103. [CrossRef]
12. Nakata, Y.; Miyanaga, N.; Momoo, K.; Hiromoto, T. Solid-liquid-solid process for forming free-standing gold nanowhisker superlattice by interfering femtosecond laser irradiation. *Appl. Surf. Sci.* **2013**, *274*, 27–32. [CrossRef]
13. Nakata, Y.; Okada, T.; Maeda, M. Nano-sized hollow bump array generated by single femtosecond laser pulse. *Jpn. J. Appl. Phys. Part 2 Lett.* **2003**, *42*, L1452–L1454. [CrossRef]
14. Nakata, Y.; Hiromoto, T.; Miyanaga, N. Mesoscopic nanomaterials generated by interfering femtosecond laser processing. *Appl. Phys. A Mater. Sci. Process.* **2010**, *101*, 471–474. [CrossRef]
15. Nakata, Y.; Miyanaga, N.; Momoo, K.; Hiromoto, T. Template free synthesis of free-standing silver nanowhisker and nanocrown superlattice by interfering femtosecond laser irradiation. *Jpn. J. Appl. Phys.* **2014**, *53*, 096701. [CrossRef]
16. Nakata, Y.; Tsuchida, K.; Miyanaga, N.; Furusho, H. Liquidly process in femtosecond laser processing. *Appl. Surf. Sci.* **2009**, *255*, 9761–9763. [CrossRef]
17. Nakata, Y.; Hayashi, E.; Tsubakimoto, K.; Miyanaga, N.; Narazaki, A.; Shoji, T.; Tsuboi, Y. Nanodot array deposition via single shot laser interference pattern using laser-induced forward transfer. *Int. J. Exreme Manuf.* **2020**, *2*, 025101. [CrossRef]
18. Nakata, Y.; Murakawa, K.; Miyanaga, N.; Narazaki, A.; Shoji, T. Local Melting of Gold Thin Films by Femtosecond Laser-Interference Processing to Generate Nanoparticles on a Source Target. *Nanomaterials* **2018**, *8*, 477. [CrossRef]
19. Nakata, Y.; Murakawa, K.; Sonoda, K.; Momoo, K.; Miyanaga, N.; Hiromoto, T. Designing of interference pattern in ultra-short pulse laser processing. *Appl. Phys. A* **2012**, *112*, 191–196. [CrossRef]

20. Zhigunov, D.M.; Evlyukhin, A.B.; Shalin, A.S.; Zywietz, U.; Chichkov, B.N. Femtosecond Laser Printing of Single Ge and SiGe Nanoparticles with Electric and Magnetic Optical Resonances. *ACS Photonics* **2018**, *5*, 977–983. [CrossRef]
21. Duocastella, M.; Kim, H.; Serra, P.; Piqué, A. Optimization of laser printing of nanoparticle suspensions for microelectronic applications. *Appl. Phys. A Mater. Sci. Process.* **2012**, *106*, 471–478. [CrossRef]
22. Nakata, Y.; Gunji, S.; Okada, T.; Maeda, M. Fabrication of LiNbO3 thin films by pulsed laser deposition and investigation of nonlinear properties. *Appl. Phys. A Mater. Sci. Process.* **2004**, *79*, 1279–1282. [CrossRef]
23. Muramoto, J.; Sakamoto, I.; Nakata, Y.; Okada, T.; Maeda, M. Influence of electric field on the behavior of Si nanoparticles generated by laser ablation. *Appl. Phys. Lett.* **1999**, *75*, 751. [CrossRef]
24. Nakata, Y.; Muramoto, J.; Okada, T.; Maeda, M. Particle dynamics during nanoparticle synthesis by laser ablation in a background gas. *J. Appl. Phys.* **2002**, *91*, 1640–1643. [CrossRef]
25. Nakata, Y.; Miyanaga, N.; Okada, T. Effect of pulse width and fluence of femtosecond laser on the size of nanobump array. *Appl. Surf. Sci.* **2007**, *253*, 6555–6557. [CrossRef]
26. Clasen, C.; Bico, J.; Entov, V.M.; McKinley, G.H. "Gobbling drops": The jettingdripping transition in flows of polymer solutions. *J. Fluid Mech.* **2009**, *636*, 5–40. [CrossRef]
27. Dinca, V.; Patrascioiu, A.; Fernández-Pradas, J.M.; Morenza, J.L.; Serra, P. Influence of solution properties in the laser forward transfer of liquids. *Appl. Surf. Sci.* **2012**, *258*, 9379–9384. [CrossRef]
28. Duocastella, M.; Fernández-Pradas, J.M.; Morenza, J.L.; Serra, P. Sessile droplet formation in the laser-induced forward transfer of liquids: A time-resolved imaging study. *Thin Solid Films* **2010**, *518*, 5321–5325. [CrossRef]
29. Flint, O. Surface tension of liquid metals. *J. Nucl. Mater.* **1965**, *16*, 233–248. [CrossRef]
30. Ofte, D. The viscosities of liquid uranium, gold and lead. *J. Nucl. Mater.* **1967**, *22*, 28–32. [CrossRef]
31. Paradis, P.F.; Ishikawa, T.; Okada, J.T. Thermophysical properties of platinum group metals in their liquid undercooled and superheated phases. *Johns. Matthey Technol. Rev.* **2014**, *58*, 124–136. [CrossRef]
32. Danilov, P.A.; Zayarny, D.A.; Ionin, A.A.; Kudryashov, S.I.; Rudenko, A.A.; Kuchmizhak, A.A.; Vitrik, O.B.; Kulchin, Y.N.; Zhakhovsky, V.V.; Inogamov, N.A. Redistribution of a Material at Femtosecond Laser Ablation of a Thin Silver Film 1. *JETP Lett.* **2016**, *104*, 759–765. [CrossRef]
33. Mukerjee, S. Particle size and structural effects in platinum electrocatalysis. *J. Appl. Electrochem.* **1990**, *20*, 537–548. [CrossRef]

nanomaterials

Article

Femtosecond Laser-Induced Periodic Surface Structures on 2D Ti-Fe Multilayer Condensates

Oleksandr V. Kuznietsov [1,2], George D. Tsibidis [3], Anatoliy V. Demchishin [4], Anatoliy A. Demchishin [4], Volodymyr Babizhetskyy [5], Ivan Saldan [5,6], Stefano Bellucci [7] and Iaroslav Gnilitskyi [1,2,*]

1   NoviNano Lab LLC, Pasternaka 5, 79015 Lviv, Ukraine; oleksandr.kuznietsov22@gmail.com
2   Department of Photonics, Lviv Polytechnic National University, 79013 Lviv, Ukraine
3   Institute of Electronic Structure and Laser (IESL), Foundation for Research and Technology (FORTH), N. Plastira 100, Vassilika Vouton, 70013 Heraklion, Crete, Greece; tsibidis@iesl.forth.gr
4   Frantsevich Institute for Problems in Materials Science of NASU, Krzhizhanovsky 3, 03142 Kyiv, Ukraine; demch@ipms.kiev.ua (A.V.D.); ntuu.kpi@ukr.net (A.A.D.)
5   Faculty of Chemistry, Ivan Franko National University of Lviv, Kyryla and Mefodia 6, 79005 Lviv, Ukraine; v.babizhetskyy@gmail.com (V.B.); saldanivan@gmail.com (I.S.)
6   Faculty of Science, P.J. Šafárik University in Košice, Šrobárova 2, 04154 Košice, Slovakia
7   INFN-Laboratori Nazionali di Frascati, Via E. Fermi 54, 00044 Frascati, Italy; bellucci@lnf.infn.it
*   Correspondence: iaroslav.gnilitskyi@novinano.com

Abstract: 2D Ti-Fe multilayer preparation has been attracting increased interest due to its ability to form intermetallic compounds between metallic titanium and metallic iron thin layers. In particular, the TiFe compound can absorb hydrogen gas at room temperature. We applied femtosecond laser pulses to heat Ti-Fe multilayer structures to promote the appearance of intermetallic compounds and generate surface nanostructuring. The surface pattern, known as Laser Induced Periodic Surface Structures (LIPSS), can accelerate the kinetics of chemical interaction between solid TiFe and gaseous hydrogen. The formation of LIPSS on Ti-Fe multilayered thin films were investigated using of scanning electron microscopy, photo-electron spectroscopy and X-ray diffraction. To explore the thermal response of the multiple layered structure and the mechanisms leading to surface patterning after irradiating the compound with single laser pulses, theoretical simulations were conducted to interpret the experimental observations.

Keywords: vacuum-arc evaporation; titanium; iron; LIPSS; multilayer structures

---

Citation: Kuznietsov, O.V.; Tsibidis, G.D.; Demchishin, A.V.; Demchishin, A.A.; Babizhetskyy, V.; Saldan, I.; Bellucci, S.; Gnilitskyi, I. Femtosecond Laser-Induced Periodic Surface Structures on 2D Ti-Fe Multilayer Condensates. *Nanomaterials* **2021**, *11*, 316. https://doi.org/10.3390/nano11020316

Received: 30 November 2020
Accepted: 19 December 2020
Published: 27 January 2021

Publisher's Note: MDPI stays neutral with regard to jurisdictional claims in published maps and institutional affiliations.

1. Introduction

The TiFe intermetallic compound (IMC) is considered a promising hydrogen storage material, but this IMC requires initial activation to effectively absorb hydrogen gas [1–6]. Using high-pressure torsion (HPT), TiFe can reversibly absorb ~1.4–1.7 wt.% H2 in practice at room temperature without thermal activation [1]. Usually, TiFe is processed by three different routes: annealing, plastic deformation using groove rolling, and severe plastic deformation using HPT [2]. Effective activation of air-exposed TiFe alloy might be performed mechanically using cold rolling [3,4] or ball milling [3,5,6]. These procedures help to restore the hydrogenation capability of TiFe. Because of the low price of production and activation of TiFe, it can be used in hydrogen storage tanks, fuel cells and secondary batteries. Recently, the operation of a bench-scale stationary hydrogen energy system comprising photovoltaic panels, a water electrolyzer and a TiFe-based tank was tested in practice [7]. Thin films of pure TiFe can be used as hydrogen sensors at room temperature, but in order to charge/discharge them faster, a more developed surface structure is needed. The nanostructuring of solid surfaces may be a reasonable way of accelerating the kinetics of their interaction with gaseous hydrogen.

Fabricating a surface pattern that can be used for the aforementioned purpose can be achieved through numerous methods including multi-beam interference-based techniques

---

and Laser Induced Periodic Surface Structures (LIPSS). LIPSS is a potentially strong method, as it is a single-step, maskless and optical patterning technique. This method of surface texturizing has been realized on metals [8], semiconductors [9], dielectric surfaces [10] and polymers [11]. Moreover, LIPSS were used in numerous applications including solar cells [12], plasmonics [13], colorizing metals [14,15], wettability [16] and tribology [17]. Recently, advanced techniques for generating highly-regular LIPSS (HR-LIPSS) have been developed to create high-quality periodic nanostructures over large area [18]. Nanolayers with a high level of regularity were obtained over a large surface area with a single-step maskless process and industrially accepted speed production. The use of the HR-LIPSS approach allows laser self-organized periodic structures with practical applications to be obtained, as they are sensitive to the optical, mechanical, adhesive and wetting properties of the surface.

The nanostructured pattern (LIPSS) of the Ti-Fe system was obtained by using femtosecond laser pulses to create a functional profile. In order to interpret the physical mechanisms that account for the surface modification, simulations were performed to highlight the thermal response of the femtosecond laser pulses during the LIPSS formation on multilayers Ti-/Fe system. The theoretical predictions, based on the 1D two-temperature model (TTM) simulations, were in good agreement with the experimental results.

2. Experimental Part

2.1. Preparation of Ti-Fe Multilayer

Made in a vacuum, thick condensates of the Ti/Fe system composed of numerous layered heterogeneous layers of the separate metals were obtained using the device "Bulat-3T" (Kharkiv, Ukraine) through the method of consecutive condensation from unfiltered plasma flows, which were respectively generated by stationary cathode-type vacuum arcs under conditions of constant ion bombardment of the formed condensates by applying an electric negative potential to the substrates [19]. The distance between the end-type cathodes and the substrates was 125 mm. Steel plates were used as a substrate made of X12Cr17 coiled steel with a size of 100 nm × 100 nm × 0.3 mm. Cathodes with a diameter of 64 mm were obtained by machining ingots of Ø70 mm from pure Ti and Fe, melted by electron beam remelting in a vacuum of 1×10^{-4} mm Hg. The condensation duration of each sublayer was 10, 20, 30, 40, 50 s, which made it possible to obtain condensates while modulating the multilayer structure in the range of 125–620 nm. The total thickness of the multilayer compositions was 60–80 microns. A negative potential of 180V was applied to the substrates. The metal deposition rate was 0.8–1.2 µm/min. The total duration of the deposition of multilayer compositions is 60 min. The arc current was tantamount to 100A and 80A for cathodes of Ti and Fe, respectively. The value of the working pressure in the installation chamber was $3–4 \times 10^{-3}$ mm Hg. The vacuum annealing of condensates without separation from the substrates was carried out in this unit at a temperature of 650–700 °C by bombarding their surfaces with argon ions for 30 min as a way of applying a negative potential of 1 kV to the substrates and a pressure of $3–4 \times 10^{-3}$ mm Hg in the chamber. The structure of the coatings was studied using a JEOL 733 scanning electron microscope (Tokyo, Japan). The phase composition of multilayer condensates was determined using X-ray diffraction analysis. XRD analysis was performed using Cu K_α radiation. The microhardness of the condensates was measured using a PMT-3 microhardness meter at a load of 50 g on transverse sections.

2.2. Processing with Ultrashort Laser

Heat-treated samples with a thickness of 60–80 µm were irradiated with a Yb: KGW laser (model Pharos 20 W, LightConversion, Vilnius, Lithuania) with 213 fs pulses and a spectrum width of 15 nm (Figure 1). The pulse repetition rate was 600 kHz. The fluence was in the range of 0.5 J/cm^2. Laser processing of the samples was carried out in air at room temperature in the regime of scanning a laser beam over their surface. The average power output was 20 W. The laser setup was equipped with lenses with a focal length of

56 mm, which ensured that a laser spot with a diameter of 10.4 μm was obtained at ℓ/e^2 intensity. The size of the lateral displacement of the laser beam was adjustable from 2 to 4 microns. Laser installation allows one to scan the surface of the samples with a linear speed of 0.5 m/s and an equivalent productivity of about 900 mm^2/min. The topography of the laser-treated sample surface was studied using an FEI Nova Nano SEM 450 electron microscope (Hillsboro, OR, USA).

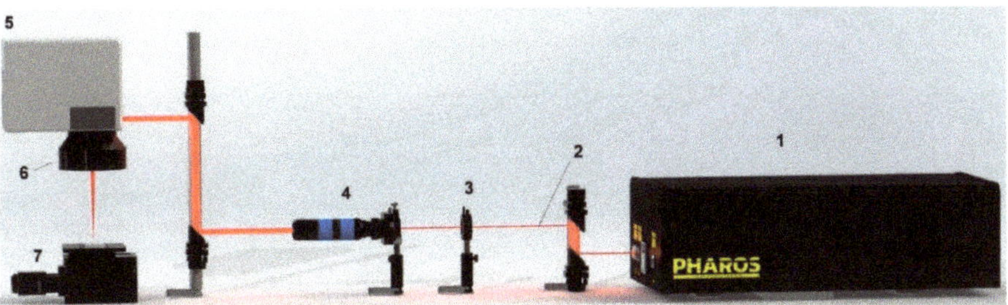

Figure 1. Laser setup for generation LIPSS: **1**—femtosecond laser "Pharos", **2**—laser beam, **3**—HWP (half wave plate), **4**—expander, **5**—galvoscaner, **6**—F-theta lens, **7**—z-axis motorized stage.

2.3. X-ray Photoelectron Spectroscopy (XPS)

XPS data were recorded using SPECS spectrometer (Berlin, Germany) with a PHOIBOS hemispherical energy analyzer and monochromatic Al Kα X-ray irradiation ($h\nu$ = 1486.74 eV, 350 W). All spectra were calibrated using the C 1s main peak located at a binding energy (BE) of 284.6 eV. The atomic concentrations of the elements were calculated taking into account corresponding relative sensitivity factors in CasaXPS software program.

2.4. X-ray Diffraction Analysis (XRD)

XRD patterns were recorded using a Rigaku diffractometer, equipped with a Miniflex goniometer and an X-ray source with Cu Kα radiation, at λ = 1.5418 A, 30 kV, and 15 mA. The samples were scanned in the 1–80°, 2θ range with a scan rate of 0.02°/s. Diffraction patterns were assigned using Joint Committee on Powder Diffraction Standards (JCPDS) cards supplied by the International Centre for Diffraction Database (ICDD).

2.5. Scanning Electron Microscope (SEM) and Focused Ion Beam (FIB) Characterization

The surface morphology of the untreated samples and those treated with ultrashort laser pulses was studied using Fa scanning electron microscope FEI Nova Nano SEM 450 with Bruker QUANTAX-200 X-EDS (Billerica, MA, USA). The cross-sections and the corresponding images were obtained by means of an FEI Strata 235 M dual beam system (Hillsboro, OR, USA). The system combines a Focused Ion Beam (FIB) equipped with a Ga Liquid Metal Ion Source (LMIS). The perpendicular cross-sections of the Ti/Fe-coated surfaces were obtained using an FIB (E-beam = 30 keV) for milling, setting 1 nA as the ion beam current for milling and 300 pA for the final polishing. In order to protect the topmost material of the coating from the ion mixing effect, the surface was protected by a thin Pt layer (Pt shield) deposited in situ using Ion Beam.

3. Research Results and Their Discussion

3.1. SEM Observation

SEM shows a typical surface morphology of metallic Ti or Fe layers after vacuum arc condensation after arc melting. The surface is chaotically covered with many inhomogeneous particles left after melting (Figure 2a). However, laser treatment with LIPSS formation leads to the structuring of the surface by creating periodic longitudinal ripples on Ti/Fe systems when the Ti layer is on the top (Figure 2b,c). LIPSS were successfully generated over a large area of 1 cm^2. Despite a lot of bifurcations that were created as a collateral effect of vacuum-arc deposition, LIPSS were homogeneously obtained over all large areas, which can be attributed to the HR-LIPSS method [20]. The quality of LIPSS was measured by a descriptor of the regularity of the periodic structures, DLOA (Dispersion of the LIPSS orientation angle), which is presented in-depth in [20]. DLOA was measured at 18.5°, thus certifying that periodic structures have a good quality despite of the numerous defects on the surface. The period of periodic structures is in the range of 650–750 nm, while the direction of the periodic structures is perpendicular to the laser polarization, which suggests an ablation mechanism for LIPSS formation.

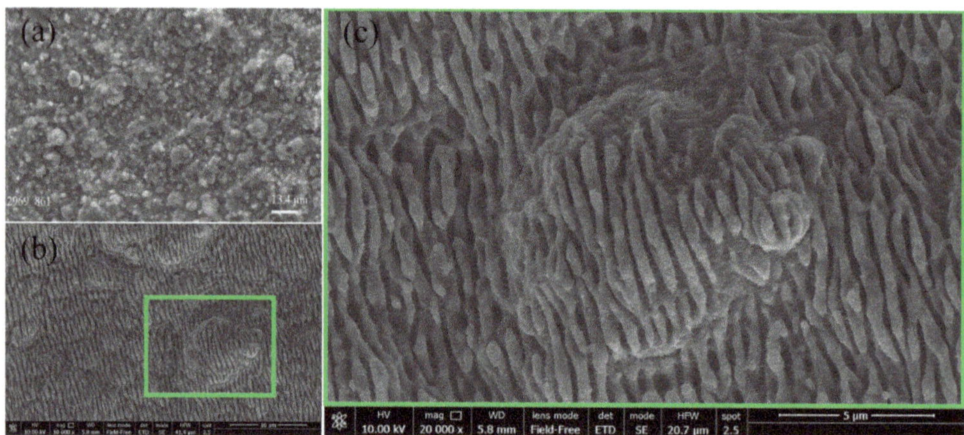

Figure 2. SEM observation of the surface of Ti/Fe (**a**) after vacuum arc condensation and after LIPSS formation on their surface (**b**) and its magnified view (**c**).

The focused ion beam (FIB) technique demonstrates the cross-sections of Ti-Fe bi-layered structures, non-irradiated as well as irradiated by laser pulses (Figure 3). This approach revealed the periodicity of metallic titanium and iron layers, suggesting a "sandwich-type" structure. The thickness of a separate layer was in the range of 90–100 nm; moreover, this value was directly proportional to the deposition time. The cross-section image on the patterned area (Figure 3b) shows that each ripple is much deeper than the thickness of the Ti or Fe layers, and consists of at least two altered layers. The width of the columnar crystallites, which was elongated in the direction of the steam flow, increased with the respective increase of the substrate temperature. It was well visible that the periodic metal layers were not completely uniform in color. That is, in addition to light and dark areas, there were also mixed shades. This suggested an alloying between two metallic layers that may occur at temperatures even lower than those in the case of bulk metals using the Ti-Fe binary phase diagram; two intermetallic compounds, Fe$_2$Ti and FeTi, may exist in the narrow regions of two components [21]. Thus, in subnano-multilayer condensates of metallic titanium and iron, the formation of the TiFe intermetallic compound at the layer interface might be observed in conformity to the Ti-Fe phase diagram.

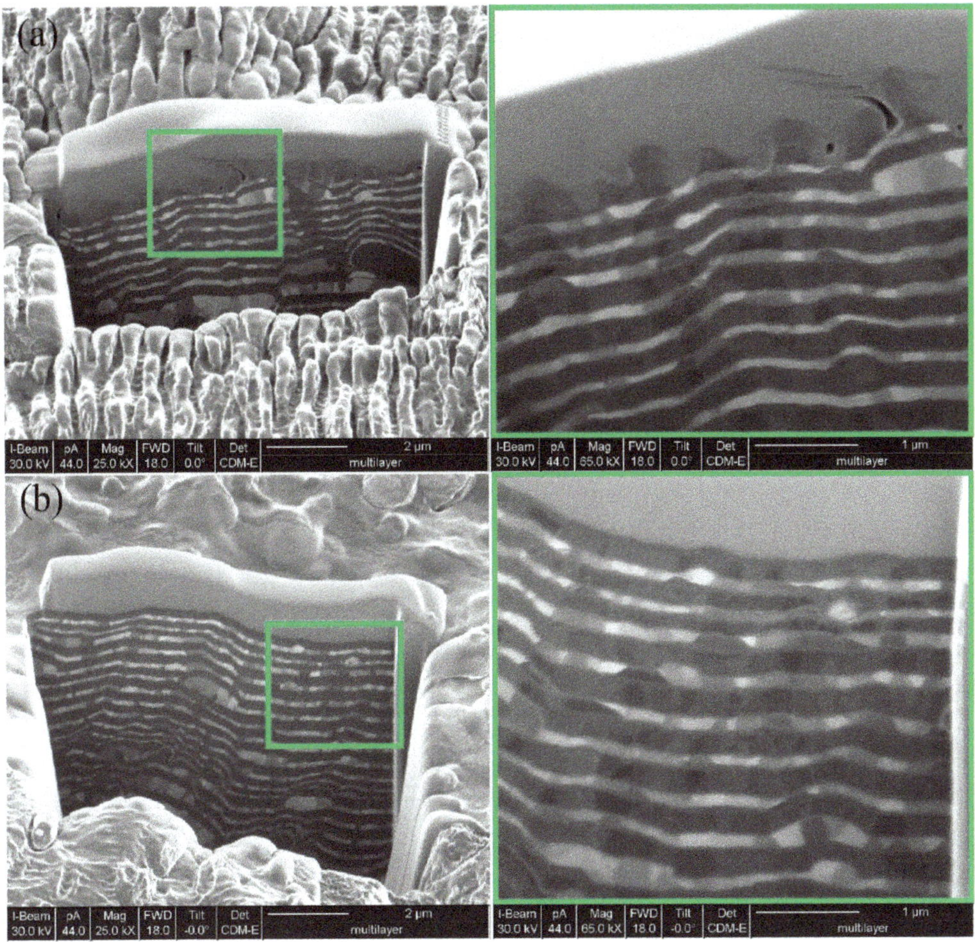

Figure 3. SEM micrographs of the vacuum-arc condensed Ti-Fe multilayer with layer formation (**b**) and after LIPSS formation (**a**) made by shot by gallium ion beam.

3.2. XPS Analysis

The surface of the Ti-Fe multilayers was analyzed by XPS. The C 1s, Fe 2p, O 1s, Ti 2p, N 1s, and Ca 2p photoelectron core level spectra of the detected elements were monitored (Table 1). The obtained XPS signals were assigned in accordance with the XPS database [22]. Along with amorphous carbon on the surface, some carbonates (C 1s at ~288.2 eV) were detected. In total, ~44 and ~35 at.% of Carbon was detected by XPS for the as-prepared and laser-treated Ti-Fe multilayer samples. These values might be considered to be surface contamination at the metallic surface [23]. The signal of O 1s demonstrates two intense peaks at 529.7 eV and 531.4 eV. The first peak suggests surface oxides on both TiO_2 and Fe_2O_3, and the second one might be associated with oxygen in the surface carbonates. The total amount of oxygen was ~45 at.%. In addition to Fe III oxide, metallic iron was also found on the surface. The Fe $2p_{3/2}$ core level in XPS spectra reveals two peaks at 706.4 and 710.4 eV related to Fe^0 and Fe^{3+} chemical states, respectively. The amount of Fe_2O_3 was higher than metallic Fe in one order. Similar to iron, titanium was also found in two chemical states: Ti^0 (Ti $2p_{3/2}$ at ~453.5 eV and Ti^{4+} (Ti $2p_{3/2}$ at ~458.4 eV) where metallic titanium is dominated by TiO_2. Calcium in the form of Ca^{2+} (Ca $2p_{3/2}$ at ~347.2 eV was also detected by XPS. Most probably, this is because of typical surface contamination $CaCO_3$ XPS signals at 400.5 and 396.4 eV might be attributed to the N 1s signal in organic compounds with C–N bonds and metal nitrides, respectively.

Table 1. Surface element analysis evaluated with XPS.

Sample	Ti, at %		Fe, at %		C, at %			O, at %		Ca, at %	N, at %	
Ti-Fe as prepared	2.3		8.2		43.5			44.4		0.3	2.9 C-N	
	2.0 TiO_2	0.3 Ti^0	7.7 F_2O_3	0.5 Fe^0	31.6 other C	11.9 CO_3^{2-}		23.2 oxide	21.2 CO_3^{2-}	Ca^{2+}		
Ti-Fe laser treated	14.6		4.5		34.6			44.9		0.2	1.2	
	11.9 TiO_2	2.7 Ti^0	4.0 F_2O_3	0.5 Fe^0	28.6 other C	2.1 carbide	3.9 CO_3^{2-}	25.5 oxide	19.4 CO_3^{2-}	Ca^{2+}	1.1 C-N	0.1 nitride

Both the as-prepared and laser-treated Ti-Fe multilayer samples were contaminated by carbon, oxygen and nitrogen, which is very common for as-prepared or laser-treated surfaces in open air. In addition, XPS data suggested a residual $CaCO_3$ on the surfaces.

3.3. XRD Analysis

XRD patterns for as-prepared and laser-treated Ti-Fe multilayer samples are shown in Figure 4. Both samples had an amorphous halo in range of 20–35°, and there was only one difference in the peak intensity of the TiFe intermetallic compound. After heat treatment of the Ti-Fe multilayer, the intermetallic compound TiFe was synthesized. The laser treatment applied to the surface resulted in a partial removal of the surface layer, since te surface of the prepared Ti-Fe multilayer samples was thoroughly "scratched". This is probably the reason why the TiFe phase, which appeared after heat treatment located between the metallic titanium and the iron layer, became more pronounced in the XRD pattern. In other words, the laser treatment of the Ti-Fe multilayer might be considered a technical tool for both surface structuring (Figure 2b,c) and opening the way to TiFe thin layers (Figure 4).

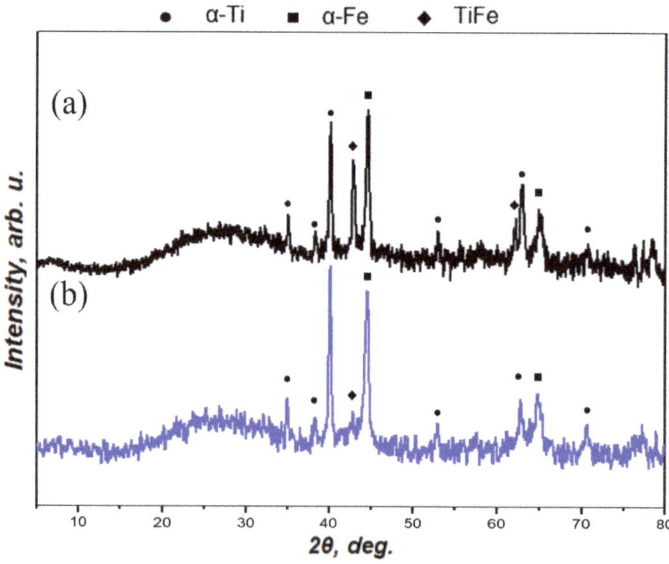

Figure 4. XRD patterns for the laser-treated (a) and as-prepared (b) Ti-Fe multilayer samples.

3.4. AFM Analysis

Atomic force microscope (AFM) analysis of the surface roughness was done for five randomly selected parts of the surface (Table 2). It is clear that all experimental values of surface area for the laser treated Ti-Fe multilayer structure were higher than those for the as-prepared Ti-Fe multilayer. This suggests that LIPSS are developing surface of Ti-Fe multilayer surface.

Table 2. Values of the surface area obtained by AFM respect to a projection area of 95.36 μm^2.

Sample	Surface Area of the Five Selected Parts of the Surface, μm^2.					
	1	2	3	4	5	Average
Ti-Fe laser treated	136.691	134.578	132.351	132.057	136.347	134.405
Ti-Fe as prepared	123.707	123.836	123.513	122.829	122.577	123.292

4. Theoretical Model-Simulation Procedure

To interpret the experimental observations, simulations were conducted to explore the thermal response of the multiple layered structure 15x(Ti/Fe) after irradiation with single laser pulses with a pulse duration of τ_p = 213 fs and a wavelength λ_L = 1030 nm. An 1D-Two Temperature Model (TTM) [24] was used to describe the relaxation process following electron excitation due to laser heating through the following equations [25,26]:

$$C_e^{(i)} \frac{\partial T_e^{(i)}}{\partial t} = \frac{\partial}{\partial z}\left(k_e^{(i)} \frac{\partial T_e^{(i)}}{\partial z} \right) - G_{eL}^{(i)}\left(T_e^{(i)} - T_L^{(i)} \right) + S^{(i)}(z,t) \quad [S^{(i)} = 0, \text{ for } i > 1]$$

$$C_L^{(i)} \frac{\partial T_L^{(i)}}{\partial t} = \frac{\partial}{\partial z}\left(k_L^{(i)} \frac{\partial T_L^{(i)}}{\partial z} \right) + G_{eL}^{(i)}\left(T_e^{(i)} - T_L^{(i)} \right) \tag{1}$$

$$S^{(1)}(z,t) = \frac{\alpha(1 - R - T)\sqrt{4\log 2}F}{\sqrt{\pi}\tau_p} \exp\left(-4\log 2\left(\frac{t - 3\tau_p}{\tau_p} \right)^2 \right) \exp(-\alpha z) \tag{2}$$

In Equations (1) and (2), $T_e^{(i)}$ ($T_L^{(i)}$) stands for the electron (lattice) temperature of layer i (i = 1, 3, 5... 2n−1 for Ti layer, i = 2, 4, 6 ... 2n for Fe layer, for n = 300 Ti/Fe layers for a multi-layered thickness equal to 60μm; each layer has a thickness equal to 100 nm). The thermophysical properties of the materials, such as electron and lattice heat capacity, ($C_e^{(i)}$, $C_L^{(i)}$), electron and lattice heat conductivity ($k_e^{(i)} \equiv k_{e0}^{(i)}\left(B^{(i)} T_L^{(i)} / \left(A^{(i)}\left(T_e^{(i)} \right)^2 + B^{(i)} T_L^{(i)} \right) \right)$, $k_L^{(i)} \sim 0.01 k_e^{(i)}$), and electron-phonon coupling strengths ($G_{eL}^{(i)}$), and the model parameters used in the simulations are listed in Table 3.

Table 3. Simulation parameters for Ti and Fe.

Parameter	Ti	Fe
G_{eL} [Wm^{-3}K^{-1}]	Fitting [26,27]	Fitting [28]
C_e [Jm^{-3}K^{-1}]	Fitting [26,27]	Fitting [28]
C_L [Jm^{-3}K^{-2}]	2.3521 × 10^6 [29]	3.675 × 10^6 [30]
k_{e0} [Jm^{-1}s^{-1}K^{-1}]	21.9 [29]	46.6 [30]
$T_{melting}$ [K]	1941 [29]	1811 [30]
$T_{boiling}$ [K]	3560 [29]	2750 [30]
$T_{critical}$ [K]	15500 [31]	8500 [30]
A [s^{-1} K^{-2}]	Fitting [26,27]	0.98 × 10^7 [30]
B [s^{-1} K^{-1}]	Fitting [26,27]	2.8 × 10^{11} [30]

While Equation (2) provides the general expression of the form of the source term due to material heating with a pulsed laser that includes the absorption coefficient α, the reflectivity R and the transmission coefficient T of the material, the Transfer Matrix Method was used to compute the optical properties of the top layer (Ti) after irradiation with pulsed laser of 1030 nm by taking into account the presence of the rest of the thin layers. Calculations yield $\alpha = 4 \times 10^5$ cm^{-1} [32], $T \cong 0$, $R = 0.54$, indicating that most of the energy will be absorbed in the first layer (Ti), while the transmitted part of the laser energy into the second layer (Fe) is essentially zero, due also to the thickness of the layer (100 nm), and it is not sufficiently high to excite the electrons in the rest of the layers (especially the second layer) and produce meaningful results. This argument justifies the use of a source term to describe laser heating only of the first layer and it is assumed that laser energy is not transmitted into the next layers. In principle, only heat transfer between the top few layers is expected, while optical excitation of the rest of the layers (except from the top one) is negligible.

Equations (1) and (2) are solved by using an iterative Crank–Nicolson scheme based on a finite-difference method. For the initial conditions, we chose thermal equilibrium at $T_e(z,t = 0) = T_L(z,t = 0) = 300$ K. Adiabatic boundary conditions are considered on the surface (at $z = 0$, $k_e^{(Ti)} \frac{\partial T_e^{(Ti)}}{\partial z} = k_L^{(Ti)} \frac{\partial T_L^{(Ti)}}{\partial z} = 0$) and the back of the complex structure. Furthermore, at the interface between the two layers, the following conditions are applied: $T_L^{(Ti)} = T_L^{(Fe)}, T_e^{(Ti)} = T_e^{(Fe)}, k_L^{(Ti)} \frac{\partial T_L^{(Ti)}}{\partial z} = k_L^{(Fe)} \frac{\partial T_L^{(Fe)}}{\partial z}, k_e^{(Ti)} \frac{\partial T_e^{(Ti)}}{\partial z} = k_e^{(Fe)} \frac{\partial T_e^{(Fe)}}{\partial z}$. The evaluation of the thermal response of the material following irradiation with single pulses was performed through the correlation of the simulation results with the measured ablation. As noted in previous reports, ablation may be associated with the lattice temperature exceeding the condition ($\sim$0.90 $T_{critical}$ where $T_{critical}$ is the critical point temperature [33–35]). Another criterion that is also usually employed is the boiling temperature of the material, $T_{boiling}$ (i.e., the region of the material that is characterized with lattice temperatures higher than $T_{boiling}$ is removed [29]). In this work, the latter condition is used, as the experimental results indicate that there is ablation, while theoretical calculations do not yield lattice temperatures ($\sim$0.90 $T_{critical}$).

5. Discussion

Despite a lot of bifurcations being created as a collateral effect of vacuum-arc deposition, the LIPSS were quite regular over all of the large areas, which can be attributed to the HR-LIPSS method [20]. Moreover, as shown by AFM, LIPSS increase the relative surface area by 9% with respect to the untreated sample, which might translate into a positive impact on kinetics for chemical interaction between solid TiFe and gaseous hydrogen. We expect that the surface properties are crucial for the acceleration of hydrogen gas adsorption on the solid surface that, together with hydrogen atom diffusion, might be a limiting stage for the hydrogenation process.

Because the laser treatment occurred in open air, the surfaces of the Ti-Fe multilayer samples were contaminated by amorphous carbon, $CaCO_3$, TiO_2 and Fe_2O_3. XPS experiments confirmed these surface contaminations, which are typical for a solid surface. AFM and XPS results suggested that because of the LIPSS treatment, the surface area of the Ti-Fe multilayers was increased; however, their surface cleanness was not affected. In practice, the laser treatment resulted in solid surface structuring and opened the way to the TiFe compound located under the surface of the prepared Ti-Fe multilayer samples. Experimental results obtained by SEM and XRD suggested that for the prepared Ti-Fe multilayers, alloying between metallic titanium and metallic iron resulted in the appearance of a TiFe compound. Usually, production of TiFe as ingot might be carried out directly by arc melting of metallic titanium and iron through the following reaction:

$$Ti + Fe = TiFe + 40.6 \text{ kJ.}$$

At the same time, chemical methods including calciothermic reduction of TiO_2 also enable the production of TiFe, but in the form of a fine powder [36]:

$$TiO_2 + Fe + 2Ca = TiFe + 2CaO + 365.7 \text{ kJ}.$$

The TiFe compound belongs to the hydrogen storage materials, in which the hydride has a similar crystal structure as the starting compound, although lattice expansion takes place during hydrogenation. More specifically, the crystal structure of the hydrogenated TiFe might be considered the derivative of that of the TiFe parent compound. The hydride stability is related to the strength of the chemical bond between metal atoms in a polyhedron in which H atoms are located. Therefore, hydrogen pressure is correlated with the ratios of the bond order between the metal atoms in the polyhedron [37,38]. On the other hand, the TiFe intermetallic compound has a CsCl-type structure, where expansion and contraction of the interatomic distances between Ti atoms during hydrogenation are 0.092 and 0.014 nm, respectively, while no changes are observed between Ti and Fe atoms. Thus, the H atom can occupy the interstitial site that is the Ti_2Fe_4 octahedron [39], resulting in a formula of hydride of $TiFeH_{1.94}$ with a calculated hydrogen content of 1.9 wt.% H_2.

Theoretical calculations of the lattice temperatures based on the scheme described above yield a spatio-temporal evolution that is illustrated in Figure 5 (in the center of the Gaussian spot). Our simulations and comparison with experimental observations indicate that nearly half of the first layer (Ti) is removed at fluence $F = 0.5 \text{ J/cm}^2$ (Figure 5a) and a second pulse is required to remove the Ti top layer completely (Figure 5b) This value corresponds to fluence that is sufficient to raise the temperature of the upper layer above $T_{boiling}$; therefore, the boiling temperature is regarded as a reliable ablation threshold criterion. It is noted that for each pulse, only the first pair of Ti/Fe layers below the upper surface thermally respond to the heat transfer. Thus, for the sake of simplicity, only a region with a small number of layers is shown to describe the affected zone.

It is important to note that due to the fact that lattice temperatures on the second layer (Fe) are lower than the boiling temperature for Ti, no material is predicted to be removed from the second layer. On the other hand, it is evident that the lattice temperature attained from a large part of the second layer (Fe) for $F = 0.5 \text{ mJ/cm}^2$ is above the melting point of the material. Finally, Figure 5c illustrates the effect of the third pulse, showing that due to the small heat capacity, large conductivity and lower boiling temperature, the energy of the laser propagates inside the Fe-layer; however, it is only sufficient to remove part of the Fe-layer, while another portion of the rest undergoes a phase transition, as the lattice temperatures lie between the melting point and the boiling temperature. The value for the reflectivity and absorption coefficient used to simulate energy absorption for Fe are 0.6173 and $\alpha = 4 \times 10^5 \text{ cm}^{-1}$, respectively, when Fe is on the upper surface. The above description indicates that fluid dynamics and resolidification processes are expected to further modify the surface profile of the assembly. Therefore, appropriate phase change-related corrections need to be incorporated into the model for a more accurate description of the surface modification processes and determination of the morphological changes. A thorough approach requires the inclusion of Navier–Stokes equations (to describe fluid dynamics) and relevant equations to account for evaporation [32,35–37]. The fact that mass displacement and mass removal is expected to lead to a corrugated profile means that appropriate conditions for surface plasmon excitation can be easily satisfied (for a detailed description, see Refs [32,40–42]). Thus, a periodic pattern can be produced, which is also confirmed by the experimental results [43,44]. This occurs irrespective of which material is on top (Fe or Ti). Although most studies have been centered on the investigation of surface plasmon (SP) excitation in dielectric–metal–dielectric multi-layered films and potential coupling of the excited SP on the dielectric–metal and metal–dielectric films in the periodicity of the periodic structures [45,46], to the best of our knowledge, no previous study has been conducted for air/multi-layered (metallic) heterostructures. Certainly, as there is no alteration of dielectric/metallic films to sustain excitation of bound SP, only SP on the top layer should be expected to have a periodicity the size of the laser wavelength (~1 µm). Nevertheless, it is known that the SP wavelength decreases at increasing energy doses [47]. The aforementioned simulations predict a sufficient ablation that induces an enhanced depth in the irradiated heterostructure, and therefore, SP wavelength should decrease. Although a consistent approach based on the employment of Finite Difference Time domain codes to solve Maxwell's equations is expected to provide a precise correlation of the ripple periodicities as a function of the produced depth (see Ref. [48] and references therein), an approximate methodology is used (see Ref. [48]). More specifically, for fourteen pulses, a SP wavelength equal to 920 nm is predicted, which yields a similar ripple periodicity on Fe illustrated in Figure 5d. Experimental measurements performed for 12.5 pulses give a ripple periodicity in the range of 650–750 nm. Certainly, a theoretical investigation assuming the surface patterning effect following a larger number of pulses could be conducted for a quantitative comparison of the ripple periodicities; however, the predominant aim of this work was focused on the thermal response of the material and the ablation-assisted modification of the surface layers. The reason why fourteen pulses were used in the simulations was to clearly illustrate both the damage induced on the complex (removal of layers) and the periodicity of the rippled zone. It is evident that there is a discrepancy between the experimentally measured value and the theoretical prediction that requires more investigation; however, the aim of the current work was to highlight the step-by-step removal of the metallic layers. Thus, the development of an improved theoretical model that yields a better agreement with the experimental observations could be the subject of a future work.

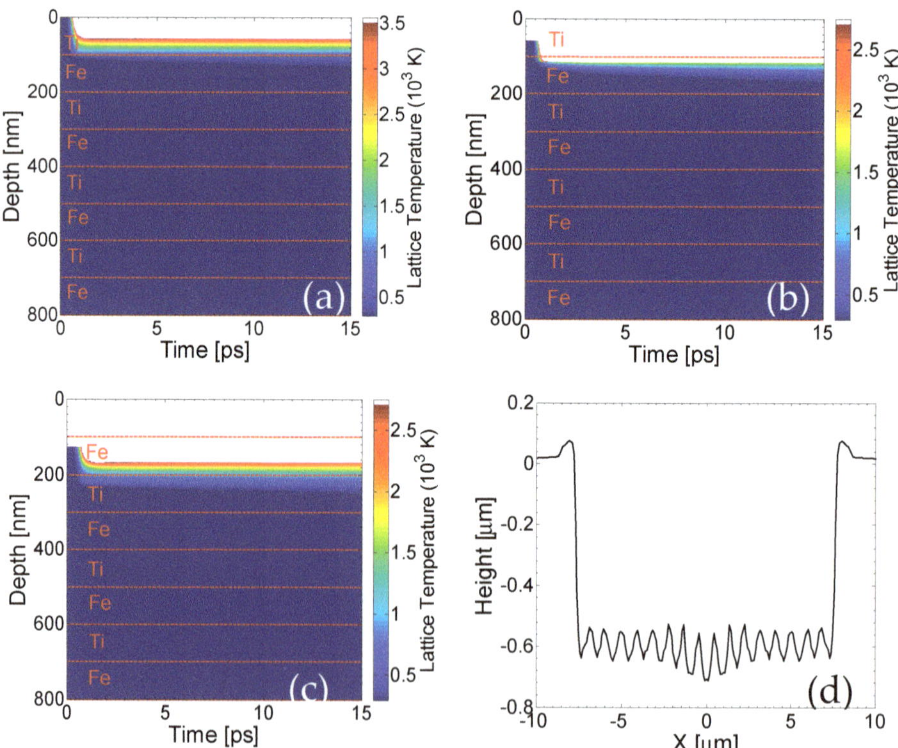

Figure 5. Lattice temperature field evolution in depth (in the center of the Gaussian spot) perpendicular to the surface of the sample after one pulse (**a**), two pulses (**b**) and three pulses (**c**) (white region indicates material removal, horizontal dashed line show the borders of the films); (**d**) periodic pattern after fourteen pulses (laser polarization is vertical to the figure plane).

6. Conclusions

The intermetallic compound TiFe was formed between multilayers made of metallic titanium and metallic iron and was detected using both XPS and XRD. The LIPSS method was applied, which generated homogeneous ripples over a large area, even on non-polished surfaces densely covered with numerous defects and bifurcations. Because of the generation of a more developed Ti-Fe multilayer surface, the absorption/desorption of gaseous hydrogen was accelerated. The thermal effects resulting from the irradiation of the compound were simulated to predict the affected region as well as to predict the periodicity of the surface pattern.

Laser-treated Ti-Fe multilayers will be tested with respect to hydrogen gas absorption and optimized as a hydrogen sensor in future publications.

Author Contributions: A.V.D., I.S. and I.G.—Conceptualization; O.V.K., A.V.D., V.B. and I.G.—methodology; G.D.T.—conducted the theoretical work and modelling of the physical processes, A.V.D., I.S. and I.G.—help writing original draft preparation; O.V.K., G.D.T., A.V.D., A.A.D., V.B., I.S., S.B. and I.G.—helped with writing review and editing; I.G.—supervised the preparation. All authors have read and agreed to the published version of the manuscript.

Funding: Financial support was provided by the Slovak Academic Information Agency (SAIA) through the National Scholarship Programme of the Slovak Republic in 2019/2020 for author I.S. as scholarship holder. G.D.T acknowledges support from COST Action TUMIEE (supported by COST-European Cooperation in Science and Technology).

Conflicts of Interest: The authors declare no conflict of interest.

References

1. Edalati, K.; Matsuda, J.; Iwaoka, H.; Toh, S.; Akiba, E.; Horita, Z. High-pressure torsion of TiFe intermetallics for activation of hydrogen storage at room temperature with heterogeneous nanostructure. *Int. J. Hydrogen Energy* **2013**, *38*, 4622–4627. [CrossRef]
2. Edalati, K.; Matsuda, J.; Yanagida, A.; Akiba, E.; Horita, Z. Activation of TiFe for hydrogen storage by plastic deformation using groove rolling and high-pressure torsion: Similarities and differences. *Int. J. Hydrogen Energy* **2014**, *39*, 15589–15594. [CrossRef]
3. Manna, J.; Tougas, B.; Huot, J. Mechanical activation of air exposed TiFe + 4 wt% Zr alloy for hydrogenation by cold rolling and ball milling. *Int. J. Hydrogen Energy* **2018**, *43*, 20795–20800. [CrossRef]
4. Vega, L.; Leiva, D.; Neto, R.L.; Silva, W.; Silva, R.; Ishikawa, T.; Kiminami, C.; Botta, W. Mechanical activation of TiFe for hydrogen storage by cold rolling under inert atmosphere. *Int. J. Hydrogen Energy* **2018**, *43*, 2913–2918. [CrossRef]
5. Emami, H.; Edalati, K.; Matsuda, J.; Akiba, E.; Horita, Z. Hydrogen storage performance of TiFe after processing by ball milling. *Acta Mater.* **2015**, *88*, 190–195. [CrossRef]
6. Vega LE, R.; Leiva, D.R.; Neto, R.L.; Silva, W.B.; Silva, R.A.; Ishikawa, T.T.; Kiminami, C.S.; Botta, W.J. Improved ball milling method for the synthesis of nanocrystalline TiFe compound ready to absorb hydrogen. *Int. J. Hydrogen Energy* **2020**, *45*, 2084–2093. [CrossRef]
7. Endo, N.; Shimoda, E.; Goshome, K.; Yamane, T.; Nozu, T.; Maeda, T. Operation of a stationary hydrogen energy system using TiFe-based alloy tanks under various weather conditions. *Int. J. Hydrogen Energy* **2020**, *45*, 207–215. [CrossRef]
8. Rotella, G.; Orazi, L.; Alfano, M.; Candamano, S.; Gnilitskyi, I. Innovative high-speed femtosecond laser nano-patterning for improved adhesive bonding of Ti6Al4V titanium alloy. *CIRP J. Manuf. Sci. Technol.* **2017**, *18*, 101–106. [CrossRef]
9. Derrien, T.J.-Y.; Itina, T.E.; Torres, R.; Sarnet, T.; Sentis, M. Possible surface plasmon polariton excitation under femtosecond laser irradiation of silicon. *J. Appl. Phys.* **2013**, *114*, 083104. [CrossRef]
10. Temple, P.; Soileau, M. Polarization charge model for laser-induced ripple patterns in dielectric materials. *IEEE J. Quantum Electron.* **1981**, *17*, 2067–2072. [CrossRef]
11. Rebollar, E.; Castillejo, M.; Ezquerra, T.A. Laser induced periodic surface structures on polymer films: From fundamentals to applications. *Eur. Polym. J.* **2015**, *73*, 162–174. [CrossRef]
12. Bonse, J.; Mann, G.; Kruger, J.; Marcinkowski, M.; Eberstein, M. Femtosecond laser-induced removal of silicon nitride layers from doped and textured silicon wafers used in photovoltaics. *Thin Solid Films* **2013**, *542*, 420–425. [CrossRef]
13. Vorobyev, A.Y.; Guo, C. Multifunctional surfaces produced by femtosecond laser pulses. *J. Appl. Phys.* **2015**, *117*, 033103. [CrossRef]
14. Gnilitskyi, I.; Gruzdev, V.; Bulgakova, N.M.; Mocek, T.; Orazi, L. Mechanisms of high-regularity periodic structuring of silicon surface by sub-MHz repetition rate ultrashort laser pulses. *Appl. Phys. Lett.* **2016**, *109*, 143101. [CrossRef]
15. Dusser, B.; Sagan, S.; Soder, H.; Faure, N.; Colombier, J.-P.; Jourlin, M.; Audouard, E. Controlled nanostructrures formation by ultra fast laser pulses for color marking. *Opt. Express* **2010**, *18*, 2913–2924. [CrossRef]

16. Gnilitskyi, I.; Pogorielov, M.; Viter, R.; Ferraria, A.M.; Carapeto, A.P.; Oleshko, O.; Orazi, L.; Mishchenko, O. Cell and tissue response to nanotextured Ti6Al4V and Zr implants using high-speed femtosecond laser-induced periodic surface structures. *Nanomed. Nanotechnol. Biol. Med.* **2019**, *21*, 102036. [CrossRef]

17. Mizuno, A.; Honda, T.; Kikuchi, J.; Iwai, Y.; Yasumaru, N.; Miyazaki, K. Friction Properties of the DLC Film with Periodic Structures in Nano-scale. *Tribol. Online* **2006**, *1*, 44–48. [CrossRef]

18. Gnilitskyi, I.; Rota, A.; Gualtieri, E.; Valeri, S.; Orazi, L. Tribological Properties of High-Speed Uniform Femtosecond Laser Patterning on Stainless Steel. *Lubricants* **2019**, *7*, 83. [CrossRef]

19. Demchishin, A.; Gnilitskyi, I.; Orazi, L.; Ascari, A. Structure, phase composition and microhardness of vacuum-arc multilayered Ti/Al, Ti/Cu, Ti/Fe, Ti/Zr nano-structures with different periods. *Appl. Surf. Sci.* **2015**, *342*, 127–135. [CrossRef]

20. Gnilitskyi, I.; Derrien, T.J.; Levy, Y.; Bugakova, N.M.; Mocek, T.; Orazi, L. High-speed, highly regularfemtosecond laser printing of laser-induced periodic surface structures on metals: Physical origin of the regularity. *Sci. Rep.* **2017**, *7*, 8485. [CrossRef]

21. Hong, B.O.; Jiang WA, N.G.; Duarte, L.; Leinenbach, C.; Liu, L.B.; Liu, H.S.; Jin, Z.P. Thermodynamic re-assessment of Fe–Ti binary system. *Trans. Nonferrous Met. Soc. China* **2012**, *22*, 2204–2211. [CrossRef]

22. Moulder, J.F.; Stickle, W.F.; Sobol, P.E.; Bomben, K.D. Perkin-Elmer Corporation, Physical Electronics Division. In *Handbook of X-ray Photoelectron Spectroscopy Eden Prairie*; Physical Electronics: Eden Prairie, MN, USA, 1992.

23. Saldan, I.; Frenzel, J.; Shekhah, O.; Chelmowski, R.; Birkner, A.; Wöll, C. Surface of Ti-Ni alloys after their preparation. *J. Alloy. Compd.* **2009**, *470*, 568–573. [CrossRef]

24. Gaković, B.; Tsibidis, G.D.; Skoulas, E.; Petrović, S.M.; Vasić, B.; Stratakis, E. Partial ablation of Ti/Al nano-layer thin film by single femtosecond laser pulse. *J. Appl. Phys.* **2017**, *122*, 223106.

25. Chen, J.K.; Beraun, J.E. Modelling of ultrashort laser ablation of gold films in vacuum. *J. Opt. A Pure Appl. Opt.* **2003**, *5*, 168–173. [CrossRef]

26. Kelly, R.; Miotello, A. Comments on explosive mechanisms of laser sputtering. *Appl. Surf. Sci.* **1996**, *96*, 205–215. [CrossRef]

27. Tsibidis, G.D.; Stratakis, E. Ripple formation on silver after irradiation with radially polarised ultrashort-pulsed lasers. *J. Appl. Phys.* **2017**, *121*, 163106. [CrossRef]

28. Tsibidis, G.D. Thermal response of double-layered metal films after ultrashort pulsed laser irradiation: The role of nonthermal electron dynamics. *Appl. Phys. Lett.* **2014**, *104*, 51603. [CrossRef]

29. Lin, Z.; Zhigilei, L.V.; Celli, V. Electron-phonon coupling and electron heat capacity of metals under conditions of strong electron-phonon nonequilibrium. *Phys. Rev. B* **2008**, *77*, 075133. [CrossRef]

30. Zhigilei, L.V. Database on 'Electron-Phonon Coupling and Electron Heat Capacity in Metals at High Electron Temperatures'. Available online: https://faculty.virginia.edu/CompMat/electron-phonon-coupling/ (accessed on 10 September 2020).

31. Lide, D.R. (Ed.) *CRC Handbook of Chemistry and Physics*; CRC Press: Boca Raton, FL, USA, 2004; Volume 85.

32. Tsibidis, G.D.; Mimidis, A.; Skoulas, E.; Kirner, S.V.; Krüger, J.; Bonse, J.; Stratakis, E. Modelling periodic structure formation on 100Cr6 steel after irradiation with femtosecond-pulsed laser beams. *Appl. Phys. A* **2018**, *124*, 27. [CrossRef]

33. Horvath, A.L. Critical temperature of elements and the periodic system. *J. Chem. Educ.* **1973**, *50*, 335. [CrossRef]

34. Johnson, P.B.; Christy, R.W. Optical constants of transition metals: Ti, V, Cr, Mn, Fe, Co, Ni, and Pd. *Phys. Rev. B* **1974**, *9*, 5056–5070. [CrossRef]

35. Tsibidis, G.D.; Barberoglou, M.; Loukakos, P.A.; Stratakis, E.; Fotakis, C. Dynamics of ripple formation on silicon surfaces by ultrashort laser pulses in subablation conditions. *Phys. Rev. B* **2012**, *86*, 115316. [CrossRef]

36. Tsuchiya, T.; Yasuda, N.; Sasaki, S.; Okinaka, N.; Akiyama, T. Combustion synthesis of TiFe-based hydrogen storage alloy from titanium oxide and iron. *Int. J. Hydrogen Energy* **2013**, *38*, 6681–6686. [CrossRef]

37. Yukawa, H.; Takahashi, Y.; Morinaga, M. Alloying effects on the electronic structures of LaNi5 containing hydrogen atoms. *Intermetallics* **1996**, *4*, S215–S224. [CrossRef]

38. Yukawa, H.; Morinaga, M. The nature of the chemical bond in hydrogen storage compounds. In *Advances in Quantum Chemistry*; Academic Press: Cambridge, MA, USA, 1998; Volume 29, pp. 83–108.

39. Nambu, T.; Ezaki, H.; Yukawa, H.; Morinaga, M. Electronic structure and hydriding property of titanium compounds with CsCl-type structure. *J. Alloy. Compd.* **1999**, *293*, 213–216. [CrossRef]

40. Tsibidis, G.D.; Skoulas, E.; Papadopoulos, A.; Stratakis, E. Convection roll-driven generation of supra-wavelength periodic surface structures on dielectrics upon irradiation with femtosecond pulsed lasers. *Phys. Rev. B* **2016**, *94*, 081305. [CrossRef]

41. Tsibidis, G.D.; Fotakis, C.; Stratakis, E. From ripples to spikes: A hydrodynamical mechanism to interpret femtosecond laser-induced self-assembled structures. *Phys. Rev. B* **2015**, *92*, 041405. [CrossRef]

42. Papadopoulos, A.; Skoulas, E.; Tsibidis, G.D.; Stratakis, E. Formation of periodic surface structures on dielectrics after irradiation with laser beams of spatially variant polarisation: A comparative study. *Appl. Phys. A* **2018**, *124*, 146. [CrossRef]

43. Fuentes-Edfuf, Y.; Sánchez-Gil, J.A.; Garcia-Pardo, M.; Serna, R.; Tsibidis, G.D.; Giannini, V.; Solis, J.; Siegel, J. Tuning the period of femtosecond laser induced surface structures in steel: From angled incidence to quill writing. *Appl. Surf. Sci.* **2019**, *493*, 948–955. [CrossRef]

44. Museur, L.; Tsibidis, G.D.; Manousaki, A.; Anglos, D.; Kanaev, A. Surface structuring of rutile TiO2(100) and (001) single crystals with femtosecond pulsed laser irradiation. *JOSA B* **2018**, *35*, 2600–2607. [CrossRef]

45. Dostovalov, A.V.; Derrien, T.J.-Y.; Lizunov, S.A.; Přeučil, F.; Okotrub, K.A.; Mocek, T.; Korolkov, V.P.; Babin, S.A.; Bulgakova, N.M. LIPSS on thin metallic films: New insights from multiplicity of laser-excited electromagnetic modes and efficiency of metal oxidation. *Appl. Surf. Sci.* **2019**, *491*, 650–658. [CrossRef]
46. Takami, A.; Nakajima, Y.; Terakawa, M. Formation of gold grating structures on fused silica substrates by femtosecond laser irradiation. *J. Appl. Phys.* **2017**, *121*, 173103. [CrossRef]
47. Huang, M.; Zhao, F.; Cheng, Y.; Xu, N.; Xu, Z. Origin of Laser-Induced Near-Subwavelength Ripples: Interference between Surface Plasmons and Incident Laser. *ACS Nano* **2009**, *3*, 4062–4070. [CrossRef] [PubMed]
48. Tsibidis, G.D.; Stratakis, E. Ionisation processes and laser induced periodic surface structures in dielectrics with mid-infrared femtosecond laser pulses. *Sci. Rep.* **2020**, *10*, 1–13. [CrossRef]

nanomaterials

Article

Influence of Sulphur Content on Structuring Dynamics during Nanosecond Pulsed Direct Laser Interference Patterning

Theresa Jähnig [1,*,†], Cornelius Demuth [1,†] and Andrés Fabián Lasagni [1,2,*]

1   Institute of Manufacturing Technology, Technische Universität Dresden, P.O. Box, 01062 Dresden, Germany; cornelius.demuth1@mailbox.tu-dresden.de
2   Fraunhofer Institute for Material and Beam Technology IWS, Winterbergstr. 28, 01277 Dresden, Germany
*   Correspondence: theresa.jaehnig@tu-dresden.de (T.J.); andres_fabian.lasagni@tu-dresden.de (A.F.L.)
†   These authors contributed equally to this work.

Abstract: The formation of melt and its spread in materials is the focus of many high temperature processes, for example, in laser welding and cutting. Surface active elements alter the surface tension gradient and therefore influence melt penetration depth and pool width. This study describes the application of direct laser interference patterning (DLIP) for structuring steel surfaces with diverse contents of the surface active element sulphur, which affects the melt convection pattern and the pool shape during the process. The laser fluence used is varied to analyse the different topographic features that can be produced depending on the absorbed laser intensity and the sulphur concentration. The results show that single peak geometries can be produced on substrates with sulphur contents lower than 300 ppm, while structures with split peaks form on higher sulphur content steels. The peak formation is explained using related conceptions of thermocapillary convection in weld pools. Numerical simulations based on a smoothed particle hydrodynamics (SPH) model are employed to further investigate the influence of the sulphur content in steel on the melt pool convection during nanosecond single-pulsed DLIP.

Keywords: direct laser interference patterning; periodic microstructure; sulphur content; nanosecond pulse; surface tension gradient; Marangoni convection; smoothed particle hydrodynamics

Citation: Jähnig, T.; Demuth, C.; Lasagni, A.F. Influence of Sulphur Content on Structuring Dynamics during Nanosecond Pulsed Direct Laser Interference Patterning. *Nanomaterials* **2021**, *11*, 855. https://doi.org/10.3390/nano11040855

Academic Editors: Jörn Bonse and Koji Sugioka

Received: 26 February 2021
Accepted: 25 March 2021
Published: 27 March 2021

Publisher's Note: MDPI stays neutral with regard to jurisdictional claims in published maps and institutional affiliations.

1. Introduction

During the process of evolution, numerous animals and plants have perfectly adapted to environmental conditions by developing repetitive surface patterns. For instance, fine riblets on their skin scales, which reduce water drag, enable sharks to swim faster and more effortlessly than other fishes [1]. In addition, hierarchical structures give rise to the structural colour of *Morpho* butterfly wings and to the hydrophobic and self-cleaning properties of rice leaves [2,3]. Learning lessons from nature and imitating these examples, surface microstructuring can be applied to improve the efficiency of ship propulsion and electricity generation in wind turbines by reducing friction, and to generate colours for decorative motifs or security marks on products as well as self-cleaning surfaces on solar panels or worktops in professional kitchens [4,5]. The success of manufacturers in a global competition relies on additional benefits to differentiate their products. Consequently, techniques allowing for the fabrication of complex surface patterns are sought to add novel functionalities to products.

Surface topographies with feature sizes in a wide range can be generated, from submicron structures made by lithography and subsequent wet chemical etching [6] or focussed beam structuring [7] to submillimetre surface textures produced by brushing, grinding, or milling. After structuring, several cleaning steps are, in general, required before the surface can be used further, with the extended production time impairing the profitability of the overall process.

In contrast, laser texturing is an effective single-step method for fabricating surface structures that dispenses with the need for material removal or cleaning processes [8,9]. More specifically, direct laser interference patterning (DLIP) provides the advantage of producing repetitive surface structures with periodicities in the submicron to micron range. This surface functionalisation technique has already been employed for diverse applications, such as antibacterial and antifouling surfaces [10], control of wettability [11], tribology [12,13], cell adhesion [14,15], and decoration and anti-counterfeit protection [16] on various materials. The DLIP process generates periodic microstructures on substrates exposed to an interference pattern, which results when at least two coherent laser beams are combined, by means of photothermal or photochemical interactions [17–19].

The intensity distribution of the interference pattern is determined by the laser wavelength λ, the number and the polar angle $\theta/2$ of the partial beams [20,21]. A two-beam configuration considered here generates line-like surface textures, where the spatial period Λ, defined as the distance between adjacent intensity maxima or minima, can be adjusted by the intersecting angle θ according to Equation (1):

$$\Lambda = \frac{\lambda}{2\sin(\theta/2)}.$$ (1)

It is evident from Equation (1) that the minimal theoretically achievable spatial period is half of the used laser wavelength, i.e., $\Lambda = \lambda/2$ for an interception angle of $\theta = \pi$. The sinusoidal laser fluence distribution of a two-beam interference pattern is given by Equation (2):

$$\Phi(x,y) = 4\Phi_0 \cos^2\left[kx\sin\left(\frac{\theta}{2}\right)\right],$$ (2)

where Φ_0 is the fluence of each partial beam and the wave number is defined as $k = 2\pi/\lambda$.

The application of short pulsed laser radiation with nanosecond pulse duration and infrared wavelengths involves photothermal interactions, in particular the formation of very localised molten zones [22,23]. In the case of metallic substrates, the local insertion of energy in a thin surface layer enables the specific shaping of well-defined melt pools associated with an alteration of the surface topography, which is influenced by the laser fluence and the material properties as well. In nanosecond pulsed DLIP of metals at moderate laser fluence, thermocapillary driven melt pool flow is interpreted as the mechanism responsible for the surface microstructure formation [24]. The in-process observation of DLIP is not possible though, owing to the short duration and the microscopic scale of the surface modification. However, the use of numerical simulation allows for an investigation and insights into the phenomena contributing to the process.

The development of the melt pool convection was originally modelled for welding processes with continuous energy input into the metallic workpiece. In detail, arc welding was considered at first, and the research was subsequently extended to laser beam welding. Accordingly, the melt pool characteristics are determined by the interaction between arc and metal, the heat transfer and fluid flow along with the thermophysical material properties, and the applied boundary conditions [25–28]. Numerical investigations on thermal fluid flow revealed that thermocapillary stresses constitute the main force driving the circulation in the weld pool [27,28]. This thermocapillary flow, or Marangoni convection, arises due to surface tension gradients along the melt pool surface [29]. Even minor additions of a surface active element, e.g., oxygen, aluminium, phosphorus, or sulphur, may significantly affect the surface tension of molten metal and, as a result, the weld penetration depth and pool width [30–32].

The effect of surface active elements on the melt convection patterns and pool shape was first investigated experimentally by Heiple and Roper [29]. Beginning at that time, mathematical models for the theoretical study of welding processes, i.e., the fluid flow [30] and heat transfer in the melt pool [33–37] were developed. While the effect of the temperature-dependent surface tension was incorporated into the models [33–37], the weld pool simulations mostly considered a constant temperature coefficient of surface tension.

On the other hand, a variable, i.e., a temperature-dependent, temperature derivative of surface tension, as required in the presence of a surfactant, was rarely included in the numerical investigations [38–42].

A very low sulphur content in steel (<30 ppm or at most 40 ppm S) involves a negative temperature coefficient of surface tension [29,32], giving rise to an outward melt flow at the surface and a wide but shallow weld pool. For a higher sulphur content in steel, the temperature derivative of surface tension is positive over a temperature interval of at least 200 K from the melting point [29]. A completely inward convection at the surface and a narrow but deep weld pool result [42], provided that the surface temperature does not exceed the aforementioned range. Furthermore, thermocapillary convection is assumed to influence the surface shape of the fusion zone [32,43]. Outward thermocapillary flow leads to a recess at the centre and bulges near the edges of the melt pool surface, as outlined in Figure 1a. On the contrary, inward Marangoni convection entails a ridge at the centre and surface depressions adjacent to the edges of the molten zone; see Figure 1b.

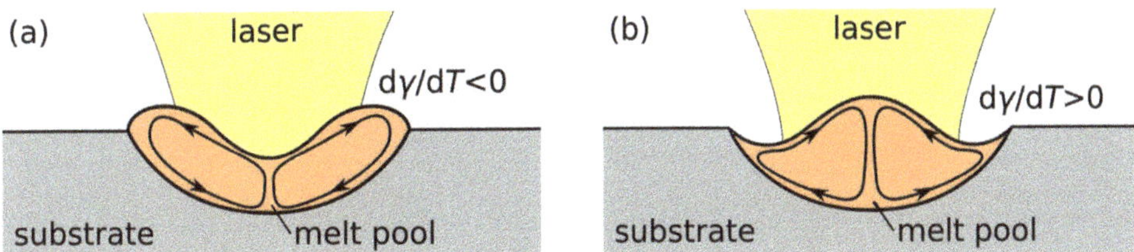

Figure 1. Surface evolution during laser processing due to temperature gradients and thermocapillary flow for (**a**) negative and (**b**) positive temperature coefficients of surface tension $d\gamma/dT$, outlines adopted from [32,43].

Notwithstanding the extensive research on thermocapillary flow in the context of welding, the effects of surface active elements in steel on the mechanisms of laser-based microprocessing are still widely unexplored. With regard to laser texturing, the melt pool flow may be employed to generate novel microstructures that provide surfaces with particular functionalities.

In this work, steel samples with different sulphur content are treated by two-beam direct laser interference patterning using a nanosecond pulsed laser to explore the possibilities of fabricating repetitive surface patterns based on thermocapillary convection due to the negative or positive temperature coefficient of surface tension. The topographies resulting from the two-beam DLIP process are analysed by confocal and scanning electron microscopy. Depending on the employed laser processing parameters and the sulphur content, individual structuring regimes are identified. In addition, numerical process simulations are performed using a smoothed particle hydrodynamics (SPH) model to understand the thermocapillary flow patterns in the melt pool and hence the microstructure formation.

2. Materials and Methods

This section presents the techniques and models employed in the combined experimental and numerical investigation. At first, the steel substrates used, the experimental setup, and the sample characterisation methods are detailed in Sections 2.1–2.3. Following this, the theoretical model and the specific parameters considered for the numerical simulations are elucidated in Sections 2.4 and 2.5.

2.1. Materials

In order to evaluate the influence of the sulphur content (S) on the structure formation with DLIP, steels with four different sulphur contents were chosen: 30 ppm, 100 ppm, <300 ppm, and 1500–3000 ppm S. The flat steel plates with a thickness of 5 mm were cut into 50×50 mm^2 pieces and polished with diamond suspension (Masterprep, Buehler, Lake Bluff, IL, USA) of 0.05 µm size to an average surface roughness (Sa) of 0.008 µm.

The root mean square roughness (*Sq*) of the same samples was 0.020 µm, with a maximum height of the assessed profile (*Sz*) at 0.450 µm. Before structuring, the surfaces were cleaned in an ultrasonic bath with ethanol (99.99% purity) for 25 min at room temperature to remove all preprocess contamination and were dried with compressed air.

2.2. Nanosecond Direct Laser Interference Patterning

The laser structuring process was accomplished on a compact DLIP system (DLIP-µFab, Fraunhofer IWS, Dresden, Germany) equipped with a Q-switched Nd:YLF laser (Tech-1053 Basic, Laser-export, Moscow, Russia) operating at 1053 nm and providing 12 ns pulses (full width at half maximum, FWHM) at 1 kHz with pulse energies up to 290 µJ. The laser emits the fundamental transverse mode (TEM_{00}) with a laser beam quality factor of $M^2 < 1.2$. This main beam is then split into two sub-beams by passing through a diffractive optical element (DOE) and the beams are subsequently parallelised by a prism. A schematic representation of the main DLIP components is shown in Figure 2a. A lens is overlapping the beams on the surface with a focal distance of 40 mm and produces an interference spot of 160 µm in diameter. Varying the angle of incidence of the partial beams by changing the distance between DOE and prism, the spatial period determined by Equation (1) is specified in the range between 1.29 µm and 7.20 µm. The latter period was chosen for the structuring on the steel samples to keep the single line-like melt pools separated from each other and to prevent thermal influence among themselves.

The Gaussian shape of the intensity profile of the overlapped beams is indicated in Figure 2b. The highest laser intensity is reached in the spot centre, whereas the intensity decreases towards the spot edges. The DLIP process was performed using laser fluences between 0.2 J/cm^2 and 1.1 J/cm^2, which correspond to 58% to 91% of laser power, respectively. The samples were positioned under the laser spot with an *x-y*-axis system (Pro115 linear stages, Aerotech Inc., Pittsburgh, PA, USA), and the focus position was controlled by vertically moving the DLIP optical head on a *z*-axis. Laser microprocessing experiments were carried out in ambient environment without posttreatments. In all cases, the structuring process was made without overlapping pulse.

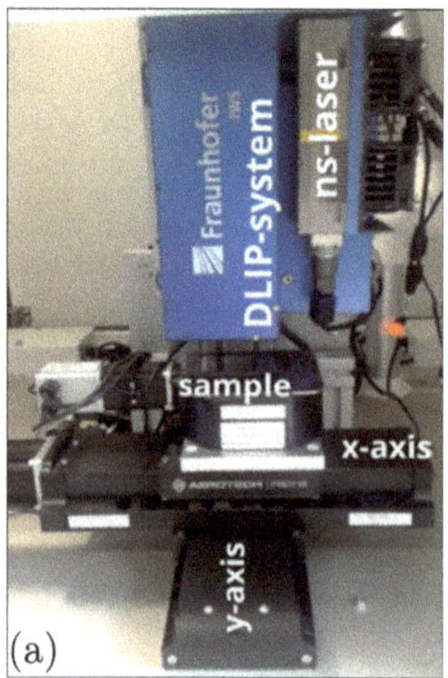

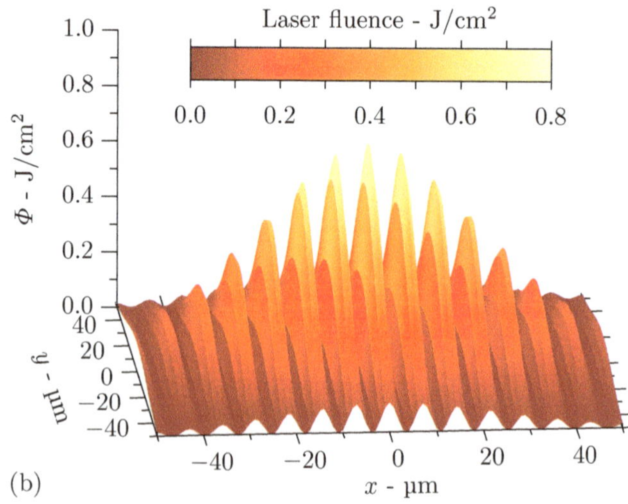

Figure 2. Setup employed for direct laser interference patterning (DLIP) experiments on steel substrates: (**a**) nanosecond pulsed infrared laser, optical head for DLIP on a *z*-stage, and sample mounted on an *x-y* positioning stage and (**b**) laser fluence distribution of interference pattern due to two coherent partial beams with a Gaussian intensity profile.

Here, the angle of incidence $\theta/2$ of the partial beams is considered small and, as a consequence, the interference spot is assumed to be circular. Therefore, the laser fluence distribution of the interference pattern exemplified in Figure 2b is given by

$$\Phi(x,y) = 2\Phi_0 \left\{ \cos\left[2kx\sin\left(\frac{\theta}{2}\right)\right] + 1 \right\} \exp\left(-\frac{2r^2}{r_0^2}\right), \tag{3}$$

where $r^2 = x^2 + y^2$ and r_0 is the Gaussian beam radius of the partial beams. Assuming that the periodicity Λ (see Equation (1)) of the interference pattern is considerably smaller than the Gaussian beam diameter, the average fluence is given by

$$\Phi_{av}(x,y) = 2\Phi_0 \exp\left(-\frac{2r^2}{r_0^2}\right) \tag{4}$$

with the central value $\Phi_{av}(r=0) = 2\Phi_0$. Furthermore, the interference spot radius r_{spot} is supposed to be smaller than the Gaussian beam radius r_0.

Consequently, the energy provided by the laser pulse is calculated by integrating the local averaged fluence in Equation (4) over the interference spot

$$E_{spot} = \int_{\phi=0}^{2\pi} \int_{r=0}^{r_{spot}} \Phi_{av}(r)r\,dr\,d\phi = \Phi_0\pi r_0^2\left[1 - \exp\left(-\frac{2r_{spot}^2}{r_0^2}\right)\right] \tag{5}$$

and divided by the spot area to obtain the fluence averaged over the interference spot

$$\Phi_{av,spot} = \frac{E_{spot}}{\pi r_{spot}^2} = \Phi_0\frac{r_0^2}{r_{spot}^2}\left[1 - \exp\left(-\frac{2r_{spot}^2}{r_0^2}\right)\right]. \tag{6}$$

Given that the averaged fluence $\Phi_{av,spot}$ in Equation (6) is known from measurement, the average fluence in the centre of the spot is determined by

$$\Phi_{av}(r=0) = 2\Phi_0 = 2\Phi_{av,spot}\frac{r_{spot}^2}{r_0^2}\left/\left[1 - \exp\left(-\frac{2r_{spot}^2}{r_0^2}\right)\right]\right.. \tag{7}$$

As mentioned above, it is assumed that $\Lambda/2 = \lambda/(4\sin(\theta/2)) \ll r_0$. Therefore, the laser fluence distribution of the period comprising the central maximum and in the central plane, i.e., for $|x| \leq \Lambda/2$ and $y = 0$, is approximately given by Equation (2), which corresponds to Equation (3) without the exponential factor.

2.3. Surface Characterisation

The polished and structured surfaces were characterised using confocal microscopy (S neox, Sensofar, Terrassa, Spain) at $150\times$ magnification with an optical resolution of $0.14\,\mu m$ and a vertical resolution of 1 nm. The field of view (FOV) is $117 \times 88\,\mu m^2$ in the x-y-plane. The recorded measurements were processed using the software MountainsMap 7.3 (Digital Surf, Besançon, France), utilising profile extraction including the step height measurement function for analysis of the peak formation and the image tools for three-dimensional images of the structured surfaces. In addition, the laser treated substrates were also evaluated using a scanning electron microscope (SEM) (Supra 40 VP, Zeiss, Jena, Germany) at an operating voltage of 5.0 kV.

2.4. Mathematical Model

The mathematical model of the material behaviour during single pulse DLIP was already described in detail by Demuth and Lasagni [44]. Here, the nonlinear temperature dependence of surface tension considered is presented along with the dimensionless form of the governing equations and the Marangoni boundary condition. The dimensional conservation equations and boundary conditions are provided in Appendix A for the sake of clarity.

In the present investigation, the surface tension of liquid steel is a nonlinear function of temperature in the presence of the surface active element sulphur. Therefore, the temperature coefficient of surface tension, i.e., $d\gamma/dT$, is a function of temperature as well, unlike the constant negative value found in the absence of a surfactant.

An expression for the temperature-dependent surface tension of liquid iron with the solute sulphur was derived by Sahoo et al. [45] in the following form:

$$\gamma = \gamma_\mathrm{m}^0 + \left.\frac{d\gamma}{dT}\right|_{\substack{\text{pure}\\\text{metal}}} (T - T_\mathrm{m}) - RT\Gamma_s \ln\left[1 + k_1 a_S \exp\left(-\frac{\Delta H_0}{RT}\right)\right], \tag{8}$$

where γ_m^0 is the surface tension of the pure metal at the melting point T_m and a_S is the sulphur activity in wt%. In Equation (8), $R = 8.3143$ J/(mol K) is the gas constant, $\Gamma_s = 1.3 \times 10^{-5}$ mol/m^2 is the surface excess at saturation, $k_1 = 3.18 \times 10^{-3}$ is a constant related to the entropy of segregation, and $\Delta H_0 = -1.88 \times 10^5$ J/mol is the standard enthalpy of adsorption [38,45]. The values of the temperature coefficient of surface tension, which is required for the melt pool convection in the boundary condition (A9) (see Appendix A) resulting for the binary system Fe–S can be used for the simulation of stainless steel melt pools [46]. Further considering the values $\gamma_\mathrm{m}^0 = 1.943$ N/m, $d\gamma/dT|_{\text{pure metal}} = -4.3 \times 10^{-4}$ N/(m K) and $T_\mathrm{m} = 1723$ K [38,45,47] in Equation (8), the surface tension is evaluated and presented in Figure 3a for liquid steel with different concentrations of the surfactant sulphur. Differentiating Equation (8) with respect to T, the relation for the temperature coefficient of surface tension is obtained as follows [38,45]:

$$\frac{d\gamma}{dT} = \left.\frac{d\gamma}{dT}\right|_{\substack{\text{pure}\\\text{metal}}} - R\Gamma_s \ln(1 + Ka_S) - \frac{Ka_S}{1 + Ka_S}\frac{\Gamma_s \Delta H_0}{T}, \tag{9}$$

where the equilibrium constant $K = k_1 \exp[-\Delta H_0/(RT)]$ is introduced. Consequently, the temperature derivative of surface tension in Equation (9) is evaluated as a function of temperature for molten stainless steel with different sulphur contents and presented in Figure 3b.

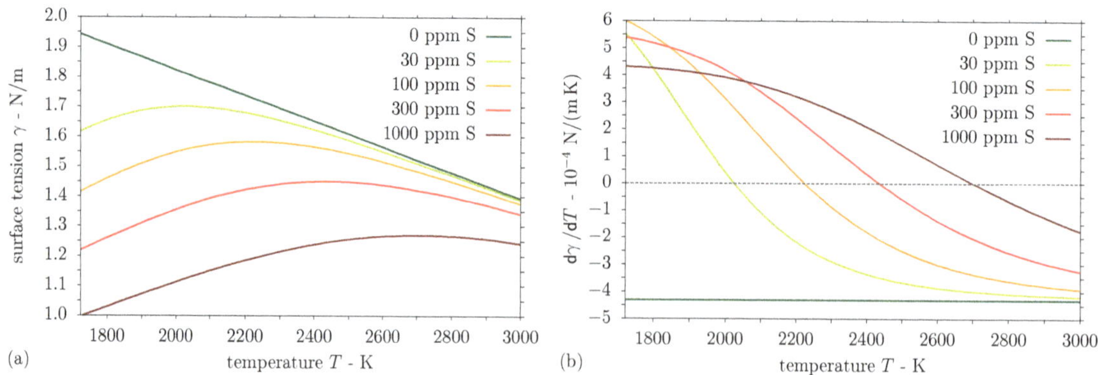

Figure 3. Temperature-dependent (**a**) surface tension and (**b**) its temperature derivative for liquid steel with different sulphur contents.

The governing equations stated in (A3)–(A6) are rewritten in dimensionless form to reveal the dimensionless numbers characterising the problem. For this purpose, the non-dimensionalisation of the variables position x, time t, velocity v, dynamic pressure p_dyn, temperature T, and specific enthalpy h is made using the scales L, L^2/a, a/L, $\rho_0 a^2/L^2$, $T_v - T_0$, and $c_p(T_v - T_0)$, respectively. In addition, the diffusion length $L = 2\sqrt{a\tau_p}$ is specified as the characteristic length scale, where the pulse width τ_p (FWHM) is selected as the laser beam dwell time. The resulting dimensionless variables are given by

$$x^* = \frac{x}{2\sqrt{a\tau_p}}, \quad t^* = \frac{t}{4\tau_p}, \quad v^* = 2v\sqrt{\frac{\tau_p}{a}}, \quad p^*_{dyn} = \frac{4\tau_p}{\rho_0 a} p_{dyn}, \quad T^* = \frac{T - T_0}{T_v - T_0} \tag{10}$$

and the dimensionless specific enthalpy assumes the shape

$$h^* = \frac{h}{c_p(T_v - T_0)} = T^* + f_m\, Ph_{s/l} + f_v\, Ph_{l/v}, \tag{11}$$

where the phase change numbers of melting and vapourisation are defined as

$$Ph_{s/l} = \frac{L_f}{c_p(T_v - T_0)}, \quad Ph_{l/v} = \frac{L_v}{c_p(T_v - T_0)}. \tag{12}$$

Employing the dimensionless variables in Equations (10) and (11), the governing equations (A3)–(A6) are rewritten in dimensionless form as

$$\frac{dh^*}{dt^*} = \Delta T^* + \frac{(1-R)\,La}{\sigma^*\sqrt{2\pi}}\left\{2\cos^2\left[\frac{\pi x^*}{\Lambda^*}\right]\right\}\exp\left(-\frac{\left(t^* - t_p^*\right)^2}{2\sigma^{*2}} + \alpha^*(z^* - z_{surf}^*)\right), \tag{13}$$

$$\nabla \cdot v^* = 0, \tag{14}$$

$$\frac{dv^*}{dt^*} = -\nabla p^*_{dyn} + Pr\Delta v^* + Pr\,Ra\frac{T^* - T_l^*}{1 - T_l^*}e_z, \tag{15}$$

$$\frac{dx^*}{dt^*} = v^* \tag{16}$$

with the dimensionless standard deviation $\sigma^* = \sigma/(4\tau_p)$ of the laser pulse, periodicity $\Lambda^* = \Lambda/L$, and absorption coefficient $\alpha^* = \alpha L$ used in Equation (13). Furthermore, the dimensionless form of the Marangoni boundary condition given in Equation (A9) is obtained as [48]

$$\frac{\partial v_x^*}{\partial z^*} = -\frac{Ma}{1 - T_l^*}\frac{\partial T^*}{\partial x^*}. \tag{17}$$

In addition to the aforementioned phase change numbers, $Ph_{s/l}$ and $Ph_{l/v}$ given in (12), and the Fourier number Fo corresponding to the considered physical time, the dimensionless numbers arising in the dimensionless Equations (13)–(17) are the laser number La

$$La = \frac{2\Phi_0\alpha}{\rho_0 c_p(T_v - T_0)}, \quad Fo = \frac{t_{end}}{4\tau_p}, \tag{18}$$

the Prandtl number Pr, the Rayleigh number Ra, and the Marangoni number Ma defined by

$$Pr = \frac{\nu}{a}, \quad Ra = 8\tau_p\frac{\beta(T_v - T_l)g\sqrt{a\tau_p}}{\nu}, \quad Ma = -\frac{d\gamma}{dT}\frac{T_v - T_l}{\rho_0\nu}2\sqrt{\frac{\tau_p}{a}}. \tag{19}$$

At first, the Marangoni number Ma in the boundary condition (17) may be evaluated using the aforementioned temperature coefficient $d\gamma/dT|_{pure\ metal}$ for liquid steel in the absence of the surfactant sulphur. In the presence of sulphur, the resulting Marangoni number Ma multiplied by the dimensionless temperature coefficient of surface tension given by (20), which is a function of temperature, is considered in the Marangoni boundary condition (17).

$$\left(\frac{d\gamma}{dT}\right)^* = \frac{d\gamma}{dT}\bigg/\frac{d\gamma}{dT}\bigg|_{\substack{pure\\metal}} = 1 - \frac{R\Gamma_s}{\frac{d\gamma}{dT}\big|_{\substack{pure\\metal}}}\ln(1 + Ka_S) - \frac{Ka_S}{1 + Ka_S}\frac{\Gamma_s\Delta H_0/T}{\frac{d\gamma}{dT}\big|_{\substack{pure\\metal}}} \tag{20}$$

2.5. Numerical Simulation

Simulation of the single pulse DLIP process considered material in the plane $y = 0$ is subject to the central period of the interference pattern comprising the fluence maximum. According to the approach presented in [44], a two-dimensional section of 7.5 µm × 4.75 µm size in the x-z-plane was discretised using particles for the numerical solution by means of the smoothed particle hydrodynamics (SPH) method. The discretisation starts with coarse particles of 1 µm diameter at the bottom of the computational domain and a consecutive reduction of the particle diameter towards the surface according to a geometric sequence with common ratio 7/9; see Figure 4a. Near the surface, an initially equidistant array of fine particles of $(7/9)^{21}$ µm $\approx$ 5.1 nm diameter is arranged for the melt pool simulation; see Figure 4b. This minimum particle size is specified to ensure a complete insertion of the energy provided by the laser irradiation in accordance with an earlier investigation [49]. In the present discretisation, the equidistant arrangement of fine particles represents a surface layer of 270 nm depth. In addition, three rows and columns of dummy particles are situated beyond the domain boundaries to complete the kernel function support of nearby interior particles and to facilitate the application of boundary conditions.

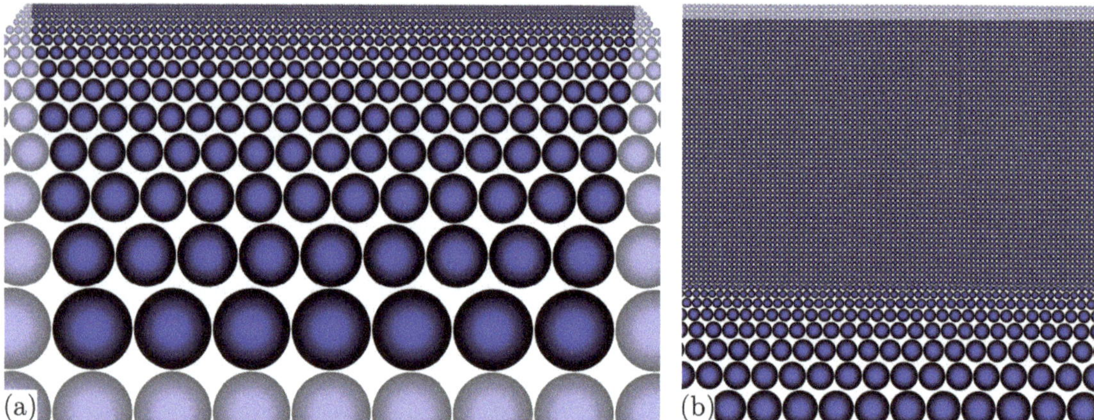

Figure 4. Particle discretisation of computational domain, dummy particles less opaque, details: (**a**) 7.5 µm × 4.4 µm, coarse particles start at the bottom, (**b**) 440 nm × 440 nm, near-surface equidistant initial array and coarsening.

The numerical investigation models the two-beam interference patterning process and setup described in Section 2.2 at moderate fluences. In detail, the process parameters considered for the numerical simulation of DLIP are presented in Table 1.

Table 1. Process parameters of single pulse DLIP on stainless steel with sulphur content.

Process Parameter	Symbol	Value
wavelength	λ	1053 nm
intersection angle between beams	θ	0.1464 rad
periodicity of interference pattern	Λ	7.2 µm
average fluence in interference spot	$\Phi_{av,spot}$	0.300 J/cm^2
fluence of interference pattern	$2\Phi_0$	0.532 J/cm^2
pulse duration (FWHM)	τ_p	12 ns
pulse time	t_p	50 ns
simulation duration	t_{end}	200 ns
initial substrate temperature	T_0	298.15 K
gravitational acceleration	g	9.81 m/s^2
Fourier number	Fo	4.1$\overline{6}$

Note that the periodicity Λ is defined by the wavelength λ and the intersection angle θ; see Equation (1). In addition, two moderate laser fluences $\Phi_{\text{av,spot}} = 0.300\ \text{J/cm}^2$ and $0.375\ \text{J/cm}^2$ were used in the numerical simulations. Considering an interference spot radius $r_{\text{spot}} = 80\ \mu\text{m}$ and a Gaussian beam radius $r_0 = 100\ \mu\text{m}$, the average fluence in the central period of the interference pattern results from Equation (7) as $2\Phi_0 = 0.532\ \text{J/cm}^2$ and $0.665\ \text{J/cm}^2$, respectively. Furthermore, simulations were performed for sulphur contents of 30 ppm, 100 ppm, 300 ppm, and 1500 ppm.

The considered material properties of the AISI 304 stainless steel substrates are compiled in Table 2. The temperature coefficient of surface tension $\mathrm{d}\gamma/\mathrm{d}T$ given in the table represents the constant negative value for stainless steel in the absence of a surfactant, which is replaced by the model introduced in Section 2.4 in the presence of sulphur. Concerning the values of the density ρ_0, specific heat c_{p}, thermal conductivity κ, and thermal diffusivity a given in Table 2, the respective temperature-dependent values, if available, are averaged over the interval from the room temperature to the vapourisation point. The thermal diffusion length is then calculated employing the determined thermal diffusivity and presented in Table 3 along with the dimensionless quantities resulting from the material properties in Table 2.

Table 2. Material properties of stainless steel substrates.

Material Property	Symbol	AISI 304 Steel	Unit	References
solidus temperature	T_{s}	1673	K	[50]
liquidus temperature	T_{l}	1727	K	[50]
vapourisation temperature	T_{v}	3273	K	[51]
density	ρ_0	7262	kg/m^3	[50]
specific heat	c_{p}	704	J/(kgK)	[50,52]
thermal conductivity	κ	26.8	W/(mK)	[50]
thermal diffusivity	a	5.24×10^{-6}	m^2/s	
enthalpy of fusion	L_{f}	251	kJ/kg	[50,51]
enthalpy of vapourisation	L_{v}	6500	kJ/kg	[51]
dynamic viscosity (at T_{l})	η	7.0×10^{-3}	Pas	[50]
kinematic viscosity (at T_{l})	ν	1.02×10^{-6}	m^2/s	
volumetric thermal expansion coefficient	β	8.5×10^{-5}	$1/\text{K}$	[53,54]
temperature coefficient of surface tension	$\mathrm{d}\gamma/\mathrm{d}T$	-4.3×10^{-4}	N/(mK)	[45]
absorption coefficient (at 1053 nm)	α	5.15×10^7	$1/\text{m}$	[55]
reflectivity (at 1053 nm)	R	0.646	1	[55]

Table 3. Diffusion length and dimensionless numbers of DLIP model for stainless steel surfaces.

Quantity	Symbol	AISI 304 Steel
thermal diffusion length	L	501.5 nm
laser number	La	18.0241
solid–liquid phase change number	$Ph_{\text{s/l}}$	0.119849
liquid–vapour phase change number	$Ph_{\text{l/v}}$	3.10367
Prandtl number	Pr	0.1947
Rayleigh number	Ra	3.04124×10^{-8}
Marangoni number	Ma	9.08939

3. Results and Discussion

This section reports and discusses the results of the DLIP experiments and the numerical simulations. In Section 3.1, the microstructures observed after laser processing, depending on the sulphur content of the substrate and the laser fluence, are presented. Furthermore, Section 3.2 provides the results of the simulations performed to comprehend the melt pool convection and the structure formation during the DLIP process.

3.1. Experimental Results

In a first set of experiments, the polished steel samples with different sulphur content were irradiated with DLIP line-like patterns by systematically changing the laser fluence. The DLIP spots showed differences in surface morphology for varying sulphur content. Figure 5 shows surface profiles of the steel samples with varied sulphur contents (30 ppm S, 100 ppm S, and 300 ppm S) treated at different laser fluences ($\Phi_{av,spot}$ = 0.46 J/cm^2, 0.63 J/cm^2, 0.82 J/cm^2, and 0.99 J/cm^2). The grey highlighted area represents the region of maximum laser intensity (interference maximum) during structuring. For instance, the steel with 30 ppm sulphur treated with a laser fluence of $\Phi_{av,spot}$ = 0.46 J/cm^2 exhibits a line-like pattern with a central peak (~40 nm height) surrounded by two valleys (~15 nm depth) below the initial surface level. Increasing the laser fluence to $\Phi_{av,spot}$ = 0.82 J/cm^2 leads to a transformation stage, where the central single peak splits into two peaks with a structure height of approximately 50 nm, forming a well-defined valley between them. By further increasing the laser fluence, both the height of the peaks and the depth of the valley increases, determined by a significant flow of the molten material during the DLIP process; see Figure 5 for 30 ppm S content and $\Phi_{av,spot}$ = 0.99 J/cm^2.

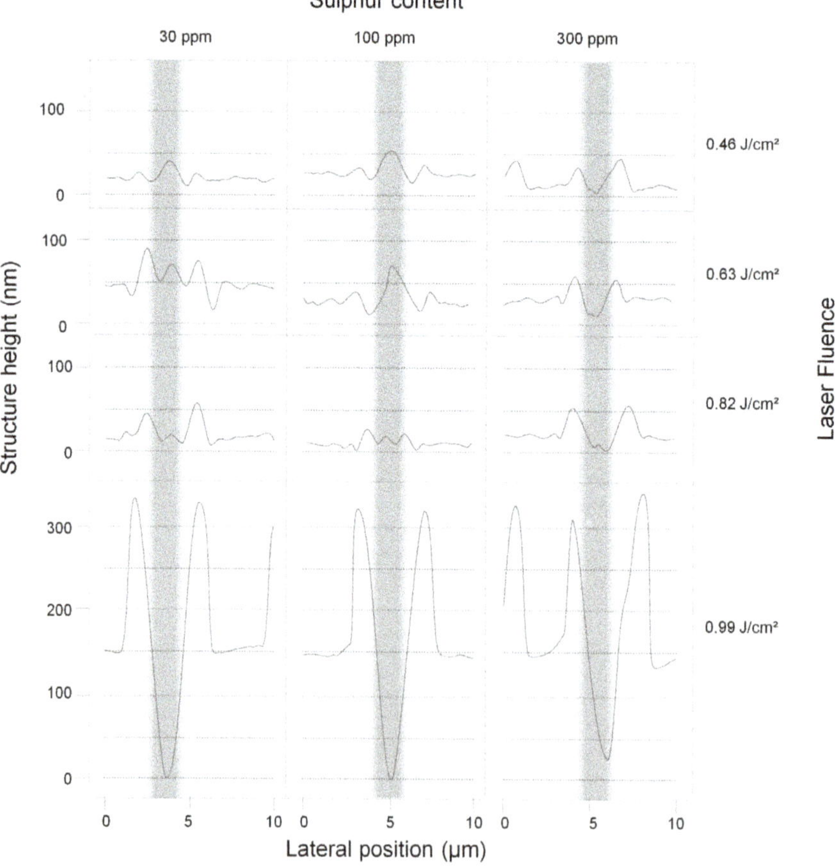

Figure 5. Overview of peak transformation of DLIP structures made on steels with varying sulphur contents. The columns show profiles of produced patterns in steels with sulphur contents of 30 ppm, 100 ppm, and 300 ppm at different laser fluences ($\Phi_{av,spot}$ = 0.46 J/cm^2, 0.63 J/cm^2, 0.82 J/cm^2, and 0.99 J/cm^2). The grey highlighted areas represent the regions of maximum laser intensity during patterning.

The line-like pattern on the steel with 100 ppm S reveals a topography similar to the processed one of 30 ppm S steel when irradiated at a low fluence level of $\Phi_{av,spot}$ = 0.46 J/cm^2, with similar peak heights and valley depths; see Figure 5. However, the central peak, with a maximum structure height of 50 nm as well, starts to split only at laser fluences above $\Phi_{av,spot}$ = 0.63 J/cm^2. By fluence increase to $\Phi_{av,spot}$ = 0.99 J/cm^2, both peak height and valley depth constantly increase further and reach a similar magnitude as that for the geometries on 30 ppm S steels irradiated with $\Phi_{av,spot}$ = 0.99 J/cm^2.

In contrast, for the steel samples with the highest concentration of sulphur (300 ppm and 1500–3000 ppm), a different behaviour was observed at low fluences. Here, two separate peaks with peak heights of up to 50 nm instead of a centre peak are found also at low laser fluences; see Figure 5 for 300 ppm S content and $\Phi_{av,spot}$ = 0.46 J/cm^2. As the laser fluence is increased, the split peaks become larger, reaching up to 50 nm absolute structure height at $\Phi_{av,spot}$ = 0.82 J/cm^2. At $\Phi_{av,spot}$ = 0.99 J/cm^2, the height of the split peaks is further increased, similar to the 30 ppm and 100 ppm samples.

The surface microstructures generated on steel samples with 30 ppm, 100 ppm, and 300 ppm S using laser fluences in the range 0.46 J/cm^2–0.99 J/cm^2 are further illustrated by the scanning electron micrographs shown in Figure 6. The small graphs inserted into the SEM images in Figure 6 indicate the areas of maximum laser intensity on each surface.

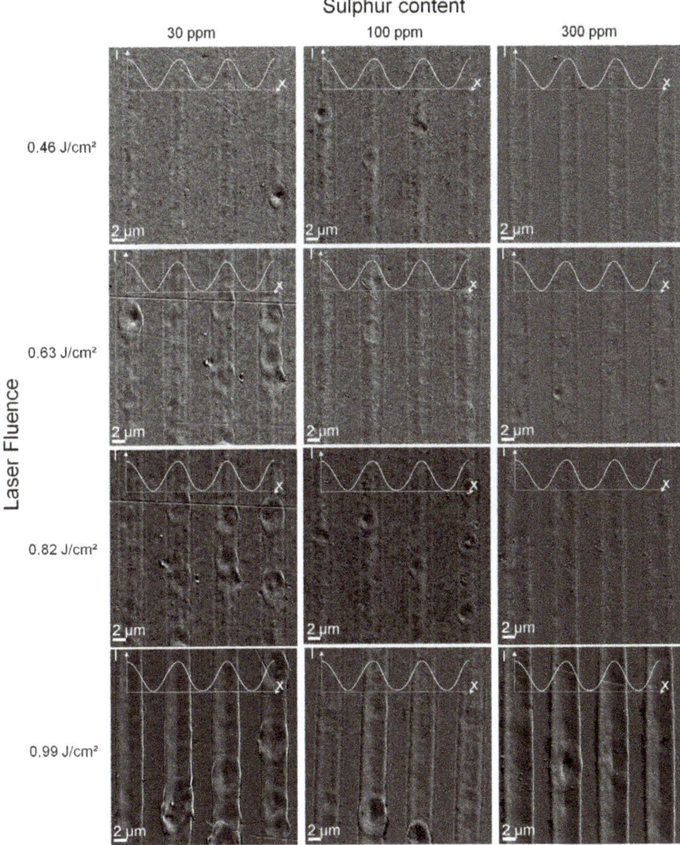

Figure 6. Scanning electron micrographs of surfaces after DLIP (λ = 1053 nm, Λ = 7.2 µm, and τ_p = 12 ns) on steel substrates with distinct sulphur contents using a single pulse and different laser fluences. Small graphs indicate areas of maximum laser intensity. The brightness and contrast of the SEM images were enhanced for better visualisation.

In agreement with the surface profiles presented in Figure 5, the micrographs of steels with 30 ppm S and 100 ppm S contents irradiated with a laser fluence of 0.46 J/cm^2 in Figure 6 depict structures with a single central peak. With an increase in laser fluence, the structure width grows, where a smaller central peak and two neighbouring subpeaks evolve instead of the single peak structure. On the other hand, microstructures fabricated using the highest laser fluence of 0.99 J/cm^2 display a split peak formation, induced by the dominant recoil pressure during the patterning process. In contrast, the SEM images of structures made on steel substrates with 300 ppm S in Figure 6 show a split peak formation for all laser fluences applied, where the structure width increases with the laser fluence up to 0.99 J/cm^2.

The differences in peak formation can also be identified in the three-dimensional images presented in Figure 7. Small, single peaks of the structures obtained on 30 ppm S steel at a laser fluence of $\Phi_{av,spot}$ = 0.46 J/cm^2 are shown in Figure 7a. The topography shown in Figure 7b on steel with 300 ppm S after DLIP treatment at $\Phi_{av,spot}$ = 0.82 J/cm^2 is characterised by split peaks and an enclosed valley below the initial surface level. The structure heights are decreasing from the DLIP spot centre towards the spot edge, following the Gaussian intensity distribution of the laser beam. Even at the spot edge, no structures in transformation stage or single peaks are visible in Figure 7b.

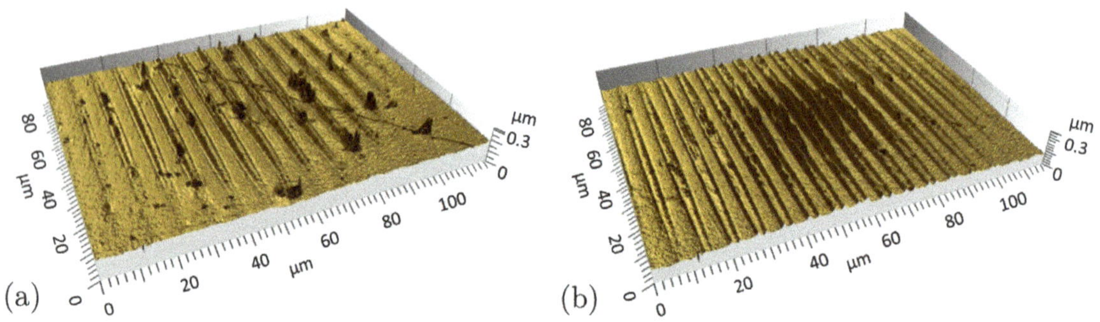

Figure 7. Three-dimensional confocal microscope images of DLIP laser spots on steel samples with different sulphur contents. (**a**) Single peak structures generated on steel with 30 ppm S content and $\Phi_{av,spot}$ = 0.46 J/cm^2, (**b**) split peak structures fabricated on steel with 300 ppm S content and $\Phi_{av,spot}$ = 0.82 J/cm^2.

An explanation of the distinct structure formation and the different peak heights at similar laser fluence is attempted on the basis of the surface tension gradient depending on the sulphur content of the melted steel [47]. As evident from Equation (9) and Figure 3b, both the surface temperature and the sulphur activity affect the temperature coefficient of surface tension, which determines the direction of thermocapillary flow and the melt pool geometry [38]. For 30 ppm S, the surface tension temperature coefficient is positive only in a small temperature range of ~300 K from the liquidus point T_l. At moderate fluences, the positive temperature derivative of surface tension induces an inward flow at the surface and the formation of a single peak structure with adjacent depressions, as shown in Figure 1b, e.g., for 30 ppm S and $\Phi_{av,spot}$ = 0.46 J/cm^2 in Figure 5. For an augmented sulphur content of 100 ppm, the magnitude and temperature range of the positive temperature coefficient of surface tension is enlarged, resulting in a stronger inward convection and more marked single peak structures at moderate fluences, as presented in Figure 5 for $\Phi_{av,spot}$ = 0.46 J/cm^2 and 0.63 J/cm^2.

With an increase in laser fluence, the higher surface temperatures are associated with a negative temperature coefficient of surface tension of the steel melt at low sulphur concentrations. Accordingly, the inward flow from the melt pool edges is temporarily suppressed by an outward flow from the centre of the melt pool surface at higher surface temperatures, which causes the evolution of outer peaks near the edges of the melt pool; see Figure 1a. In addition, small central peaks may form due to inward convection in the late stages of the melt pool presence at lower melt surface temperatures; see the microstructures for 30 ppm S at $\Phi_{av,spot} = 0.63$ J/cm² and 0.82 J/cm² and for 100 ppm S at $\Phi_{av,spot} = 0.82$ J/cm² in Figure 5. On the contrary, it is evident from the split peak structures observed in Figure 5 for steel with 300 ppm S at fluences up to $\Phi_{av,spot} = 0.82$ J/cm² that the conception of an even stronger inward convection at higher sulphur contents is not confirmed by the experiments. This deviation suggests that, in the presence of higher impurity levels, the expression for the temperature-dependent surface tension in Equation (8) is no longer appropriate or that effects other than thermocapillary convection are responsible for the microstructure evolution. For a high fluence of $\Phi_{av,spot} = 0.99$ J/cm², split peak structures of similar height are produced on all steel samples, independent of the sulphur content; see Figure 5. This observation may be attributed to the dominance of the recoil pressure induced by vapourisation of molten metal at high laser intensity, which causes a lateral ejection of melt during the patterning process [56].

It is important to mention the different behaviour observed at the fluences below $\Phi_{av,spot} = 0.46$ J/cm² or at the edges of the DLIP spots. In these cases, single peaks are observed without the surrounding valleys, as compared to the centre peaks of the 30 ppm and 100 ppm S samples treated with $\Phi_{av,spot} = 0.46$ J/cm². In other words, this structure has been produced without the flow of molten metal and thus can only be explained by the transformation of the crystalline structure of the used steel. When an austenitic steel is heated, depending on the temperature reached and the cooling rate, the structure can transform to martensite, which has a lower density (austenite: 7.9 g/cm³; martensite: 6.5 g/cm³ [57]). Taking into consideration that the measured height of the peak was 15 nm, it is estimated that the austenitic steel was affected up to a depth of 69 nm.

The dependence of structure heights and peak splitting on sulphur content and laser fluence is shown in Figure 8. The change in structure height between centre peak and initial surface level is depicted in Figure 8a. Fluences below $\Phi_{av,spot} = 0.46$ J/cm² produce slight surface elevations related to phase changes austenite–martensite in the steel. A fluence increase from $\Phi_{av,spot} = 0.46$ J/cm² to $\Phi_{av,spot} = 0.9$ J/cm² results in a gain in structure heights, up to 49 nm for single peaks with surrounding valleys for <100 ppm S steels. A further rise in fluence only results in a decrease in structure height due to a transformation stage from single peak to split peaks. Structures on steel with higher sulphur content (>100 ppm S) show negative structure heights, that is, valleys at the centre, for fluences above $\Phi_{av,spot} = 0.46$ J/cm² due to the split peak formation. With increasing fluence to $\Phi_{av,spot} = 0.99$ J/cm², the valley enclosed by the split peaks becomes slightly deeper and increases further for fluences above $\Phi_{av,spot} = 0.99$ J/cm². Maximum peak heights of 150 nm are found for $\Phi_{av,spot} = 1.1$ J/cm² on all steel samples, irrespective of the sulphur content.

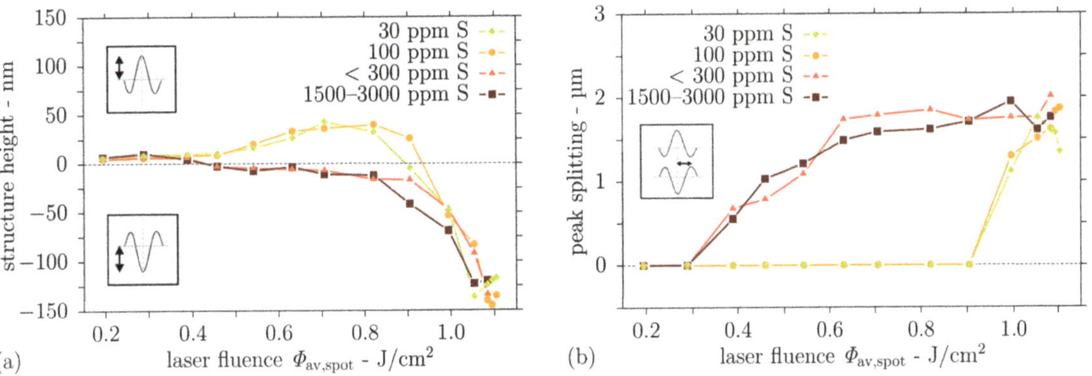

Figure 8. Surface topography observed after single pulse DLIP ($\lambda = 1053$ nm, $\Lambda = 7.2$ μm, $\tau_p = 12$ ns) on steel with different sulphur contents as a function of fluence: (**a**) structure height and (**b**) distance of peak(s) from centre.

The topic of peak splitting is also addressed in Figure 8b. Here, the distance between the split peaks is analysed. No peak splitting is detected for fluences below $\Phi_{av,spot} = 0.46$ J/cm^2. Starting from $\Phi_{av,spot} = 0.46$ J/cm^2, a separation distance (half distance between both peaks) increasing with the laser fluence up to a maximum of 2 µm is measured for high sulphur content steels.

3.2. Simulation Results

Simulations of single pulse DLIP were performed for stainless steel substrates with different sulphur contents. As mentioned in Section 2.5, the numerical simulations considered the process parameters and material properties given in Tables 1 and 2, respectively, as well as the resulting diffusion length and dimensionless quantities in Table 3. The dimensionless form (20) of the surface tension temperature coefficient in Equation (9) was incorporated into the Marangoni boundary condition in Equation (17). Furthermore, simulations were carried out for the central period of the interference pattern with an average fluence of either $2\Phi_0 = 0.532$ J/cm^2 or $2\Phi_0 = 0.665$ J/cm^2, corresponding to a laser number of $La = 18.0241$ or $La = 22.5301$.

The evolution of the maximum surface temperature predicted in the simulation for the moderate laser fluence $2\Phi_0 = 0.532$ J/cm^2 is presented in Figure 9 along with the temporal variation of the laser pulse intensity. The maximum surface temperature trend in Figure 9 is in line with earlier simulation results in [44], although $T_{max} \sim 2570$ K is lower here, for DLIP of stainless steel at a laser wavelength of $\lambda = 355$ nm and a lower fluence of $2\Phi_0 = 0.4$ J/cm^2, which is attributed to the higher reflectivity at the present wavelength. It is evident from Figure 9 that the substrate surface is heated up to a temperature significantly above the liquidus point at the interference maximum due to the action of the laser pulse. Accordingly, a melt layer with the dimensions computed by the SPH model and presented in Figure 10 develops near the interference maximum after the onset of the laser pulse. Concerning the trends in Figure 10, the discrete values of the melt pool dimensions are employed only once at a central point in time to avoid a stair-step appearance of the graphs. The melt pool depth and the duration of the melt presence are compatible with the aforementioned numerical results in [44], whereas the melt pool width in Figure 10 exceeds the earlier calculations owing to the larger periodicity Λ of the interference pattern considered here.

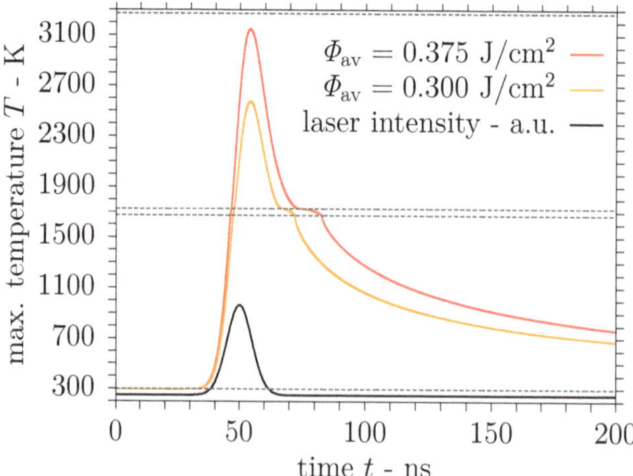

Figure 9. Maximum surface temperature during single pulse DLIP of stainless steel with Gaussian beams using the process parameters $\lambda = 1053$ nm, $\Lambda = 7.2$ µm, $\tau_p = 12$ ns and $2\Phi_0 = 0.532$ J/cm^2 or $2\Phi_0 = 0.665$ J/cm^2.

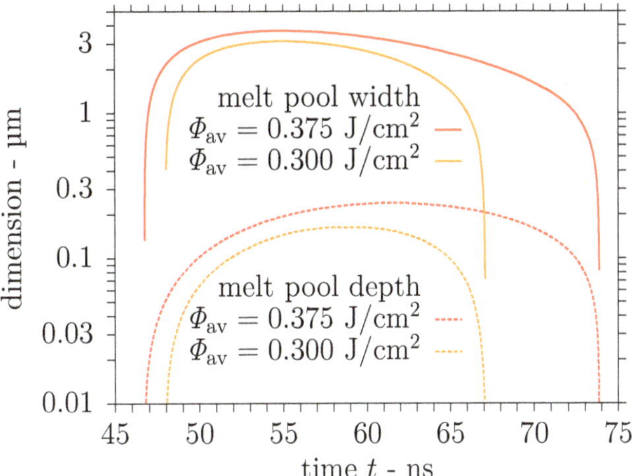

Figure 10. Transient melt pool dimensions predicted by the smoothed particle hydrodynamics (SPH) model of DLIP (λ = 1053 nm, Λ = 7.2 µm, τ_p = 12 ns) on stainless steel using a laser fluence of either $2\Phi_0$ = 0.532 J/cm^2 or $2\Phi_0$ = 0.665 J/cm^2.

Furthermore, simulations were performed for an elevated laser fluence $2\Phi_0 = 0.665$ J/cm^2 to investigate the influence of the laser fluence on the material behaviour, in particular the nonlinear effect on the melt pool dimensions and the velocity field, and the structure formation reported in Section 3.1. Consequently, the evolution of the maximum surface temperature computed by the SPH model is also shown in Figure 9. The temporal maximum of the surface temperature $T_{max} \sim 3150$ K at the interference maximum is below the vapourisation point, unlike the simulation results in [44] for the comparable fluence $2\Phi_0 = 0.5$ J/cm^2 and the laser wavelength $\lambda = 355$ nm with a lower reflectivity of the stainless steel substrate. These differences from the earlier results are attributed to the slightly larger width of the laser pulse considered in the present investigation. The computed dimensions of the melt layer developing after the beginning of the laser pulse are shown in Figure 10 for the elevated fluence $2\Phi_0 = 0.665$ J/cm^2 as well. In accordance with the numerical results in [44], the maximum melt pool dimensions are roughly 20% wider and even 50% deeper than the calculations for the moderate fluence, i.e., $2\Phi_0 = 0.532$ J/cm^2 here. Again, the duration of the melt presence and the determined melt pool depth conform with the previous results in [44] unlike the increased melt pool width due to the larger periodicity Λ employed.

As set out in the Introduction, the nonuniform surface temperature of the melt layer gives rise to surface tension gradients, entailing thermocapillary convection in the direction of high surface tension. Considering the presence of the surfactant sulphur in liquid steel, the temperature coefficient of surface tension is positive near the melting point, i.e., there is inward convection from the melt pool edges towards regions of higher surface tension and temperature. The computed velocity magnitude of this inward flow at the melt pool surface is indicated by the solid lines in Figure 11a for different sulphur contents and the moderate laser fluence $2\Phi_0 = 0.532$ J/cm^2. If the surface of the melt pool is heated to higher temperatures, the temperature coefficient of surface tension is negative in the proximity of the interference maximum and an additional outward flow from the centre of the melt pool surface towards regions of maximum surface tension develops. The predicted velocity magnitude of this outward convection is indicated by the dashed lines in Figure 11a for different sulphur contents.

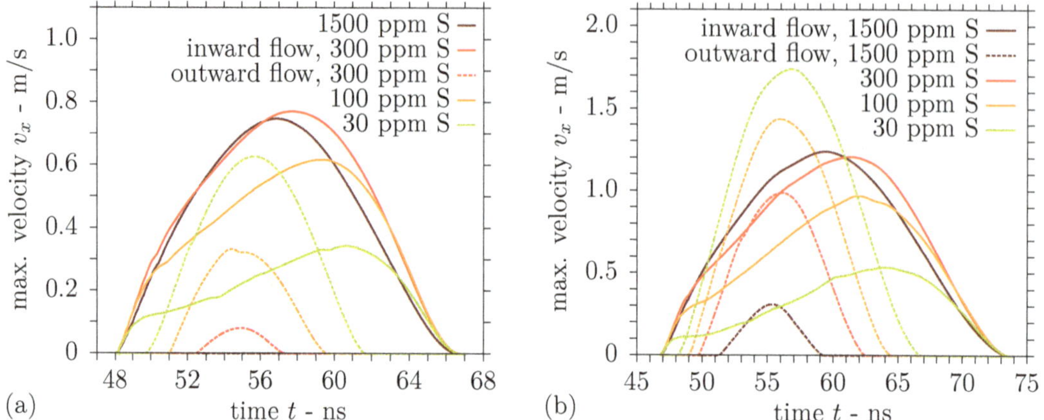

Figure 11. Horizontal velocity magnitudes at melt pool surface during DLIP with $\lambda = 1053$ nm, $\Lambda = 7.2$ μm, and $\tau_p = 12$ ns using a fluence of (**a**) $2\Phi_0 = 0.532$ J/cm^2 or (**b**) $2\Phi_0 = 0.665$ J/cm^2 on steel with sulphur content.

It is observed from Figure 11a that the magnitude of the outward flow is particularly prominent for a low sulphur concentration of 30 ppm and decreases with an increase in sulphur content. On the contrary, the inward flow becomes more pronounced for a higher surfactant concentration, as can be identified from Figure 11a. For 1500 ppm sulphur, the highest sulphur content investigated, the thermocapillary convection is completely inward at the melt pool surface for the moderate laser fluence, as the maximum surface temperature does not exceed the point of maximum surface tension; see also Figures 3 and 9. For a clearer idea of the substrate behaviour in the interaction zone, Figure 12a presents the predicted temperature and velocity fields in a section of the computational domain comprising the melt pool half width for different sulphur contents at several times. In addition to the velocity magnitudes shown in Figure 11a for the moderate fluence $2\Phi_0 = 0.532$ J/cm^2, Figure 12a illustrates the strong outward convection for a low sulphur content of 30 ppm, which gradually decreases and is dominated by the inward convection for higher sulphur concentrations of 100 ppm and 300 ppm. Compared with the results in [44], lower horizontal velocity magnitudes are obtained in this work, which is attributed to the generally smaller magnitude of the surface tension temperature coefficient in the presence of a surfactant and the lower surface temperature gradients due to the larger periodicity Λ and longer pulse duration τ_p.

The magnitudes of the horizontal velocity extrema at the melt pool surface predicted for the elevated laser fluence $2\Phi_0 = 0.665$ J/cm^2 are presented in Figure 11b. It is noted in Figure 11b that the velocity magnitude of the inward flow from the melt pool edges is consistently higher than the results in Figure 11a for the moderate laser fluence. As the surface temperature largely exceeds the point of maximum surface tension in the central region due to the elevated fluence $2\Phi_0 = 0.665$ J/cm^2, the outward flow from the centre of the melt pool surface is significantly enhanced; see Figure 11b. Considering a low sulphur content of 30 ppm, Figure 11b shows that the outward flow clearly dominates the melt pool convection. For a concentration of 100 ppm sulphur in liquid steel, outward and inward convection both contribute considerably to the melt pool flow pattern. On the contrary, the graphs in Figure 11b suggest that the inward flow still dominates the melt pool convection for a higher sulphur content of 300 ppm in spite of the augmented outward flow. The foregoing statements are further illustrated in Figure 12b, which depicts the temperature and velocity fields computed for DLIP of stainless steel employing the elevated fluence in a section comprising the melt pool half width. In particular, a comparison of the results for 100 ppm sulphur presented in the central columns of Figure 12a,b reveals that a slight increase in the laser fluence changes the melt pool convection from a predominant inward flow to competing outward and inward flow. This changed melt flow character may explain the transition from a single peak microstructure to split peaks in Figure 5 when augmenting the laser fluence.

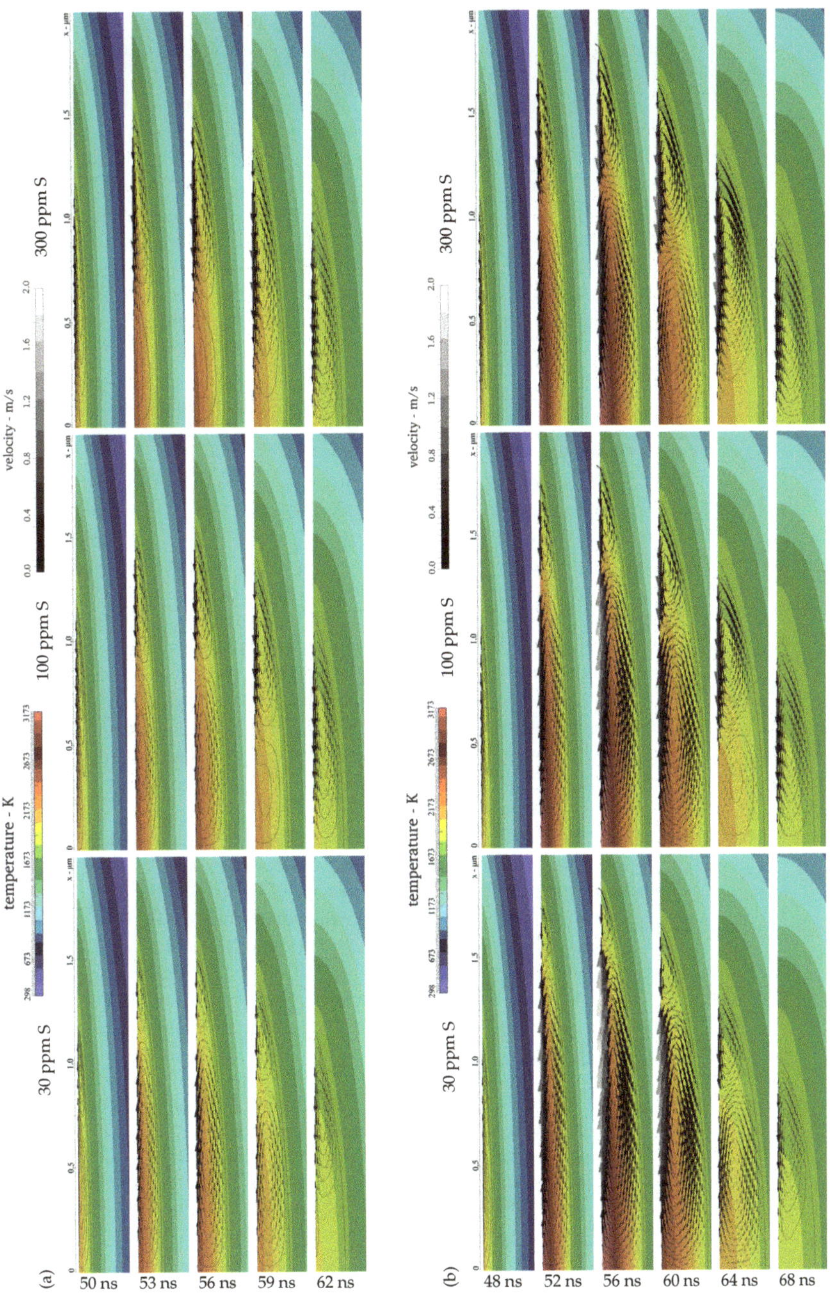

Figure 12. Simulation of melt pool flow during DLIP (λ = 1053 nm, Λ = 7.2 µm, and τ_p = 12 ns) of steel with (l) 30 ppm, (c) 100 ppm, and (r) 300 ppm sulphur at a laser fluence of (**a**) $2\Phi_0$ = 0.532 J/cm^2 and (**b**) $2\Phi_0$ = 0.665 J/cm^2 with detail of 2 µm × 250 nm including isotherms, streamlines, and velocity vectors in melt pool half width at times (**a**) t = 50 ns, 53 ns, 56 ns, 59 ns, and 62 ns, and (**b**) t = 48 ns, 52 ns, 56 ns, 60 ns, 64 ns, and 68 ns.

4. Conclusions

Direct laser interference patterning was employed on steel substrates with varied sulphur content to study the influence of melt dynamics on the produced surface topographies. For the structuring process, a laser source with a wavelength of 1053 nm and a pulse duration of 12 ns was selected to permit a photothermal interaction at the steel surface and thus the melt pool formation. At laser fluences below $\Phi_{av,spot} = 0.46$ J/cm^2, low peak heights were produced independently of the sulphur content. An explanation can be found in the phase change in the steel microstructure from austenite to martensite, which involves a volume expansion due to the lower density of martensite. A slight enhancement of the laser fluence resulted in an increase in height of the single peak structures formed on low sulphur steels, which can be related to the inverse Marangoni convection in the melt pool. With a further increase in laser fluence, the structures on low sulphur steels change from single peak to split peaks. In contrast, for steels with high sulphur content, structures with split peaks were produced also at low laser fluences. Structures generated employing laser fluences above $\Phi_{av,spot} = 0.90$ J/cm^2 exhibited split peaks, irrespective of the sulphur content, due to the dominance of the recoil pressure during the DLIP process. The difference in the surface topography observed as a function of the laser parameters and sulphur content can be used in the future to create pattern geometries different from the yet achievable with two-beam or three-beam DLIP and thus to produce surfaces with novel functions.

As the process is not accessible to measurement, numerical simulation is used to gain insight into the mechanisms effective during nanosecond pulsed DLIP of stainless steel with sulphur content. For this purpose, an SPH model for heat transfer and fluid flow during the patterning process is extended to take the nonlinear temperature dependence of surface tension in the presence of a surfactant into account. Considering the process parameters and physical properties of the substrates employed in the experiments, process simulations of DLIP on stainless steel with different sulphur content were performed. The computational results provide detailed information on the temperature field, the melt pool size, and flow pattern in the interaction zone. In particular, the influence of sulphur concentration on the thermocapillary convection and the effect of laser fluence on the melt pool dimensions and flow pattern were investigated numerically. Consequently, the simulations support the explanation of the microstructure formation observed in the experiments. For instance, the evolution of both single peak structures at moderate fluence and split peak structures at elevated fluence found after DLIP on stainless steel with 100 ppm sulphur is confirmed by the dominating inward flow and the competing outward flow, respectively, in the SPH simulations.

The simulation results hint at the relevance of the melt pool flow in the microstructure evolution due to nanosecond pulsed DLIP of metals at moderate fluences, although the melt displacement and the structure formation cannot be reproduced at present. Prospective numerical investigations of nanosecond pulsed DLIP on metal substrates will include the modelling of the surface deformation and the vapourisation recoil pressure at higher fluences.

Author Contributions: Conceptualisation, T.J. and A.F.L.; methodology, T.J., C.D., and A.F.L.; software, C.D.; validation, T.J. and C.D.; formal analysis, T.J. and C.D.; investigation, T.J.; resources, A.F.L.; data curation, T.J. and C.D.; writing—original draft preparation, T.J. and C.D.; writing—review and editing, T.J., C.D., and A.F.L.; visualisation, T.J. and C.D.; supervision, A.F.L.; project administration, A.F.L.; funding acquisition, A.F.L. All authors have read and agreed to the published version of the manuscript.

Funding: The experimental part of this research was carried out in a project in the framework of the priority programme SPP 1676 "Nachhaltige Produktion durch Trockenbearbeitung in der Umformtechnik" funded by the German Research Foundation (Deutsche Forschungsgemeinschaft DFG) grant No F-004442-541-003-1132104. The work of A.F.L. was further supported under the Excellence Initiative programme by the German federal and state governments to promote top-level research at German universities grant No F-003661-553-71A-1132104. The APC was discounted thanks to the participation in the Institutional Open Access Programme and borne in the frame of Open Access Funding by the Publication Fund of the TU Dresden and the Saxon State and University Library Dresden (SLUB).

Institutional Review Board Statement: Not applicable.

Informed Consent Statement: Not applicable.

Data Availability Statement: The data presented in this article are available upon request from the authors.

Acknowledgments: The authors acknowledge Achim Mahrle (Fraunhofer IWS Dresden) for the helpful discussions.

Conflicts of Interest: The authors declare no conflict of interest. The funders had no role in the design of the study; in the collection, analyses, or interpretation of data; in the writing of the manuscript; or in the decision to publish the results.

Abbreviations

The following abbreviations are used in this manuscript:

AISI	American Iron and Steel Institute
DLIP	direct laser interference patterning
DOE	diffractive optical element
FOV	field of view
FWHM	full width at half maximum
LASER	light amplification by stimulated emission of radiation
Nd:YLF	neodymium-doped yttrium lithium fluoride
ppm	parts per million
SEM	scanning electron microscope
SPH	smoothed particle hydrodynamics
TEM_{00}	fundamental transverse electromagnetic mode

Appendix A. Governing Equations and Boundary Conditions

The conservation of energy determines the state of the material throughout the patterning process. Here, the interference irradiation constitutes a volumetric heat source:

$$\dot{q}'''(x,y,z,t) = \alpha(1-R)\frac{\Phi(x,y)}{\sigma\sqrt{2\pi}}\exp\left[-\frac{(t-t_\mathrm{p})^2}{2\sigma^2}+\alpha(z-z_\mathrm{surf})\right], \tag{A1}$$

where $\Phi(x,y)$ is the fluence distribution of the interference pattern (see Equation (2)) and where a single laser pulse of Gaussian temporal shape with the standard deviation $\sigma = \tau_\mathrm{p}/(2\sqrt{2\ln 2})$ and the pulse duration τ_p (FWHM) at the pulse time t_p is considered along with the Beer–Lambert absorption law. Furthermore, in Equation (A1), α denotes the absorption coefficient of the substrate and R is its reflectivity, both evaluated at the laser wavelength λ. In addition, the mixed enthalpy-temperature formulation is employed in the heat transfer Equation (A3) below to account for the latent heat of involved phase changes. For this purpose, the specific enthalpy comprises sensible and latent amounts:

$$h = h_\mathrm{sens} + h_\mathrm{lat} = c_\mathrm{p}(T-T_0) + f_\mathrm{m}L_\mathrm{f} + f_\mathrm{v}L_\mathrm{v}, \tag{A2}$$

where f_m and f_v represent the molten and vaporised mass fractions and the specific latent heats of fusion and vapourisation are denoted by L_f and L_v, respectively.

As soon as the material is molten due to the absorption of the incident radiation, the melt pool convection starts, which is particularly governed by the conservation of mass and momentum. The liquid metal is conceived as an incompressible fluid with constant density and viscosity, except for the density in the body force term of the momentum Equation (A5) within the scope of the Oberbeck–Boussinesq approximation. Furthermore, as the conservation equations are solved by a Lagrangian particle method, an equation describing the evolution of particle positions is required. Accordingly, the governing equations are considered in the following form:

$$\rho_0 \frac{dh}{dt} = \kappa \Delta T + \dot{q}''', \tag{A3}$$

$$\nabla \cdot v = 0, \tag{A4}$$

$$\rho_0 \frac{dv}{dt} = -\nabla p_{\mathrm{dyn}} + \eta \Delta v - \beta (T - T_1)\rho_0 g, \tag{A5}$$

$$\frac{dx}{dt} = v. \tag{A6}$$

In Equations (A3), (A5), and (A6), it should be noted that the left-hand side is written in terms of the substantial derivative $\frac{d}{dt} = \frac{\partial}{\partial t} + v \cdot \nabla$.

To solve the above equations, suitable boundary conditions have to be prescribed. Given the short time scale of the nanosecond laser interference patterning process, heat losses to the environment are neglected and adiabatic boundary conditions are employed:

$$-\kappa \frac{\partial T}{\partial n} = 0, \quad \frac{\partial h}{\partial n} = 0. \tag{A7}$$

Concerning the liquid metal flow, the dynamic pressure is subject to homogeneous Neumann boundary conditions and the no-slip condition is applied at the bottom of the melt pool:

$$\frac{\partial p_{\mathrm{dyn}}}{\partial n} = 0, \quad v = 0. \tag{A8}$$

Apart from that, an inhomogeneous Neumann condition, the Marangoni boundary condition, on the horizontal velocity [48,58] and a vanishing vertical velocity are imposed at the free surface:

$$\eta \frac{\partial v_x}{\partial z} = \frac{\partial \gamma}{\partial x} = \frac{d\gamma}{dT}\frac{\partial T}{\partial x}, \quad v_z = 0. \tag{A9}$$

References

1. Ball, P. Shark skin and other solutions. *Nature* **1999**, *400*, 507–509. [CrossRef]
2. Dusser, B.; Sagan, Z.; Soder, H.; Faure, N.; Colombier, J.P.; Jourlin, M.; Audouard, E. Controlled nanostructures formation by ultra fast laser pulses for color marking. *Opt. Express* **2010**, *18*, 2913–2924. [CrossRef] [PubMed]
3. Lee, S.G.; Lim, H.S.; Lee, D.Y.; Kwak, D.; Cho, K. Tunable anisotropic wettability of rice leaf-like wavy surfaces. *Adv. Funct. Mater.* **2013**, *23*, 547–553. [CrossRef]
4. Ahsan, M.S.; Ahmed, F.; Kim, Y.G.; Lee, M.S.; Jun, M.B.G. Colorizing stainless steel surface by femtosecond laser induced micro/nano-structures. *Appl. Surf. Sci.* **2011**, *257*, 7771–7777. [CrossRef]
5. Verma, L.K.; Sakhuja, M.; Son, J.; Danner, A.J.; Yang, H.; Zeng, H.C.; Bhatia, C.S. Self-cleaning and antireflective packaging glass for solar modules. *Renew. Energy* **2011**, *36*, 2489–2493. [CrossRef]
6. Madou, M.J. *Manufacturing Techniques for Microfabrication and Nanotechnology*, 3rd ed.; Fundamentals of microfabrication and nanotechnology; CRC Press: Boca Raton, FL, USA, 2011; Volume II.
7. Matsui, S. Three-dimensional nanostructure fabrication by focused ion beam chemical vapor deposition. In *Springer Handbook of Nanotechnology*, 2nd ed.; Bhushan, B., Ed.; Springer: Berlin/Heidelberg, Germany, 2007; Chapter 6, pp. 179–195. [CrossRef]
8. Etsion, I. State of the Art in Laser Surface Texturing. *J. Tribol.* **2005**, *127*, 248–253. [CrossRef]
9. Bonse, J.; Kirner, S.V.; Griepentrog, M.; Spaltmann, D.; Krüger, J. Femtosecond Laser Texturing of Surfaces for Tribological Applications. *Materials* **2018**, *11*, 801. [CrossRef]
10. Tripathy, A.; Sen, P.; Su, B.; Briscoe, W.H. Natural and bioinspired nanostructured bactericidal surfaces. *Adv. Colloid Interface Sci.* **2017**, *248*, 85–104. [CrossRef]
11. Alamri, S.; Aguilar-Morales, A.I.; Lasagni, A.F. Controlling the wettability of polycarbonate substrates by producing hierarchical structures using direct laser interference patterning. *Eur. Polym. J.* **2018**, *99*, 27–37. [CrossRef]
12. Roch, T.; Benke, D.; Milles, S.; Roch, A.; Kunze, T.; Lasagni, A. Dependence between friction of laser interference patterned carbon and the thin film morphology. *Diamond Relat. Mater.* **2015**, *55*, 16–21. [CrossRef]
13. Baharin, A.F.S.; Ghazali, M.J.; Wahab, J.A. Laser surface texturing and its contribution to friction and wear reduction: A brief review. *Ind. Lubr. Tribol.* **2016**, *68*, 57–66. [CrossRef]

14. Bremus-Koebberling, E.A.; Beckemper, S.; Koch, B.; Gillner, A. Nano structures via laser interference patterning for guided cell growth of neuronal cells. *J. Laser Appl.* **2012**, *24*, 042013. [CrossRef]
15. Langheinrich, D.; Yslas, E.; Broglia, M.; Rivarola, V.; Acevedo, D.; Lasagni, A. Control of cell growth direction by direct fabrication of periodic micro- and submicrometer arrays on polymers. *J. Polym. Sci. Part B Polym. Phys.* **2012**, *50*, 415–422. [CrossRef]
16. Rößler, F.; Kunze, T.; Lasagni, A.F. Fabrication of diffraction based security elements using direct laser interference patterning. *Opt. Express* **2017**, *25*, 22959–22970. [CrossRef]
17. Bieda, M.; Schmädicke, C.; Roch, T.; Lasagni, A. Ultra-low friction on 100Cr6-steel surfaces after direct laser interference patterning. *Adv. Eng. Mater.* **2015**, *17*, 102–108. [CrossRef]
18. Lasagni, A.F.; Beyer, E. Fabrication of periodic submicrometer and micrometer arrays using laser interference-based methods. In *Laser Surface Engineering: Processes and Applications*, 1st ed.; Lawrence, J., Waugh, D.G., Eds.; Woodhead Publishing Series in Electronic and Optical Materials; Woodhead Publishing: Cambridge, UK, 2015; Chapter 17, Volume 65, pp. 423–439. [CrossRef]
19. Bieda, M.; Siebold, M.; Lasagni, A.F. Fabrication of sub-micron surface structures on copper, stainless steel and titanium using picosecond laser interference patterning. *Appl. Surf. Sci.* **2016**, *387*, 175–182. [CrossRef]
20. Nebel, C.; Dahlheimer, B.; Schöniger, S.; Stutzmann, M. Sub-micron silicon structures for thin film solar cells. *Phys. Status Solidi B* **1996**, *194*, 55–67. [CrossRef]
21. Lasagni, A.; Holzapfel, C.; Mücklich, F. Periodic pattern formation of intermetallic phases with long range order by laser interference metallurgy. *Adv. Eng. Mater.* **2005**, *7*, 487–492. [CrossRef]
22. Chichkov, B.N.; Momma, C.; Nolte, S.; von Alvensleben, F.; Tünnermann, A. Femtosecond, picosecond and nanosecond laser ablation of solids. *Appl. Phys. A: Mater. Sci. Process.* **1996**, *63*, 109–115. [CrossRef]
23. Semerok, A.; Chaléard, C.; Detalle, V.; Lacour, J.L.; Mauchien, P.; Meynadier, P.; Nouvellon, C.; Sallé, B.; Palianov, P.; Perdrix, M.; et al. Experimental investigations of laser ablation efficiency of pure metals with femto, pico and nanosecond pulses. *Appl. Surf. Sci.* **1999**, *138–139*, 311–314. [CrossRef]
24. D'Alessandria, M.; Lasagni, A.; Mücklich, F. Direct micropatterning of aluminum substrates via laser interference metallurgy. *Appl. Surf. Sci.* **2008**, *255*, 3210–3216. [CrossRef]
25. Rosenthal, D. Mathematical theory of heat distribution during welding and cutting. *Weld. J.* **1941**, *20*, 220–234.
26. Ishizaki, K.; Murai, K.; Kanbe, Y. *Penetration in Arc Welding and Convection in Molten Metal*; Technical Report, Study Group 212, Document 212-77-66; International Institute of Welding: Genoa, Italy, 1966.
27. Kou, S.; Wang, Y.H. Weld pool convection and its effect. *Weld. J. Res. Suppl.* **1986**, *65*, 63–70.
28. Mundra, K.; Debroy, T.; Zacharia, T.; David, S.A. Role of thermophysical properties in weld pool modeling. *Weld. J. Res. Suppl.* **1992**, *71*, 313–320.
29. Heiple, C.R.; Roper, J.R. Mechanism for minor element effect on GTA fusion zone geometry. *Weld. J. Res. Suppl.* **1982**, *61*, 97–102.
30. Atthey, D.R. A mathematical model for fluid flow in a weld pool at high currents. *J. Fluid Mech.* **1980**, *98*, 787–801. [CrossRef]
31. Aidun, D.K.; Martin, S.A. Effect of sulfur and oxygen on weld penetration of high-purity austenitic stainless steels. *J. Mater. Eng. Perform.* **1997**, *6*, 496–502. [CrossRef]
32. Mills, K.C.; Keene, B.J.; Brooks, R.F.; Shirali, A. Marangoni effects in welding. *Phil. Trans. R. Soc. Lond. A* **1998**, *356*, 911–925. [CrossRef]
33. Oreper, G.M.; Eagar, T.W.; Szekely, J. Convection in arc weld pools. *Weld. J. Res. Suppl.* **1983**, *62*, 307–312.
34. Oreper, G.M.; Szekely, J. Heat- and fluid-flow phenomena in weld pools. *J. Fluid Mech.* **1984**, *147*, 53–79. [CrossRef]
35. Chan, C.; Mazumder, J.; Chen, M.M. A two-dimensional transient model for convection in laser melted pool. *Metall. Trans. A* **1984**, *15*, 2175–2184. [CrossRef]
36. Kou, S.; Sun, D.K. Fluid flow and weld penetration in stationary arc welds. *Metall. Trans. A* **1985**, *16*, 203–213. [CrossRef]
37. Kou, S.; Wang, Y.H. Three-dimensional convection in laser melted pools. *Metall. Trans. A* **1986**, *17*, 2265–2270. [CrossRef]
38. Zacharia, T.; David, S.A.; Vitek, J.M.; Debroy, T. Weld pool development during GTA and laser beam welding of type 304 stainless steel, Part I—Theoretical analysis and Part II—Experimental correlation. *Weld. J. Res. Suppl.* **1989**, *68*, 499–509; 510–519.
39. Zacharia, T.; David, S.A.; Vitek, J.M.; Kraus, H.G. Computational modeling of stationary gas-tungsten-arc weld pools and comparison to stainless steel 304 experimental results. *Metall. Trans. B* **1991**, *22*, 243–257. [CrossRef]
40. Choo, R.T.C.; Szekely, J. Vaporization kinetics and surface temperature in a mutually coupled spot gas tungsten arc weld and weld pool. *Weld. J. Res. Suppl.* **1992**, *71*, 77–93.
41. Choo, R.T.C.; Szekely, J.; David, S.A. On the calculation of the free surface temperature of gas-tungsten-arc weld pools from first principles: Part II. Modeling the weld pool and comparison with experiments. *Metall. Trans. B* **1992**, *23*, 371–384. [CrossRef]
42. Lei, Y.; Shi, Y.; Murakawa, H.; Ueda, Y. Numerical analysis of the effect of sulfur content upon fluid flow and weld pool geometry for type 304 stainless steel. *Trans. JWRI* **1997**, *26*, 1–8.
43. Bäuerle, D. *Laser Processing and Chemistry*, 4th ed.; Springer: Berlin/Heidelberg, Germany, 2011. [CrossRef]
44. Demuth, C.; Lasagni, A.F. An incompressible smoothed particle hydrodynamics (ISPH) model of direct laser interference patterning. *Computation* **2020**, *8*, 9. [CrossRef]
45. Sahoo, P.; Debroy, T.; McNallan, M.J. Surface tension of binary metal—surface active solute systems under conditions relevant to welding metallurgy. *Metall. Trans. B* **1988**, *19*, 483–491. [CrossRef]
46. McNallan, M.J.; Debroy, T. Effect of temperature and composition on surface tension in Fe-Ni-Cr alloys containing sulfur. *Metall. Trans. B* **1991**, *22*, 557–560. [CrossRef]

47. Wang, Y.; Tsai, H.L. Effects of surface active elements on weld pool fluid flow and weld penetration in gas metal arc welding. *Metall. Mater. Trans. B* **2001**, *32*, 501–515. [CrossRef]

48. Nepomnyashchy, A.; Simanovskii, I.; Legros, J.C. *Interfacial Convection in Multilayer Systems*; Springer Monographs in Mathematics; Springer: New York, NY, USA, 2006.

49. Demuth, C.; Bieda, M.; Lasagni, A.F.; Mahrle, A.; Wetzig, A.; Beyer, E. Thermal simulation of pulsed direct laser interference patterning of metallic substrates using the smoothed particle hydrodynamics approach. *J. Mater. Process. Technol.* **2012**, *212*, 689–699. [CrossRef]

50. Mills, K.C. *Recommended Values of Thermophysical Properties for Selected Commercial Alloys*; Woodhead Publishing: Cambridge, UK, 2002.

51. Steen, W.M.; Mazumder, J. *Laser Material Processing*, 4th ed.; Springer: London, UK, 2010.

52. Valencia, J.J.; Quested, P.N. Thermophysical properties. In *Casting, ASM Handbook*; ASM International: Materials Park, OH, USA, 2008; Volume 15, pp. 468–481.

53. Kaptay, G. A unified model for the cohesive enthalpy, critical temperature, surface tension and volumetric thermal expansion coefficient of liquid metals of bcc, fcc and hcp crystals. *Mater. Sci. Eng. A* **2008**, *495*, 19–26. [CrossRef]

54. Nasch, P.M.; Steinemann, S.G. Density and thermal expansion of molten manganese, iron, nickel, copper, aluminum and tin by means of the gamma-ray attenuation technique. *Phys. Chem. Liq.* **1995**, *29*, 43–58. [CrossRef]

55. Lide, D.R. (Ed.) *Handbook of Chemistry and Physics*, 85th ed.; CRC Press: Boca Raton, FL, USA, 2004.

56. Semak, V.; Matsunawa, A. The role of recoil pressure in energy balance during laser materials processing. *J. Phys. D Appl. Phys.* **1997**, *30*, 2541–2552. [CrossRef]

57. Abbaschian, R.; Abbaschian, L.; Reed-Hill, R.E. *Physical Metallurgy Principles*, 4th ed.; The Hardening of Steel, Section 19.11 Dimensional Changes Associated with Transformation of Martensite; Cengage Learning: Stamford, CT, USA, 2010; Chapter 19, p. 631.

58. Landau, L.D.; Lifshitz, E.M. *Fluid Mechanics*; Course of Theoretical Physics; Pergamon Press: London, UK, 1959; Volume 6.

nanomaterials

MDPI

Article

The Tuning of LIPSS Wettability during Laser Machining and through Post-Processing

Michael J. Wood, Phillip Servio and Anne-Marie Kietzig *

Department of Chemical Engineering, McGill University, Montréal, QC H3A 0C5, Canada; michael.wood3@mail.mcgill.ca (M.J.W.); phillip.servio@mcgill.ca (P.S.)
* Correspondence: anne.kietzig@mcgill.ca

Abstract: In this work, we investigate the fabrication of stainless-steel substrates decorated with laser-induced periodic surface structures (LIPSS) of both hydrophilic and hydrophobic wettability through different post-processing manipulation. In carrying out these experiments, we have found that while a CO_2-rich atmosphere during irradiation does not affect final wettability, residence in such an atmosphere after irradiation does indeed increase hydrophobicity. Contrarily, residence in a boiling water bath will instead lead to a hydrophilic surface. Further, our experiments show the importance of removing non-sintered nanoparticles and agglomerates after laser micromachining. If they are not removed, we demonstrate that the nanoparticle agglomerates themselves become hydrophobic, creating a Cassie air-trapping layer on the surface which presents with water contact angles of 180°. However, such a surface lacks robustness; the particles are removed with the contacting water. What is left behind are LIPSS which are integral to the surface and have largely been blocked from reacting with the surrounding atmosphere. The actual surface presents with a water contact angle of approximately 80°. Finally, we show that chemical reactions on these metallic surfaces decorated with only LIPSS are comparatively slower than the reactions on metals irradiated to have hierarchical roughness. This is shown to be an important consideration to achieve the highest degree of hydro-philicity/phobicity possible. For example, repeated contact with water from goniometric measurements over the first 30 days following laser micromachining is shown to reduce the ultimate wettability of the surface to approximately 65°, compared to 135° when the surface is left undisturbed for 30 days.

Keywords: fs-laser; micromachining; wetting; hydrophobicity; LIPSS

Citation: Wood, M.J.; Servio, P.; Kietzig, A.-M. The Tuning of LIPSS Wettability during Laser Machining and through Post-Processing. *Nanomaterials* **2021**, *11*, 973. https://doi.org/10.3390/nano11040973

Academic Editor: Peter Simon, Jürgen Ihlemann, Jörn Bonse

Received: 12 March 2021
Accepted: 8 April 2021
Published: 10 April 2021

Publisher's Note: MDPI stays neutral with regard to jurisdictional claims in published maps and institutional affiliations.

1. Introduction

The irradiation of solids with linearly-polarized laser radiation has long been known to result in the formation of laser-induced periodic surface structures (LIPSS) [1]. Because of the contactless, maskless, single-step nature of laser irradiation—coupled with advances in high repetition rate, low pulse duration laser technology—the processing of LIPSS has become a scalable method of surface functionalization [2]. Specifically, the wettability of a surface can be controlled by combining the regular nanoscale roughness imparted through laser irradiation with favourable surface chemistry. A highly hydrophilic surface, for example, can be fabricated through the creation of an increased hydrophilic surface area—leading to the Wenzel wetting state [3]. Contrarily, a highly hydrophobic surface can be fabricated through the creation of trapped air pockets—leading to the Cassie wetting state [4].

The typical method of rendering laser-irradiated solids highly hydrophobic had been to deposit a low-energy coating onto the micromachined surface [5]. However, Kietzig et al. (2009) reported the creation of superhydrophobic surfaces from smooth, hydrophilic, steel alloys some time after laser irradiation without the need for such chemical coatings. They noted that the water contact angle measured on their micromachined samples evolved from hydrophilic to highly hydrophobic over the first 2–3 weeks following exposure to the femtosecond laser beam, with the samples exposed simply to lab air. It was noted via surface chemistry analyses that the increase in hydrophobicity correlated with an increase in the relative amount of carbon—with the surface being solely made up of iron oxides and carbon after some time [6]. Kietzig et al. (2011) performed a follow-up study in which they showed that pure metals exhibit this same hydrophilic-to-highly hydrophobic wetting evolution upon laser irradiation, similarly following an increase in surface carbon content [7]. Indeed, in the years since these results were published, there have been a considerable number of studies carried out which demonstrate similar wetting evolutions for laser micromachined metallic surfaces, generating much interest around the ability to create superhydrophobic metals in a single processing step [8–14].

A specific mechanism or reaction scheme for the increase in surface carbon content after irradiating metals with ultrashort laser pulses remains unknown. However, it is clear that a sort of cleaning of the metallic surface occurs during the ablation process whereas the outermost oxidized layers are removed, exposing a hydrophilic reactive surface on which a new chemical layer can form [15]. Kietzig et al. (2009) presented their original hypothesis that the progressive dissociative adsorption of CO_2 onto the reactive surface was the cause of the increase in non-polar carbon—and increased hydrophobicity. They tested this hypothesis by exposing a freshly irradiated surface to a CO_2-rich environment. Indeed, this exposure led to a surface with a higher carbon content and a higher water contact angle than a surface exposed to lab air [6]. This technique of exposing a micromachined metal to a CO_2-rich environment has since been reported as an experimental method to fabricate superhydrophobic surfaces [16,17]. On the contrary, Long et al. (2015) were the first to report that exposure of a laser-irradiated metal to pure CO_2 actually led to a restraint in the hydrophilic-to-highly hydrophobic wetting transition. Rather, they report that residence in a vacuum atmosphere accelerated the transition due to adsorption of organic matter, originating from the mechanical vacuum pump, which had contaminated the chamber. Their theory of the progressive adsorption of hydrocarbons being the origin of the non-polar carbon surface chemistry was tested by exposing a laser-irradiated sample to an atmosphere rich in organic matter, which again led to accelerated transitions [10]. This contrary theory that airborne hydrocarbons are the source of the non-polar carbon which progressively adsorbs onto laser-irradiated metals—rendering them highly hydrophobic—has been the basis of discussion in many subsequent studies [12,13,15,18,19].

The creation of a stable highly hydrophilic metallic surface upon laser micromachining had also been reported by Kietzig et al. (2009). Instead of exposing the freshly irradiated surface to lab air or CO_2, it was immersed in boiling water for two days. This residence in the water bath led to a surface chemistry rich in hydrophilic oxides, hypothesized to be formed through dissociative adsorption of the water molecules. This surface did not undergo the typical hydrophilic-to-highly hydrophobic wetting transition after exposure to lab air [6].

The manipulation of the micromachined substrate in the early times after irradiation is evidently paramount to the final wettability obtained. This has led to interest into the effect of the atmosphere during the irradiation process itself on final wettability of the surface. Pou et al. (2019) performed a study in which they streamed: air, Ar, N_2, O_2, and CO_2 onto the substrate during nanosecond laser irradiation. After irradiation, the samples were individually sealed in an air-filled bag for a week before analytical tests were performed. What they report is that streaming N_2, O_2, CO_2, or even simply streaming air during the irradiation process ultimately results in a hydrophilic surface rich in metal lattice oxides and hydroxides [20]. The origins of the air used in these experiments is not explicitly stated, thus it is difficult to determine if these results fit with the theory that airborne hydrocarbons are the source of the non-polar carbon which progressively adsorbs onto laser-irradiated metals. Further, the negative effect that the CO_2 stream has on ultimate wettability of the surface contradicts what has been reported by those already mentioned.

Pou et al. (2019) make a point of mentioning in their experimental section that after their texturing process, the samples were rinsed with ethanol to remove any debris that could affect subsequent measurements [20]. This is an important procedural point considering that the waste generated during laser micromachining are reactive nanoparticles. Analytical techniques such as water contact angle or surface chemistry measurements could be greatly influenced by the presence of non-sintered nanoparticles or agglomerates. Nonetheless, it is often unclear how these nanoparticles are dealt with in published studies—or if they are considered at all [7,8,11–15].

The present work takes another look at the effect of CO_2 exposure on the final wettability of laser micromachined metals. Specifically, we explore the effect of this gas on the wettability of LIPSS-decorated stainless-steel in an attempt to rigorously describe what we have empirically known—that CO_2 exposure does in fact promote hydrophobicity of our laser micromachined metals. We also take this opportunity to discuss otherwise how the manipulation of LIPSS-decorated stainless-steel at early times following irradiation greatly affects the final wettability of the surface. In doing so, we note the effect of non-sintered nanoparticles generated during micromachining and the effect of disturbance of the nascent surface chemistry on the ultimate wettability. These are all important considerations as metal laser micromachining technologies get scaled and transferred to industrial applications.

2. Materials and Methods

2.1. Fs-Laser Micromachining

The substrate used in this study was a 0.036″-thick 316 stainless-steel sheet (McMaster-Carr). The manufacturer has supplied us with an ASTM A751/14a chemical analysis of the substrate, as presented in Table 1. This material came pre-polished with a mirror-like ASTM #8 finish. This sheet was cut into small coupons, and was then ultrasonically cleaned in acetone for 5 min to remove any oils from the surface prior to micromachining.

Table 1. ASTM A751/14a chemical analysis of 316 stainless-steel substrate.

C%	Cr%	Cu%	Mn%	Mo%	N%	Ni%	P%	S%	Si%
0.0205	16.5820	0.4420	1.1830	2.0205	0.0465	10.0700	0.0310	0.0010	0.2750

The cleaned surfaces were in turn decorated with laser-induced periodic surface structures (LIPSS) by exposure to ultrashort laser pulses. The laser system used was a *Libra* amplified Ti:sapphire laser (Coherent, Inc., Santa Clara, CA, USA) with a central wavelength of 800 nm, output power of 4 W, pulse duration of <100 fs, and a 1 kHz repetition rate. A variable attenuator comprised of a half-wave plate and a polarizing beam splitter was used to reduce the laser output power. A plano-convex lens with a focal length of 200 mm was used to focus the laser beam. The pristine substrates were mounted horizontally onto XY linear translation stages, with their top surface being 3 mm closer to the focusing lens than the focal point, as shown in Figure 1a. The XY-stages were actuated by an *XPS* universal high-performance motion/driver controller (Newport Corp., Irvine, CA, USA). Successive pulses were overlapped in lines irradiated at 10 mm/s. Using the methods of Liu [21], the ablation threshold fluence of the 316 stainless-steel at these conditions was measured to be 0.23 J·cm^{-2}, matching well what is reported in the literature [22]. Further, the effective beam diameter at these conditions is measured to be 80 μm. Through trial-and-error, optimal conditions for the formation of LIPSS were determined to be at a peak intensity of 696 W·cm^{-2}, pulse fluence of 0.70 J·cm^{-2}, and $N_{eff,2D}$ of 200 pulses-per-spot. In total, a 10×10 mm square patch was irradiated on each stainless-steel coupon on which to perform the subsequent goniometric measurements.

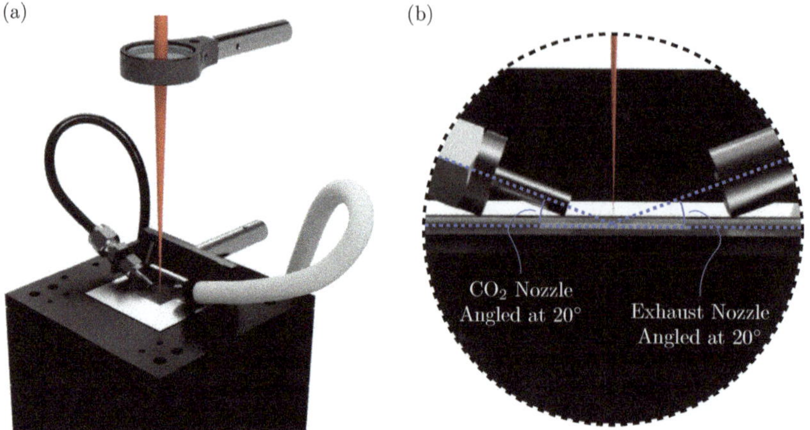

Figure 1. (**a**) Rendering of the fs-laser micromachining setup. Stainless-steel coupons are mounted horizontally on XY translation stages, 3 mm closer to the laser focusing lens than the focal point. A stationary nozzle is used to introduce CO_2 gas at the point of laser irradiation. (**b**) Magnified rendering of the CO_2 nozzle and exhaust line in relation to the stainless-steel substrate and laser beam spot.

A gas nozzle with an inner diameter of 2 mm was mounted stationary with the laser beam at an angle of 20° to the stainless-steel substrate, as shown in Figure 1b. This nozzle allowed flowing of carbon dioxide directly on the laser spot during irradiation, and thus a CO_2-rich environment to be formed over the irradiated patch. A needle valve and calibrated in-line rotameter allowed for the carbon dioxide volumetric flow rate to be set as an experimental parameter (from 0 to 10,000 mL·min^{-1}). An exhaust line was mounted opposite the CO_2 nozzle as a safety measure to collect nanoparticles blown away by the gas stream.

2.2. Post-Processing Steps

The samples in the present work were prepared in three ways: (1) no further processing after laser micromachining—referred to hereafter as *no-post*; (2) residence in a CO_2-rich reactor after micromachining—referred to hereafter as *CO_2-reactor*; and (3) residence in boiling water after micromachining—referred to hereafter as *boiling-water*.

CO_2-reactor samples were immediately removed from the XY translation stages after the micromachining routine had finished. They were placed in a small reactor with an ambient temperature of 60 °C, held constant via a PID-controlled heating element. Carbon dioxide was introduced to the reactor with its release valve initially open to flush the volume of air. The release valve was closed, the CO_2 pressure inside the reactor was increased to 20 PSI, and the reactor was sealed. These samples had a residence time of 48 h inside the reactor before further tests were performed.

Boiling-water samples were immediately removed from the XY translation stages after the micromachining routine had finished. They were placed in a beaker of boiling deionized water which was continually topped off to maintain an adequate level of water. These samples had a residence time of 48 h in the boiling water before further tests were performed.

2.3. Sample Characterization

Carefully timed dynamic water contact angle measurements were performed on each of the micromachined stainless-steel coupons. These measurements were taken at 0-, 10-, 20-, and 30-days after micromachining was performed. The 0-day measurements were started within 10 min of the experimental surface being created (after finishing micromachining, or after finishing the post-processing step, as applicable). These water contact angle measurements were performed using a lab-built goniometer comprised of an *Infinity 3* microscope camera (Teledyne Lumenera, Inc., Ottawa, ON, Canada.) equipped with a *VZM 200i* zoom imaging lens (Edmund Optics, Inc., Barrington, NJ, USA). High backlighting is provided by an ultra-bright LED spotlight (Optikon Corp, Kitchener, ON, Canada). A 70-2203 syringe pump module (Harvard Apparatus, Inc., Holliston, MA, USA), controlled by a custom *Arduino* code, pumps degassed reverse osmosis water onto the studied surface. $\theta_{sessile}$ was measured by gently placing a 5 µL droplet onto the studied surface. $\theta_{advancing}$ and $\theta_{receding}$ were measured by first gently placing a 2 µL droplet onto the studied surface, then pumping 5 µL to and from said droplet, respectively, at a rate of 0.25 µL/s.

The goniometric measurements on each stainless-steel coupon were carried out as follows. First, $\theta_{sessile}$ was measured on five different locations within the laser-irradiated patch. In the case of CO_2-*reactor* samples, a second round of $\theta_{sessile}$ measurements was performed on the same five locations within the patch, after allowing the surface to dry. Next, $\theta_{advancing}$ and $\theta_{receding}$ were measured on the same five locations within the patch, after allowing the surface to dry. The error bars in the data presented herein represent the bounds of the 95% confidence intervals on the mean of these goniometric measurements.

After the 30-day goniometric measurements were completed all of the substrates described in the previous sections were imaged using a *Quanta 450* scanning electron microscope (ThermoFisher, Inc., Waltham, MA, USA). An accelerating voltage of 5.0 kV and spot size of 2.5 nm were used to image the stainless-steel samples.

3. Results and Discussion

3.1. Effect of CO_2 Stream Flow Rate

We begin the discussion with an overview presentation of the bulk of the data collected. Figure 2a–c presents the average sessile water contact angles measured over the 30-day period on the laser micromachined surfaces prepared with increasing flow rate of CO_2 during irradiation, and the three separate post-processing steps. We note that the dynamic water contact angles are typically better suited to describe the wettability of a surface, and indeed we include this data as Supplementary Materials, however the nature of our experimental method leads us to centre our discussion around both datasets, as explained later.

The data in Figure 2a–c show that the flow rate of CO_2 gas during fs-laser irradiation has no appreciable effect on the wettability of the resulting surface. There is no statistically-significant difference in contact angles measured on those surfaces prepared with no CO_2 stream, and those surfaces prepared with increasing volumetric flow rates of CO_2. This lack of effect persists over the 30 days of goniometric measurements following surface preparation. It is important to highlight that in this data we do not see a negative effect on wettability of the surface from streaming CO_2 during irradiation, as was reported by Pou et al. [20]. A possible explanation for these results is that the rate of the reaction which re-forms the surface chemistry on the laser irradiated stainless-steel is far too slow to be affected by the CO_2 surrounding the micromachining step itself. Very little is known about the kinetics of this chemical reaction, yet it is often reported that water contact angles continue to increase (and hence the surface chemistry continues to change) for around 30 days following laser irradiation [6,15].

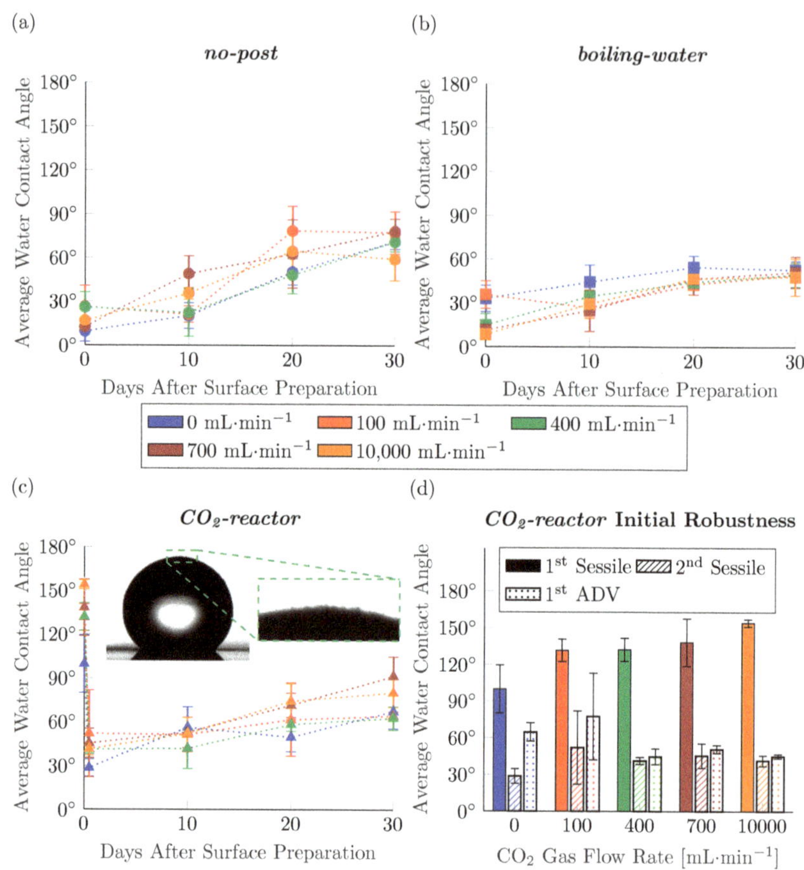

Figure 2. Average sessile water contact angle measurements taken over the 30 days following laser micromachining/post-processing on surfaces prepared with: (**a**) no post-processing step; (**b**) 48 h in boiling water; and (**c**) 48 h in the CO_2 reactor. Inset shows nanoparticle agglomerates on periphery of first sessile drop placed on *CO₂-reactor* surfaces. (**d**) Comparison of the average: contact angle of the first sessile drop, contact angle of the second sessile drop, and advancing contact angle at day 0 on *CO₂-reactor* surfaces.

An interesting result that was evident during experimentation was the exceptionally high sessile contact angles presented by the *CO₂-reactor* surfaces immediately following their 48-hour residences (Figure 2c). Contrary to the rest of the data collected in the present work, the flow rate of CO_2 during laser irradiation seemed to have a positive effect on the resulting contact angles. However, the first sessile droplets placed on these surfaces clearly had agglomerated nanoparticles at their periphery, as exemplified by the inset images of Figure 2c. Evidently, nanoparticles generated during laser micromachining agglomerate and remain, non-sintered, atop the laser-induced periodic surface surface structures. These agglomerates are hydrophobic following the residence in the CO_2 reactor and lead to the air-trapping characteristic of the Cassie wetting state. The coverage of the LIPSS-decorated surfaces with nanoparticles and nanoparticle agglomerates is an important consideration both in the research context, and as laser-irradiated surface technology starts being transferred to industry. Observations such as the inset image of Figure 2c bring into question whence the positive characteristics of a laser-irradiated surface arrive. For in effect, the first sessile droplet placed on the *CO₂-reactor* surfaces is measuring the hydrophobicity of the compound LIPSS–nanoparticle agglomerate system. As the surface is covered with more and more nanoparticle agglomerates, a sessile drop coming into contact with it will resemble more and more a Janus particle. Surfaces such as these inherently lack robustness, as rolling droplets of water carry away the nanoparticle agglomerates, leaving behind the underlying substrate.

To illustrate the lack of anti-wetting robustness, we placed a second sessile droplet at approximately the same locations on the *CO_2-reactor* surfaces at Day 0, allowing the surface sufficient time to dry after the first sessile droplet. Figure 2d shows that while these surfaces present as hydrophobic (even highly-hydrophobic with increased CO_2 flow rate during micromachining) with the first sessile droplet placed on the surface, they present as hydrophilic with the second sessile droplet. Further, the increased volumetric flow rate of the CO_2 jet has not lead to a statistically-significant increase in the average contact angles of the second sessile droplet. This is further evidence that the rate of the reaction which re-forms the surface chemistry of the LIPSS-decorated stainless-steel is too slow to be affected by the composition of the atmosphere surrounding the micromachining process itself. Figure 2d also presents the average advancing contact angle measured at approximately the same locations as the first and second sessile droplets at Day 0. The data from these measurements are interesting in that they describe surfaces which are hydrophilic, however, in some cases the advancing contact angle is especially higher than the sessile contact angle previously measured and in some cases they are statistically-equivalent. The variability between the average contact angles measured by the second sessile droplets and first advancing droplets to touch the *CO_2-reactor* surfaces likely stems from the exact location where these droplets were placed. That is, the wettability of the system depends greatly on the ratio of hydrophobic nanoparticle agglomerates to hydrophilic LIPSS.

To explore the non-sintered nanoparticle agglomerate phenomenon further, we micromachined two more LIPSS-decorated surfaces: one fabricated without CO_2 streaming during micromachining and one fabricated with a 10,000 mL·min^{-1} stream of CO_2. After micromachining, these samples were sputter-coated with 5 nm of platinum to afix any nanoparticles in place for scanning electron microscopy. The resulting scanning electron micrographs are presented in Figure 3.

The images in Figure 3 show clearly that nanoparticle agglomerates are formed and redeposited homogeneously on the surface during the fs-laser micromachining process, even at the low-fluence conditions necessary for the formation of only LIPSS. Micromachining with increased laser fluence—such as for the formation of bumpy self-assembled structures or inscribed geometric patterns—will only exacerbate this issue of nanoparticle agglomerate redeposition. The inset photographs in the top row of Figure 3 show the macroscopic results of streaming CO_2 during irradiation. With no CO_2 stream during micromachining, much of the perimeter surrounding the irradiated patch tends to become covered in a layer of non-sintered nanoparticles/agglomerates, giving the pristine stainless steel a golden brown hue. However, the streaming of CO_2 gas during micromachining has confined the redeposition of nanoparticles and agglomerates to only downstream from the nozzle. Comparing the left and right columns of Figure 3, it is obvious that the jet of CO_2 gas serves to increase the degree to which the underlying structured surface is covered with large nanoparticle agglomerates. As the flow rate of gas is increased, the expanding nanoparticle plume is confined to a greater degree and nanoparticles ejected from the substrate tend to agglomerate. These nanoparticle agglomerates are redirected back down to the surface where they homogeneously blanket the LIPSS. We theorize that the angle between the CO_2 nozzle and the surface will also affect the degree of nanoparticle plume confinement. Whereas if the jet of CO_2 is directed to the substrate from a higher incident angle, we would expect to see more coverage of the laser-irradiated patch with nanoparticle agglomerates, and therefore a positive effect on the initial hydrophobicity of the sample. The relationship between CO_2 stream angle of incidence and initial surface hydrophobicity could be the topic of interesting future research.

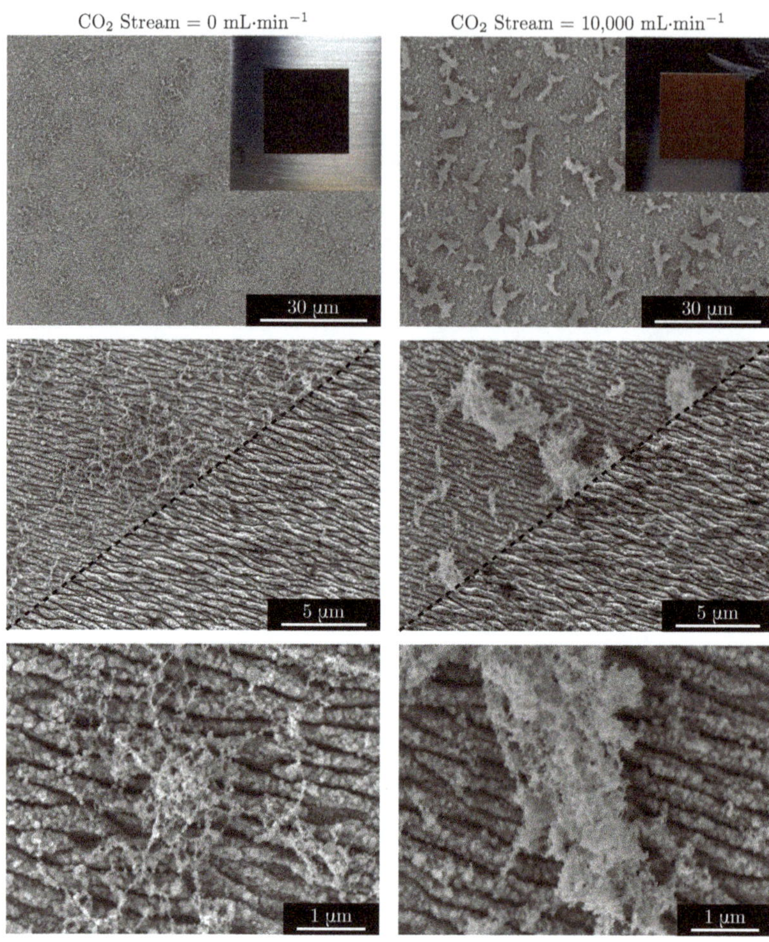

Figure 3. Scanning electron micrographs of the surfaces produced by fs-laser micromachining with (right column) and without (left column) a CO_2 stream. Inset photographs of the top row demonstrate the CO_2 jet serves to confine the area of nanoparticle (agglomerate) redeposition to downstream from the nozzle. Middle row shows a comparison between the surfaces after micromachining (upper left) and after ultrasonication (lower right).

The higher magnification micrographs of Figure 3 show that with increased CO_2 flow rate, the nanoparticle agglomerates transition from spider web–like formations to large cloud–like flakes. That is, further confinement of the nanoparticle plume with higher stream flow rates means that more ejected hot nanoparticles come into contact and agglomerate. In addition to the large agglomerates, the LIPSS are left covered with a blanket of nanoparticles. The middle row of Figure 3 presents a side-by-side comparison of the surfaces before and after ultrasonication in acetone for 5 min. Clearly, ultrasonication has removed not only the large agglomerates from the irradiated patch, but also many of the nanoparticles—suggesting that they are redeposited without becoming sintered. Inspection of the micrographs from the ultrasonicated samples confirms that the stream of CO_2—and the nanoparticle plume confinement which results—does not affect the appearance of LIPSS. In fact, as expected, both the surfaces prepared with no CO_2 stream and a stream of 10,000 mL·min^{-1} possess the characteristic nanoripples with no appreciable difference.

As the volumetric flow rate of the CO_2 streamed during the laser micromachining process was seen to have little effect on the hydrophobicity of the resulting surfaces—save for the initial hydrophobicity of *CO_2-reactor* surfaces—we plot in Figure 4a the sessile contact angle data as an average of all surfaces prepared with the same post-processing step, regardless of stream flow rate. The very small 95% confidence intervals of the contact angle data presented this way is further evidence that the volumetric flow rate of the CO_2 stream has no effect on the wettability of the resulting LIPSS-decorated surfaces. The variance in the data analyzed this way can also elucidate the homogeneity of the fabricated surfaces. Figure 4b presents the variance of each data point graphed in Figure 4a. In general the data of the present work is of low, consistent, variance with the exception of four outliers as determined by statistical F-tests. The two datasets with statistically *greater* variances are not surprisingly those contact angle measurements taken on the compound hydrophobic nanoparticle agglomerate—hydrophilic LIPSS systems. The variance in the contact angle measurements of the first sessile drops to touch the *CO_2-reactor* surfaces is high because of the varying degree of nanoparticle agglomerate coverage with increasing CO_2 stream flow rate. Thus, these surfaces manifest the Cassie wetting state to varying degrees, with more nanoparticle agglomerates leading to more of the air-trapping effect. The advancing contact angle data taken on the *CO_2-reactor* surfaces on Day 0 also has statistically greater variability than those of the other datasets, backing up our hypothesis that as the footprint of the droplet is increased beyond the contact area of the original sessile droplets, it is in contact with a compound hydrophillic LIPSS–hydrophobic nanoparticle agglomerate system. What we find especially interesting in Figure 4b is that the contact angle data of the second sessile droplets to touch the *CO_2-reactor* surfaces have a variance which matches well the variances of the bulk of the dataset. This low variability in the measured contact angles of the second sessile droplets to touch the *CO_2-reactor* surfaces shows that the underlying micromachined substrate—left behind once the hydrophobic agglomerated nanoparticles are removed by the first sessile droplet—possesses homogeneous wettability, like the *no-post* and *boiling-water* surfaces.

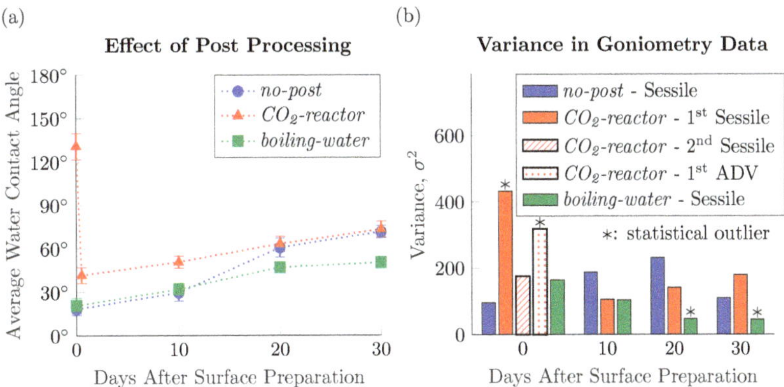

Figure 4. (**a**) Average sessile contact angle measured on the *no-post*, *CO_2-reactor*, and *boiling-water* surfaces regardless of CO_2 jet volumetric flow rate during micromachining over the 30 days following surface preparation. (**b**) Variance in the water contact angle data presented in part (**a**). Asterisk denotes a statistically-significant difference in variance compared to the rest of the dataset, as determined through F-tests.

The final two datasets presented in Figure 4 with statistically different variances are the measured sessile contact angles on the *boiling-water* surfaces 20 and 30 days after their initial preparation. Evidently, the residence in boiling water for 48 h has yielded surfaces with exceptionally homogeneous wettability. The high level of surface chemistry homogeneity could be the result of the washing away of any nanoparticle agglomerates which are on these surfaces after the micromachining step as well as a higher reaction rate for the formation of hydrophilic hydroxides in the water bath than for the formation of hydrophobic oxides in the CO_2 chamber or lab air. This point is discussed further in the next section.

3.2. Effect of Water Contact

The temporal evolution of the studied surfaces' wettabilities, as presented in Figure 4a, warrants further discussion. Kietzig et al. (2009), in their seminal work on the surface chemistry changes induced when metals are irradiated with ultrashort laser pulses, showed that hierarchically-rough fs-laser micromachined stainless-steels can be made to remain hydrophilic, in the Wenzel wetting state, if boiled in water for 2 days following the laser treatment [6]. We see similar, yet distinct results here for surfaces decorated with LIPSS alone. In Figure 4a, it is shown that the average sessile water contact angle measured on the *boiling-water* surfaces increases from $21 \pm 5°$ immediately after removal from the boiling water bath to $50 \pm 3°$ 30 days later. Whereas Kietzig et al. (2009) measured no appreciable change in water contact angle on their boiled surfaces, we see a nearly $30°$ increase. This suggests that the same 2-day residence in the boiling water bath was not sufficient to allow for the hot water to fully react with the laser irradiated surface to form a hydrophilic iron oxide layer. Rather, as the *boiling-water* surface sat exposed to lab air for 30 days, either some CO_2 from the air dissociatively adsorbed to the surface [6] or airborne hydrocarbons adsorbed to the surface [15], forming some hydrophoboic carbonaceous compounds. The weakened reactivity of the surfaces in the present study—those decorated with only LIPSS—compared to the surfaces prepared by Kietzig et al. (2009) stems from the lessened fluence used to induce only the formation of the nanoripples. With lowered irradiation fluence comes a weakened cleaning effect where less of the outermost, more oxidized layers of the material are removed [15]. Therefore, there are less Fe(metal) and Fe(II) oxide components at the surface to evolve to more oxidized forms and hence the slower reaction rate [6,15]. Further, the single-scale roughness of the surfaces decorated with only LIPSS inherently possess less total reactive surface area than the hierarchically-rough surfaces created by the higher fluences employed by Kietzig et al.

Figure 4a shows that the wettability transition of the *no-post* surfaces do not match the typical exponential growths of those metals reported in the literature to be hierarchically-roughened by laser micromachining [6,7,9,11–15]. We suspected that these contradictory findings are the result of repeated contact of the surface with water during the goniometric measurements. Similar to how the slowed reaction rate of these LIPSS-decorated surfaces allowed for hydrophobic carbonaceous compounds to form on the *boiling-water* preparations over the 30 days following treatment. The increased time required to form the hydrophobic carbonaceous surface chemistry allowed for the water to form some hydrophilic iron oxide on the surface as contact angles were measured. What is presented instead of an exponential growth is a linear increase in water contact angles over the 30 days following the post-processing step. Further, while we may not expect the nearly $150°$ sessile water contact angles reported by others on their hierarchically-roughened laser micromachined surfaces, we would expect to obtain decidedly hydrophobic surfaces from the *no-post* preparations after 30 days.

Further, the results presented in Figure 4a seem to contradict the hypothesis that the hydrophobic carbonaceous layer is the result of dissociative adsorption of CO_2 onto the irradiated surface. If that were the case, the 48-h residence in the elevated temperature/pressure CO_2 reactor should have resulted in *CO_2-reactor* surfaces with higher hydrophobicity than the *no-post* surfaces, based on previously reported findings [6,13,17]. From Figure 4a, it would appear that the hypothesis of adsorbed hydrocarbons being the source of the hydrophobic carbonaceous surface chemistry is the more likely explanation [12,13,15,18,19]. Whereas both the *no-post* and *CO_2-reactor* surface present with equivalent average contact contact angles after their residences in lab air. However, as mentioned above, we suspected that these wettability results were being affected by the goniometric measurements themselves.

This hypothesis of the water contact angle measurements affecting the final wettability of the surfaces was tested by preparing four more LIPSS-decorated samples: (i) 0 mL·min^{-1} *no-post*; (ii) 0 mL·min^{-1} *CO$_2$-reactor*; (iii) 10,000 mL·min^{-1} *no-post*; and (iv) 10,000 mL·min^{-1} *CO$_2$-reactor* and only taking goniometric measurements on their surfaces after 30 days. Figure 5a compares the average water contact angle measured at day 30 on these "undisturbed" preparations with the day 30 measurement taken on those surfaces prepared the same way, but repeatedly tested on the goniometer (i.e., every 10 days for 30 days). It is immediately evident that the average sessile contact angle measured on any of the four undisturbed surfaces is much greater than the average sessile contact angle measured on the same preparation which has come into repeated contact with water over the 30 days of experimentation. In the case of the *no-post* surfaces fabricated with no CO$_2$ stream during the micromachining step, the average static contact angle has increased from $71 \pm 5°$ to $116 \pm 2°$ by avoiding any contact with water for the 30 days following fabrication. The hydrophobic sessile contact angle manifested by the undisturbed surface matches well what is reported in the literature for LIPSS-decorated metals [9,15,23,24].

(a)

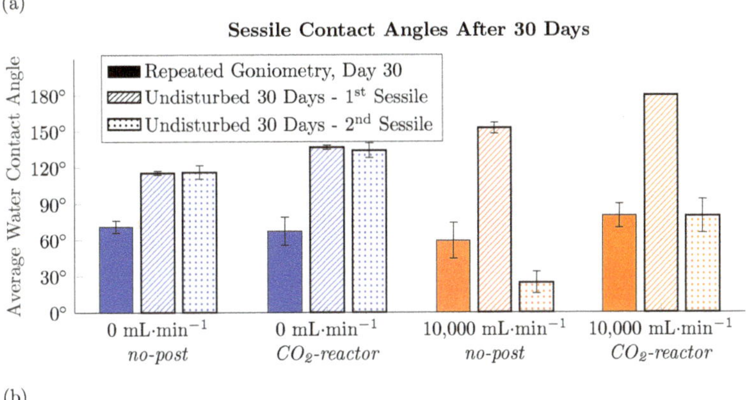

(b)

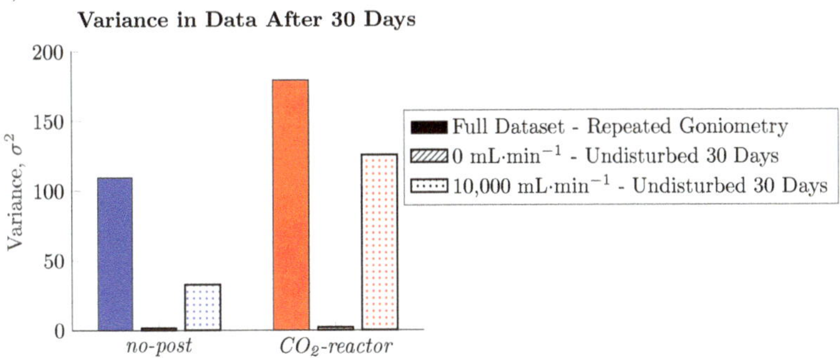

Figure 5. (**a**) Average sessile contact angle measured on *no-post* and *CO$_2$-reactor* surfaces left undisturbed for 30 days after fabrication before goniometric measurements were taken. (**b**) Variance in the sessile water contact angle data presented in part (**a**).

It is interesting to note that Figure 5a now shows that residence in the warm, elevated pressure CO$_2$ reactor positively affects the hydrophobicity of the laser micromachined metal sample. Once the 0 mL·min^{-1} *CO$_2$-reactor* sample is left undisturbed for 30 days following its fabrication, it matches the hypothesis that the dissociative absorption of CO$_2$ is (at least partly) responsible for the hydrophobic carbonaceous surface chemistry found on laser micromachined metals. Whereas the average contact angle of the first sessile drop to touch the undisturbed *CO$_2$-reactor* surface prepared with no CO$_2$ stream during irradiation is $137 \pm 2°$ compared to $116 \pm 2°$ for the *no-post* preparation (and compared to $67 \pm 12°$ for the *CO_2-reactor* surface with repeated water contact). These drastic changes in the wettability of the surfaces with avoidance of water contact for the 30 days following their fabrication confirms our hypothesis. What is remarkable is just how affected the hydrophobicity of the nascent surface chemistry is by the relatively short contact time with water during the goniometric measurements at 10-day intervals. Micromachining with low fluences, such as in the present work, evidently requires added consideration of avoiding any water contact before the hydrophobic carbonaceous chemical layer forms on the laser irradiated metal, if one desires a highly hydrophobic surface.

The results presented in Figure 5a are also a good recall to our discussion in Section 3.1. Similarly to the surfaces prepared with no streaming CO_2 during the micromachining step, avoiding contact with water for the first 30 days after fabrication has yielded much higher water contact angles for the first sessile drop placed on the surfaces prepared with CO_2 flowing at 10,000 mL·min^{-1}. In fact, the *CO_2-reactor* surface, prepared with a 10,000 mL·min^{-1} gas stream during micromachining and no water contact for the first 30 days after fabrication, presented as so hydrophobic that a droplet could not be placed on the surface at all. A video showing a water droplet immediately rolling off the surface is included as Supplementary Materials. Where the behaviours of these undisturbed surfaces diverge is in their wetting performance against the second sessile droplet placed on them. The two samples prepared without the flow of CO_2 gas during the micromachining steps present as equally hydrophobic to the second sessile drop as to the first sessile drop to touch the surface. On the contrary, the two samples created with a high flow rate of CO_2 gas present with a much lower average static contact angle with the second sessile drops to touch their surfaces. The blanketing of the surface with non-sintered nanoparticles and agglomerated nanoparticle flakes when micromachining in the presence of the high flow rate CO_2 stream has led to the same result on the undisturbed surfaces as the surfaces repeatedly tested on the goniometer, as described in Section 3.1. That is, the hydrophobicity of the non-sintered nanoparticles themselves is what imparts the Cassie state non-wetting behaviour of the surface against the first sessile droplet. These hydrophobic nanoparticles/agglomerates are removed with the first sessile drop, exposing the LIPSS for contact with subsequent water.

What is especially interesting to note in Figure 5a is that even after 30 days of aging in the lab air with no water contact, the LIPSS themselves which decorate the *no-post* sample prepared with a high flow rate of CO_2 gas during micromachining remain hydrophilic. In fact, the second sessile water droplet to touch this undisturbed surface—the first to touch the LIPSS themselves—presents with an average contact angle of $25 \pm 9°$, statistically equivalent to the average sessile contact angle measured at day 0 on the same surface preparation (see Figure 2a). Figure 3 shows LIPSS visible underneath the nanoparticle agglomerates which homogeneously cover the surface when CO_2 gas is streamed during micromachining. However, we hypothesize that the chemical reaction at the surface is blocked from occuring by a coverage of small non-sintered nanoparticle clusters. Crystallographic techniques could be used in future studies to distinguish the laser-induced nanoripples, integral to the surface, from non-sintered nanoparticles sitting atop them. Figure 5a shows that even the elevated pressure, temperature, and carbon dioxide-rich environment used to prepare the 10,000 mL·min^{-1} *CO_2-reactor* sample was not sufficient to completely overcome the blocking of the reactive surface by nanoparticles and nanoparticle agglomerates. Even though this *CO_2-reactor* sample is more hydrophobic than the *no-post* sample prepared with the same 10,000 mL·min^{-1} CO_2 stream during the micromachining step, both the high flow rate surfaces are much less hydrophobic than the two undisturbed samples prepared with no gas stream during irradiation.

The two major observations made in this work with regards to fabricating LIPSS-decorated surfaces with tuned wettability bear repeating. The first is that any non-sintered nanoparticles and nanoparticle agglomerates should be removed from the surface prior to any post processing steps to completely expose the reactive surface. However, it should be noted that removal of non-sintered nanoparticles is accomplished by ultrasonication, which exposes the reactive irradiated surface to the solvent itself. This has recently been shown to also affect the final wettability of laser micromachined metals [25]. Other techniques such as laser cleaning with photoacoustic monitoring may also prove effective in removing non-sintered nanoparticle agglomerates from laser-irradiated surfaces [26]. If non-sintered nanoparticle agglomerates are not removed, then the wettability measured is not that of the LIPSS themselves, but rather the compound LIPSS-nanoparticle agglomerate system, which inherently lacks robustness. The second observation is that the reformation of the surface chemistry on a LIPSS-only surface is much slower than hierarchically-roughened laser irradiated surfaces due to the lowered fluence employed for micromachining. As such, care must be taken to not disturb the nascent surface chemistry before it has sufficiently reacted to form either a robust hydrophilic oxide layer or a robust hydrophobic carbonaceous layer. For the fabrication of a hydrophilic surface, this means that the freshly irradiated substrate must be kept in a boiling water bath for longer than the two days reported necessary for metals irradiated with higher fluences [6]; For the fabrication of a hydrophobic surface, this means that the irradiated substrate must be kept away from even brief contact with water. The high level of homogeneity that is achieved when both recommendations are followed is exemplified by the very small variances in the contact angle data measured on the undisturbed surfaces. Figure 5b presents the variance in the sessile contact angles measured on those surfaces prepared with no stream of gas flowing during the micromachining step (i.e., low amount of non-sintered nanoparticles/agglomerates on the surface) and left undisturbed by water contact for the first 30 days following fabrication. The comparison with the contrary surface preparations is staggering. Both repeated contact with water through goniometric measurements and failure to remove non-sintered nanoparticles/agglomerates results in much more heterogeneous surfaces.

4. Conclusions

In this work, we have taken a systematic look at the effect of surface handling in the early times after irradiation on the final wettability of stainless-steel decorated with laser-induced periodic surface structures. In doing so we have been able to rigorously demonstrate that exposure to a CO_2-rich environment after micromachining does, in fact, lead to increased hydrophobicity. This proves what we have empirically known from our lab's work with laser irradiated metals: that CO_2 as well as airborne hydrocarbons can be the source of the hydrophobic carbonaceous layer which forms in the weeks following micromachining.

While performing this work, we have made a few interesting observations which we see as pertinent to metal laser micromachining getting transferred to industrial applications. The first observation is that the waste product of ultrafast laser micromachining—nanoparticles and nanoparticle agglomerates—can affect the wettability of the surface. We have shown that as these non-sintered particles sit on top of the irradiated surface, they become hydrophobic with residence in a CO_2-rich environment (or indeed with aging in lab air), while blocking this reaction from proceeding on the irradiated surface itself. Hence, contact angle measurements taken on such a surface will be measuring the wettability of the water-nanoparticle (agglomerate) system, and not the wettability of the actual micromachined metal surface. The substrate will behave much differently to subsequent contact with water as hydrophobic nanoparticle (agglomerates) are carried away, exposing the underlying hydrophillic micromachined surface. The second pertinent observation is that the metal surfaces prepared in the present study to possess only laser-induced periodic surface structures underwent much slower surface chemical reactions than those previously reported on metals hierarchically roughened by laser micromachining. This point is important with regards to industrial scale-up as we have shown that fabrication of a stable hydrophilic LIPSS-decorated stainless-steel requires a longer residence in boiling water after irradiation than a hierarchically-roughened substrate. Similarly, fabrication of a stable hydrophobic LIPSS-decorated stainless-steel requires that water contact be avoided at early times while a carbonaceous layer forms on the surface.

The observations and recommendations made in the present work can hopefully help guide the handling procedures for laser irradiated metal substrates in the lab and industry alike. Taking into account the need to remove non-sintered nanoparticles and nanoparticle agglomerates as well as requisite surface reaction time requirements can greatly aid in achieving desired wettabilities on laser micromachined metals.

Supplementary Materials: The following are available online at www.mdpi.com/xxx/s1, Figure S1: Effect of CO_2 flow rate on advancing contact angles, Figure S2: Effect of post-processing on advancing contact angles, Video S1: Droplet_Rolloff.

Author Contributions: Conceptualization, M.J.W. and A.-M.K.; methodology, M.J.W.; formal analysis, M.J.W.; writing—original draft preparation, M.J.W.; writing—review and editing, A.-M.K.; supervision, A.-M.K. and P.S. All authors have read and agreed to the published version of the manuscript.

Funding: This research was funded by a Natural Sciences and Engineering Research Council of Canada (NSERC) Discovery Grant, grant number: DG PT#69301.

Data Availability Statement: The data presented in this study are available on request from the corresponding author.

Conflicts of Interest: The authors declare no conflict of interest.

References

1. Van Driel, H.M.; Sipe, J.E.; Young, J.F. Laser-Induced Periodic Surface Structure on Solids: A Universal Phenomenon. *Phys. Rev. Lett.* **1982**, *49*, 1955–1958. [CrossRef]
2. Bonse, J.; Hoehm, S.; Kirner, S.V.; Rosenfeld, A.; Krueger, J. Laser-Induced Periodic Surface Structures—A Scientific Evergreen. *IEEE J. Sel. Top. Quantum Electron.* **2017**, *23*, 9000615. [CrossRef]
3. Wenzel, R.N. Resistance of Solid Surfaces to Wetting by Water. *Ind. Eng. Chem.* **1936**, *28*, 988–994. [CrossRef]
4. Johnson, R.E.; Dettre, R.H. Contact Angle Hysteresis. In *Contact Angle, Wettability, and Adhesion*; Advances in Chemistry, Book Section 7; American Chemical Society: Washington, DC, USA, 1964; Volume 43, pp. 112–135. [CrossRef]
5. Baldacchini, T.; Carey, J.E.; Zhou, M.; Mazur, E. Superhydrophobic surfaces prepared by microstructuring of silicon using a femtosecond laser. *Langmuir* **2006**, *22*, 4917–4919. [CrossRef] [PubMed]
6. Kietzig, A.M.; Hatzikiriakos, S.G.; Englezos, P. Patterned Superhydrophobic Metallic Surfaces. *Langmuir* **2009**, *25*, 4821–4827. [CrossRef] [PubMed]
7. Kietzig, A.M.; Negar Mirvakili, M.; Kamal, S.; Englezos, P.; Hatzikiriakos, S.G. Laser-Patterned Super-Hydrophobic Pure Metallic Substrates: Cassie to Wenzel Wetting Transitions. *J. Adhes. Sci. Technol.* **2011**, *25*, 2789–2809. [CrossRef]
8. Tang, M.; Shim, V.; Pan, Z.Y.; Choo, Y.S.; Hong, M.H. Laser Ablation of Metal Substrates for Super-hydrophobic Effect. *J. Laser Micro Nanoeng.* **2011**, *6*, 6–9. [CrossRef]
9. Bizi-bandoki, P.; Valette, S.; Audouard, E.; Benayoun, S. Time dependency of the hydrophilicity and hydrophobicity of metallic alloys subjected to femtosecond laser irradiations. *Appl. Surf. Sci.* **2013**, *273*, 399–407. [CrossRef]
10. Long, J.; Zhong, M.; Fan, P.; Gong, D.; Zhang, H. Wettability conversion of ultrafast laser structured copper surface. *J. Laser Appl.* **2015**, *27*, S29107. [CrossRef]
11. Ta, D.V.; Dunn, A.; Wasley, T.J.; Kay, R.W.; Stringer, J.; Smith, P.J.; Connaughton, C.; Shephard, J.D. Nanosecond laser textured superhydrophobic metallic surfaces and their chemical sensing applications. *Appl. Surf. Sci.* **2015**, *357*, 248–254. [CrossRef]
12. Skoulas, E.; Manousaki, A.; Fotakis, C.; Stratakis, E. Biomimetic surface structuring using cylindrical vector femtosecond laser beams. *Sci. Rep.* **2017**, *7*, 45114. [CrossRef] [PubMed]
13. Yan, H.; Abdul Rashid, M.R.B.; Khew, S.Y.; Li, F.; Hong, M. Wettability transition of laser textured brass surfaces inside different mediums. *Appl. Surf. Sci.* **2018**, *427*, 369–375. [CrossRef]
14. Gregorčič, P.; Conradi, M.; Hribar, L.; Hočevar, M. Long-Term Influence of Laser-Processing Parameters on (Super)hydrophobicity Development and Stability of Stainless-Steel Surfaces. *Materials* **2018**, *11*, 2240. [CrossRef] [PubMed]
15. Giannuzzi, G.; Gaudiuso, C.; Di Mundo, R.; Mirenghi, L.; Fraggelakis, F.; Kling, R.; Lugarà, P.M.; Ancona, A. Short and long term surface chemistry and wetting behaviour of stainless steel with 1D and 2D periodic structures induced by bursts of femtosecond laser pulses. *Appl. Surf. Sci.* **2019**, *494*, 1055–1065. [CrossRef]
16. Tanvir Ahmmed, K.M.; Kietzig, A.M. Drag reduction on laser-patterned hierarchical superhydrophobic surfaces. *Soft Matter* **2016**, *12*, 4912–4922. [CrossRef] [PubMed]
17. Ling, E.J.Y.; Uong, V.; Renault-Crispo, J.S.; Kietzig, A.M.; Servio, P. Reducing Ice Adhesion on Nonsmooth Metallic Surfaces: Wettability and Topography Effects. *ACS Appl. Mater. Interfaces* **2016**, *8*, 8789–8800. [CrossRef] [PubMed]

18. Boinovich, L.B.; Emelyanenko, A.M.; Emelyanenko, K.A.; Domantovsky, A.G.; Shiryaev, A.A. Comment on "Nanosecond laser textured superhydrophobic metallic surfaces and their chemical sensing applications" by Duong V. Ta, Andrew Dunn, Thomas J. Wasley, Robert W. Kay, Jonathan Stringer, Patrick J. Smith, Colm Connaughton, Jonathan D. Shephard (Appl. Surf. Sci. 357 (2015) 248–254). *Appl. Surf. Sci.* **2016**, *379*, 111–113. [CrossRef]
19. Yasumaru, N.; Sentoku, E.; Kiuchi, J. Formation of organic layer on femtosecond laser-induced periodic surface structures. *Appl. Surf. Sci.* **2017**, *405*, 267–272. [CrossRef]
20. Pou, P.; del Val, J.; Riveiro, A.; Comesaña, R.; Arias-González, F.; Lusquiños, F.; Bountinguiza, M.; Quintero, F.; Pou, J. Laser texturing of stainless steel under different processing atmospheres: From superhydrophilic to superhydrophobic surfaces. *Appl. Surf. Sci.* **2019**, *475*, 896–905. [CrossRef]
21. Liu, J.M. Simple technique for measurements of pulsed Gaussian-beam spot sizes. *Opt. Lett.* **1982**, *7*, 196–198. [CrossRef]
22. Villerius, V.; Kooiker, H.; Post, J.; Pei, Y.T. Ultrashort pulsed laser ablation of stainless steels. *Int. J. Mach. Tools Manuf.* **2019**, *138*, 27–35. [CrossRef]
23. Florian, C.; Skoulas, E.; Puerto, D.; Mimidis, A.; Stratakis, E.; Solis, J.; Siegel, J. Controlling the Wettability of Steel Surfaces Processed with Femtosecond Laser Pulses. *ACS Appl. Mater. Interfaces* **2018**, *10*, 36564–36571. [CrossRef] [PubMed]
24. Raimbault, O.; Benayoun, S.; Anselme, K.; Mauclair, C.; Bourgade, T.; Kietzig, A.M.; Girard-Lauriault, P.L.; Valette, S.; Donnet, C. The effects of femtosecond laser-textured Ti-6Al-4V on wettability and cell response. *Mater. Sci. Eng. C* **2016**, *69*, 311–320. [CrossRef] [PubMed]
25. Schnell, G.; Polley, C.; Bartling, S.; Seitz, H. Effect of Chemical Solvents on the Wetting Behavior Over Time of Femtosecond Laser Structured Ti6Al4V Surfaces. *Nanomaterials* **2020**, *10*, 1241. [CrossRef] [PubMed]
26. Athanasia, P.; George, J.T.; Kristalia, M.; Costas, F.; Giannis, Z.; Paraskevi, P. Development of a hybrid photoacoustic and optical monitoring system for the study of laser ablation processes upon the removal of encrustation from stonework. *Opto Electron. Adv.* **2020**, *3*, 190037. [CrossRef]

nanomaterials

MDPI

Article

Self-Organization Regimes Induced by Ultrafast Laser on Surfaces in the Tens of Nanometer Scales

Anthony Nakhoul [1], Claire Maurice [2], Marion Agoyan [1], Anton Rudenko [3], Florence Garrelie [1], Florent Pigeon [1] and Jean-Philippe Colombier [1,*]

[1] UJM-St-Etienne, CNRS, Laboratoire Hubert Curien UMR 5516, Institute of Optics Graduate School, Univ Lyon, F-42023 Saint-Etienne, France; anthony.nakhoul@univ-st-etienne.fr (A.N.); marion.agoyan@univ-st-etienne.fr (M.A.); garrelie@univ-st-etienne.fr (F.G.); florent.pigeon@univ-st-etienne.fr (F.P.)

[2] Mines Saint-Etienne, CNRS, Univ Lyon, UMR 5307 LGF, Centre SMS, F-42023 Saint-Etienne, France; maurice@emse.fr

[3] Arizona Center for Mathematical Sciences and College of Optical Sciences, University of Arizona, Tucson, AZ 85721, USA; antmipt@gmail.com

* Correspondence: jean.philippe.colombier@univ-st-etienne.fr

Abstract: A laser-irradiated surface is the paradigm of a self-organizing system, as coherent, aligned, chaotic, and complex patterns emerge at the microscale and even the nanoscale. A spectacular manifestation of dissipative structures consists of different types of randomly and periodically distributed nanostructures that arise from a homogeneous metal surface. The noninstantaneous response of the material reorganizes local surface topography down to tens of nanometers scale modifying long-range surface morphology on the impact scale. Under ultrafast laser irradiation with a regulated energy dose, the formation of nanopeaks, nanobumps, nanohumps and nanocavities patterns with 20–80 nm transverse size unit and up to 100 nm height are reported. We show that the use of crossed-polarized double laser pulse adds an extra dimension to the nanostructuring process as laser energy dose and multi-pulse feedback tune the energy gradient distribution, crossing critical values for surface self-organization regimes. The tiny dimensions of complex patterns are defined by the competition between the evolution of transient liquid structures generated in a cavitation process and the rapid resolidification of the surface region. Strongly influencing the light coupling, we reveal that initial surface roughness and type of roughness both play a crucial role in controlling the transient emergence of nanostructures during laser irradiation.

Keywords: ultrafast laser nanostructuring; self-organization; nanobumps; femtosecond laser; LIPSS; nanopatterns

Citation: Nakhoul, A.; Maurice, C.; Agoyan, M.; Rudenko, A.; Garrelie F.; Pigeon F.; Colombier J.-P. Self-Organization Regimes Induced by Ultrafast Laser on Surfaces in the Tens of Nanometer Scales. *Nanomaterials* **2021**, *11*, 1020. https://doi.org/10.3390/nano11041020

Academic Editors: Jörn Bonse, Peter Simon and Jürgen Ihlemann

Received: 29 March 2021
Accepted: 13 April 2021
Published: 16 April 2021

Publisher's Note: MDPI stays neutral with regard to jurisdictional claims in published maps and institutional affiliations.

1. Introduction

Femtosecond laser sources allow us to explore the material's behavior under swift photoexcitation. This represents a privileged way to bring material surfaces in extreme states of temperature and pressure far from equilibrium. During the relaxation stage, out-of-equilibrium systems usually exhibit pattern formation as a result of spontaneous spatial breaking symmetry of the initial homogeneous state [1]. In particular, ultrashort laser energy deposition induces local thermal stresses and transient phase changes which modifies materials microstructures and topography [2,3]. The fast interaction with material surfaces results in the formation of anisotropic surface morphology, usually recognized as laser-induced periodic surface structures (LIPSS) or ripples. LIPSS are a special type of arranged nanopatterns, known for their periodicity varying from near laser wavelength down to sub-100 nm scale [4,5]. Their formation generally occurs in a multi-pulse irradiation regime with a fluence below the ablation threshold. Consequently, the first laser pulse modifies the surface topography randomly by creating roughness centers that are perceived by the next pulses. All the consecutive laser pulses impact on a surface being constantly modified by the previous pulses and being roughened constantly, from an initially quasi-flat surface to transiently growing disturbances [6–8]. LIPSS can be generated with different periodic scales on the irradiated surface by virtue of the radiative and nonradiative modes of light coupling, which involves polarization effects on the freshly formed roughness centers [9]. Far beyond the diffraction limit, most of the sub-wavelength structures were oriented by the laser polarization, commonly spaced between 50–200 nm in the literature and a priori provoked by local field enhancement on the local roughness [6,10]. They are generally called high spatial frequency LIPSS (HSFL) of type *r* in the literature [11]. The light coupling is composed of scattered waves but also of nonradiative fields that emerge from nanoreliefs and that generate coherent exchanges of optical near fields [9]. Strong enhancement of the oriented electromagnetic field at surface defects trigger local heat confinement that can destabilize the thin laser-induced melt layer. During the sub-surface rarefaction process, periodic nanovoids enclosed under the surface and nanocavities emerge at the surface by cavitation [12]. During the acceleration of the molten surface by the rarefaction wave, a Marangoni-convection instability can be triggered by hydrothermal waves perpendicular to these temperature gradients [13], forming high spatial frequency periodic surface structures.

Furthermore, it was proven that under multi-pulse feedback regimes, there is a strong dependence of the generation of surface roughness and finally periodic nanoripples on the crystallographic surface orientation [2]. More refined nanostructure was obtained on Cr (100) by single-pulse irradiation and a more pronounced LIPSS after two laser pulses as compared to (110) surfaces [14]. The competition between the evolution of transient liquid structures generated in the spallation process and the rapid resolidification of the surface region were proposed to define the types, sizes and dimensions of the nanoscale surface features [14]. Thus, the formation of sub-100nm nanopatterns are clearly dependent on crystallographic orientation and initial surface roughness. Recent works show that it is possible to emancipate from the polarization-dependency by combining two orthogonal linear polarization states. The control of the ultrafast laser pulse polarization, inter-pulse delay and laser fluence leads to the generation of innovative 2D surface morphologies on different scales. Several works focused on the fabrication of 2D-LIPSS in the near-submicron length [15–17]. These microstructures can lead to strong modifications on surface wetting [18], tribological and optical properties [19]. Laser-induced textures enable surface functionalities, such as hydrophobicity [20], antibacterial properties [21], coloring [22,23] and wear resistance [24].

Driven by near-field light enhancement, periodic patterns recently approached ultimate scales down to tens of nanometers [25]. A self-organization of unusual arrays of nanocavities of 20 nm diameter with a periodicity down to 60 nm was created on a Nickel surface oriented in (100) plan. Obtained by overcoming the anisotropic polarization response of the surface by a delayed action of cross-polarized laser pulses, this process opens the route for self-arranged topography at the nanoscale highly demanded in advanced technology sectors. Analogous structures can be produced by using other advanced technologies such as electrochemical etching, but they suffer from low throughput and accuracy. Instead, ultrafast laser–surface interactions reveal sub-100 nm self-arranged hexagonal arrays with a high throughout and the highest precision. As the resulting morphology is energy-driven, the purpose of this paper is to explore the formation of unrevealed nanopatterns under regulated energy dose. Beyond the already reported spatial instability triggering hexagonal arrays, we discuss the laser irradiation parameters as well as the material conditions apt to generate a self-organization regime of nanostructuring. We show that a slight modification of the amount and the timescale of the energy delivery perturbs the system and a stable self-organization regime shifts towards another one with new properties. The initial surface roughness features, in particular, the absence of inordinately high peaks and deep valleys is also shown to be a crucial feature to be checked to control the self-organization regime. A large variety of nanostructuring regimes are reported here with an abrupt transition between them. This is a characteristic feature of dissipative structures, including random patterns with concentrated nanoreliefs, 1D nanostripes, and 2D hexagonal arrays. Disordered, labyrinthine and bumpy nanopatterns that arise from a homogeneous metal surface under ultrafast excitation are originally reported.

2. Materials and Methods

2.1. Material Preparation

A mono-crystalline Nickel (Ni) sample oriented in (100) direction was used during this study. Single crystals are preferred to insure the uniformity of the structuring process. The Ni (100) bar was grown by directional solidification and cut into $10 \times 10 \times 10$ mm^3 cubes by using a wire saw. Two types of polishing were performed during experiments, the mechanical and electrochemical polishing. The automatic polishing was performed on "Buehler Automet 250" by using a coarse paper of P180, moving successively to P320, P600, P1200 and P2400 followed by a diamond 3 µm and 1 µm and vibratory polishing on "Buehler Vibromet 2" with the colloidal silica 0.05 µm for 17 h, prior to laser irradiations. Electropolishing was performed after automatic polishing on "Struers LectroPol-5" by using stainless steel electrolyte at 25 Volts for 60 s. Both polishing procedures assure mirror-zero scrach samples with an initial arithmetical mean surface roughness (Ra) below 5 nm. The Ra threshold Ra$_{th}$ for nanostructures formation was found to be 5 nm on the Atomic Force Microscopy (AFM), on a scan of 5×5 µm. Crystal orientations were checked by X-ray diffraction prior to laser irradiation to ensure the uniform cutting direction.

2.2. Laser Setup

A Ti: Sapphire laser from Coherent (Legend Elite Series) was used in this setup. Powered by an integrated Revolution pump laser and seeded by a Vitara-S oscillator, the ultrafast amplifier delivers pulse durations down to 25 fs, a 1 kHz repletion rate, an output power up to 7 W and a $\lambda = 800$ nm central wavelength. A thin film polarizer at the output of the laser allows us to separate s and p-polarized components of the beam. A half-wave plate is devoted to tuning the incoming energy on the double-pulse irradiation setup.

To force an isotropic energy deposition on the surface that fosters self-organization free from polarization orientation, a cross-polarized irradiation strategy is followed. The double-pulse irradiation setup consists of a modified Mach–Zehnder interferometer combining the effect of a crossed-polarization with an inter-pulse delay Δt. The aim of this setup is to help to control the laser-induced structure formation process. A first beam splitter divides the incoming laser beam into two beams. These beams take two different optical paths and then encounter a second beam splitter which recombines the two of them. We call the "fixed arm" the optical path of the "reference beam" and "mobile arm" the optical path of the "sample beam". The length of the mobile arm can be adjusted by the use of a highly accurate motorized stage. From a distance allowing a path shorter to the fixed arm to another one a little bit longer, the optical path difference induces a temporal delay. The "zero delay" $\Delta t = 0$ corresponds to a spatial and temporal overlap. Temporal overlap is obtained when the two optical paths have the same length and spatial overlap needs a good orientation of the beam splitters. The "zero delay" configuration is obtained when the pattern of interference is the most contrasted. In each arm, we introduce a polarizer in order to obtain the cross-polarization configuration after recombination. An half-wave plate is also introduced in each arm so we get the same energy for the reference and the sample beam, always checked by a power-meter measurement.

After recombination, the laser beam is then collimated by an optical lens with a focal length of 250 mm and irradiates the surface of a sample with a normal incidence. The surface of the sample is set in the focal plane of the lens independently of the thickness of the sample thanks to the sample holder. The size of the laser spot is always measured before irradiation by the D^2 method [26] and the spot size of the Gaussian profile at $(1/e^2)$ was $2\omega_0 = 56$ µm where ω_0 is the beam waist.

2.3. Characterization

Surface topography was visualized by using scanning electron microscopy (SEM) of type "Nova NanoSEM" equipped with a field emission gun. The images were taken at 15 kV with Everhart–Thornley detector (ETD), at different magnitudes. Atomic force microscopy (AFM) of type "Bruker Dimension ICON" was used for characterization in 2D and 3D. Images were taken in SCANASYST-AIR imaging mode, by silicon tip on nitride lever, at different scales and at a resolution of 1024 lines per scan. This microscope has a local probe allowing to visualize the surface topography of a sample with a resolution of about 1 nm. Topological investigations and roughness calculation were performed by using NanoScope Analysis software.

3. Results

3.1. Advanced Surface Topography Control as a Function of Time Delay and Laser Fluence

Femtosecond laser irradiations are performed by controlling several influential parameters. The two main parameters are the laser peak fluence and the time delay between the double pulses. Other substantial parameters are the polarization angle between the two delayed pulses, number of pulses, pulse duration and initial surface roughness. Based on previous experiments, the chosen number of double-pulse sequences (N_{DPS}) was 25 pulses [25]. A cross-polarized ultrashort double pulses are applied in order to break the surface isotropy, avoiding the formation of LIPSS. The relative polarization of the two pulses was fixed by an angle of 90°. The chosen pulse duration was 150 fs and the initial arithmetical mean roughness (Ra) was below 5 nm.

Figure 1 presents 2D AFM images of laser-induced nanopatterns formation by controlling the laser peak fluence from 0.18 J/cm² to 0.24 J/cm² and the time delay (Δt) from 8 ps to 25 ps. Laser parameters are chosen to be below the LIPSS formation threshold for a single pulse, which requires a combined action of the double pulses. At low Δt (8 ps and 10 ps) and at a peak fluence above 0.20 J/cm², chaotic nanostructures were observed. They are also observed at (15 ps; 0.24 J/cm²) and they are formed for a small time-delay between pulses. Energetically, it is somehow acting similarly to a single pulse process, at a cumulated absorbed fluence higher than the single-pulse absorbed threshold since the small delay leads to a slight decrease of Ni reflectivity with electronic thermal excitation before the second pulse [27]. Up to 15% of reflectivity decrease can be expected for ultrafast photoexcited nickel under electron–phonon nonequilibrium. At 0.18 J/cm², three different types of nanopatterns were observed at 8 ps, 10 ps and 15 ps. The tallest nanostructures were at Δt = 8 ps and have a shape comparable to a karst peak, so they are named as "nanopeaks". Nanostructures shape has changed at a larger time delay of 10 ps. Their diameter has increased and their shape was similar to a bump, so they are named as "nanobumps". We note that the formation of microbumps and nanojets on the metal targets was observed upon single pulse tight focusing ultrashort laser excitation of thin gold films and attributed to Marangoni effect [28]. Our investigation shows that similar structures can be obtained also on irradiated bulks. At Δt = 15 ps, nanostructures height was extensively decreased and they are named as "nanohumps". Nanostructures were progressively disappeared while increasing the time delay from 15 ps to 25 ps. In this case, the cumulated energy was lower and it can be reasonably assumed that the thermal gradients resulting from local polarization effects were strongly softened by thermal dissipation before the second pulse arrived. The transverse thermal dissipation time due to gradients between two structures at a characteristic distance of 100 nm can be roughly estimated as $\tau_L = \frac{\rho C_i x^2}{2k_i}$ where $\rho = 7.90 \times 10^3$ kg/m³ is the liquid nickel density, $C_i = 630$ J/kg/K is the heat capacity, $k_i = 50$ W/m/K is the thermal conductivity. Considering a distance of $x = 50$ nm, $\tau_L \sim 10$ ps. In these conditions, we can estimate that the local polarization memory corresponding to transverse temperature gradients can be rapidly erased by thermal dissipation before the second pulse arrived in the ten of picosecond timescale. At 0.20 and 0.22 J/cm², we can observe the transition from chaotic structures to stripes, passing by a transition regime at (0.20 J/cm²; 15 ps and 0.22 J/cm²; 20 ps). This transition region has a short-range order, which consists of locally nanobumps and chaotic nanostructures. However, at Δt = 25 ps , labyrinthine patterns were observed [1], which are disordered spatial structures that show a short-range order. At 0.24 J/cm^2, we can observe the switch from disordered nanostructures to a pattern of nanocavities when Δt increases, passing a by transition regime from 15 to 20 ps. The hexagonal array of nanocavities were observed previously at similar parameters [25].

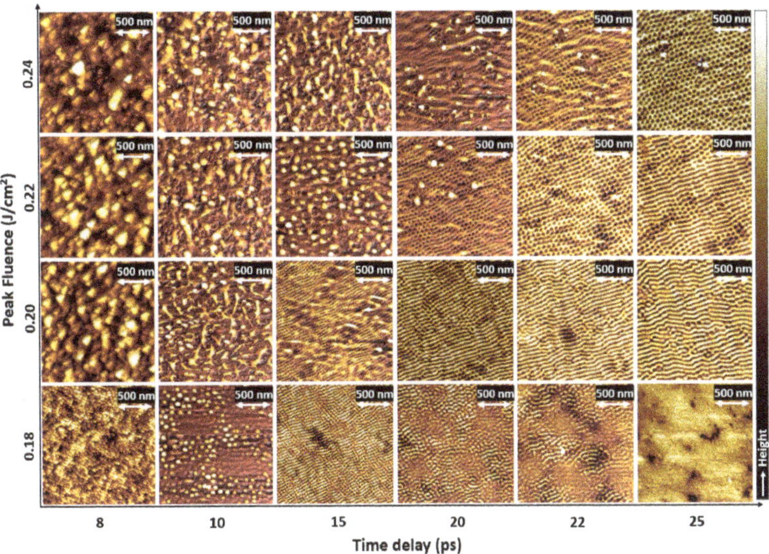

Figure 1. 2D AFM images of laser-induced nanopatterns formation on Ni(100) as a function of time delay and peak fluence at a fixed N_{DPS} of 25. The different zones of interest are nanopeaks, nanobumps, nanohumps and nanocavities. They are created progressively at different doses: (0.18 J/cm^2; 8 ps), (0.18 J/cm^2; 10 ps), (0.18 J/cm^2; 15 ps), (0.24 J/cm^2; 25 ps).

3.2. Wide Variety of Nanostructure Regimes

Figure 2a–c present 3D AFM images of the principal nanostructures, which correspond to the laser parameters of Figure 1 (a(0.18 J/cm^2; 8 ps), b(0.18 J/cm^2; 10 ps), c(0.18 J/cm^2; 15 ps) and d(0.24 J/cm^2; 25 ps)). Figure 2e presents the maximum nanostructure height as a function of time delay and laser fluence. The maximum height was characterized on AFM for each parameter. It was classified into two different categories. One category of nanostructures with a maximum height of almost 110 nm presented in (a) and (b), marked by black and green circles respectively. The other category has a maximum height of almost 20 nm presented in (c) and (d), marked by blue and orange circles respectively. The material surface exhibits a significant swelling suggesting a frustrated ablation regime [29,30].

It can be observed that (a) and (b) have respectively the shape of nanopeaks and nanobumps but with different widths, organization and concentrations. However, (c) and (d) present hexagonal self-organized nanohumps and nanocavities on the surface. Interestingly, (c) exhibits a state with persistent dynamics with oriented stripes independent from the local polarization that may correspond to "spiral defect chaos" [31].

The transition regimes between the two different nanostructures categories of (110 nm and 20 nm) are shown in an almost diagonal way in Figures 1 and 2e. It can be understood that the lost energy by increasing the time delay was covered by the intensification of laser fluence to maintain the instability of the critical phase leading to the presence of these nanostructures. The laser spots images (a, b and c) confirm the occurrence of a spallation regime that exists above the melting threshold, while remaining below phase explosion threshold.

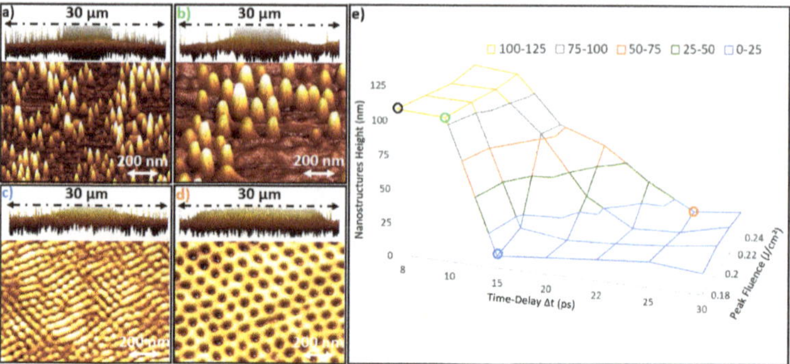

Figure 2. (**a–d**) The 3D AFM images of the laser spot topography in the spallation regime and the principal nanostructures types (nanopeaks, nanobumps, nanohumps and nanocavities). (**e**) Maximum nanostructures height as a function of time delay and laser fluence. The colored circles in (**e**) (black, green, blue and orange) present respectively the regions of (**a–d**) nanostructures.

3.3. Wide Variety Nanopatterns Morphologies

Figure 3 presents the 3D AFM images of the principal nanopatterns (a–d) at different scanning scales. (a) presents a nanobump shape comparable to a karst peak. It has a width of ≈20 nm and a height of ≈100 nm and is named a nanopeak. At the same laser peak fluence of 0.18 J/cm^2, and just by increasing the time delay from 8 ps to 10 ps, the nanostructures have transformed from (a) to (b). In (b), it has a shape of a nanobump but totally different than in (a). The diameter of the bump is ≈60 nm and the height is ≈100 nm. Moreover, by a slight increase of the time delay from 10 to 15 ps, the nanobump scale was reduced to reach a height less than 10 nm in Figure 3c, which are named nanohumps. Those small hexagonal nanohumps have a periodicity of ≈50 nm. On the other hand, (d) presents self-arranged hexagonal nanocavities with a periodicity of ≈85 nm and a depth of ≈25 nm.

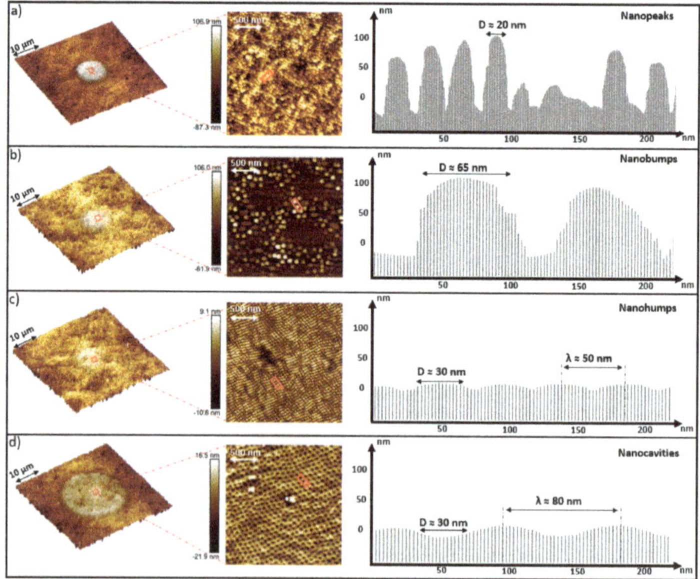

Figure 3. 3D AFM images of the principal nanopatterns (a–d) at different scales, presenting the laser spot region. A scan profile was performed for each type of the principal nanostructures presenting the shape and the periodicity of each type.

In addition, the laser spot diameter for the nanopatterns (a, b and c) was ≤ 9 μm. However, the laser spot diameter for the nanocavities shown in (d) was ≤ 22 μm. Thus, laser fluence is a critical factor in determining the self-organized regime of nanostructures growth.

3.4. Nanopatterns Control by Laser Dose

As presented previously, surface topography and nanopatterns shapes are controlled by laser parameters. Laser dose has a significant role in creating these different patterns. Figure 4a emphasizes the essential role of the N_{DPS} in controlling surface topography and enhancing nanostructures, by fixing the peak fluence to 0.18 J/cm^2 and the time delay to 10 ps. Therefore, the surface roughness created by the last pulse is playing an important role in the appearance of different nanopatterns. However, Figure 4b presents SEM images at a fixed N_{DPS} of 25 pulses and a peak fluence of 0.24 J/cm^2. As presented in Figure 1, we can observe the transformation from chaotic surface to nanocavities by only inceasing the time delay between the two cross-polarized pulses, passing by a transition regime between 15 and 25 ps. On the other hand, Figure 4c shows the laser spot at the same laser parameters of Figure 4a, but at $N_{DPS} = 44$ pulses. Finally, Figure 4d exhibits the left region of the gaussian laser spot of Figure 4c, showing how nanopatterns can be simply regulated by setting a specific laser peak fluence. The presence of different nanostructures on the laser spot periphery stresses the significant role of the laser peak fluence on creating different surface morphologies.

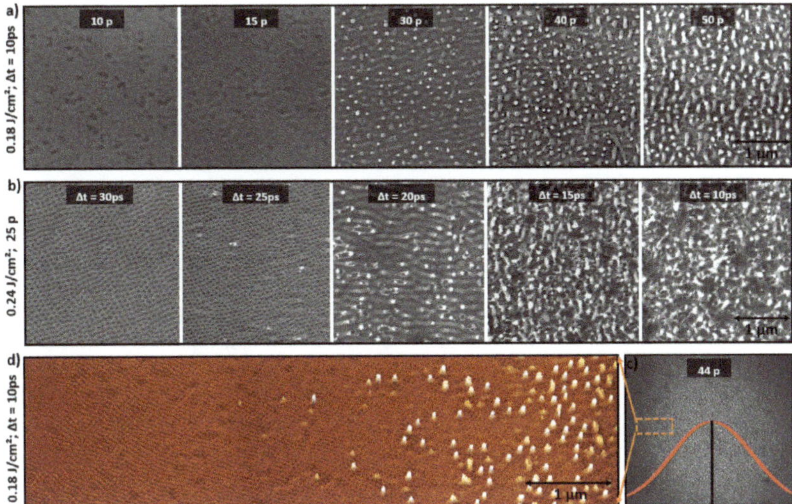

Figure 4. (**a**) 2D scanning electron microscopy (SEM) images presenting the influence of the N_{DPS} in enhancing nanostructures at a fixed time delay. (**b**) 2D SEM images of growing nanopatterns by decreasing the time delay at a fixed peak fluence of 0.24 J/cm^2 and $N_{DPS} = 25$. (**c**) 2D SEM image of the gaussian laser spot at a fixed laser parameters (0.18 J/cm^2; 10 ps; $N_{DPS} = 44$). (**d**) The left region of the gaussian laser spot, showing the significant role of laser fluence in controlling different types of nanostructures.

3.5. Initial Roughness Effect on Nanostructures Formation

Initial surface topography plays an essential role in nanostructures formation. Several polishing procedures were developed in order to obtain different initial surface topographies, with various initial arithmetic surface roughness (Ra). A Ra > 5 nm, will not permit the formation of different types of previously reported nanohumps and nanocavities.

As presented in Figure 5a,c, sample 1 and sample 2 have an initial Ra < Ra_{th}. Mechanical polishing was performed for sample 1 by using a coarse paper up to P2400 followed by a diamond polishing of 3 μm and 1 μm and vibratory polishing with colloidal silica of 0.05 μm for several hours, which guarantee a mirror surface with a very low arithmetic roughness. Kurtosis (Ku) statistical parameters was used to characterize the type of initial roughness and sharpness of surface spikes. If Ku > 3 the surface is considered "spiky" and if Ku < 3 the surface is considered "bumpy". If Ku = 3, the surface has completely random surface roughness. For sample 1, the Ku = 7.54 which is considered spiky surface with a Ra < Ra_{th}.

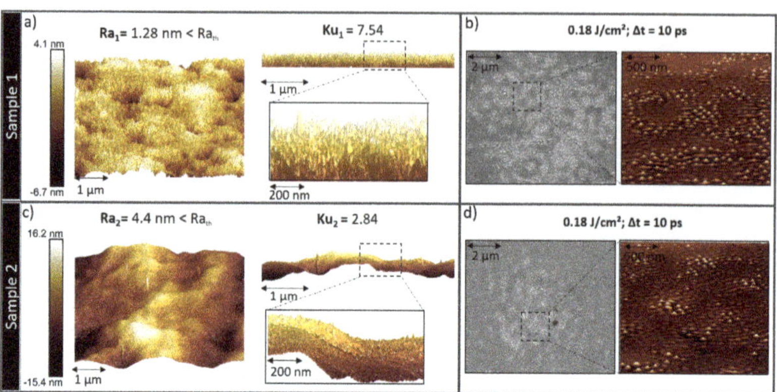

Figure 5. Initial surface topography of two different samples polished by different procedures (Mechanical polishing in (**a**) and Electrochemical in (**c**)). Arithmetic roughness (Ra) and Kurtosis (Ku) were measured and compared for both samples. (**b,d**) 2D and 3D SEM images of (**a,c**) respectively after laser irradiation. The crucial role of initial type of roughness is observed by comparing the nanostructures concentration in the SEM.

On the other hand, an electrochemical polishing was performed for sample 2 by using stainless steel electrolyte at 25 Volts for 60 s. The surface has an arithmetic roughness Ra = 4.4 nm < Ra_{th} and a Ku of 2.84 which is considered bumpy surface.

By comparing the SEM images of Figure 5b,d, the difference in nanobumps concentration is clearly observed. The spiky surface of sample 1 has a higher surface absorbance compared to the bumpy surface of sample 2, which increased the concentration of nanostructures. The initial type of roughness plays an essential role in controlling surface absorbance and reflectivity. Moreover, AFM characterizations show the difference in the scale of the obtained nanostructures in the two different samples. For sample 1, the maximum nanobumps height was 89 nm and for sample 2 was only 22.3 nm. Thus, the initial surface type of roughness has a prominent role besides the laser parameters, in controlling surface topography at the nanoscale.

4. Discussion

Self-organization regimes may result from complex flows conditions depending on competing destabilization and dissipation forces that occur in the transiently laser-induced molten surface. At the considered fluence below ablation threshold, the electrons are excited up to an estimated electronic temperature of 1–2 eV after each laser pulse of the double-pulse sequence. On the electron–phonon relaxation timescale, a thin liquid layer of 10–20 nm thickness is expected to be formed at the surface a few ps after each pulse. During this relaxation stage, the confined temperature gradient of femtosecond laser irradiation induces a pressure wave on the top surface which is followed by a rarefaction wave [12]. Consequently, the melt layer is driven toward the free surface by the rarefaction wave in the same direction of the temperature gradient. This mechanism is vectorially the opposite of the destabilizing effect of the gravitational force acting on a liquid heated from below [32]. In classical Rayleigh–Bénard instability, the gravitational force is directed against the density gradient, pushing down the heavier liquid. In the case of ultrashort laser excitation, the rarefaction below the surface might play a destabilizing role due to the force that results from different pressures in rarefied or non-rarefied liquid, being directed towards the surface and pushing colder liquid inside the warmer upper layer.

Guided by surface tension and rarefaction forces, hydrothermal flows develop a thermoconvective instability at the nanoscale, similar to well-known Rayleigh–Bénard-Marangoni instabilities [33,34]. This mechanism represents a convincing scenario for generating hexagonal nanostructures (nanocavities or nanohumps on the surface) [33]. The Rayleigh–Bénard-like instability occurs since there is a density gradient between the top and the bottom surface, rarefaction wave acts trying to pull the cooler, denser liquid from the bottom to the top [35,36]. Therefore, a Bénard–Marangoni convection occurs along the unstable fluids interface, initiated by small flow perturbations due to recurring surface tension gradients. This Marangoni flow transports thermal energy, that can be characterized by the Marangoni number, comparing the rate at which thermal energy is transported by this flow to the rate at which thermal energy diffuses. Modulated by near field coupling on local roughness, transverse temperature gradients are built parallel to the surface that turns to be unstable [25]. The process starts with a roughness-triggered inhomogeneous energy absorption featured by electromagnetic patterns. They induce a surface thermomechanical fingerprint persisting after electron–ion thermal equilibrium at $t = 10$ ps. Later on, at 15 ps $< t < 25$ ps, the lattice density is strongly influenced by the inhomogeneous shock and rarefaction waves. The surface starts to melt inhomogeneously and the destabilization occurs. The liquid flow is then driven by transverse surface tension forces, rearranging the material from hot spots to colder regions on the surface [13]. Anisotropic temperature gradients generate hydrothermal waves and convection rolls in a disturbed thin liquid layer [37,38]. This generates Marangoni forces balanced by thermal diffusion that reorganize the material over the surface. If isotropic conditions are fulfilled, this can be done in the most compact hexagonal way. Finally, the dimensions of the cell are correlated to the Marangoni number which is equivalent to the ratio between the destabilizing Marangoni force and the viscous restrictive force.

The nature and the orientation of the resulting standing waves are defined by the dimensionless number of Prandtl, which expresses the ratio between the kinematic viscosity and thermal diffusivity and which is leading to hydrothermal waves perpendicular to transverse temperature gradients [37,39]. It makes it possible to understand whether the thermal phenomenon is more or less rapid than the hydrodynamic phenomenon. If the Prandtl number is small, then it should be understood that the phenomenon of thermal conduction is so fast that the velocity profile has little or no effect on the temperature profile. However, for a large Prandtl number, the temperature gradient in the fluid will be strongly influenced by the local velocity of the fluid. It is demonstrated that there exists a critical Prandtl number [40] $P_r^c = 0.25$ such as: For $P_r < P_r^c$, the production of kinetic energy is intensified by isoenergetic redistributions of vorticity that lead to a concentration of the vertical velocity in the vicinity of the center of the cell. Therefore, the convection sets in as a pattern of hexagonal cells with downward motion in the center (g-hexagons) which are similar to the obtained nanocavities. On the other hand, for $P_r^c < P_r$, the intensification of kinetic energy production by the expansion of the energy-producing temperature gradient constitutes the reason for the preference of l hexagons in high-Prandtl-number convection. Therefore, the conventional hexagonal cells with upward motion in the center (l-hexagons) appear at the onset of instability [41,42], which are similar to the obtained nanohumps.

In our case, the time delay plays a major role in controlling the viscosity of the liquid layer. For example, when the time delay is higher, the second pulse shock wave reacts with a surface that already was cooling down, promoting the formation of nanohumps at low irradiation fluence. The mechanism ends up with resolidification of the liquid fluid, when the surface lattice is cooling and the Marangoni forces decrease (after $t = 100$ ps) [13].

On the other hand, it was observed that the cavitation-induced roughness is likely to play a key role in triggering the generation of high-frequency LIPSS upon irradiation by multiple laser pulses. Large-scale molecular dynamics simulation demonstrated that the competition between the evolution of transient liquid structures generated in the spallation process and the rapid resolidification of the surface region are defining the types, sizes and dimensions of the nanoscale surface features [14]. Moreover, associated to void formation and growth, liquid filaments with a high aspect-ratio perpendicular to the surface were reported when the spallation process starts in atomistic computations targeting the nanohydrodynamics phenomena [43]. The observed nanospikes on the previous study closely resemble the obtained nanopeaks in our study. To go further ahead in this analysis, complementary microstructural investigations are needed to enlarge the study of the physical mechanisms leading to the formation of the observed nanopatterns.

5. Conclusions

Various nanopatterns corresponding to distinct self-organization regimes induced by convection mechanism at the nanoscale were investigated on femtosecond laser-irradiated nickel surfaces. Forming far from equilibrium, complex patterns of structures are revealed caused by intensive light energy exchange with a dissipative environment. An hydrothermal mechanism is highlighted by the appearance of two competing nanostructures, nanohumps and hexagonal nanocavities. The nanorelief is characterized by mounds and depressions where the unbalance and the respective predominance is shown to be tunable by the time-delay and the fluence. Observed for (0.18 J/cm^2; $\Delta t = 15$ ps) and (0.24 J/cm^2; $\Delta t = 25$ ps) both kind of nanostructures have 30 nm diameter and tens of nanometers height. Nanopeaks and nanobumps were observed at (0.18 J/cm^2; $\Delta t = 8$ ps) and (0.18 J/cm^2; $\Delta t = 10$ ps). The nanopeaks have 20 nm width and 110 nm height whereas the nanobumps exhibit a 65 nm width and 100 nm height. Potentially resulting from thermo-convective instabilities, they were formed during the competition of the pressure-gradient (via surface tension and cavitation) forces and the rapid resolidification of the molten surface region. Successively implying emergence, growth, amplification and regulation of the pattern formation, laser parameters as peak fluence, time delay, and number of double-pulses sequences play a significant role on modifying surface morphology and topography. Furthermore, the initial surface roughness and type of roughness is another essential feature to be controlled to switch from a regime of self-organization to another one. The involved instabilities being expected in most of ultrafast-laser irradiated materials, the process reported in this paper is a priori generalizable to any metals in order to design a predefined surface morphology at the nanoscale. The production of these nanostructures can enable unique surface functionalizations toward the control of mechanical, biomedical, optical, or chemical surface properties on a nanometric scale [44].

Author Contributions: F.P., F.G., C.M. and J.-P.C. designed and directed the project. A.N. performed the samples preparations experiments. M.A. and A.N. performed the laser irradiation experiments. A.N. performed the characterizations experiments on SEM and AFM. A.R. and J.-P.C. developed the theoretical framework. A.N. and J.-P.C. wrote the manuscript. All authors discussed the results and contributed to the final manuscript. All authors have read and agreed to the published version of the manuscript.

Funding: This work was carried out thanks to the financial support of the Project ANR-17-EURE-0026 of the University of Lyon within the framework of the Programme Investissement d'Avenir (ANR-16-IDEX-0005).

Institutional Review Board Statement: Not applicable.

Informed Consent Statement: Not applicable.

Data Availability Statement: Not applicable.

Conflicts of Interest: The authors declare no conflict of interest.

Abbreviations

The following abbreviations are used in this manuscript:

LIPSS	Laser-Induced Periodic Surface Structures
HSFL	High Spatial Frequency LIPSS
SEM	Scanning Electron Microscopy
AFM	Atomic Force Microscopy
Ra	Arithmetic Roughness
Ra_{th}	Arithmetic Roughness Threshold
Ku	Kurtosis
P_r	Prandtl Number
P_r^c	Critical Prandtl Number
N_{DPS}	Number of Double-Pulses Sequences
Δt	Time-Delay Between the Double-Pulses

References

1. Echeverría-Alar, S.; Clerc, M.G. Labyrinthine patterns transitions. *Phys. Rev. Res.* **2020**, *2*, 042036. [CrossRef]
2. Sedao, X.; Maurice, C.; Garrelie, F.; Colombier, J.-P.; Reynaud, S.; Quey, R.; Pigeon, F. Influence of crystal orientation on the formation of femtosecond laser-induced periodic surface structures and lattice defects accumulation. *Appl. Phys. Lett.* **2014**, *104*, 171605. [CrossRef]
3. Abou-Saleh, A.; Colombier, J.-P.; Garrelie, F. Relation Entre Auto-Organisation et Création/Résorption de Défauts Microsturaux Sous Irradiation Laser Ultrabrèves. Ph.D. Thesis, Université de Lyon, Lyon, France, 2019. Available online: http://theses.fr/2019LYSES001 (accessed on 27 January 2020).
4. Bonse, J.; Krüger, J.; Höhm, S.; Rosenfeld, A. Femtosecond Laser-Induced Periodic Surface Structures. *J. Laser Appl.* **2012**, *24*, 042006. [CrossRef]
5. Wang, S.; Jiang, L.; Han, W.; Liu, W.; Hu, J.; Wang, S.; Lu, Y. Controllable formation of laser-induced periodic surface structures on ZnO film by temporally shaped femtosecond laser scanning. *Opt. Lett.* **2020**, *45*, 2411–2414. [CrossRef]
6. Zhang, H.; Colombier, J.-P.; Li, C.; Faure, N.; Cheng, G.; Stoian, R. Coherence in Ultrafast Laser-Induced Periodic Surface Structures. *Phys. Rev. B* **2015**, *92*, 174109. [CrossRef]
7. Déziel, J.-L.; Dumont, J.; Gagnon, D.; Dubé, L.J.; Messaddeq, S.H.; Messaddeq, Y. Constructive feedback for the growth of laser-induced periodic surface structures. *Phys. Status Solidi C* **2016**, *13*, 121–124. [CrossRef]
8. Skolski, J.Z.P.; Römer, G.R.B.E.; Vincenc Obona, J.; Huis in't Veld, A.J. Modeling laser-induced periodic surface structures: Finite-difference time-domain feedback simulations. *J. Appl. Phys.* **2014**, *115*, 103102. [CrossRef]
9. Rudenko, A.; Mauclair, C.; Garrelie, F.; Stoian, R.; Colombier, J.P. Light absorption by surface nanoholes and nanobumps. *Appl. Surf. Sci.* **2019**, *470*, 228–233. [CrossRef]
10. Rudenko, A.; Colombier, J.-P.; Höhm, S.; Rosenfeld, A.; Krüger, J.; Bonse, J.; Itina, T.E. Spontaneous Periodic Ordering on the Surface and in the Bulk of Dielectrics Irradiated by Ultrafast Laser: A Shared Electromagnetic Origin. *Sci Rep.* **2017**, *7*, 12306. [CrossRef] [PubMed]
11. Skolski, J.Z.P.; Römer, G.R.B.E.; Obona, J.V.; Ocelik, V.; Huis in't Veld, A.J.; De Hosson, J.T.M. Laser-Induced Periodic Surface Structures: Fingerprints of Light Localization. *Phys. Rev. B* **2012**, *85*, 075320. [CrossRef]
12. Sedao, X.; Abou Saleh, A.; Rudenko, A.; Douillard, T.; Esnouf, C.; Reynaud, S.; Maurice, C.; Pigeon, F.; Garrelie, F.; Colombier, J.-P. Self-Arranged Periodic Nanovoids by Ultrafast Laser-Induced Near-Field Enhancement. *ACS Photonics* **2018**, *5*, 1418–1426. [CrossRef]
13. Rudenko, A.; Saleh, A.A.; Pigeon, F.; Mauclair, C.; Garrelie, F.; Stoian, R.; Colombier, J.-P. High-frequency periodic patterns driven by non-radiative fields coupled with Marangoni convection instabilities on laser-excited surfaces. *ACS Photonics* **2020**, *194*, 93–105. [CrossRef]
14. Abou-Saleh, A.; Karim, E. T.; Maurice, C.; Reynaud, S.; Pigeon, F.; Garrelie, F.; Zhigilei, L.V.; Colombier, J.P. Spallation-Induced Roughness Promoting High Spatial Frequency Nanostructure Formation on Cr. *Appl. Phys.* **2018**, *124*, 308. [CrossRef]
15. Jalil, S.A.; Yang, J.; ElKabbash, M.; Cong, C.; Guo, C. Formation of controllable 1D and 2D periodic surface structures on cobalt by femtosecond double pulse laser irradiation. *Appl. Phys. Lett.* **2019**, *115*, 031601. [CrossRef]
16. Fraggelakis, F.; Mincuzzi, G.; Lopez, J.; Manek-Hönninger, I.; Kling, R. Controlling 2D Laser Nano Structuring over Large Area with Double Femtosecond Pulses. *Appl. Surf. Sci.* **2019**, *470*, 677–686. [CrossRef]
17. Fraggelakis, F.; Giannuzzi, G.; Gaudiuso, C.; Manek-Hönninger, I.; Mincuzzi, G.; Ancona, A.; Kling, R. Double- and Multi-Femtosecond Pulses Produced by Birefringent Crystals for the Generation of 2D Laser-Induced Structures on a Stainless Steel Surface. *Materials* **2019**, *12*, 1257. [CrossRef] [PubMed]
18. Wu, Z.; Yin, K.; Wu, J.; Zhu, Z.; Duan, J.-A.; He, J. Recent advances in femtosecond laser-structured Janus membranes with asymmetric surface wettability. *Nanoscale* **2021**, *13*, 2209–2226. [CrossRef] [PubMed]
19. Bonse, J. Quo Vadis LIPSS?—Recent and Future Trends on Laser-Induced Periodic Surface Structures. *Nanomaterials* **2020**, *10*, 1950. [CrossRef]

20. Zorba, V.; Stratakis, E.; Barberoglou, M.; Spanakis, E.; Tzanetakis, P.; Anastasiadis, S.H.; Fotakis, C. Biomimetic Artificial Surfaces Quantitatively Reproduce the Water Repellency of a Lotus Leaf. *Adv. Mater.* **2008**, *20*, 4049–4054. [CrossRef]

21. Elbourne, A.; Crawford, R.J.; Ivanova E.P. Nano-structured antimicrobial surfaces: From nature to synthetic analogues *J. Colloid Interface Sci.* **2017**, *508*, 603–616. [CrossRef] [PubMed]

22. Vorobyev, A.Y.; Guo, C. Colorizing Metals with Femtosecond Laser Pulses. *Appl. Phys.* **2008**, *4*. [CrossRef]

23. Guay, J.-M.; Calà Lesina, A.; Côté, G.; Charron, M.; Poitras, D.; Ramunno, L.; Berini, P.; Weck, A. Laser-Induced Plasmonic Colours on Metals. *Nat Commun.* **2017**, *8*, 16095. [CrossRef] [PubMed]

24. Bonse, J.; Koter, R.; Hartelt, M.; Spaltmann, D.; Pentzien, S.; Höhm, S.; Rosenfeld, A.; Krüger, J. Tribological Performance of Femtosecond Laser-Induced Periodic Surface Structures on Titanium and a High Toughness Bearing Steel. *Appl. Surf. Sci.* **2015**, *336*, 21–27. [CrossRef]

25. Abou Saleh, A.; Rudenko, A.; Reynaud, S.; Pigeon, F.; Garrelie, F.; Colombier, J.-P. Sub-100 Nm 2D Nanopatterning on a Large Scale by Ultrafast Laser Energy Regulation. *Nanoscale* **2020**, *12*, 6609–6616. [CrossRef]

26. Liu, J.M. Simple technique for measurements of pulsed Gaussian-beam spot sizes. *Opt. Lett.* **1982**, *7*, 196–198. [CrossRef] [PubMed]

27. Bévillon, E.; Stoian, R.; Colombier, J.P. Nonequilibrium optical properties of transition metals upon ultrafast electron heating. *J. Phys. Condens. Matter.* **2018**, *30*, 385401. [CrossRef]

28. Korte, F.; Koch, J.; Chichkov, B.N. Formation of microbumps and nanojets on gold targets by femtosecond laser pulses. *Appl. Phys. A* **2004**, *79*, 879–881. [CrossRef]

29. Savolainen, J.M.; Christensen, M.S.; ; Balling, P. Material swelling as the first step in the ablation of metals by ultrashort laser pulses. *Phys. Rev. B* **2011**, *84*, 193410. [CrossRef]

30. Baset, F.; Popov, K.; Villafranca, A.; Alshehri, A.M.; Guay, J.M.; Ramunno, L.; Bhardwaj, V.R. Nanopillar formation from two-shot femtosecond laser ablation of poly-methyl methacrylate. *Appl. Surf. Sci.* **2015**, *357*, 273–281. [CrossRef]

31. Vitral, E.; Mukherjee, S.; Leo, P.H.; Viñals, J.; Paul, M.R.; Huang, Z.F. Spiral defect chaos in Rayleigh-Bénard convection: Asymptotic and numerical studies of azimuthal flows induced by rotating spirals. *Phys. Rev. Fluids* **2020**, *5*, 093501. [CrossRef]

32. Bodenschatz, E.; Pesch, W.; Ahlers, G. Recent Developments in Rayleigh-Bénard Convection. *Phys. Rev. Fluids Annu. Rev. Fluid Mech.* **2000**, *32*, 709–778. [CrossRef]

33. Thess, A.; Bestehorn, M. Planform selection in Bénard-Marangoni convection: l hexagons versus g hexagons. *Phys. Rev. E* **1995**, *52*, 6358–6367. [CrossRef] [PubMed]

34. Pearson, J. On convection cells induced by surface tension. *J. Fluid Mech.* **1958**, *4*, 489–500. [CrossRef]

35. Sharp, D.H. An overview of rayleigh-taylor instability. Presented at the International Conference on "Fronts, Interfaces and Patterns", Los Alamos, NM, USA, 2–6 May 1983.

36. Morgan, R.V.; Cabot, W.H.; Greenough, J.A.; Jacobs, J.W. Rarefaction-driven Rayleigh–Taylor instability. Part 2. Experiments and simulations in the nonlinear regime. *J. Fluid Mech.* **2018**, *838*, 320–355. [CrossRef]

37. Smith, M.K.; Davis, S.H. Instabilities of dynamic thermocapillary liquid layers. Part 1. Convective instabilities. *J. Fluid Mech.* **1983**, *132*, 119–144. [CrossRef]

38. Smith, M. Instability mechanisms in dynamic thermocapillary liquid layers. *Phys. Fluids* **1986**, *29*, 3182–3186. [CrossRef]

39. Tsibidis, G.D.; Fotakis, C.; Stratakis, E. From ripples to spikes: A hydrodynamical mechanism to interpret femtosecond laser-induced self-assembled structures. *Phys. Rev. B* **2015**, *92*, 041405. [CrossRef]

40. Busse, F.H. Remarks on the critical value Pc = 0.25 of the Prandtl number for internally heated convection found by Tveitereid and Palm. *Eur. J. Mech. B/Fluids* **2014**, *47*, 32–34. [CrossRef]

41. Boeck, T.; Thess, A. Bénard–Marangoni convection at low Prandtl number. *J. Fluid Mech.* **1999**, *399*, 251–275. [CrossRef]

42. Bragard, J.; Velarde, M.G. Bénard–Marangoni convection: planforms and related theoretical predictions. *J. Fluid Mech.* **1998**, *368*, 165–194. [CrossRef]

43. Starikov, S.V.; Pisarev, V.V. Atomistic simulation of laser-pulse surface modification: predictions of models with various length and time scales. *J. Appl. Phys.* **2015**, *117*, 135901. [CrossRef]

44. Bonse, J.; Kirner, S.V.; Krüger, J. Laser-Induced Periodic Surface Structures (LIPSS). In *Handbook of Laser Micro- and Nano-Engineering*; Springer International Publishing: Cham, Switzerland, 2020; pp. 1–59. [CrossRef]

 *nanomaterials*

Article

Polystyrene Thin Films Nanostructuring by UV Femtosecond Laser Beam: From One Spot to Large Surface

Olga Shavdina [1],*, Hervé Rabat [1], Marylène Vayer [2], Agnès Petit [1], Christophe Sinturel [2] and Nadjib Semmar [1]

[1] GREMI (Groupe de Recherches sur l'Energétique des Milieux Ionisés)-UMR (Unité Mixte de Recherche) 7344-CNRS, University of Orleans, 45067 Orléans, France; herve.rabat@univ-orleans.fr (H.R.); agnes.petit@univ-orleans.fr (A.P.); nadjib.semmar@univ-orleans.fr (N.S.)

[2] ICMN (Interfaces, Confinement, Matériaux et Nanostructures)-UMR (Unité Mixte de Recherche) 7374-CNRS, Université d'Orleans, 45071 Orléans, France; marylene.vayer@cnrs-orleans.fr (M.V.); christophe.sinturel@univ-orleans.fr (C.S.)

* Correspondence: olga.shavdina@univ-orleans.fr

Abstract: In this work, direct irradiation by a Ti:Sapphire (100 fs) femtosecond laser beam at third harmonic (266 nm), with a moderate repetition rate (50 and 1000 Hz), was used to create regular periodic nanostructures upon polystyrene (PS) thin films. Typical Low Spatial Frequency LIPSSs (LSFLs) were obtained for 50 Hz, as well as for 1 kHz, in cases of one spot zone, and also using a line scanning irradiation. Laser beam fluence, repetition rate, number of pulses (or irradiation time), and scan velocity were optimized to lead to the formation of various periodic nanostructures. It was found that the surface morphology of PS strongly depends on the accumulation of a high number of pulses (10^3 to 10^7 pulses) at low energy (1 to 20 µJ/pulse). Additionally, heating the substrate from room temperature up to 97 °C during the laser irradiation modified the ripples' morphology, particularly their amplitude enhancement from 12 nm (RT) to 20 nm. Scanning electron microscopy and atomic force microscopy were used to image the morphological features of the surface structures. Laser-beam scanning at a chosen speed allowed for the generation of well-resolved ripples on the polymer film and homogeneity over a large area.

Keywords: polymer thin films; femtosecond beam; laser texturing; LSFL; 2D LIPSS; substrate temperature effect; scanning processing

Citation: Shavdina, O.; Rabat, H.; Vayer, M.; Petit, A.; Sinturel, C.; Semmar, N. Polystyrene Thin Films Nanostructuring by UV Femtosecond Laser Beam: From One Spot to Large Surface. *Nanomaterials* **2021**, *11*, 1060. https://doi.org/10.3390/nano11051060

Academic Editors: Peter Simon, Jürgen Ihlemann and Jörn Bonse

Received: 23 March 2021
Accepted: 16 April 2021
Published: 21 April 2021

Publisher's Note: MDPI stays neutral with regard to jurisdictional claims in published maps and institutional affiliations.

1. Introduction

Periodic structured surfaces at micro- or nanometric scales find applications in a wide variety of fields, such as microelectronics [1], optics [2], photovoltaics [3], tribology [4], biological sensors [5], and anti-counterfeiting [6]. The generation of periodic surface structures can be performed by using different methods, for instance, nanoimprint [7], which is limited by the necessity to manufacture the micro- or nanostructured molds. The lithographic method requires clean-room facilities and multi-step procedure involving complex mask fabrication [8]. Laser patterning can be a powerful technique for high-resolution surface nano-engineering [9–12]. It is a non-contact, mask-less, and flexible/versatile method with direct generation of micro- and nano-structures called laser-induced periodic surface structures (LIPSSs) observed on a wide range of materials, such as semiconductors [13,14], metals and alloys [15–19], metal oxides [20–23], inorganic compounds [24,25], and polymers [26]. Micro- and nano-structures could also be simultaneously observed in hierarchical patterns [19,24].

The formation mechanisms of LIPSS are still under investigation and have been described by several theories, including interference between the incident and scattered waves [27], self-organization [25], second-harmonic generation [28], and excitation of surface plasmon polaritons [29]. Redistribution of the energy through a feedback mechanism enhances surface structuring [30]. The dimensions of the structures, as well as their morphology, depend on the laser-beam texturing parameters. The periods of the generated structures are strongly correlated to the wavelength of the laser source [31], and the linear polarization allows us to form regular ripple structures parallel to the polarization [32].

Some polymers were structured in the bulk state by femtosecond laser, namely poly(methyl methacrylate) PMMA [33], polystyrene PS [34], polydimethylsiloxane PDMS [35], polytetrafluoroethylene PTFE [36], and polyether ether ketone PEEK [37]. Potential applications of structured polymer surfaces are nano-templating, storage media applications [3], and some specific solutions for biosystems [38]. By direct irradiation of PTFE plate [36], micro–nano dual-scale composite structures were created. The micro-grooves morphology of LIPSS was formed thanks to Ti:Sapphire laser on PMMA microchip [33]. Due to the particular properties of the polymers, such as low glass transition temperature (T_g), the LIPSS formations on polymer foils are often achieved with low beam fluencies (with a fluence of 10.5 and 12.5 mJ/cm^2 for PET and PS irradiated by nanosecond laser beam at 248 nm) [26].

Only a few polymer thin films, such as polyethylene terephthalate PET, polytrimethylene terephthalate PTT, polyvinylidene fluoride PVDF, and polycarbonate bisphenol A PC, were irradiated by nanosecond laser in order to create periodic structures with a period (Λ) similar to the laser wavelength ($\Lambda{\sim}\lambda$) and parallel to the laser polarization direction [39]. Moreover, several studies on LIPSS generation in polymer films were conducted by using solid-state laser-beam sources from the fundamental wavelength, typically 1064 nm with Nd:YAG sources down to the fourth harmonic at 266 nm [32], and in other works, with excimer sources from 193 to 308 nm [40].

Considering the laser regime in the literature, femtosecond beams are less employed for polymer-thin-films processing [41,42], and particularly with UV beams, as investigated in the present work. The advantage of the femtosecond (fs) regime compared to the nanosecond one is that the LIPSS could be formed despite the low absorption coefficient of the polymers [43]. In the fs regime, non-linear absorption and ionization processes mediate the coupling of laser light with the thin films, as reported in Reference [44].

In this work, we first studied femtosecond multi-pulse laser irradiation at 266 nm on spin-coated PS thin films within a single spot zone. More specifically, using low energy values (few μJ/pulse), we studied the influence of laser frequency (50 and 1000 Hz) and number of laser pulses (from 5×10^3 to 800×10^3 pulses) on the surface morphology and on the LIPSS main features (period and amplitude). The influence of the substrate temperature on LIPSS morphology, rarely explored to our knowledge, was also investigated thanks to a controlled heating Peltier module.

Using a XY-translation stage, we created ripples over large area (few mm length and width) and optimized the different parameters for this process. The surface morphology of the irradiated films was systematically analyzed by scanning electron microscopy (SEM) and atomic force microscopy (AFM).

2. Materials and Methods

The schematic of the procedure leading to LIPSS formation on polymer films is shown in Figure 1. Firstly, polymer solution was prepared by dissolution of polystyrene in dichloroethane (30 mg/mL) (PS with a molecular weight of 170 kg·mol^{-1} was purchased from Sigma-Aldrich (Merck KGaA, Saint-Louis, MO, USA)). Silicon wafers were cut into pieces of 1.5×1.5 cm^2. These substrates were cleaned with water, ethanol and acetone. Spin-coating of polymer solution was carried out at the speed of 4000 rpm, during 30 s, on silicon substrates (Figure 1b). The deposited polymer film's thickness (around 180 nm) was determined by using a DektakXT Stylus profilometer (Bruker Corporation, Billerica, MA, USA).

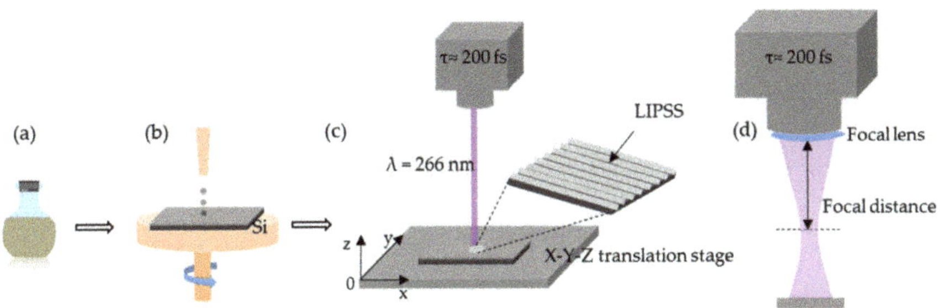

Figure 1. Schematics of PS laser surface structuring steps: (**a**) preparation of polymer solution by dissolution of the polymer in solvent, (**b**) spin-coating of polymer solution onto silicon substrate to obtain a supported polymer film, (**c**) irradiation of polymer film by fs laser, and (**d**) laser-beam focus using a focal lens.

PS-thin-film irradiations were carried out with femtosecond laser beam at a wavelength of 266 nm, thanks to the third harmonic generator of a Ti:Sapphire source (800 nm, 100 fs). An estimation of the temporal duration of these UV pulses is about 200 fs, as reported in Reference [45]. The Spectra Physics Mai Tai laser system (Mai Tai Diode-Pumped, Mode-Locked Ti:Sapphire Laser, Spectra-Physics Inc., Mountain View, CA, USA) used in our experiment consists of two parts: a CW diode-pumped Nd:YVO$_4$ laser and a mode-locked Ti:Sapphire pulsed laser. We used linearly polarized laser pulses with a Gaussian beam spatial profile. The beam spot size and, thus, the fluence were adjusted by the distance between the focal lens and the sample XY-translation stage, at normal incidence parallel to the Z-axis (Figure 1d). The working distance (from focusing lens to substrate surface) was determined to be close to 78.5 mm in the single spot case. The control of the pulse number (N) for 50 Hz and 1 kHz repetition rates was achieved thanks to an electronic shutter. The experiments were performed under standard laboratory conditions, ambient atmosphere, and room temperature (RT).

To study the effect of substrate temperature on the structure morphology, the substrate was heated during the irradiation, from RT to 40, 60, and 97 °C. Then the Peltier module (RS 490-1525, Adaptive, Russia) was used. In the scanning mode, only the 1 kHz repetition rate condition was applied. The same focal lens as in single spot processing was used to focalize the laser beam. The velocity of the XY-translation stage displacement was between 20 and 100 µm/s. The effective number of pulses, N$_{eff}$, was directly derived from the overlap of the spots and depends on the spot size and the scanning velocity. The working distance was determined to be close to 68.5 mm in the scanning mode, leading to the best process parameters in terms of spot size, laser fluence, and process time.

An Atomic Force Microscope (AFM) Dimension ICON from Bruker with ScanAsyst in Air mode (Bruker Corporation, Billerica, MA, USA) was used to study the surface morphologies and to compare sample roughness before and after irradiation. The morphology of the thin film was also observed after irradiation, using a Zeiss SUPRA 40 Scanning Electron Microscope (SEM) (German Carl Zeiss Company, Oberkochen, Germany).

3. Results

3.1. Single Spot Processing

Figure 2 shows SEM micrographs of PS film before (Figure 2a) and after laser irradiation (Figure 2b–f) for 50 Hz of repetition rate, 8 µJ of pulse energy, and 11.8 mJ/cm^2 of fluence per pulse. The generation of LIPSS structures in PS requires a minimum number of pulses to start the modification of the surface morphology. Within our experimental parameters, the first structures started to emerge from 5×10^3 pulses (Figure 2b). These structures are probably formed around microscopic defects generation (hot points) and film roughness, as previously suggested [33]. These few-micrometers hot points grow and merge into larger areas. At N corresponding to 10×10^3 pulses, the micrographs reveal the formation of 1D periodic parallel structures (Figure 2c) that can be attributed to a low spatial-repetition rate LIPSS (LSFL). During the subsequent pulses, probably through the radiation–structuration feedback mechanisms, LSFL structures emerged as dominant structures. By further increasing the number of shots up to 15×10^3 shots, we observed nanodots structures with a 2D periodicity (2D-LIPSS) from Figure 2d. For the number of pulses between 15×10^3 and 35×10^3, LSFL and 2D-LIPSS were observed simultaneously within the same irradiation spot. From AFM analyses for the same samples characterized by SEM, the 2D Fast Fourier Transformation (FFT) was used in order to analyze the periodicities of LIPSS (Figure 2). From FFT analysis, we found that, for 10×10^3 pulses, LSFL has a period of $\Lambda{\sim}165$ nm, while above 15×10^3 pulses, the observed 2D-LIPSS exhibit a 2D periodicity of $\Lambda_1{\sim}165$ nm and $\Lambda_2{\sim}230$ nm. For 25×10^3 and 35×10^3 pulses (Figure 2e,f), from 2D-FFT analysis, the same 2D orientation and periodicity are observed.

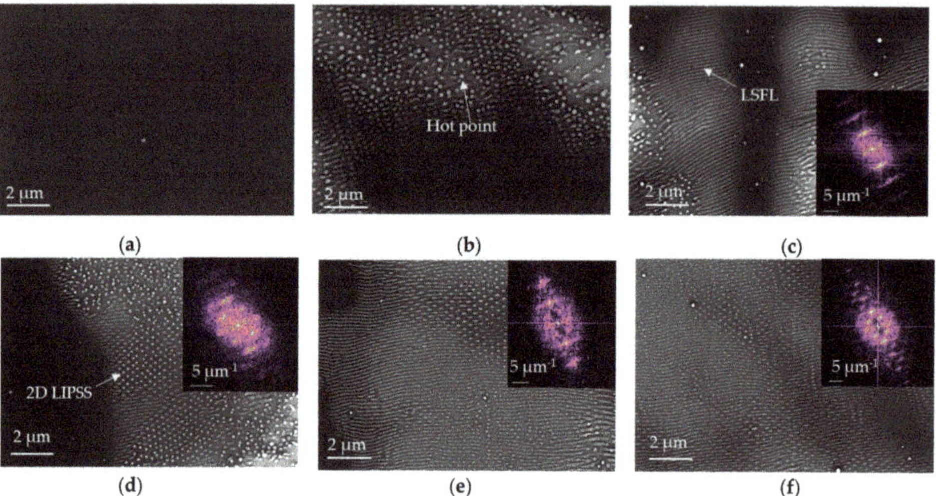

Figure 2. SEM micrographs of PS film with corresponding FFT images from AFM analysis: (**a**) before and after laser irradiation at pulse fluence F = 11.8 mJ/cm^2, repetition rate of 50 Hz for pulse number N: (**b**) 5×10^3, (**c**) 10×10^3, (**d**) 15×10^3, (**e**) 25×10^3, and (**f**) 35×10^3.

From classical electromagnetic theory, it is known that the period, Λ, of the ripples depends on the laser wavelength, λ, effective refractive index, n, and the angle of incidence of the radiation, θ:

$$\Lambda = \frac{\lambda}{n \pm \sin(\theta)} \tag{1}$$

Thanks to the ellipsometric measurement, we found a refractive index of 1.57 for PS film at 266 nm, using Cauchy model. PS film was irradiated without tilting, and the calculated period of LIPSS corresponds to $\Lambda{\sim}170$ nm, confirming theoretical approach. This type of structure is induced by interference between incident wave and scattered and reflected waves that creates a modulation of the intensity of the beam in thin film.

Figure 3a,b presents AFM images of PS film after 35×10^3 pulses of laser irradiation and the corresponding cross-section profiles. Figure 3c displays a SEM micrograph of the tilted sample exhibiting ripples in the foreground and 2D-LIPSS in the background.

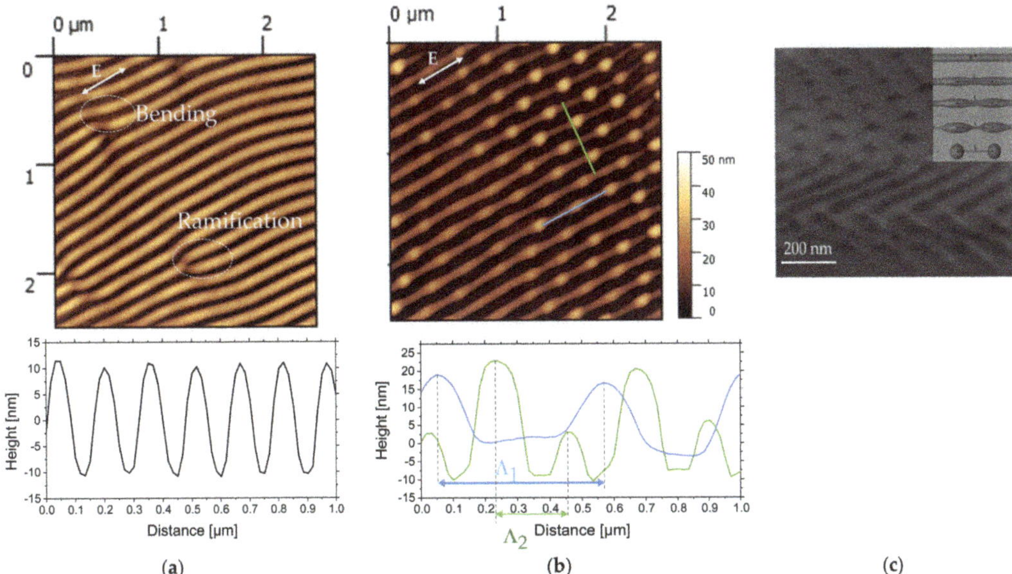

Figure 3. AFM height (amplitude) images and cross-section profiles along the corresponding lines (2.5 × 2.5 μm^2 size, Z scale is 50 nm) of PS film irradiated at 266 nm, at a repetition rate of 50 Hz, a fluence of 11.8 mJ/cm^2, after 35 × 10^3 pulses: (**a**) area 1, (**b**) area 2, and (**c**) SEM image. The polarization direction of the laser beam is indicated by double arrows.

From AFM images on area 1 (Figure 3a), ripples with a typical height of 20 nm can be observed with some bendings and ramifications. The ramifications are characterized by the separation of the ripples into two or three separated ripples. These types of structures are associated with a spatial modulation of absorbed energy by defects on polystyrene thin film and superposition of the LIPSS wave vectors. The same types of structures were observed on silicon surface [46] and were explained by combining two-temperature model, free-carrier dynamics and Sipe's theory [47].

From Figure 3b, another type of structure is observed on area 2, characterized by the formation of 2D-LIPSS within the ripples. From the AFM image, 2 periodicities can be observed with Λ_1~340 nm and Λ_2~165 nm. 2D LIPSSs with a typical height of 30 nm are formed inside parallel LSFL structures with a typical height of 13 nm. We suggest that formation of these nanodots is dominated by hydrodynamic instabilities, resulting in material redistribution from parallel ripples. We suppose that local instabilities are induced by surface-tension minimization that results in the ripple breaking into smaller parts. The viscosity of polymers is decreasing during irradiation, and hydrodynamic phenomenon starts to play a dominant role. The presence of two types of structures, namely ripples and nanodots, is the result of viscosity variations and a combination of broken and unbroken LSFL. Hexagonal lattice types have also been studied by M. Fermigier et al. [48], where the nonlinear growth of the Rayleigh–Taylor instability leads to the formation of two-dimensional patterns with hexagonal shape.

In order to produce the periodic structures at higher laser repetition rate and reduce the duration between pulses down to 1 ms, the irradiations were also carried out with the repetition rate of 1 kHz. Figure 4 displays AFM height images of 2.5 × 2.5 μm^2 area of PS films irradiated with 50 × 10^3, 100 × 10^3, and 800 × 10^3 pulses for 0.6 mJ/cm^2 of pulse fluence. For the lowest number of pulses (Figure 4a), first morphological changes are observed with the creation of periodic parallel structures. The period is Λ~170 nm, and the height of the structures is up to 10 nm. With the increasing of the pulses number up to 100 × 10^3, nanodots start to appear with the decreasing of the period Λ down to 150 nm.

The height of these nanodots is around 10 nm. Finally, for an even higher number of pulses corresponding to 800×10^3 (Figure 4c), the parallel structures start to be distorted with the irradiated area exhibiting an array of like-groove morphologies. The distance between two consecutive grooves is around $\Lambda \sim 250$ nm, and their depth is around 25 nm.

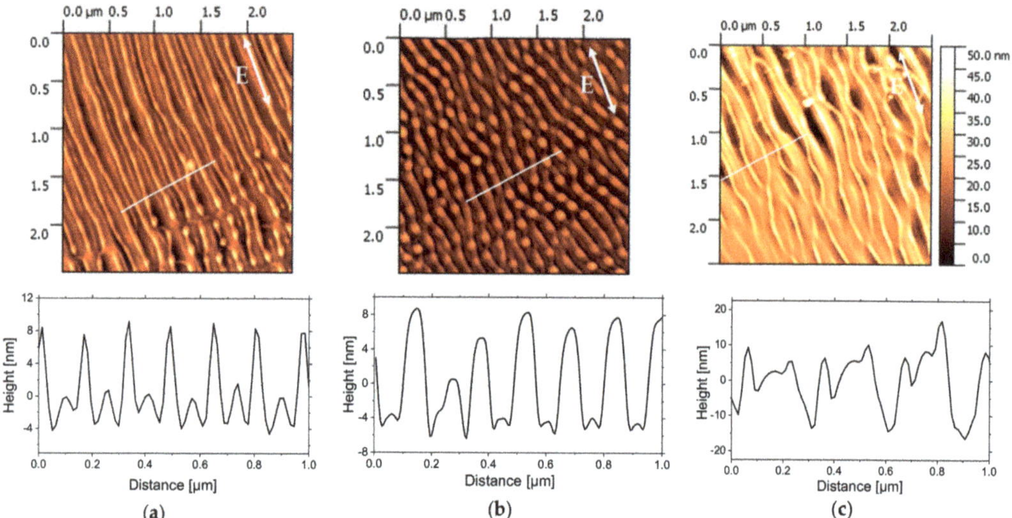

Figure 4. AFM height images (2.5×2.5 μm^2 size) and cross-sections profiles along the corresponding white lines of PS films irradiated at 266 nm, 1 kHz of repetition rate, and at F = 0.6 mJ/cm^2, with number of pulses, N, (**a**) 50×10^3, (**b**) 100×10^3, and (**c**) 800×10^3. The polarization direction of the laser beam is indicated by double arrows.

It was proposed by He and al. [46] that the quasi-periodic pattern of stripes on silicon is related to a spatial redistribution of the absorbed energy and to the formation of the regions where the absorbed fluence is not enough to induce effective ablation and could fuse the surface nanostructures leading to the progressive generation of the groove stripes covering the underlying ripples. We suppose that the same phenomenon was observed for PS film and reactions of photolysis are generated and induced the material loss. These photochemical reactions are characterized by much shorter time scales than thermal-related processes.

Based on Sipe's theory [47], wave vectors are possible from LIPSS, depending on surface parameters (surface roughness and dielectric permittivity) and laser irradiation parameters. Once the LIPSSs are formed, surface roughness will induce the further enhancement of the ripples' formation characterized by the redistribution of optical energy from subsequent laser pulses. In other words, we find the feedback effect, which explains the increase on the surface of the ripples. It is important to underline that the ripple formation is extended over a large area, and this is related to repeat scattering during multi-pulse irradiation, which can propagate the pattern.

Irradiation at 50 Hz can be compared with irradiation at 1 kHz, using the accumulated dose, which is the fluence per pulse times the number of pulses. The accumulated dose (60 J/cm^2) at 50 Hz for 5×10^3 pulses is the same as the one at 1 kHz for 100×10^3 pulses, and the accumulated dose (413 J/cm^2) at 50 Hz for 35×10^3 pulses is similar as the one (480 J/cm^2) at 1 kHz for 800×10^3 pulses.

3.2. Influence of the Substrate Temperature on the LIPSS Morphology

It is known that the behavior of polymers strongly depends on the temperature [49]. From a microscopic point of view, their behavior is associated with molecular relaxations and the elastic modulus decreases above T_g (here measured for PS bulk by DSC at 95 °C, in accordance with the value estimated by Vignaud et al. [50] at 97 °C for a PS 135 nm–thick layer). To investigate the influence of the substrate temperature, the ripples were generated with number of pulses $N = 10^5$, which is optimal for the formation of organized structures at a fixed fs-laser pulse fluence ($F = 0.6$ mJ/cm^2) and laser repetition rate (1 kHz). This allowed us to create ripples parallel to the polarization vector. We varied the temperature of the Peltier module from RT to 97 °C, heating the polymer film during the irradiation. Figure 5 displays the corresponding SEM images. LSFLs are observed with the same period between ripples $\Lambda{\sim}170$ nm whatever the temperature of the substrate. However, the width of the ripples decreases from 80 nm at RT (Figure 5a) to 55 nm for 97 °C (Figure 5d). From AFM images, it could be seen that the height of the ripples was increased from 10 nm for RT (Figure 4b) to 20 nm for 97 °C (Figure 5e). The LIPSSs are formed independently of PS thickness, indicating that structuration process is significant only in the surface layer, which absorbs the light of the laser beam.

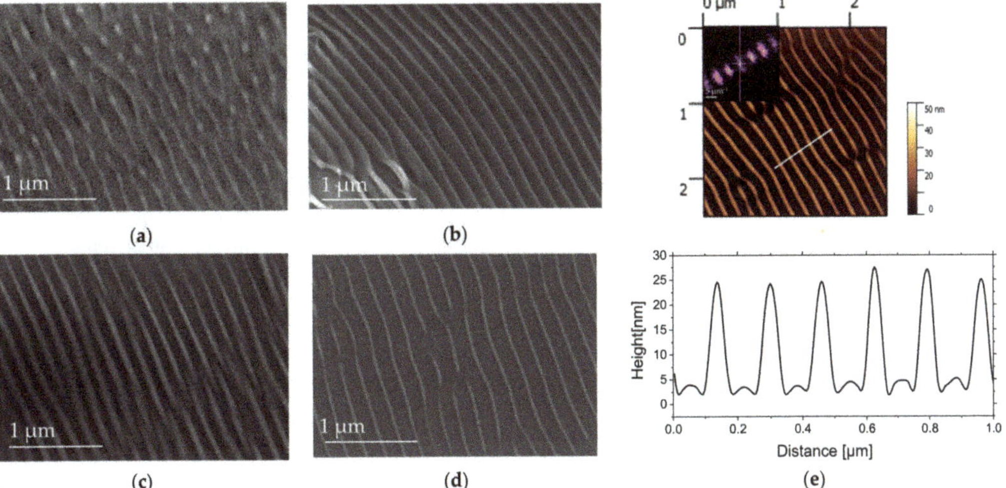

Figure 5. SEM images of PS film irradiated at $\lambda = 266$ nm, 1 kHz of repetition rate and pulse fluence $F = 0.6$ mJ/cm^2 with 10^5 pulses at (**a**) 23.5 °C, (**b**) 40 °C, (**c**) 60 °C, and (**d**) 97 °C. (**e**) AFM height image (2.5×2.5 µm^2 size) of PS film irradiated at 266 nm, 1 kHz of repetition rate with number of pulses $N = 100 \times 10^3$ at 97 °C, and cross-section profile along the white line.

At low temperature corresponding to 23.5 °C, chains of the polymer are in a glassy state. Increasing of the temperature causes an increase in the mobility of macromolecular chains within thin film. The application of the laser energy at a temperature higher than RT can promote the passage of PS in soften state (Figure 5b) [51]. The softening of the PS makes hydrodynamic contributions, including thermo-capillary flow to the LSFL mechanisms, dominant, leading to higher and narrower ripples.

3.3. Scanning Processing

We also proceed to the LIPSS generation in scanning mode in order to create periodic structures on large area. In order to optimize the scanning time, we approached the substrate towards the FoF position to $z = 68.5$ mm from the focusing laser lens. Figure 6a illustrates the beam overlap during the scanning process. There are two zones in the laser spot to be consider, as shown in Figure 6b: The first one in the middle of the laser spot, with a diameter equal to 2r, is the zone where there are many photons interacting with the PS layer and where soft ablation can take place in relation to the Gaussian beam shape.

There is also the weak zone of interaction of the laser beam with the layer, with the outer diameter 2R.

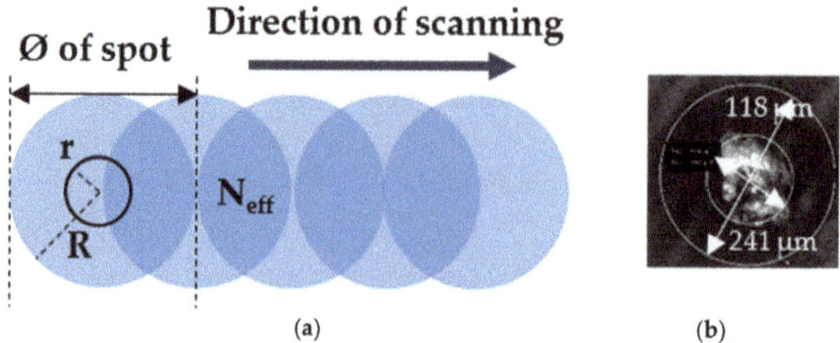

Figure 6. (**a**) Schematic of the beam overlap during a scanning process. (**b**) SEM image of laser impact on PS film irradiated at 266 nm at the repetition rate of 1 kHz at z = 68.5 mm.

For this, the scanning was carried out at different speeds, v, for a repetition rate, ω, corresponding to 1 kHz, which then defines a number of effectives pulses, N_{eff}, that applied during the overlap of two laser spots with their diameters, Ø, corresponding to 241 µm:

$$N_{eff} = O\frac{\omega}{v} \tag{2}$$

The average number of laser pulses in overlapping area was thus ranged from 2410 to 12,050 by varying the scanning speed of the laser beam from 20 to 100 µm/s at fixed fluence corresponding to 2.2 mJ/cm^2.

Figure 7 presents SEM images of the formed structures. At the microscopic scale, the periodic parallel structures for 20 µm/s are created with $\Lambda \sim 170$ nm (Figure 7a). At low scanning speed, a soft ablative regime dominates in the center of the spot with the LIPSS generation around this area. By increasing the speed until 40 µm/s, the polymer is less affected in central part of laser impact (Figure 7b). For a scanning speed of 60 µm/s, the quasi-parallel ripples are dominant and cover a large area of scanning line (Figure 7c). Higher speeds (80 and 100 µm/s) lead to the same result, with observed quasi-periodic structures with higher period (Figure 7d. This is related to the fact that the effective number of pulses of the laser beam in the overlap zone is not sufficient to induce the LIPSS.

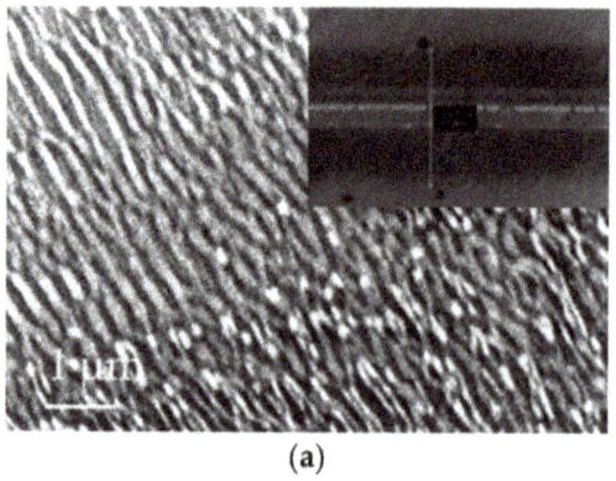

(a)

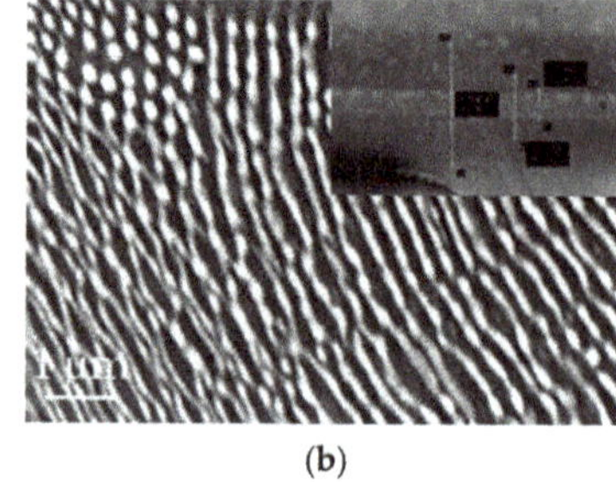

(b)

Figure 7. *Cont.*

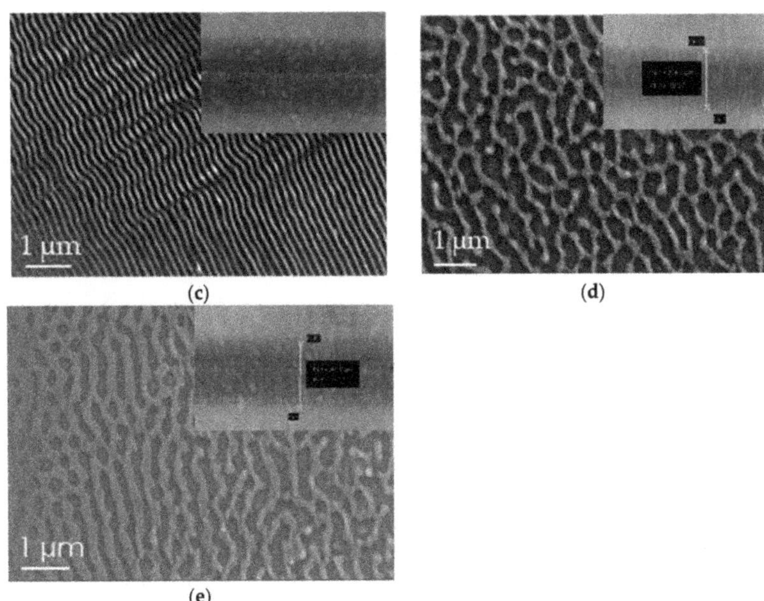

Figure 7. SEM images of PS film irradiated at λ = 266 nm at frequency of 1 kHz and pulse fluence of 2.2 mJ/cm^2 at different scanning speeds: (**a**) 20, (**b**) 40, (**c**) 60, (**d**) 80, and (**e**) 100 μm/s.

From AFM images (Figure 8a–c) and corresponding 2D-FFT image analyses, it can be seen that, at a lower scanning speed corresponding to 20 μm/s (with N_{eff} = 12,050), the structures are linearly oriented with height H~130 nm but partially destructed. This could relate to the material loss due to the photochemical reactions initiated by the absorption of energy of laser beam with the creation of transient excited states and future consequence of bond breaking and radicalization. For a scanning speed of 60 μm/s, well-defined and periodic LSFLs are observed. The height of ripples is about 35 nm, which could be related to the material redistribution with modulated energy deposition but not material loss. When radiation order couples with the surface structure order, one enhances the other, and periodic structures grow faster than disordered ones and emerge as a dominant surface feature. The generation of higher-order harmonics from 2D-FFT analysis highlights the good quality and homogeneity of the ripples on large areas for this scanning speed with the period of Λ~170 nm and the height of H~35 nm. A scanning speed of 80 μm/s leads mostly to an elliptic profile of 2D-FFT, with a higher period corresponding to 400 nm.

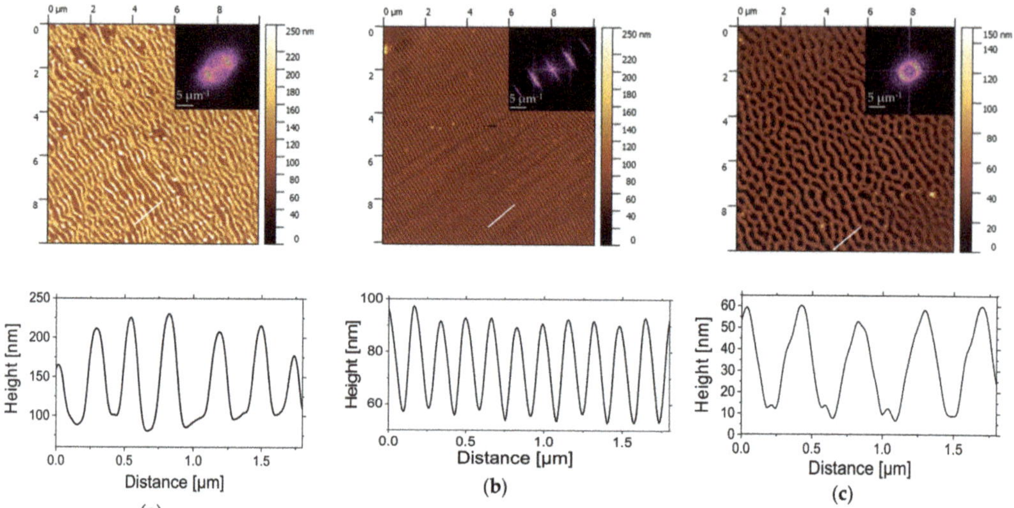

Figure 8. AFM height images (10 × 10 μm² size) and corresponding 2D-FFT scattering graphs of PS film irradiated at λ of 266 nm with repetition rate of 1 kHz and a pulse fluence of 2.2 mJ/cm², at scanning speeds (**a**) 20, (**b**) 60, and (**c**) 80 μm/s.

As shown in Figure 9a, the effective number of pulses decreases from 2410 to 12,050 when the scanning speed increases from 20 to 100 μm/s. The conditions for well-defined and organized LIPSS were optimal for an effective number of pulses of 4000 for a pulse fluence of 2.2 mJ/cm². These conditions allowed us to obtain LIPSS without any defect, except for bending over a large area, as shown in Figure 9b. The size of the area is limited by the size of the scanning area. The test was performed over 4 × 4 mm².

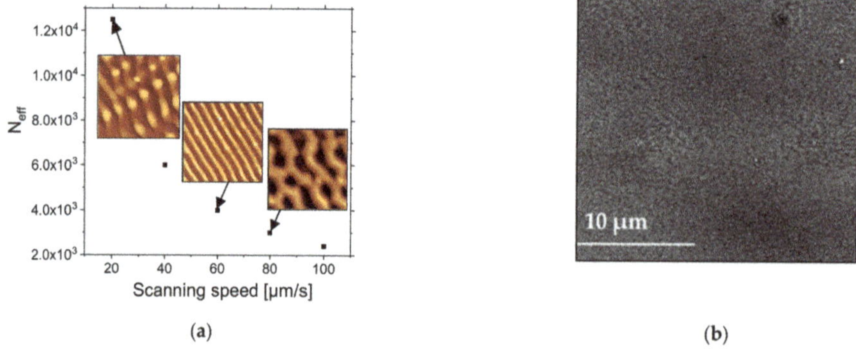

Figure 9. (**a**) Average effective number N_{eff} of pulses at overlapping area as a function of scanning speed (from 20 to 100 μm/s). (**b**) SEM image of PS film irradiated at λ 266 nm, with a repetition rate of 1 kHz and a pulse fluence of 2.2 mJ/cm² at a 60 μm/s scanning speed.

4. Conclusions

We performed PS thin films nanostructuring by UV femtosecond laser beam at 266 nm, with two repetitions rates corresponding to 50 and 1000 Hz. For 50 Hz of repetition rate and 11.8 mJ/cm² of fluence per pulse, parallel LSFL structures were observed above 10^3 pulses structures with a period Λ smaller than the laser irradiation wavelength λ. Ramifications and bendings of ripples were also observed, and they are probably associated with the spatial modulation of the absorbed energy by PS-thin-film defects on polystyrene thin film. Two-dimensional LIPSS structures started to appear for pulse numbers greater than 15×10^3. For 1 kHz of repetition rate, 0.6 mJ/cm^2 of fluence per pulse and a number of pulses between 50×10^3 and 800×10^3, we found parallel structures LSFL with the period close to 170 nm and less than the beam wavelength. For 800×10^3 pulses, the ripples started to be destroyed with the formation of the groove-like morphology with a spatial period of 260 nm. Scanning the PS film with the laser beam also generated LIPSS on large areas, and we found that 60 µm/s was the optimal speed for the creation of well-resolved ripples for a 2.2 mJ/cm^2 of fluence per pulse.

Finally, we demonstrated that heating the PS films from RT up to 97 °C upon femtosecond laser irradiation simultaneously induced the ripples' width decrease and amplitude increase.

Author Contributions: O.S., H.R., M.V., A.P., C.S. and N.S., conceived of and realized the experiments, and wrote the manuscript. All authors have read and agreed to the published version of the manuscript.

Funding: This research received funding from Region Centre-Val-de-Loire (French Government) through the national project APR-IA number 201900134921.

Data Availability Statement: The data is included in the main text.

Acknowledgments: This work is part of a National Project entitled "APR-IA CoMeMAT" funded by the Region Centre-Val-de-Loire.

Conflicts of Interest: The authors declare no conflict of interest.

References

1. Bucknal, D.G. *Nanolithography and Patterning Techniques in Microelectronics*; CRC Press: Boca Raton, FL, USA, 2005.
2. Dewan, R.; Knipp, D. Light trapping in thin-film silicon solar cells with integrated diffraction grating. *J. Appl. Phys.* **2009**, *106*, 23058–23065. [CrossRef]
3. Hauser, H.; Tucher, N.; Tokai, K.; Schneider, P.; Wellens, C.; Volk, A.K.; Seitz, S.; Benick, J.; Barke, S.; Dimroth, F.; et al. Development of nanoimprint processes for photovoltaic applications. *J. Micro/Nanolithography MEMS MOEMS* **2015**, *14*, 031210. [CrossRef]
4. Ranc, H.; Servais, C.; Chauvy, P.-F.; Debaud, S.; Mischler, S. Effect of surface structure on frictional behaviour of a tongue/palate tribological system. *Tribol. Int.* **2006**, *39*, 1518–1526. [CrossRef]
5. Bienaime, A.; Elie-Caille, C.; Leblois, T. Micro Structuration of GaAs Surface by Wet Etching: Towards a Specific Surface Behavior. *J. Nanosci. Nanotechnol.* **2012**, *12*, 6855–6863. [CrossRef]
6. Cui, Y.; Phang, I.Y.; Lee, Y.H.; Lee, M.R.; Zhang, Q.; Ling, X.Y. Multiplex plasmonic anti-counterfeiting security labels based on surface-enhanced Raman scattering. *Chem. Commun.* **2015**, *51*, 5363–5366. [CrossRef]
7. Cattoni, A.; Chen, J.; Decanini, D.; Shi, J.; Haghiri-Gosnet, A.-M. Soft UV Nanoimprint Lithography: A Versatile Tool for Nanostructuration at the 20nm Scale. In *Recent Advances in Nanofabrication Techniques and Applications*; IntechOpen: London, UK, 2011. [CrossRef]
8. Pimpin, A.; Srituravanich, W. Review on Micro- and Nanolithography Techniques and their Applications. *Eng. J.* **2012**, *16*, 37–55. [CrossRef]
9. Yang, L.; Wei, J.; Ma, Z.; Song, P.; Ma, J.; Zhao, Y.; Huang, Z.; Zhang, M.; Yang, F.; Wang, X. The Fabrication of Micro/Nano Structures by Laser Machining. *Nanomaterials* **2019**, *9*, 1789. [CrossRef]
10. Ruffino, F.; Grimaldi, M.G. Nanostructuration of Thin Metal Films by Pulsed Laser Irradiations: A Review. *Nanomaterials* **2019**, *9*, 1133. [CrossRef]
11. Peter, A.; Lutey, A.H.; Faas, S.; Romoli, L.; Onuseit, V.; Graf, T. Direct laser interference patterning of stainless steel by ultrashort pulses for antibacterial surfaces. *Opt. Laser Technol.* **2020**, *123*, 105954. [CrossRef]
12. Gnilitskyi, I.; Derrien, T.J.-Y.; Levy, Y.; Bulgakova, N.M.; Mocek, T.; Orazi, L. High-speed manufacturing of highly regular femtosecond laser-induced periodic surface structures: Physical origin of regularity. *Sci. Rep.* **2017**, *7*, 8485. [CrossRef]
13. Bonse, J. Quo Vadis LIPSS?—Recent and Future Trends on Laser-Induced Periodic Surface Structures. *Nanomaterials* **2020**, *10*, 1950. [CrossRef]
14. Alves-Lopes, I.; Almeida, A.; Oliveira, V.; Vilar, R. Influence of Femtosecond Laser Surface Nanotexturing on the Friction Behavior of Silicon Sliding Against PTFE. *Nanomaterials* **2019**, *9*, 1237. [CrossRef]
15. Liu, Y.-H.; Yeh, S.-C.; Cheng, C.-W. Two-Dimensional Periodic Nanostructure Fabricated on Titanium by Femtosecond Green Laser. *Nanomaterials* **2020**, *10*, 1820. [CrossRef]
16. Huynh, T.T.D.; Vayer, M.; Sauldubois, A.; Petit, A.; Semmar, N. Evidence of liquid phase during laser-induced periodic surface structures formation induced by accumulative ultraviolet picosecond laser beam. *Appl. Phys. Lett.* **2015**, *107*, 193105. [CrossRef]
17. Huynh, T.T.D.; Petit, A.; Semmar, N. Picosecond laser induced periodic surface structure on copper thin films. *Appl. Surf. Sci.* **2014**, *302*, 109–113. [CrossRef]

18. Huynh, T.T.D.; Semmar, N. Dependence of ablation threshold and LIPSS formation on copper thin films by accumulative UV picosecond laser shots. *Appl. Phys. A* **2014**, *116*, 1429–1435. [CrossRef]

19. Yang, Z.; Zhu, C.; Zheng, N.; Le, D.; Zhou, J. Superhydrophobic Surface Preparation and Wettability Transition of Titanium Alloy with Micro/Nano Hierarchical Texture. *Materials* **2018**, *11*, 2210. [CrossRef] [PubMed]

20. Talbi, A.; Semmar, N.; Tabbal, M.; Connor, G.O.; Coddet, P.; Thomann, A.-L.; Stolz, A.; Leborgne, C.; Millon, E. Femtosecond laser irradiation of titanium oxide thin films: Accumulation effect under IR beam. *Appl. Phys. A* **2020**, *126*, 390. [CrossRef]

21. Talbi, A.; Tameko, C.T.; Stolz, A.; Millon, E.; Boulmer-Leborgne, C.; Semmar, N. Nanostructuring of titanium oxide thin film by UV femtosecond laser beam: From one spot to large surfaces. *Appl. Surf. Sci.* **2017**, *418*, 425–429. [CrossRef]

22. Talbi, A.; Kaya-Boussougou, S.; Sauldubois, A.; Stolz, A.; Boulmer-Leborgne, C.; Semmar, N. Laser-induced periodic surface structures formation on mesoporous silicon from nanoparticles produced by picosecond and femtosecond laser shots. *Appl. Phys. A* **2017**, *123*. [CrossRef]

23. Talbi, A.; Petit, A.; Melhem, A.; Stolz, A.; Boulmer-Leborgne, C.; Gautier, G.; Defforge, T.; Semmar, N. Nanoparticles based laser-induced surface structures formation on mesoporous silicon by picosecond laser beam interaction. *Appl. Surf. Sci.* **2016**, *374*, 31–35. [CrossRef]

24. Liu, M.-N.; Wang, L.; Yu, Y.-H.; Li, A.-W. Biomimetic construction of hierarchical structures via laser processing. *Opt. Mater. Express* **2017**, *7*, 2208. [CrossRef]

25. Costache, F.; Henyk, M.; Reif, J. Surface patterning on insulators upon femtosecond laser ablation. *Appl. Surf. Sci.* **2003**, *208–209*, 486–491. [CrossRef]

26. Rebollar, E.; Ezquerra, T.A.; Nogales, A. Laser-induced periodic surface structures (Lipss) on polymer surfaces. *Wrinkled Polym. Surf. Strateg. Methods Appl.* **2019**, 143–155. [CrossRef]

27. Emmony, D.C.; Howson, R.P.; Willis, L.J. Laser mirror damage in germanium at 10.6 μm. *Appl. Phys. Lett.* **1973**, *23*, 598–600. [CrossRef]

28. Le Harzic, R.; Dörr, D.; Sauer, D.; Stracke, F.; Zimmermann, H. Generation of high spatial frequency ripples on silicon under ultrashort laser pulses irradiation. *Appl. Phys. Lett.* **2011**, *98*, 2009–2012. [CrossRef]

29. Miyaji, G.; Miyazaki, K. Origin of periodicity in nanostructuring on thin film surfaces ablated with femtosecond laser pulses. *Opt. Express* **2008**, *16*, 16265. [CrossRef]

30. Young, J.F.; Sipe, J.E.; Van Driel, H.M. Laser-induced periodic surface structure. III. Fluence regimes, the role of feedback, and details of the induced topography in germanium. *Phys. Rev. B* **1984**, *30*, 2001. [CrossRef]

31. Rebollar, E.; Castillejo, M.; Ezquerra, T.A. Laser induced periodic surface structures on polymer films: From fundamentals to applications. *Eur. Polym. J.* **2015**, *73*, 162–174. [CrossRef]

32. Rebollar, E.; Hernández, M.; Sanz, M.; Pérez, S.; Ezquerra, T.A.; Castillejo, M. Laser-induced surface structures on gold-coated polymers: Influence of morphology on surface-enhanced Raman scattering enhancement. *J. Appl. Polym. Sci.* **2015**, *132*, 2–7. [CrossRef]

33. Goya, K.; Yamachoshi, Y.; Fuchiwaki, Y.; Tanaka, M.; Ooie, T.; Abe, K.; Kataoka, M. Femtosecond laser direct fabrication of micro-grooved textures on a capillary flow immunoassay microchip for spatially-selected antibody immobilization. *Sens. Actuators B* **2017**, *239*, 1275–1281. [CrossRef]

34. Wang, B.; Wang, X.C.; Zheng, H.Y.; Lam, Y.C. Femtosecond laser-induced surface wettability modification of polystyrene surface. *Sci. China Phys. Mech. Astron.* **2016**, *59*, 124211. [CrossRef]

35. Sarbada, S.; Shin, Y.C. Superhydrophobic contoured surfaces created on metal and polymer using a femtosecond laser. *Appl. Surf. Sci.* **2017**, *405*, 465–475. [CrossRef]

36. Fan, W.; Qian, J.; Bai, F.; Li, Y.; Wang, C.; Zhao, Q.-Z. A facile method to fabricate superamphiphobic polytetrafluoroethylene surface by femtosecond laser pulses. *Chem. Phys. Lett.* **2016**, *644*, 261–266. [CrossRef]

37. Riveiro, A.; Soto, R.; Comesaña, R.; Boutinguiza, M.; del Val, J.; Quintero, F.; Lusquiños, F.; Pou, J. Laser surface modification of PEEK. *Appl. Surf. Sci.* **2012**, *258*, 9437–9442. [CrossRef]

38. Riveiro, A.; Maçon, A.L.B.; Val, J.; Comesaña, R.; Pou, J. Laser Surface Texturing of Polymers for Biomedical Applications. *Front. Phys.* **2018**, *6*. [CrossRef]

39. Rebollar, E.; Pérez, S.; Hernández, J.J.; Martín-Fabiani, I.; Rueda, D.R.; Ezquerra, T.A.; Castillejo, M. Assessment and Formation Mechanism of Laser-Induced Periodic Surface Structures on Polymer Spin-Coated Films in Real and Reciprocal Space. *Langmuir* **2011**, *27*, 5596–5606. [CrossRef]

40. Barb, R.-A.; Hrelescu, C.; Dong, L.; Heitz, J.; Siegel, J.; Slepička, P.; Vosmanská, V.; Svorcik, V.; Magnus, B.; Marksteiner, R.; et al. Laser-induced periodic surface structures on polymers for formation of gold nanowires and activation of human cells. *Appl. Phys. A* **2014**, *117*, 295–300. [CrossRef]

41. Rebollar, E.; De Aldana, J.R.V.; Pérez-Hernández, J.A.; Ezquerra, T.A.; Moreno, P.; Castillejo, M. Ultraviolet and infrared femtosecond laser induced periodic surface structures on thin polymer films. *Appl. Phys. Lett.* **2012**, *100*, 041106. [CrossRef]

42. Daskalova, A.; Bliznakova, I.; Angelova, L.; Trifonov, A.; Declercq, H.; Buchvarov, I. Femtosecond Laser Fabrication of Engineered Functional Surfaces based on Biodegradable Polymer and Biopolymer/Ceramic Composite Thin Films. *Polymers* **2019**, *11*, 378. [CrossRef]

43. Rebollar, E.; De Aldana, J.R.V.; Martín-Fabiani, I.; Hernández, M.; Rueda, D.R.; Ezquerra, T.A.; Domingo, C.; Moreno, P.; Castillejo, M. Assessment of femtosecond laser induced periodic surface structures on polymer films. *Phys. Chem. Chem. Phys.* **2013**, *15*, 11287–11298. [CrossRef] [PubMed]
44. Ionin, A.A.; Kudryashov, S.I.; Ponomarev, Y.N.; Seleznev, L.V.; Sinitsyn, D.V.; Tikhomirov, B.A.; Phipps, C. Non-linear Absorption and Ionization of Gases by Intense Femtosecond Laser Pulses. *AIP Conf. Proc.* **2010**, *1278*, 354–363. [CrossRef]
45. Talbi, A. Effect Of Ultra-Short Laser Nanostructuring Of Material Surfaces On The Evolution Of Their Thermoelectric Properties. Ph.D. Thesis, Université d'Orléans, Orléans, France, 2018.
46. He, S.; Nivas, J.J.; Anoop, K.; Vecchione, A.; Hu, M.; Bruzzese, R.; Amoruso, S. Surface structures induced by ultrashort laser pulses: Formation mechanisms of ripples and grooves. *Appl. Surf. Sci.* **2015**, *353*, 1214–1222. [CrossRef]
47. Sipe, J.E.; Young, J.F.; Preston, J.S. Laser-induced periodic surface structure. I. Theory. *Phys. Rev. B* **1983**, *27*, 1141–1154. [CrossRef]
48. Fermigier, M.; Limat, L.; Wesfreid, J.E.; Boudinet, P.; Quilliet, C. Two-dimensional patterns in Rayleigh-Taylor instability of a thin layer. *J. Fluid Mech.* **1992**, *236*, 83. [CrossRef]
49. Richeton, J.; Ahzi, S.; Vecchio, K.; Jiang, F.; Adharapurapu, R. Influence of temperature and strain rate on the mechanical behavior of three amorphous polymers: Characterization and modeling of the compressive yield stress. *Int. J. Solids Struct.* **2006**, *43*, 2318–2335. [CrossRef]
50. Vignaud, G.; Chebil, M.S.; Bal, J.K.; Delorme, N.; Beuvier, T.; Grohens, Y.; Gibaud, A. Densification and Depression in Glass Transition Temperature in Polystyrene Thin Films. *Langmuir* **2014**, *30*, 11599–11608. [CrossRef]
51. Bodiguel, H.; Frétigny, C. Viscoelastic Properties of Ultrathin Polystyrene Films. *Macromolecules* **2007**, *40*, 7291–7298. [CrossRef]

 *nanomaterials*

Article

On the Insignificant Role of the Oxidation Process on Ultrafast High-Spatial-Frequency LIPSS Formation on Tungsten

Priya Dominic [1,2], Florent Bourquard [1], Stéphanie Reynaud [1], Arnaud Weck [2,3,4], Jean-Philippe Colombier [1] and Florence Garrelie [1,*]

[1] UJM-Saint-Etienne, CNRS, Laboratoire Hubert Curien UMR 5516, Institute of Optics Graduate School, University Lyon, F-42023 St-Etienne, France; priya.dominic@univ-st-etienne.fr (P.D.); Florent.Bourquard@univ-st-etienne.fr (F.B.); stephanie.reynaud@univ-st-etienne.fr (S.R.); jean.philippe.colombier@univ-st-etienne.fr (J.-P.C.)

[2] Department of Physics, University of Ottawa, STEM Complex, 150 Louis-Pasteur, Ottawa, ON K1N 6N5, Canada; aweck@uOttawa.ca

[3] Department of Mechanical Engineering, University of Ottawa, 161 Louis Pasteur, Ottawa, ON K1N 6N5, Canada

[4] Centre for Research in Photonics, University of Ottawa, 25 Templeton St, Ottawa, ON K1N 6N, Canada

* Correspondence: florence.garrelie@univ-st-etienne.fr

Abstract: The presence of surface oxides on the formation of laser-induced periodic surface structures (LIPSS) is regularly advocated to favor or even trigger the formation of high-spatial-frequency LIPSS (HSFL) during ultrafast laser-induced nano-structuring. This paper reports the effect of the laser texturing environment on the resulting surface oxides and its consequence for HSFLs formation. Nanoripples are produced on tungsten samples using a Ti:sapphire femtosecond laser under atmospheres with varying oxygen contents. Specifically, ambient, 10 mbar pressure of air, nitrogen and argon, and 10^{-7} mbar vacuum pressure are used. In addition, removal of any native oxide layer is achieved using plasma sputtering prior to laser irradiation. The resulting HSFLs have a sub-100 nm periodicity and sub 20 nm amplitude. The experiments reveal the negligible role of oxygen during the HSFL formation and clarifies the significant role of ambient pressure in the resulting HSFLs period.

Keywords: ultrafast laser nanostructuring; femtosecond laser; oxidation; sputtering; laser-induced periodic surface structures; high-spatial-frequency LIPSS; vacuum; Marangoni flow

Citation: Dominic, P.; Bourquard, F.; Reynaud, S.; Weck, A.; Colombier, J.-P.; Garrelie, F. On the Insignificant Role of the Oxidation Process on Ultrafast High-Spatial-Frequency LIPSS Formation on Tungsten. *Nanomaterials* **2021**, *11*, 1069. https://doi.org/10.3390/nano11051069

Academic Editors: Jörn Bonse, Peter Simon and Jürgen Ihlemann

Received: 29 March 2021
Accepted: 18 April 2021
Published: 22 April 2021

Publisher's Note: MDPI stays neutral with regard to jurisdictional claims in published maps and institutional affiliations.

1. Introduction

Ultrafast laser machining on material surfaces including metals [1,2], semiconductors [3–5], and polymers [6,7] to generate laser-induced periodic surface structures (LIPSS) has a broad spectrum of applications in biomedical surface engineering [8–10], tribology [11–14], color marking [15,16], memory devices fabrication [17–19], etc. Surface nanostructure formation can be mainly divided into (i) low-spatial-frequency LIPSS (**LSFLs**), with period $\lambda/2 < \Lambda < \lambda$ (where λ is the laser wavelength) and (ii) high-spatial-frequency LIPSS (**HSFLs**), with period $\Lambda < \lambda/2$. Since the generation of LIPSS using a ruby laser by Birnbaum in 1965 [20], controversial theories have emerged over time to explain the formation of these nanostructures. One of the theories that explains the formation of LSFLs is the scattered wave model [21], where LSFLs are explained by the interference between a scattered surface wave or a polariton for metals (electromagnetic modes bound to and propagating along the surface) and the incoming laser wave. Sipe's efficacy theory [22] analytically summarizes the generation of LSFLs on surface-roughness-induced inhomogeneous subsurface energy absorption and was confirmed by numerical simulations involving 3D finite-difference time-domain (3D FDTD) simulations [23]. These approaches can reproduce LSFLs formation based on an initial roughness layer, where topography centers induce a coherent superposition between the refracted wave and the

275

far field scattered wave or the surface polaritons. Although these theories provide some insight into the formation of LSFLs, several theories have been proposed to explain all aspects associated with HSFLs, including second harmonic generation [24,25], changes in optical properties of the irradiated surface during the pulse [6], nascent plasma theory [26], surface instabilities leading to self-reorganization of matter upon laser irradiation [27], or hydrodynamics instabilities driven by electromagnetics [28].

One of the notable attempts in recent years to understand the formation of HSFLs is recognizing the associated role of oxidation. Öktem et al. [29] proposed a theory based on a feedback mechanism between oxide and nanostructure formation, where an initially positive feedback for nanostructure formation occurs by oxygen incorporation into the surface and is followed by a negative feedback as the thickness of this oxide layer increases, slowly breaking the nanostructure growth. Dostovalov et al. [30] introduced thermochemical LIPSS (TLIPSS), which are LIPSS formed due to metal oxidation rather than ablation. These TLIPSS have unique characteristics, including a rise in relief height, high degree of order, and an orientation parallel to the incident beam direction. The formation of TLIPSS on Ti, Ni, and Cr is explained based on a thermodynamic theory called Wagner theory [30,31]. The same group later recognized that HSFLs cannot be explained by Sipe's theory and proposed an HSFLs formation mechanism based on different modes of propagation associated with scattered electromagnetic waves (SEWs), which evolve with a different percentage fraction of oxide [32].

Zuhlke et al. [33] and Peng et al. [34] demonstrated that structures like mounds formed on Ni and Ti surfaces after femtosecond laser irradiation have an oxide layer thickness ranging from nanometers to micrometers as confirmed by EDX and TEM characterizations. Later studies by Kirner et al. [35] showed that on Ti, hundreds of nanometers of oxide layer were found in association with both LSFLs and HSFLs, confirmed by μ-Raman, AES, and XPS analyses. Another investigation by Florian et al. [36] on a titanium alloy using glow discharge optical emission spectroscopy (GD-OES) for depth profiling the oxygen content in LSFLs showed a 200 nm thick oxide layer. One of the interesting conclusions of Florian et al. [37], based on FDTD simulations, is the requirement of 100 nm thick rough oxide layer for the formation of certain types of LSFL (with anomalous orientation) on CrN. However, the existence of sub 100 nm HSFLs with amplitude as small as a few tens of nanometers [26,38] may appear contradictory to previously mentioned theories.

Several HSFL formation theories point toward surface oxidation as a key parameter in understanding HSFL formation during ultrafast laser texturing. Oxygen can be incorporated from either one of the following sources, or both, during laser irradiation: (i) the native oxide layer, i.e., the adsorbed oxygen layer on the surface of materials with a thickness as low as 5 nm; (ii) the ambient air, which one might think is a major contributor.

In this work, the effect of laser processing environment on surface morphology was investigated to better understand HSFLs formation on tungsten. Various environments were used including high vacuum (10^{-7} mbar), argon (10 mbar), nitrogen (10 mbar), air (10 mbar), as well as ambient. Experiments were also carried out where the native oxide layer was removed by plasma sputtering, directly followed by laser irradiation. This study provides insight into the effect of oxygen on the formation of HSFLs. To date, only a very few attempts have been made to link processing environment to LSFLs or microspikes formation [39–41], conical microstructures [42], and HSFLs [43]. Here, HSFLs formation on tungsten (W) was investigated, because W is an important industrial metal (high hardness and high melting point) [44,45] and because it belongs to the same group as chromium in the periodic table, and thus W might be expected to undergo a similar oxidation-based HSFLs formation mechanism as discussed above. Furthermore, an investigation into the incorporation of oxygen during laser mater interaction is relevant not only to better understand HSFLs formation, but also because surface chemistry can influence many surface properties, including wettability [46–48], color [15,49,50], catalytic properties [51,52], and heat transfer potential [53,54], as seen for LSFLs.

2. Experimental Details

Polycrystalline tungsten samples (Goodfellow SARL, Lille, France) with dimensions $10 \times 10 \times 1$ mm were electrochemically polished, resulting in an average roughness Ra of ~6 nm as measured with an AFM (Burker Dimension ICON, Billerica, MA, USA). Polished tungsten samples were irradiated with a linearly polarized laser beam (Titanium-Sapphire, Legend Coherent Inc., Santa Jose, CA, USA) with a central wavelength of 800 nm, pulse duration of 60 fs, and 1 kHz pulse repetition rate, focused with a converging lens of focal length 30 cm to a focal spot size diameter of 100 μm into an ultrahigh vacuum chamber (Turbo pump, VINCI Technologies, Nanterre, France) equipped with a sputtering apparatus (DC pinnacle plus plasma generator, Denver, CO, USA), which can generate an argon ion plasma, as shown in Figure 1.

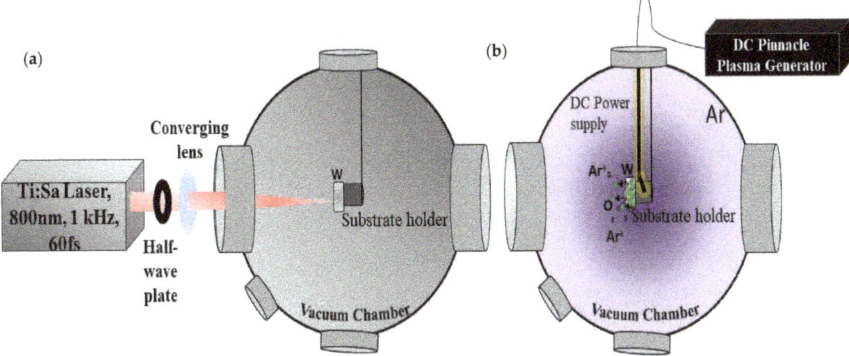

Figure 1. Experimental setup with the femtosecond laser beam focused into the vacuum chamber equipped with a plasma generator. (**a**) Irradiation of the W target inside a vacuum chamber. (**b**) Sputtering of the target surface prior to laser irradiation, in which the inert argon ion plasma created by polarizing the substrate holder using a continuous power supply provided by the pinnacle plus and the accelerated argon ions spray off the adsorbed oxide layer.

W surfaces were irradiated in different atmospheres including ambient, 10 mbar pressure of air, nitrogen (Air Products, Paris, France), argon (Air Products, Paris, France), and under high vacuum (10^{-7} mbar.). Tungsten peak ablation threshold fluence for a single pulse was 0.6 J/cm^2, as calculated by Liu's method [55]. The presented laser impacts were achieved with a peak fluence F_p = 0.35 J/cm^2 and a number of pulses N = 25 in all environmental conditions, and the focal spot diameter of 100 μm was also evaluated near to these laser parameters by Liu's method [55]. In order to eliminate the undesired native oxide layer, which could alter the photon-metal absorption process, argon ion sputtering was used (current = 0.25 A, frequency = 120 kHz, power = 16 W, voltage = 26 V, argon gas pressure = 10^{-2} mbar, and time of exposure to sputtering = 3 min). The process ensured the sputtering of any native oxide layer before irradiation with the laser and would thus reveal the role of the native oxide layer on HSFL formation. Sputtering was performed in the same vacuum chamber as the laser irradiation, thus ensuring no ambient air contamination of the sample surface. Surface morphology was characterized by SEM (FEI Nova NanoSEM 200, Hillsboro, OR, USA) and AFM.

3. Results and Discussion

High-resolution SEM images of laser-irradiated samples without sputtering, taken at the center of the laser irradiated spots along with the corresponding FFT transforms (inset), are shown in Figure 2a.1,b.1,c.1,d.1,e.1,a.2,b.2,c.2,d.2,e.2. In the figure, HSFLs are observed to form parallel to the horizontally polarized femtosecond laser beam for all processing atmospheres. FFT analysis provided an HSFL period as small as 100 nm and AFM data, as presented in Figure 2a.4,b.4,c.4,d.4,e.4, confirmed this periodicity, and revealed the small amplitudes of the HSFL, varying from 10 to 20 nm. The scanning transmission electron microscope (STEM) annular dark field (ADF) cross-sectional images (Figure 2e.5,e.6) of the HSFLs formed in vacuum conditions agrees with the sub 20 nm amplitude obtained by AFM. The HSFL period (Λ) was 81 ± 2 nm, 92 ± 3 nm, 109 ± 3 nm, 95 ± 3 nm, and 123 ± 2 nm for ambient, air (10 mbar), N (10 mbar), Ar (10 mbar), and vacuum (10^{-7} mbar), respectively.

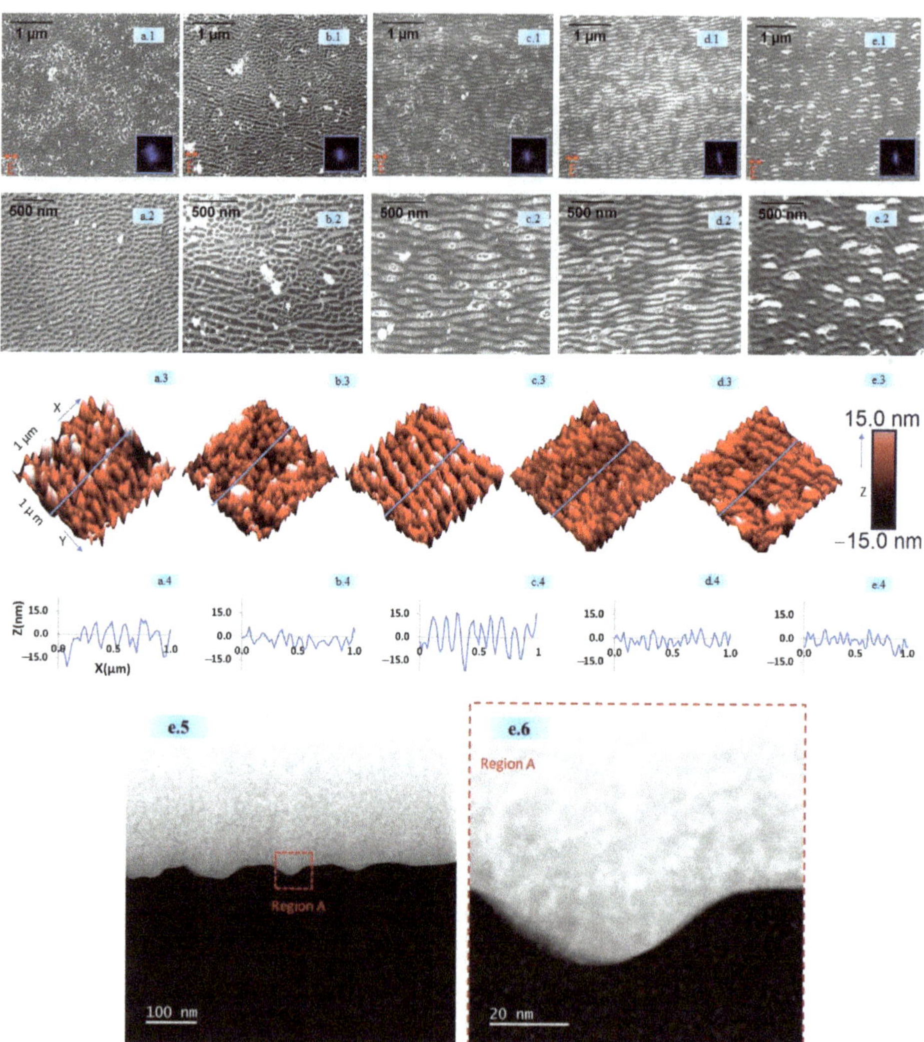

Figure 2. (**i**) **Top** (**a.1–e.2**) SEM images of HSFLs generated with no sputtering with FFT for (**a.1,a.2**) ambient, (**b.1,b.2**) air (10 mbar), (**c.1,c.2**) nitrogen (10 mbar), (**d.1,d.2**) argon (10 mbar), and (**e.1,e.2**) vacuum (10^{-7} mbar). (**ii**) **Middle** (**a.3–e.3**) AFM images (1×1 µm area) for (**a.3**) ambient, (**b.3**) air (10 mbar), (**c.3**) nitrogen (10 mbar), (**d.3**) argon (10 mbar), and (**e.3**) vacuum (10^{-7} mbar). (**iii**) (**a.4–e.4**) Line profiles corresponding to the AFM images indicated by the blue line (**a.4,b.4,c.4,d.4,e.4**). (**iv**) **Bottom** (**e.5,e.6**) STEM ADF cross-sectional images obtained for HSFLs formed in the vacuum. The ADF image shows several crests and valleys of HSFLs (**e.5**). A high-resolution image of Region A (**e.6**) confirms the amplitude of the HSFLs being sub 20 nm.

Figure 3a.1,b.1,c.1,d.1,e.1,a.2,b.2,c.2,d.2,e.2 shows the topography of the HSFLs formed after sputtering under the same various environments as those used in Figure 2. Here, the variation in periodicity was 51 ± 4 nm, 72 ± 5nm, 79 ± 3 nm, 88 ± 4 nm to 152 ± 5 nm, for ambient, 10 mbar air, 10 mbar N, 10 mbar Ar, and vacuum (10^{-7} mbar), respectively, which agrees with the periodicity obtained in the AFM profiles reported in Figure 3a.4,b.4,c.4,d.4,e.4. The amplitude of HSFL varied between 10 and 20 nm.

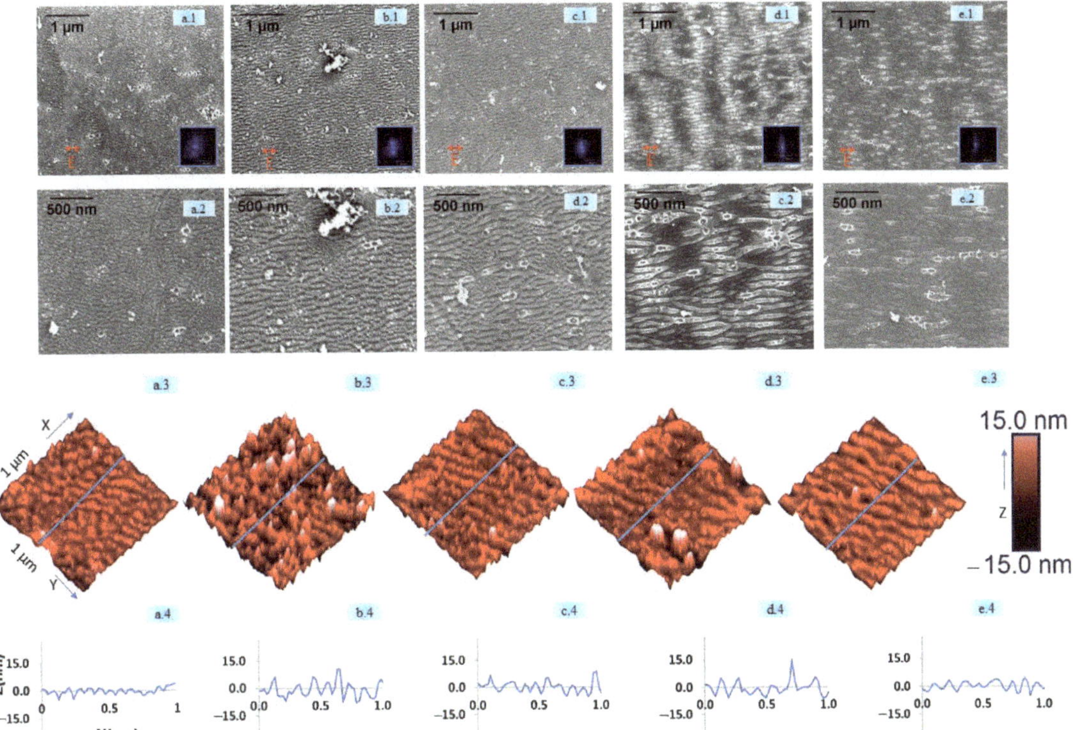

Figure 3. **(i) Top (a.1–e.2)** SEM images (row 1 and row 2, different magnifications) of HSFLs generated with sputtering with FFT for **(a.1,a.2)** ambient, **(b.1,b.2)** air (10 mbar), **(c.1,c.2)** nitrogen (10 mbar), **(d.1,d.2)** argon (10 mbar), and **(e.1,e.2)** vacuum (10^{-7} mbar). **(ii) Middle (a.3–e.3)** AFM images (1×1 μm area) for **(a.3)** ambient, **(b.3)** air (10 mbar), **(c.3)** nitrogen (10 mbar), **(d.3)** argon (10 mbar), and **(e.3)** vacuum (10^{-7} mbar). **(iii) Bottom (a.4–e.4)** Line profiles corresponding to the AFM images indicated by the blue line **(a.4,b.4,c.4,d.4,e.4)**.

As evident from Figures 2 and 3, the presence of HSFLs in high vacuum conditions and after sputtering confirms the insignificant role of atmospheric oxygen or any native oxide layer for the formation of these kind of laser-induced high-frequency structures. Upon analyzing Figure 4, we can observe that at a fixed pressure of 10 mbar, different processing atmospheres, whether air or in the presence of a non-reactive gas like argon or a reactive gas like nitrogen, yield HSFLs with almost the same features, thus again confirming the negligible contribution of the laser processing environment on the formation of HSFLs.

The variation in the HSFLs periods and amplitudes with different processing atmospheres, with and without sputtering, is shown in Figure 4. The general observed trend, with and without sputtering, is that the period of the HSFLs increases as the pressure decreases. As the ambient air was controlled to a pressure of 10 mbar (i.e., the condition of 10 mbar of air), the period increased compared with that of atmospheric pressure, and it further continued to increase as the pressure was reduced as low as 10^{-7} mbar (under vacuum). The decrease in ambient pressure can affect the hydrodynamical aspects of laser–matter interaction by influencing the surface tension of the molten liquid, thus leading to an increase in the period, as observed experimentally.

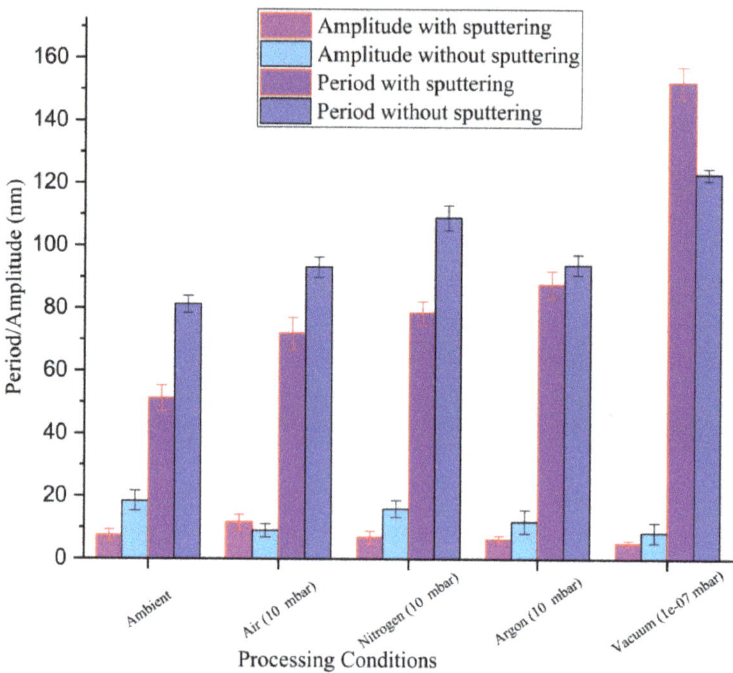

Figure 4. Variation in the period and amplitude of the HSFLs formed under different processing environments.

Rather than originating from SEWs in the laser-induced oxide layer, HSFLs are proposed to be established by molten material reorganization, driven by surface-tension-dependent transverse Marangoni gradient process [56]. The observed periodicity of the HSFLs (Λ) is related to the Marangoni instability as [28,55] $\Lambda = \frac{2pL}{\sqrt{Ma/8}}$, where L is the liquid layer thickness and Ma is a dimensionless quantity called the Marangoni number, as given by $Ma = -\frac{d\gamma}{dT}\frac{L\Delta T}{\mu D}$ [28,55], where $d\gamma/dT$ is the temperature-dependent surface tension gradient, ΔT is the temperature difference, μ is the viscosity, and D is the thermal diffusivity. The dependence of the surface tension gradient on pressure (P) is given by $\frac{d\gamma}{dT} \propto -P$ [57]. From these equations, we obtain the HSFL period, $\Lambda \propto P^{-1/2}$. This equation agrees with our experimental observations where smaller HSFL periods were observed at higher pressures, and supports a molten material reorganization process. The small HSFL period of approximately Λ–$\lambda/8$ can also be correlated to the confinement of the light absorption, as Marangoni instability provides $\Lambda \propto \sqrt{L}$, where L is the liquid layer depth. Assuming that at this low fluence and high electron–phonon coupling strength, the electron thermal energy diffusion is weak, so the energy confinement can roughly be estimated by the optical penetration depth for tungsten at 800 nm. This is given by $\lambda/4pk$ = 25 nm, where k = 2.9 is the extinction coefficient for photoexcited W [58]. For this small penetration depth, a small liquid layer, and thus small HSFL period, is expected.

The sputtering prior to laser irradiation tended to induce small changes in the periodicity for structures generated, as shown in Figure 4. This may be due to the change in roughness on the tungsten target induced by the sputtering process. It is known that the initial roughness can highly influence laser coupling and hence the resulting topography [59].

In particular, the concentration and kind of nanoreliefs as bumps and cavities resulting from the sputtering process affect the final period. It is expected that the introduction of additional scattering inhomogeneities on the surface with an average distance lower than the initial roughness would result in a higher concentration of scattering centers. This favors dipole–dipole coupling for nonradiative fields, reducing the pattern period [58]. Topographical features in the SEM (Figures 1 and 2), for different gaseous atmospheres are different arising from the polycrystallinity of the samples. Because of the variation in energy absorption, phase transition, and lattice defects storage for different crystal orientations, their interaction with the laser beam can be different [60]. Because of the lower fluence used for the generation of HSFLs, these changes can be more intense.

All observations in this work signal that the formation of HSFLs for a given fluence and number of pulses under different laser processing atmospheres is mostly independent of any supply of oxygen. These results strongly suggest that HSFLs are more likely to be explained by theories based on electromagnetics coupling with hydrodynamical concepts of laser matter rather than the propagation modes implying oxidation models. More specifically, inhomogeneous subsurface energy absorption arising from the local non-radiative electromagnetic response can trigger nanoconvective instabilities. These thermoconvective instabilities mediated by temperature-gradient-dependent flow of capillary waves and their further solidification producing nanostructures [28,55,61] can explain the HSFL formation observed in the absence of oxygen and the decrease in HSFLs period with the ambient pressure.

4. Conclusions

Femtosecond-laser-induced HSFLs with sub 100 nm period and sub 20 nm amplitude were achieved on tungsten with a given fluence (Fp = 0.35 J/cm^2) and number of laser pulses (N = 25) under different processing atmospheres, namely ambient, air (10 mbar), nitrogen (10 mbar), argon (10 mbar), and vacuum (10^{-7} mbar) with and without sputtering. This points toward neither ambient oxygen nor the native oxide layer playing a significant role in the formation of HSFLs. The experiment under different atmospheres showed how HSFLs can be obtained without any external supply of oxygen. They can be achieved in a chamber almost devoid of any gases with a pressure as low as 10^{-7} mbar, or in the presence of a nonreactive gaseous atmosphere like argon, or even on the presence of a reactive gas like nitrogen. Laser processing experiments immediately after plasma sputtering of the tungsten surface yielded HSFLs with no native oxide layer. The generation of HSFLs is mostly dependent upon material properties like surface roughness, grain orientation, and laser parameters; hence, oxygen as a necessary condition can be neglected. SEM, FFT, and AFM data provided complementary information regarding the topography of HSFLs. The decrease in HSFL periodicity with increasing pressure may be understood in terms of the influence of pressure on the Marangoni flow of the molten liquid. This supports a hydrodynamic origin of HSFLs formation where the observed period can be tuned by the pressure dependence of the thermo-capillarity process. This work paves the way toward the control of 100 nm features by changing the pressure and chemical environment.

Author Contributions: P.D., F.B., S.R. and A.W. performed conceptualization. F.G. and F.B. designed and directed the project. P.D. performed the samples preparations, the laser irradiation experiments and the characterizations experiments on SEM and AFM. S.R. performed FIB and STEM analysis. J.-P.C. developed the theoretical framework. P.D. wrote the original draft manuscript. All authors discussed the results and contributed to the final manuscript. All authors have read and agreed to the published version of the manuscript.

Funding: This research was funded by Auvergne Rhône Alpes region.

Institutional Review Board Statement: Not applicable.

Informed Consent Statement: Not applicable.

Data Availability Statement: Not applicable.

Acknowledgments: The authors acknowledge Anthony Nakhoul for providing support with the polishing requirements of the surface at Ecole Nationale Superieure des Mines de Saint Etienne, France. The authors acknowledge the Auvergne Rhône Alpes region for funding.

Conflicts of Interest: The authors declare no conflict of interest.

References

1. Dar, M.H.; Kuladeep, R.; Saikiran, V. Femtosecond laser nanostructuring of titanium metal towards fabrication of low-reflective surfaces over broad wavelength range. *Appl. Surf. Sci.* **2016**, *371*, 479–487. [CrossRef]
2. Bonse, J.; Höhm, S.; Rosenfeld, A.; Krüger, J. Sub-100-nm laser-induced periodic surface structures upon irradiation of titanium by Ti:sapphire femtosecond laser pulses in air. *Appl. Phys. A* **2013**, *110*, 547–551. [CrossRef]
3. Hsu, E.M.; Crawford, T.H.R.; Maunders, C.; Botton, G.A.; Haugen, H.K. Cross-sectional study of periodic surface structures on gallium phosphide induced by ultrashort laser pulse irradiation. *Appl. Phys. Lett.* **2008**, *92*, 221112. [CrossRef]
4. Casquero, N.; Fuentes-Edfuf, Y.; Zazo, R.; Solis, J.; Siegel, J. Generation, control and erasure of dual LIPSS in germanium with fs and ns laser pulses. *J. Phys. Appl. Phys.* **2020**, *53*, 485106. [CrossRef]
5. Tsibidis, G.D.; Stratakis, E.; Loukakos, P.A.; Fotakis, C. Controlled ultrashort-pulse laser-induced ripple formation on semiconductors. *Appl. Phys. A* **2014**, *114*, 57–68. [CrossRef]
6. Wu, Q.; Ma, Y.; Fang, R.; Liao, Y.; Yu, Q.; Chen, X.; Wang, K. Femtosecond laser-induced periodic surface structure on diamond film. *Appl. Phys. Lett.* **2003**, *82*, 1703–1705. [CrossRef]
7. Rebollar, E.; de Aldana, J.R.; Martín-Fabiani, I.; Hernández, M.; Rueda, D.R.; Ezquerra, T.A.; Domingo, C.; Moreno, P.; Castillejo, M. Assessment of femtosecond laser induced periodic surface structures on polymer films. *Phys. Chem. Chem. Phys.* **2013**, *15*, 11287. [CrossRef] [PubMed]
8. Kunz, C.; Bartolomé, J.F.; Gnecco, E.; Müller, F.A.; Gräf, S. Selective generation of laser-induced periodic surface structures on Al_2O_3-ZrO_2-Nb composites. *Appl. Surf. Sci.* **2018**, *434*, 582–587. [CrossRef]
9. Gnilitskyi, I.; Pogorielov, M.; Viter, R.; Ferraria, A.M.; Carapeto, A.P.; Oleshko, O.; Orazi, L.; Mishchenko, O. Cell and tissue response to nanotextured Ti6Al4V and Zr implants using high-speed femtosecond laser-induced periodic surface structures. *Nanomed. Nanotechnol. Biol. Med.* **2019**, *21*, 102036. [CrossRef] [PubMed]
10. Santos, A.D.; Campanelli, L.C.; Silva, P.S.; Vilar, R.; Almeida, M.A.; Kuznetsov, A.; Achete, C.A.; Bolfarini, C. Influence of a Femtosecond Laser Surface Modification on the Fatigue Behavior of Ti-6Al-4V ELI Alloy. *Mater. Res.* **2019**, *22*, e20190118. [CrossRef]
11. Bonse, J.; Koter, R.; Hartelt, M.; Spaltmann, D.; Pentzien, S.; Höhm, S.; Rosenfeld, A.; Krüger, J. Tribological performance of femtosecond laser-induced periodic surface structures on titanium and a high toughness bearing steel. *Appl. Surf. Sci.* **2015**, *336*, 21–27. [CrossRef]
12. Duarte, M.; Lasagni, A.; Giovanelli, R.; Narciso, J.; Louis, E.; Mücklich, F. Increasing Lubricant Film Lifetime by Grooving Periodical Patterns Using Laser Interference Metallurgy. *Adv. Eng. Mater.* **2008**, *10*, 554–558. [CrossRef]
13. Bonse, J.; Kirner, S.V.; Koter, R.; Pentzien, S.; Spaltmann, D.; Krüger, J. Femtosecond laser-induced periodic surface structures on titanium nitride coatings for tribological applications. *Appl. Surf. Sci.* **2017**, *418*, 572–579. [CrossRef]
14. Bonse, J.; Koter, R.; Hartelt, M.; Spaltmann, D.; Pentzien, S.; Höhm, S.; Rosenfeld, A.; Krüger, J. Femtosecond laser-induced periodic surface structures on steel and titanium alloy for tribological applications. *Appl. Phys. A* **2014**, *117*, 103–110. [CrossRef]
15. Dusser, B.; Sagan, Z.; Soder, H.; Faure, N.; Colombier, J.P.; Jourlin, M.; Audouard, E. Controlled nanostructrures formation by ultra fast laser pulses for color marking. *Opt. Express* **2010**, *18*, 2913. [CrossRef] [PubMed]
16. Vorobyev, Y.; Guo, C. Colorizing metals with femtosecond laser pulses. *Appl. Phys. Lett.* **2008**, *92*, 41914. [CrossRef]
17. Martínez-Tong, D.E.; Rodríguez-Rodríguez, Á.; Nogales, A.; Garcia-Gutierrez, M.C.; Pérez-Murano, F.; Llobet, J.; Ezquerra, T.A.; Rebollar, E. Laser Fabrication of Polymer Ferroelectric Nanostructures for Nonvolatile Organic Memory Devices. *ACS Appl. Mater. Interfaces* **2015**, *7*, 19611–19618. [CrossRef] [PubMed]
18. He, X.; Datta, A.; Nam, W.; Traverso, L.M.; Xu, X. Sub-Diffraction Limited Writing based on Laser Induced Periodic Surface Structures (LIPSS). *Sci. Rep.* **2016**, *6*, 35035. [CrossRef]
19. Yu, Y.; Bai, S.; Wang, S.; Hu, A. Ultra-Short Pulsed Laser Manufacturing and Surface Processing of Microdevices. *Engineering* **2018**, *4*, 779–786. [CrossRef]
20. Birnbaum, M. Modulation of the Reflectivity of Semiconductors. *J. Appl. Phys.* **1965**, *36*, 657–658. [CrossRef]
21. Keilmann, F.; Bai, Y.H. Periodic surface structures frozen into CO_2 laser-melted quartz. *Appl. Phys. Solids Surf.* **1982**, *29*, 9–18. [CrossRef]
22. Sipe, J.E.; Young, J.F.; Preston, J.S.; van Driel, H.M. Laser-induced periodic surface structure. I. Theory. *Phys. Rev. B* **1983**, *27*, 1141–1154. [CrossRef]
23. Skolski, J.Z.; Römer, G.R.; Vincenc Obona, J.; Huis in 't Veld, A.J. Modeling laser-induced periodic surface structures: Finite-difference time-domain feedback simulations. *J. Appl. Phys.* **2014**, *115*, 103102. [CrossRef]
24. Borowiec, A.; Haugen, H.K. Subwavelength ripple formation on the surfaces of compound semiconductors irradiated with femtosecond laser pulses. *Appl. Phys. Lett.* **2003**, *82*, 4462–4464. [CrossRef]
25. Bonse, J.; Munz, M.; Sturm, H. Structure formation on the surface of indium phosphide irradiated by femtosecond laser pulses. *J. Appl. Phys.* **2005**, *97*, 013538. [CrossRef]

26. Pan, F.; Wang, W.J.; Mei, X.S.; Yang, H.Z.; Sun, X.F. The formation mechanism and evolution of ps-laser-induced high-spatial-frequency periodic surface structures on titanium. *Appl. Phys. B* **2017**, *123*, 21. [CrossRef]

27. Reif, J.; Varlamova, O.; Varlamov, S.; Bestehorn, M. The role of asymmetric excitation in self-organized nanostructure formation upon femtosecond laser ablation. *Appl. Phys. A* **2011**, *104*, 969–973. [CrossRef]

28. Rudenko, A.; Abou-Saleh, A.; Pigeon, F.; Mauclair, C.; Garrelie, F.; Stoian, R.; Colombier, J.P. High-frequency periodic patterns driven by non-radiative fields coupled with Marangoni convection instabilities on laser-excited metal surfaces. *Acta Mater.* **2020**, *194*, 93–105. [CrossRef]

29. Öktem, B.; Pavlov, I.; Ilday, S.; Kalaycıoğlu, H.; Rybak, A.; Yavaş, S.; Erdoğan, M.; Ilday, F.Ö. Nonlinear laser lithography for indefinitely large-area nanostructuring with femtosecond pulses. *Nat. Photonics* **2013**, *7*, 879–901. [CrossRef]

30. Dostovalov, V.; Korolkov, V.P.; Okotrub, K.A.; Bronnikov, K.A.; Babin, S.A. Oxide composition and period variation of thermo-chemical LIPSS on chromium films with different thickness. *Opt. Express* **2018**, *26*, 7712. [CrossRef] [PubMed]

31. Dostovalov, V.; Korolkov, V.P.; Terentyev, V.S.; Okotrub, K.A.; Dultsev, F.N.; Babin, S.A. Study of the formation of thermochemical laser-induced periodic surface structures on Cr, Ti, Ni and NiCr films under femtosecond irradiation. *Quantum Electron.* **2017**, *47*, 631–637. [CrossRef]

32. Dostovalov, A.V.; Derrien, T.J.; Lizunov, S.A.; Přeučil, F.; Okotrub, K.A.; Mocek, T.; Korolkov, V.P.; Babin, S.A.; Bulgakova, N.M. LIPSS on thin metallic films: New insights from multiplicity of laser-excited electromagnetic modes and efficiency of metal oxidation. *Appl. Surf. Sci.* **2019**, *491*, 650–658. [CrossRef]

33. Zuhlke, C.A.; Anderson, T.P.; Alexander, D.R. Comparison of the structural and chemical composition of two unique micro/nanostructures produced by femtosecond laser interactions on nickel. *Appl. Phys. Lett.* **2013**, *103*, 121603. [CrossRef]

34. Peng, E.; Bell, R.; Zuhlke, C.A.; Wang, M.; Alexander, D.R.; Gogos, G.; Shield, J.E. Growth mechanisms of multiscale, mound-like surface structures on titanium by femtosecond laser processing. *J. Appl. Phys.* **2017**, *122*, 133108. [CrossRef]

35. Kirner, S.V.; Wirth, T.; Sturm, H.; Krüger, J.; Bonse, J. Nanometer-resolved chemical analyses of femtosecond laser-induced periodic surface structures on titanium. *J. Appl. Phys.* **2017**, *122*, 104901. [CrossRef]

36. Florian, C.; Wonneberger, R.; Undisz, A.; Kirner, S.V.; Wasmuth, K.; Spaltmann, D.; Krüger, J.; Bonse, J. Chemical effects during the formation of various types of femtosecond laser-generated surface structures on titanium alloy. *Appl. Phys. A* **2020**, *126*, 266. [CrossRef]

37. Florian, C.; Déziel, J.L.; Kirner, S.V.; Siegel, J.; Bonse, J. The Role of the Laser-Induced Oxide Layer in the Formation of Laser-Induced Periodic Surface Structures. *Nanomaterials* **2020**, *10*, 147. [CrossRef] [PubMed]

38. Sedao, X.; Shugaev, M.V.; Wu, C.; Douillard, T.; Esnouf, C.; Maurice, C.; Reynaud, S.; Pigeon, F.; Garrelie, F.; Zhigilei, L.V.; et al. Growth Twinning and Generation of High-Frequency Surface Nanostructures in Ultrafast Laser-Induced Transient Melting and Resolidification. *ACS Nano* **2016**, *10*, 6995–7007. [CrossRef] [PubMed]

39. Her, T.-H.; Finlay, R.J.; Wu, C.; Deliwala, S.; Mazur, E. Microstructuring of silicon with femtosecond laser pulses. *Appl. Phys. Lett.* **1998**, *73*, 1673–1675. [CrossRef]

40. Schade, M.; Varlamova, O.; Reif, J.; Blumtritt, H.; Erfurth, W.; Leipner, H.S. High-resolution investigations of ripple structures formed by femtosecond laser irradiation of silicon. *Anal. Bioanal. Chem.* **2010**, *396*, 1905–1911. [CrossRef] [PubMed]

41. Gesuele, F.; Nivas, J.J.; Fittipaldi, R.; Altucci, C.; Bruzzese, R.; Maddalena, P.; Amoruso, S. Analysis of nascent silicon phase-change gratings induced by femtosecond laser irradiation in vacuum. *Sci. Rep.* **2018**, *8*, 12498. [CrossRef] [PubMed]

42. Nayak, K.; Gupta, M.C.; Kolasinski, K.W. Formation of nano-textured conical microstructures in titanium metal surface by femtosecond laser irradiation. *Appl. Phys. A* **2008**, *90*, 399–402. [CrossRef]

43. Girolami, M.; Bellucci, A.; Mastellone, M.; Orlando, S.; Serpente, V.; Valentini, V.; Polini, R.; Sani, E.; De Caro, T.; Trucchi, D.M. Femtosecond-Laser Nanostructuring of Black Diamond Films under Different Gas Environments. *Materials* **2020**, *13*, 5761. [CrossRef]

44. Smid, I.; Akiba, M.; Vieider, G.; Plöchl, L. Development of tungsten armor and bonding to copper for plasma-interactive components. *J. Nucl. Mater.* **1998**, *258–263*, 160–172. [CrossRef]

45. Tobar, M.J.; Álvarez, C.; Amado, J.M.; Rodríguez, G.; Yáñez, A. Morphology and characterization of laser clad composite NiCrBSi–WC coatings on stainless steel. *Surf. Coat. Technol.* **2006**, *200*, 22–23. [CrossRef]

46. Samanta, A.; Wang, Q.; Shaw, S.K.; Ding, H. Roles of chemistry modification for laser textured metal alloys to achieve extreme surface wetting behaviors. *Mater. Des.* **2020**. [CrossRef]

47. Giannuzzi, G.; Gaudiuso, C.; Di Mundo, R.; Mirenghi, L.; Fraggelakis, F.; Kling, R.; Lugarà, P.M.; Ancona, A. Short and long term surface chemistry and wetting behaviour of stainless steel with 1D and 2D periodic structures induced by bursts of femtosecond laser pulses. *Appl. Surf. Sci.* **2019**, *494*, 1055–1065. [CrossRef]

48. Pou, P.; Del Val, J.; Riveiro, A.; Comesaña, R.; Arias-González, F.; Lusquiños, F.; Bountinguiza, M.; Quintero, F.; Pou, J. Laser texturing of stainless steel under different processing atmospheres: From superhydrophilic to superhydrophobic surfaces. *Appl. Surf. Sci.* **2019**, *457*, 896–905. [CrossRef]

49. Guay, J.M.; Lesina, A.C.; Côté, G.; Charron, M.; Poitras, D.; Ramunno, L.; Berini, P.; Weck, A. Laser-induced plasmonic colours on metals. *Nat. Commun.* **2017**, *8*, 16095. [CrossRef]

50. Guay, J.-M.; Killaire, G.; Gordon, P.G.; Barry, S.T.; Berini, P.; Weck, A. Passivation of Plasmonic Colors on Bulk Silver by Atomic Layer Deposition of Aluminum Oxide. *Langmuir* **2018**, *34*, 4998–5010. [CrossRef]

51. Lange, K.; Schulz-Ruhtenberg, M.; Caro, J. Platinum Electrodes for Oxygen Reduction Catalysis Designed by Ultrashort Pulse Laser Structuring. *Chem. Electro Chem.* **2017**, *4*, 570–576. [CrossRef]
52. Satapathy, P.; Pfuch, A.; Grunwald, R.; Das, S.K. Enhancement of photocatalytic activity by femtosecond-laser induced periodic surface structures of Si. *J. Semicond.* **2020**, *41*, 032303. [CrossRef]
53. Kruse, C.; Lucis, M.; Shield, J.E.; Anderson, T.; Zuhlke, C.; Alexander, D.; Gogos, G.; Ndao, S. Effects of Femtosecond Laser Surface Processed Nanoparticle Layers on Pool Boiling Heat Transfer Performance. *J. Therm. Sci. Eng. Appl.* **2018**, *10*, 031009. [CrossRef] [PubMed]
54. Kruse, C.; Tsubaki, A.; Zuhlke, C.; Alexander, D.; Anderson, M.; Peng, E.; Shield, J.; Ndao, S.; Gogos, G. Influence of Copper Oxide on Femtosecond Laser Surface Processed Copper Pool Boiling Heat Transfer Surfaces. *J. Heat Transf.* **2019**, *141*, 051503. [CrossRef]
55. Liu, J.M. Simple technique for measurements of pulsed Gaussian-beam spot sizes. *Opt. Lett.* **1982**, *7*, 196. [CrossRef] [PubMed]
56. Zhang, H.; Colombier, J.P.; Li, C.; Faure, N.; Cheng, G.; Stoian, R. Coherence in ultrafast laser-induced periodic surface structures. *Phys. Rev. B* **2015**, *92*, 174109. [CrossRef]
57. Slowinski, J.; Gates, E.E.; Waring, C.E. The Effect of Pressure on the Surface Tensions of Liquids. *J. Phys. Chem.* **1957**, *61*, 808–810. [CrossRef]
58. Bévillon, E.; Colombier, J.P.; Recoules, V.; Zhang, H.; Li, C.; Stoian, R. Ultrafast switching of surface plasmonic conditions in nonplasmonic metals. *Phys. Rev. B* **2016**, *93*, 165416. [CrossRef]
59. Colombier, J.P.; Rudenko, A.; Silaeva, E.; Zhang, H.; Sedao, X.; Bévillon, E.; Reynaud, S.; Maurice, C.; Pigeon, F.; Garrelie, F.; et al. Mixing periodic topographies and structural patterns on silicon surfaces mediated by ultrafast photoexcited charge carriers. *Phys. Rev. Res.* **2020**, *2*, 043080. [CrossRef]
60. Sedao, X.; Maurice, C.; Garrelie, F.; Colombier, J.P.; Reynaud, S.; Quey, R.; Pigeon, F. Influence of crystal orientation on the formation of femtosecond laser-induced periodic surface structures and lattice defects accumulation. *Appl. Phys. Lett.* **2014**, *104*, 171605. [CrossRef]
61. Tsibidis, D.; Fotakis, C.; Stratakis, E. From ripples to spikes: A hydrodynamical mechanism to interpret femtosecond laser-induced self-assembled structures. *Phys. Rev. B* **2015**, *92*, 041405. [CrossRef]

 *nanomaterials*

Article

Initial Morphology and Feedback Effects on Laser-Induced Periodic Nanostructuring of Thin-Film Metallic Glasses

Mathilde Prudent [1], Florent Bourquard [1], Alejandro Borroto [2], Jean-François Pierson [2], Florence Garrelie [1] and Jean-Philippe Colombier [1,*]

[1] Univ Lyon, UJM-Saint-Etienne, CNRS, Institute of Optics Graduate School, Laboratoire Hubert Curien UMR CNRS 5516, F-42023 St-Etienne, France; mathilde.prudent@univ-st-etienne.fr (M.P.); florent.bourquard@univ-st-etienne.fr (F.B.); florence.garrelie@univ-st-etienne.fr (F.G.)

[2] Université de Lorraine, CNRS, IJL, F-54000 Nancy, France; alejandro.borroto@univ-lorraine.fr (A.B.); jean-francois.pierson@univ-lorraine.fr (J.-F.P.)

* Correspondence: jean.philippe.colombier@univ-st-etienne.fr

Abstract: Surface nanostructuring by femtosecond laser is an efficient way to manipulate surface topography, creating advanced functionalities of irradiated materials. Thin-film metallic glasses obtained by physical vapor deposition exhibit microstructures free from grain boundaries, crystallites and dislocations but also characterized by a nanometric surface roughness. These singular properties make them more resilient to other metals to form laser-induced nanopatterns. Here we investigate the morphological response of $Zr_{65}Cu_{35}$ alloys under ultrafast irradiation with multipulse feedback. We experimentally demonstrate that the initial columnar microstructure affects the surface topography evolution and conditions the required energy dose to reach desired structures in the nanoscale domain. Double pulses femtosecond laser irradiation is also shown to be an efficient strategy to force materials to form uniform nanostructures even when their thermomechanical properties have a poor predisposition to generate them.

Keywords: thin film metallic glasses; femtosecond laser; LIPSS; surface functionalization; nanostructuring

Citation: Prudent, M.; Bourquard, F.; Borroto, A.; Pierson, J.-F.; Garrelie, F.; Colombier, J.-P. Initial Morphology and Feedback Effects on Laser-Induced Periodic Nanostructuring of Thin-Film Metallic Glasses. *Nanomaterials* **2021**, *10*, 1076. https://doi.org/10.3390/nano11051076

Academic Editor: Maria Grazia Grimaldi

Received: 30 March 2021
Accepted: 20 April 2021
Published: 22 April 2021

Publisher's Note: MDPI stays neutral with regard to jurisdictional claims in published maps and institutional affiliations.

1. Introduction

For several decades, the femtosecond laser has been considered as a universal one-step procedure allowing the functionalization of solid materials. Benefitting from a high flexibility, ultrashort laser-induced surface nanostructuring has a panel of applications in various domains such as biomedical technologies, nanofluidics, renewable energies or aeronautics, to name a few [1–3]. The preparation and structuring of material surfaces on the nanometer scale are of prime importance for the advancement of these applications. The fast and repeated cycles of melting and resolidification upon multiple pulse irradiation progressively build up unique surface morphologies. Main topography features can be dictated by tuning the laser wavelength, polarization states, pulse duration or the fluence conditions, able to statistically promote specific shapes, scales, depths and orientations. These structures can transform or enhance surface properties such as wettability [4,5], mechanical and corrosion behavior, optical properties [3] or tribological performances [6], giving new functionalities to the structured material. More specifically, laser-induced periodic surface structures (LIPSS) with various characteristics can be generated on transparent materials, semi-conductors or metals [7,8]. For laser–metal interaction, several class of micro- and nano-structures are classically encountered. Formed near the ablation fluence regime, low spatial frequency LIPSS (LSFL) exhibit a periodicity marginally lower than the laser wavelength λ. LSFL are perpendicular to the electric field polarization, and are triggered by the interference of the incident light with the radiative scattered fields from laser-induced surface heterogeneities as bumps, holes or nanoparticles. Parallel or perpendicular to the laser polarization, high spatial frequency LIPSS (HSFL) have a period

285

lower than $\lambda/2$ and are strongly suspected to be initiated by local field enhancement on nanoreliefs whereas their growth is sustained by hydrodynamic mechanisms [9]. They are often categorized according to features depending on their localization and their orientation parallel or perpendicular to the polarization [10]. Parallel to the polarization and with a periodicity $\Lambda > \lambda$, a groove-like structures can also be generated on metals by high feedback femtosecond laser irradiation [11]. Finally, nanocavities or nanobumps were recently disclosed on the nanoscale with original hexagonal symmetries [9,12–14].

Even if high regularity process has been reported [15], typical LSFL generated on metals used to present many heterogeneities and bifurcations, limiting the process to be competitive for advanced industrial applications as nanolithography. Contrary to the crystalline metals, bulk amorphous metals, also named bulk metallic glasses, have the particularity to form very regular LSFL with few bifurcations [16]. This may be due to the excellent surface state of these amorphous materials, free from any topographic inhomogeneities. Indeed, the absence of crystalline defects such as grain boundaries or dislocations reduces the roughness and the number of surface heterogeneities [15]. However, few works deal with the HSFL creation on amorphous metals, limiting the main structures to LSFL and rings [16–19]. Although these bulk metallic glasses are already used in niche markets, they are very expensive to fabricate and, since the 1980s, thin-film metallic glasses (TFMG) have been elaborated varying the composition systems. The physical vapour deposition (PVD) method, and in particular magnetron sputtering, constitutes one of the most popular elaboration techniques, allowing the control of the composition, the density and the morphological aspect of the films. Zirconium-based TFMG can considerably increase the mechanical properties and the tribological behaviour of the covered material. Thereby, a Zr-Cu system constitutes a TFMG thoroughly studied because of its good properties and also for its good glass forming ability [20].

The aim of this work is to exploit the special structural nature of Zr-Cu thin film to investigate the susceptibility of this metallic glass to form nanopatterns exhibiting periods down to 100 nm. To understand the mechanisms underlying their emergence, we have investigated the effects of different initial morphologies. Two TFMG with the same composition and different surface aspects were irradiated varying the irradiation conditions. The difference in terms of feedback was revealed by electron and atomic microscopies.

2. Material Surface Features and Ultrafast Irradiation Process

Two thin films of metallic glass with the composition $Zr_{65}Cu_{35}$ have been deposited for the experiments. They were both manufactured by PVD magnetron sputtering from two targets of zirconium and copper. The thin films were synthesized on a silicon wafer. The argon working pressure was modified between the two samples, inducing two different kinds of film morphology [20,21]. The first sample, named "Tight Columns" (TC sample), was manufactured with a working pressure of 1 Pa while the second one, named "Coarse Columns" (CC sample), was deposited under 0.5 Pa. The substrates were not intentionally heated during the film deposition and the sample temperature was lower than 80 °C. The surface morphologies were observed by scanning electron microscopy ("SEM", Nova NanoSEM 200, FEI, Hillsboro, OR, USA) and atomic force microscopy ("AFM", Dimension ICON, Bruker, Billerica, MA, USA) in ScanAsyst-Air mode with a silicon nitride probe, coupled with image analysis. These analyses are described in part 3.

To perform periodic structuring of the surface, femtosecond laser irradiations were carried out using a commercial Ti: sapphire laser system (Legend Elite, Coherent Inc. San Jose, US), producing pulses centered at $\lambda = 800$ nm wavelength with a pulse duration fixed at 60 fs. The beam was focused on the samples by a 25 cm achromatic lens. Double pulses with the possibility to adjust the delay between them were produced by splitting and recombining the beam using a Mach–Zehnder interferometer configuration with a mobile arm. In each optical path the combination of a half-wave plate and a polarizer allowed for a perfect control of the pulses polarizations and energies [14]

Using the D^2 technique [22], the theoretical spot size and the damage threshold fluence for 1 laser pulse were determined for both samples. Figure 1 shows the evolution of the squared diameter increasing the irradiation energies for both samples for one single shot. Figure 1 also displays two SEM pictures of the impacts after one pulse obtained after irradiation for a fluence of 0.25 J/cm^2. The diameters of the spots were determined by an accurate measurement of the observed damage area. The D^2 technique yields the damage threshold fluences for one laser pulse for both samples. The TC sample presents a threshold fluence of $F_{th\text{-}TC}$ = 0.08 J/cm^2 lower than those of the CC sample with $F_{th\text{-}CC}$ = 0.12 J/cm^2.

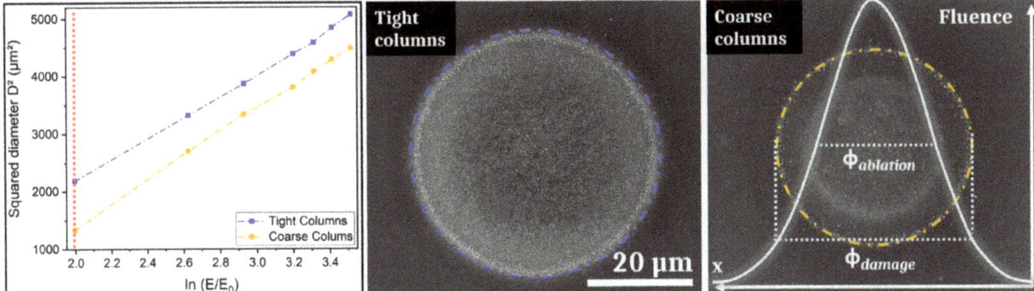

Figure 1. Evolution of the squared diameter versus the neperian logarithm of the energy of irradiation for a single pulse irradiation. The scanning electron microscope (SEM) images show the impact aspect for both samples for the lowest irradiation fluence (F_{1p} = 0.25 J/cm^2) corresponding to the value displayed by the red line: ln(E/E$_0$) = 2 (with E$_0$ = 1 µJ).

For single pulse experiments, different fluence conditions of irradiation were tested on both samples and three of them are reported and analyzed in the paper: 0.12, 0.08 and 0.06 J/cm^2. Different numbers of laser pulses were investigated from 1 to 50 pulses. For double pulse experiments, the polarizations of both pulses were horizontal and the lowest fluence condition (0.06 J/cm^2) is shown here. The burst number was fixed to 50 pulses and the time delay between the two pulses was increased from 0 ps to 70 ps. All of the irradiations were performed at atmospheric pressure in ambient air.

3. Experimental Results

3.1. Single Pulse Laser-Induced Nanostructuring: Initial Morphology Dependence on the Feedback

As mentioned in part 2, both samples used for the irradiation experiments have different initial morphologies resulting from different pressure conditions during PVD elaboration. Figure 2a,d show SEM pictures of the surface of, respectively, the TC and CC samples. The surface morphology of both samples displays a high density of interstices between so-called columns, which are classically formed during the PVD process. In an interesting way, the grain size observed on the surface plane varies with the pressure conditions. The TC sample presents a "granular" morphology with a size distribution denser and more regular than the CC sample. An image analysis reveals that the transverse size of the TC columns vary between ~10 and ~50 nm of diameter. The CC sample exhibits larger columns on the surface, with a typical width between ~30 and ~140 nm of diameter. If SEM provides high-quality images of the prepared surfaces, this remains limited to provide quantitative information of those morphologies and require complementary microscopy to define precise roughness profiles.

To complete the information visible on the SEM pictures, AFM measurements were carried out for both samples. Figure 2b,c respectively present an AFM analysis of the surface of the Tight Columns sample and a height profile of the white line visible on Figure 2b. The "peak-to-peak" roughness of the TC sample is measured at 10.1 nm, for an average RMS (root mean square) roughness of 1.44 nm. This profile shows a low regular distribution and confirms the average column's size found with the image analysis. Figure 2e,f display the AFM analysis of the initial morphology surface of the CC sample and the height profile of the white line displayed on the AFM picture. The roughness range is marginally higher than for TC as the measure gives a "peak-to-peak" roughness of 13.2 nm and a RMS roughness of 1.98 nm for CC. The height profile reveals an average column size in accordance with the observations done on the SEM picture. Figure 2f also shows small bumps on the top of many large columns, confirming the heterogeneity of roughness distribution for this sample. Thereby, both samples exhibit very low roughness, with optical surfaces of high quality with irregularities of the dimension $<\lambda/100$.

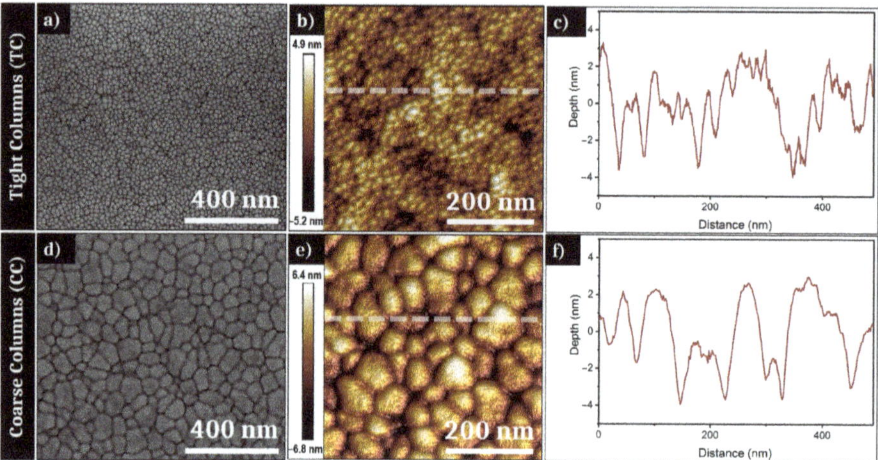

Figure 2. (**a**) Scanning electron microscope (SEM) picture and (**b**) atomic force microscope (AFM) 3D-projection of the surface of the Tight Columns (TC) sample; (**c**) height profile of the white line in (**b**); (**d**) SEM picture and (**e**) AFM 3D-projection of the surface of the Coarse Columns (CC) sample; (**f**) height profile of the white line in (**e**).

The first laser experiments were done with a fluence of 0.12 J/cm^2 corresponding to the CC sample's one pulse damage threshold, which is higher than that of the TC sample. After the irradiations with an average fluence of 0.12 J/cm^2, LSFL are obtained for high numbers of laser pulses for both samples. Figure 3 presents the SEM pictures corresponding to the evolution of structures obtained on both samples for different pulse numbers N. The red arrow shows the electric field polarization. Colored 3D-projections of SEM images correspond to the white box visible on each large SEM pictures.

Regarding the evolution of structures for the TC sample, it is remarkable that for the first pulse, frozen liquid bridges are already present on the impact. They are characteristic of the spallation mechanism. After two pulses, these filaments orientate preferentially in parallel with the laser polarization direction. They are observable up to 4 pulses and, for a higher feedback, nascent HSFL are oriented along this preferred direction. However, HSFL are superposed to underlying LSFL, revealing a crossed structure that mixes both types on the center of the impact. LSFL are clearly present from 15 to 50 pulses with HSFL perpendicular to them and mostly visible in the hollows. The irradiation fluence used in this case (0.12 J/cm^2) is 50% more elevated than the damage threshold of the TC sample. The thin film experienced a photomechanical ablation regime with identifiable specificities as surface structures revealed on Figure 3. After heating, the amorphous material turns to a molten state that affects and partially erases the initial columnar morphology. During this liquid state, spallation induced by generation and coalescence of voids in the liquid layer generates filaments that remain on the surface after the solidification process [23]. Near-field localized in the cavities between the filaments lead to the apparition of HSFL parallel to the electric field direction. Resulting from far-field scattering on the roughened surface, LSFL start to appear at the same time perpendicularly to the HSFL and their definition increase for high numbers of pulses [7,9,24].

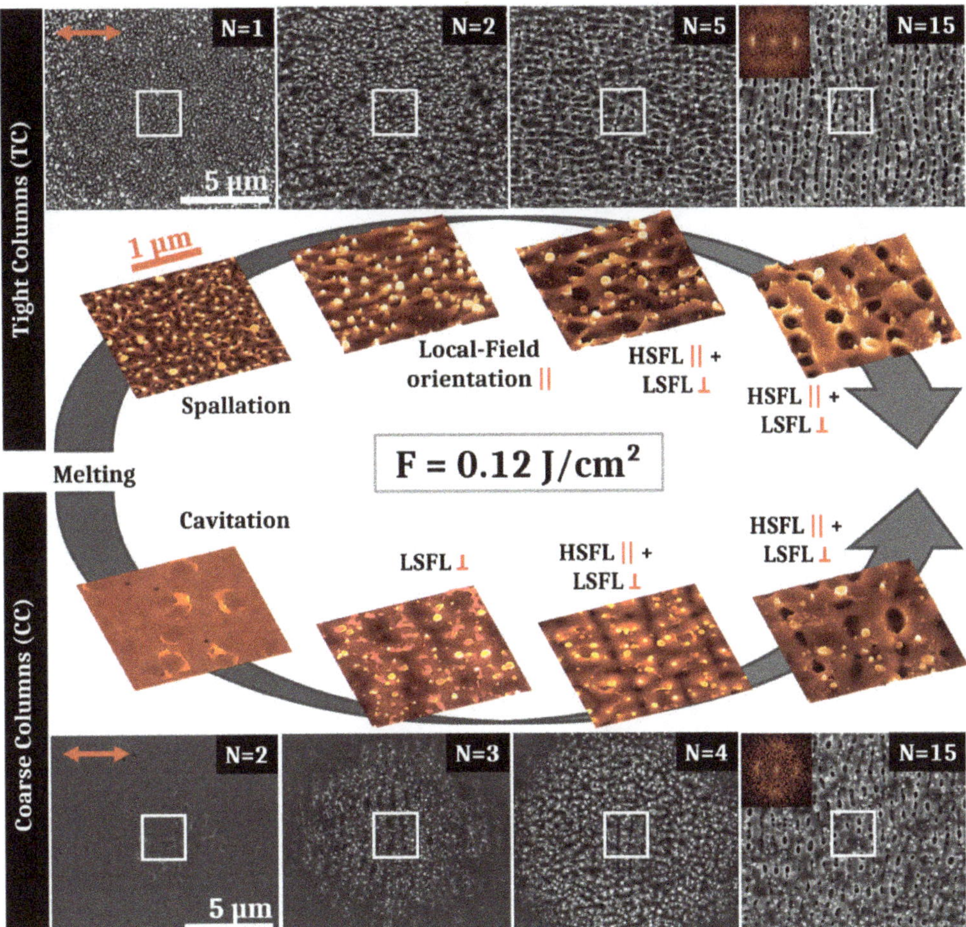

Figure 3. SEM and 3D-projections of SEM pictures of surface structures obtained after irradiation of the Tight Columns and the Coarse Columns samples with an average fluence of 0.12 J/cm² and different numbers of pulses. The evolution of structures is shown for different numbers of pulses. For the TC sample, successively appear disorganized filaments induced by spallation (N = 1), orientated filaments (N = 2), high spatial frequency laser-induced periodic surface structures (HSFL) parallel to the polarization intertwined between low spatial frequency laser-induced periodic surface structures (LSFL) perpendicular to the polarization (N = 5 and N =15). For the CC sample, successively appear residual cavitation bubbles (N = 2), LSFL (N = 3), horizontal HSFL and vertical LSFL (N = 4 and N = 15). The red arrow represents the electric field polarization. The symbols (| |) and (⊥) respectively mean "parallel" and "perpendicular" to the polarization. 2D-Fourier transforms shown as insets support the presence of a crossed structure of HSFL and LSFL for the highest pulse numbers.

For the Coarse Columns sample irradiated by one single shot at 0.12 J/cm² (not presented on Figure 3), the laser impact is barely visible on the surface and no structure can be evidenced. However, the initial columnar surface morphology disappears. For N = 2, structures resulting from exploded cavitation bubbles are visible on the surface. For N = 3, periodic structures perpendicular to the polarization are already discernible at the impact centre. They present a period marginally lower than the wavelength of the laser, and constitute the precursor of low spatial frequency LIPSS. From 4 to 10 pulses, HSFL formation, parallel to the electric field, emerge between the LSFL structures. Increasing the number of pulses, LSFL are increasingly defined. The definitive crossed structure with mixed HSFL and LSFL is observable on Figure 3 for 15 pulses for the CC sample. This crossed structure is present up to 50 pulses.

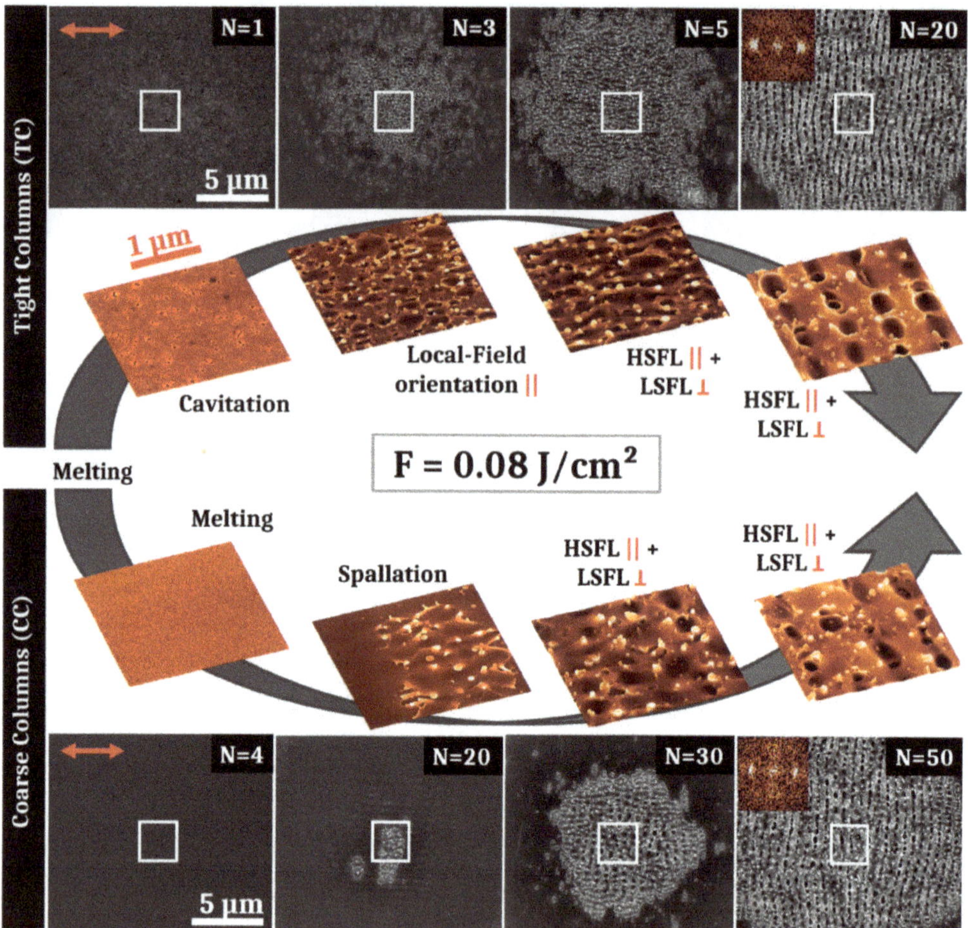

Figure 4. SEM and 3D-projections of SEM pictures of surface structures obtained after irradiation of both samples with an average fluence of 0.08 J/cm² and different numbers of pulses. The evolution of structures is shown for different numbers of pulses. For the TC sample, successively appear residual cavitation bubbles (N = 1), disorganized filaments (N = 3), HSFL parallel to the polarization intertwined between LSFL perpendicular to the polarization (N = 5 and N = 20). For the CC sample, successively appear a smooth surface (N = 4), spallation effects (N = 20), horizontal HSFL and vertical LSFL (N = 30 and N = 50). The red arrow represents the electric field polarization. The symbols (| |) and (⊥) respectively mean "parallel" and "perpendicular" to the polarization. 2D-Fourier transforms shown as insets support the presence of a crossed structure of HSFL and LSFL for the highest pulse numbers.

Although the final crossed structures mixing HSFL and LSFL are similar for both samples, the structure evolution differs between the CC sample and the TC sample. The average roughness for both samples is very low (less than 2 nm), but the different roughness morphology affects considerably the damage threshold fluence leading to a delayed surface structures apparition and evolution in terms of feedback. For the CC sample, after the surface melting stage, a cavitation mechanism occurs leading to the apparition of surface bubbles but spallation filaments are less pronounced than for the TC sample. Large cavitation bubbles appear with a low density and constitute larger precursors than the filaments for light scattering compared to the TC sample. These frozen cavitation structures trigger the apparition of low spatial frequency LIPSS perpendicular to the electric field. For a higher number of pulses, the LSFL hollows undergo local melt flow that initiates high spatial frequency LIPSS parallel to the polarization [9]. While they are less regular than for the TC sample, LSFL contrast increases for higher number of pulses. For N = 15 pulses, the period of the LSFL on the TC sample is ~630 nm versus ~650 nm for the LSFL of the CC sample. Comparing LIPSS on this TFMG with that already reported on bulk metallic glass, the LSFL obtained here present many heterogeneities, bifurcations and redeposited particles [25,26]. As they are triggered by sharp nanoreliefs in the photomechanical regime of ablation, the HSFL present also bifurcations and irregularities for both samples [16,17]. Moreover, in both cases, HSFL are superposed with LSFL structures. To favor the HSFL apparition and prevent the formation of LSFL, the fluence has been reduced to 0.08 J/cm^2.

The results of these second irradiations performed at a lower fluence of 0.08 J/cm^2 are presented in Figure 4, showing the surface structures obtained on both samples after different pulse numbers. Similar periodic surfaces structures are visible for the TC sample in order of appearance: oriented filaments, HSFL and LSFL. However, at this reduced fluence, spallation effects are not visible for N = 1 and the material surface has undergone a localized cavitation process with a high concentration. These surface cavities are not leading to the apparition of LSFL but the smaller scattering centers have stimulated localized field enhancement that orientates filaments along the electric field direction. For the CC thin film, no surface structure is visible for N < 10 pulses as the fluence was lower than the damage threshold of the sample. At 10 pulses, structures parallel to the polarization with a period similar as LSFL are present with a very low contrast. On the top of these structures, some cracks seem to be the precursors of spallation effects, visible for N > 20. These cracks may appear in response of the particular deformation modes of metallic glasses that generate high stresses [20,27]. After spallation mechanisms (N = 20), HSFL mixed with LSFL appear for N = 30 pulses. Finally, for higher numbers of pulses, LSFL are observable on the CC sample with a better contrast and more regularly than for the higher fluence.

For both fluences (0.12 J/cm^2 and 0.08 J/cm^2), HSFL are obtained concurrently with LSFL. These two types of structure are always superposed with a better definition of either one or the other depending on the pulse number. The HSFL formed are always less regular than those reported on crystalline metals [28,29]. In order to create homogeneous and regular high spatial frequency LIPSS, irradiations in a sub-ablation regime were performed with a fluence lower, at 0.06 J/cm^2, and increasing the feedback.

3.2. High Feedback with Single and Double Pulses Strategy to Foster High Spatial Frequency Laser-Induced Periodic Surface Structures (HSFL)

Using a low fluence of 0.06 J/cm^2 with single pulses experiments, the samples are submitted to a sub-ablation regime promoting solid–liquid phase transition that may involve hydrodynamic instabilities due to a capillary process [9]. These instabilities are correlated to time-dependent liquid properties that can be extended using a collinear double pulses approach [9]. Therefore, a second set of irradiations were done at the same fluence of 0.06 J/cm^2 in a double pulse configuration, with an equally distributed fluence in each arm, with an elevated number of pulses in order to enhance the feedback. Figure 5 shows comparatively the results obtained for both 50 single-pulse and 50 double-pulse sequences. For single pulses (Figure 5a), the TC sample exhibits a reduced zone where low spatial frequency LIPSS are formed at the center of the irradiation area. Two zooms at remote locations surrounding the LSFL are presented in the Figure 5a (left and right) with their respective 2D-Fourier transform (2D-FT) as insets. On the border of the observed laser impact area, weakly contrasted nanostructures are present. They consist of periodic arrangement of material, parallel to the polarization, with a period approximately equal to 100 nm. These periodic structures are only observed at the outside region of the Gaussian energy distribution and their contrast is similar to structures associated to annealing mechanisms [13,30]. In addition, in the intermediate region between these structures and the impact center, highly-concentrated dark contrasted points are visible at the interstices of the initial columnar morphology. These dark points indicate the presence of cavitation-induced gaps between the columns. Low contrasted wavy patterns, confirmed by vertical bright dots in the 2D-Fourier transform, reveal the starting of HSFL generation that preserve the microstructural morphology.

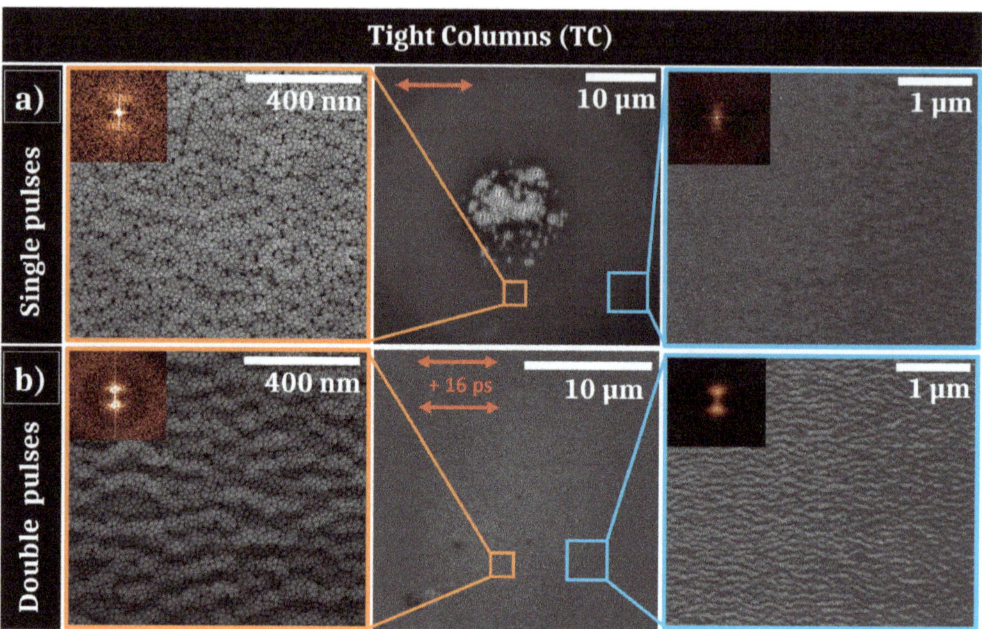

Figure 5. (a) SEM pictures of irradiated area of the Tight Columns sample after 50 single pulses of laser irradiation with 0.06 J/cm^2 with 2D-Fourier transform as insets; (b) SEM pictures of irradiated area after 50 double pulses of laser irradiation separated by 16 ps with a total fluence of 0.06 J/cm^2 with 2D-Fourier transform as insets. The red arrows represent the electric field polarizations.

Figure 5b displays the results obtained in a similar energy dose but with a different fluence feedthrough brought by two consecutive pulses, delayed by a time $\Delta t = 16$ ps. This corresponds to the optimal time separation between both horizontally polarized pulses to form HSFL. These nanostructures are, however, observed in the range $\Delta t = 2 - 30$ ps, with a maximum contrast at $\Delta t = 16$ ps. For this delay condition, the impact is homogeneously covered by this original kind of HSFL and two different magnitude of zooms are presented at the border of the impact in Figure 5b with the associated 2D-FT as insets. The distributed HSFL are noticeably regular and present a period $\Lambda \sim 100$ nm at the center and also at the border of the laser impact. They present an amplitude between 15 and 20 nm. Once again, the persistence of grains is remarkable and questions on the thermodynamic conditions underlying this phenomenon.

The same irradiations at 50 single- and double-pulse sequences were applied to the CC sample at 0.06 J/cm^2. These results are shown on Figure 6. As previously noted for the TC sample, for single pulse irradiation shown in the middle of Figure 6a, the border of the irradiated zone of the CC sample presents horizontally periodic surface structures that may be generated by an annealing process. However, these formed structures are larger than those observed on the TC sample, that may be due to the larger size of coarse columns or to a different thickness of heated/liquid layer. On the center of the impact, LSFL are not observed as for the TC sample, indicating that the energetic dose condition is not sufficient to initiate them. Instead, some big cracks are present at the center of the Gaussian fluence distribution, probably resulting from intense tensile stresses at the surface of the thin-film metallic glass. Around these cracks, a SEM zoomed picture reveals that the initial columnar morphology of the TFMG has disappeared. The interstices are less contrasted than before irradiation and the roughness is likely to be lower. The 2D-Fourier analysis performed in this irradiated region confirms the absence of order.

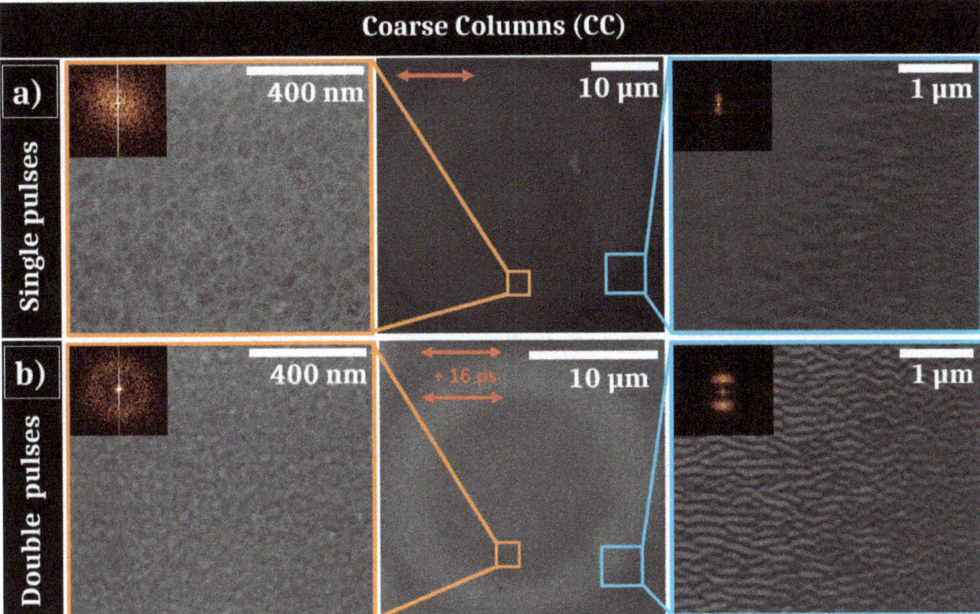

Figure 6. (a) SEM pictures of irradiated area of the Coarse Columns sample after 50 single pulses of laser irradiation with 0.06 J/cm^2 with 2D-Fourier transform as insets; (b) SEM pictures of irradiated area after 50 double pulses of laser irradiation separated by 16 ps with a total fluence of 0.06 J/cm^2 with 2D-Fourier transform as insets. The red arrows represent the electric field polarizations.

For the double pulse experiments presented in Figure 6b (middle), the material response is not as homogeneous as that observed for the TC sample and a bright corona with a 10 µm radius is visible at the inner edge of the laser impact. A zoomed picture of this zone (Figure 6b right) reveals the presence of high spatial frequency LIPSS, parallel to the polarizations. The impacts are similar for time delays $\Delta t = 2 - 30$ ps with more contrasted HSFL at $\Delta t = 16$ ps, as observed for the TC sample. The HSFL have a period $\Lambda \sim 150$ nm, slightly smaller than the periodic structures created on the border of the single-pulse impact visible on Figure 6a. Furthermore, they are more contrasted, as evidenced on the SEM picture, that is confirmed by the 2D-FT. The center zone does not exhibit HSFL and the "columnar" morphology of the film has disappeared as well as for single pulse experiments; 2D-FT analysis validates the absence of periodic structures near the center (Figure 6b left).

As a resume, a double-pulse sequence enhances the regularity and contrast of HSFL formation. This has been optimized by a specific time delay between the two collinear pulses that could correspond to the thermomechanical characteristic time of the ultrafast irradiated surface. They emerge as the material is heated near the melting threshold, enabling the preservation of the initial columnar morphology for the TC sample, but counter-intuitively, the initial columns for the CC sample disappear. Even if the one-pulse damage threshold is lower for TC, this may not the case for a higher number of shots for which the CC sample damage threshold could be lower. Thereby, this confirms that the feedback dynamics is strongly morphology-dependent.

4. Discussion

Ultrashort-laser irradiation of $Zr_{65}Cu_{35}$ thin film metallic glass shows that standard LIPSS formation can be achieved with a fluence adjusted to the initial morphology. In particular HSFL and LSFL with polarization-driven orientations propose features similar to what is usually observed in crystalline metals [8,31,32]. It is important to point out that behind the generic term of HSFL, nanostructures show two kinds of very distinct morphologies for situations where spallation process occur on the surface, as shown on conditions of Figures 3 and 4, and when the surface seems to conserve its general integrity, in particular conserving columnar growth patterns as observed for Figures 5 and 6.

Both kinds of PVD prepared thin film, with distinct roughness shown in Figure 2, allow progressively the establishment of LSFL when the number of laser shots increases as topography-driven feedback offers all the requirements for surface waves scattering [33]. This is expected in the classic scenario stipulating an interference process between the incident pulse with the surface scattered waves [34]. SEM images of the impact after tens of impacts, presented in Figures 3 and 4 for two different fluences, confirm that the conditions required to form LSFL are poorly correlated to the initial roughness even though we need to adjust energetic dose conditions. This observation was anticipated as the coherent far-field scattering is weakly dependent on the topography morphologies, supporting the "surface-scattered wave" as a robust concept not very demanding on the initial conditions [34,35]. More interestingly, the transient evolution of both surfaces is slightly different and the CC sample exhibits delayed dynamics in terms of feedback or requires a higher fluence to achieve LSFL, in accordance with the threshold of material response. We can note that this different response between both initial morphologies subsides as the number of pulses increase, i.e., that the cumulative dynamics differ.

At 0.12 J/cm^2 and 0.08 J/cm^2 fluences, the photomechanical regime of ablation is gradually set up. This induces a spallation process with coalescence of sub-surface voids associated with breakup and fast freezing of the transient liquid structures remaining on the surface [23]. The retraction dynamics of liquid droplets, driven by capillary and viscosity forces, play a key role for the formation of further reliefs at the nanoscale. Pulse after pulse, the progressively formed roughness centers experience non-radiative light coupling that will trigger orientation of energy deposition and transverse thermal gradients. The polarization-oriented structures around asperities are clearly visible on Figure 3. A high concentration can favor multiple scattering whereas the size and shape of a relief unit determines the strength of the local field enhancement. This complexity and diversity of nanoreliefs prevent or limit the effective control of this kind of HSFL. Moreover, initiating and associating with an ablation regime, these structures are superposed to more resilient LSFL, which lessen the potential properties of this kind of HSFL. In our irradiating strategy, a particular emphasis was thereby placed on the apparition of uniform nanostructures well below the wavelength, parallel to the laser polarization, and not superposed to LSFL. This requires the fluence to decrease from 0.12 J/cm^2 to 0.06 J/cm^2 whereas the number of pulses has to increase as revealed by SEM images shown in Figures 3, 5 and 6. In absolute terms, as previously reported on other materials, a high feedback regime for fluence conditions close to melting threshold fosters the emergence of wavy structures with structural modifications combined with a periodic topography while limiting the flatness alteration of the thin film [13,36]. This point is crucial for targeted mechanical and tribological applications.

Finally, in a sub-ablation regime, hydrothermal waves driving the hotter melt flow above the initial surface level were shown to be at the origin of HSFL growth for metals and semi-conductors [9,13]. Regular nanopatterns inherit the light polarization response on the local roughness with a formation resulting from the competition between destabilizing gradients as surface tension and a rarefaction wave and thermal dissipation on the surface. In this Bénard–Rayleigh–Marangoni-like hydrodynamic instability resides the lifetime of the molten layer as well as an efficient displacement of the liquid interface [9]. This suggests that the viscous flow is facilitated by the non-equilibrium properties of the system. In particular, the related characteristic time for Marangoni flow against the viscous forces is estimated as $\tau_M = \frac{\mu L^2}{4h\gamma\Delta T}$ where μ is the material viscosity, L the expected distance, h the liquid thickness and $\gamma\Delta T$ the surface tension dependence. At the melting temperature estimated to $T_m = 1300$ K, the viscosity is around 50 mPa s which is one order of magnitude higher than the values reported for most crystalline metals [37]. However, at higher liquid temperatures in the range T = 2000–2500 K, μ decreases to values similar to pure metals and reaches 2–6 mPa s, enabling a strongly more pronounced capillary-driven melted flow [37]. Reducing this Marangoni timescale, the liquid-displacement from hot to cold regions can modulate surface topography before thermal dissipation and solidification. To exploit this high-T dependence of viscosity for amorphous metals, a collinear double pulse irradiation strategy has been followed varying the pulse delay. This way, the material is brought to a higher temperature as long as possible with the opportunity to control cavitation formation by the time separation between two pulses. Figures 5 and 6 clearly show that the double-pulse sequence promotes HSFL formation preventing LSFL apparition at the low fluence regime. Double-pulse irradiation provides this way an additional degree for controlling the desired material properties far from equilibrium. This easily implementable beam engineering opens the route to functionalize materials, as thin film amorphous metallic alloys, unwilling to form uniform HSFL.

5. Conclusions

Since Sipe's theory, it has been clearly established that the roughness layer has a central role in the formation of LIPSS, in particular for LSFL development. Thin film deposition by PVD offers a remarkable way to change the initial morphologies conserving other alloy properties. In particular, high-quality surface metallic glass exhibits granular columns that can affect the transient ultrafast light coupling under multipulse feedback irradiation. Although the initial average roughness of the film is very low, the columnar morphology constitutes roughness heterogeneities inducing more bifurcations than for polished bulk amorphous metals in laser-induced surface structures. These heterogeneities can be comparable to crystalline defects as grain boundaries or dislocations and represent privileged precursors for local light absorption and scattering. This is a valuable asset to favor non-radiative field structures and subsequent HSFL creation despite the resilience of metallic glass to form such nanopatterns. We show that the high viscosity of an amorphous alloy can be bypassed by double-pulse irradiation sequences that bring the material into a higher temperature and a presumed lower viscosity state more favorable to undergo required capillary flows. This temporal optimization of the energy feedthrough promotes the development of homogeneous high spatial frequency LIPSS with a high regularity and preserving the surface integrity. Therefore, ultrashort laser pulses can functionalize thin-film metallic glasses generating patterns with various periodicities, down to 100 nanometers. These results pave the way for the use of a femtosecond laser process for nanostructuring and further functionalization of metallic glass thin film, with dimensions and accuracies relevant for mechanical, tribological and biomedical applications.

Author Contributions: J.-P.C., F.B. and F.G. directed the project. A.B. and J.-F.P. performed the elaboration of the thin films. M.P. operated the laser experiments and the characterizations (SEM, AFM and image analyses). M.P. and J.-P.C. wrote the manuscript. All authors examined the results and ideas and reviewed the manuscript. All authors have read and agreed to the published version of the manuscript.

Funding: This work is supported by the ANR project MEGALIT (ANR-18-CE08-0018) of the French Agence Nationale de la Recherche and the project FORMEL of the Pack Ambition Research program of the Auvergne Rhône-Alpes Region.

Institutional Review Board Statement: Not applicable.

Informed Consent Statement: Not applicable.

Data Availability Statement: Not applicable.

Conflicts of Interest: The authors declare no conflict of interest.

References

1. Bonse, J.; Höhm, S.; Kirner, S.V.; Rosenfeld, A.; Krüger, J. Laser-Induced Periodic Surface Structures—A Scientific Evergreen. *IEEE J. Sel. Top. Quantum Electron.* **2017**, *23*, 1–15. [CrossRef]
2. Sugioka, K.; Cheng, Y. Ultrafast lasers—Reliable tools for advanced materials processing. *Light Sci. Appl.* **2014**, *3*, e149. [CrossRef]
3. Papadopoulos, A.; Skoulas, E.; Mimidis, A.; Perrakis, G.; Kenanakis, G.; Tsibidis, G.D.; Stratakis, E. Biomimetic Omnidirectional Antireflective Glass via Direct Ultrafast Laser Nanostructuring. *Adv. Mater.* **2019**, *31*, e1901123. [CrossRef] [PubMed]
4. Krishna, H.; Favazza, C.; Gangopadhyay, A.K.; Kalyanaraman, R. Functional nanostructures through nanosecond laser dewetting of thin metal films. *JOM* **2008**, *60*, 37–42. [CrossRef]
5. Bizi-Bandoki, P.; Benayoun, S.; Valette, S.; Beaugiraud, B.; Audouard, E. Modifications of roughness and wettability properties of metals induced by femtosecond laser treatment. *Appl. Surf. Sci.* **2011**, *257*, 5213–5218. [CrossRef]
6. Woloszynski, T.; Touche, T.; Podsiadlo, P.; Stachowiak, G.W.; Cayer-Barrioz, J.; Mazuyer, D. Effects of Nanoscale Ripple Texture on Friction and Film Thickness in EHL Contacts. *Tribol. Lett.* **2019**, *67*, 16. [CrossRef]
7. Rudenko, A.; Colombier, J.-P.; Höhm, S.; Rosenfeld, A.; Krüger, J.; Bonse, J.; Itina, T.E. Spontaneous periodic ordering on the surface and in the bulk of dielectrics irradiated by ultrafast laser: A shared electromagnetic origin. *Sci. Rep.* **2017**, *7*, 12306. [CrossRef] [PubMed]
8. Bonse, J.; Krüger, J.; Höhm, S.; Rosenfeld, A. Femtosecond laser-induced periodic surface structures. *J. Laser Appl.* **2012**, *24*, 042006. [CrossRef]
9. Rudenko, A.; Abou-Saleh, A.; Pigeon, F.; Mauclair, C.; Garrelie, F.; Stoian, R.; Colombier, J.P. High-frequency periodic patterns driven by non-radiative fields coupled with Marangoni convection instabilities on laser-excited metal surfaces. *Acta Mater.* **2020**, *194*, 93–105. [CrossRef]
10. Skolski, J.Z.P.; Römer, G.R.B.E.; Vincenc Obona, J.; Huis in 't Veld, A.J. Modeling laser-induced periodic surface structures: Finite-difference time-domain feedback simulations. *J. Appl. Phys.* **2014**, *115*, 103102. [CrossRef]
11. Tsibidis, G.D.; Fotakis, C.; Stratakis, E. From ripples to spikes: A hydrodynamical mechanism to interpret femtosecond laser-induced self-assembled structures. *Phys. Rev. B* **2015**, *92*, 041405. [CrossRef]
12. Giannuzzi, G.; Gaudiuso, C.; Franco, C.D.; Scamarcio, G.; Lugarà, P.M.; Ancona, A. Large area laser-induced periodic surface structures on steel by bursts of femtosecond pulses with picosecond delays. *Opt. Lasers Eng.* **2019**, *114*, 15–21. [CrossRef]
13. Colombier, J.-P.; Rudenko, A.; Silaeva, E.; Zhang, H.; Sedao, X.; Bévillon, E.; Reynaud, S.; Maurice, C.; Pigeon, F.; Garrelie, F.; et al. Mixing periodic topographies and structural patterns on silicon surfaces mediated by ultrafast photoexcited charge carriers. *Phys. Rev. Res.* **2020**, *2*, 043080. [CrossRef]
14. Abou Saleh, A.; Rudenko, A.; Reynaud, S.; Pigeon, F.; Garrelie, F.; Colombier, J.-P. Sub-100 nm 2D nanopatterning on a large scale by ultrafast laser energy regulation. *Nanoscale* **2020**, *12*, 6609–6616. [CrossRef] [PubMed]
15. Gnilitskyi, I.; Derrien, T.J.-Y.; Levy, Y.; Bulgakova, N.M.; Mocek, T.; Orazi, L. High-speed manufacturing of highly regular femtosecond laser-induced periodic surface structures: Physical origin of regularity. *Sci. Rep.* **2017**, *7*, 8485. [CrossRef] [PubMed]
16. Li, C.; Cheng, G.; Sedao, X.; Zhang, W.; Zhang, H.; Faure, N.; Jamon, D.; Colombier, J.-P.; Stoian, R. Scattering effects and high-spatial-frequency nanostructures on ultrafast laser irradiated surfaces of zirconium metallic alloys with nano-scaled topographies. *Opt. Express* **2016**, *24*, 11558. [CrossRef] [PubMed]
17. Ran, L.; Qu, S. Femtosecond laser induced surface structures on amorphous alloys. In Proceedings of the 2011 Academic International Symposium on Optoelectronics and Microelectronics Technology, Harbin, China, 12–16 October 2011; pp. 134–137.
18. Hoppius, J.S.; Bialuschewski, D.; Mathur, S.; Ostendorf, A.; Gurevich, E.L. Femtosecond laser crystallization of amorphous titanium oxide thin films. *Appl. Phys. Lett.* **2018**, *113*, 071904. [CrossRef]
19. Ma, F.; Yang, J.; Zhu, X.; Liang, C.; Wang, H. Femtosecond laser-induced concentric ring microstructures on Zr-based metallic glass. *Appl. Surf. Sci.* **2010**, *256*, 3653–3660. [CrossRef]
20. Apreutesei, M.; Steyer, P.; Joly-Pottuz, L.; Billard, A.; Qiao, J.; Cardinal, S.; Sanchette, F.; Pelletier, J.M.; Esnouf, C. Microstructural, thermal and mechanical behavior of co-sputtered binary Zr–Cu thin film metallic glasses. *Thin Solid Film.* **2014**, *561*, 53–59. [CrossRef]
21. Apreutesei, M.; Steyer, P.; Billard, A.; Joly-Pottuz, L.; Esnouf, C. Zr–Cu thin film metallic glasses: An assessment of the thermal stability and phases' transformation mechanisms. *J. Alloys Compd.* **2015**, *619*, 284–292. [CrossRef]
22. Liu, J.M. Simple technique for measurements of pulsed Gaussian-beam spot sizes. *Opt. Lett.* **1982**, *7*, 196. [CrossRef]
23. Abou-Saleh, A.; Karim, E.T.; Maurice, C.; Reynaud, S.; Pigeon, F.; Garrelie, F.; Zhigilei, L.V.; Colombier, J.P. Spallation-induced roughness promoting high spatial frequency nanostructure formation on Cr. *Appl. Phys. A* **2018**, *124*, 308. [CrossRef]
24. Aguilar, A.; Mauclair, C.; Faure, N.; Colombier, J.-P.; Stoian, R. In-situ high-resolution visualization of laser-induced periodic nanostructures driven by optical feedback. *Sci. Rep.* **2017**, *7*, 16509. [CrossRef]
25. Zhang, W.; Cheng, G.; Hui, X.D.; Feng, Q. Abnormal ripple patterns with enhanced regularity and continuity in a bulk metallic glass induced by femtosecond laser irradiation. *Appl. Phys. A* **2014**, *115*, 1451–1455. [CrossRef]
26. Lei, Y.; Yang, J.; Cong, C.; Guo, C. Fabrication of homogenous subwavelength grating structures on metallic glass using double-pulsed femtosecond lasers. *Opt. Lasers Eng.* **2020**, *134*, 106273. [CrossRef]
27. Etiemble, A.; Der Loughian, C.; Apreutesei, M.; Langlois, C.; Cardinal, S.; Pelletier, J.M.; Pierson, J.-F.; Steyer, P. Innovative Zr-Cu-Ag thin film metallic glass deposed by magnetron PVD sputtering for antibacterial applications. *J. Alloys Compd.* **2017**, *707*, 155–161. [CrossRef]

28. Žemaitis, A.; Mimidis, A.; Papadopoulos, A.; Gečys, P.; Račiukaitis, G.; Stratakis, E.; Gedvilas, M. Controlling the wettability of stainless steel from highly-hydrophilic to super-hydrophobic by femtosecond laser-induced ripples and nanospikes. *RSC Adv.* **2020**, *10*, 37956–37961. [CrossRef]

29. Liu, Y.-H.; Yeh, S.-C.; Cheng, C.-W. Two-Dimensional Periodic Nanostructure Fabricated on Titanium by Femtosecond Green Laser. *Nanomaterials* **2020**, *10*, 1820. [CrossRef]

30. Bonse, J.; Baudach, S.; Krüger, J.; Kautek, W.; Lenzner, M. Femtosecond laser ablation of silicon–modification thresholds and morphology. *Appl. Phys. A* **2002**, *74*, 19–25. [CrossRef]

31. Buividas, R.; Mikutis, M.; Juodkazis, S. Surface and bulk structuring of materials by ripples with long and short laser pulses: Recent advances. *Prog. Quantum Electron.* **2014**, *38*, 119–156. [CrossRef]

32. Zhang, H.; Colombier, J.-P.; Li, C.; Faure, N.; Cheng, G.; Stoian, R. Coherence in ultrafast laser-induced periodic surface structures. *Phys. Rev. B* **2015**, *92*. [CrossRef]

33. Rudenko, A.; Mauclair, C.; Garrelie, F.; Stoian, R.; Colombier, J.-P. Self-organization of surfaces on the nanoscale by topography-mediated selection of quasi-cylindrical and plasmonic waves. *Nanophotonics* **2019**, *8*, 459–465. [CrossRef]

34. Sipe, J.E.; Young, J.F.; Preston, J.S.; van Driel, H.M. Laser-induced periodic surface structure. I. Theory. *Phys. Rev. B* **1983**, *27*, 1141–1154. [CrossRef]

35. Young, J.F.; Preston, J.S.; van Driel, H.M.; Sipe, J.E. Laser-induced periodic surface structure. II. Experiments on Ge, Si, Al, and brass. *Phys. Rev. B* **1983**, *27*, 1155–1172. [CrossRef]

36. Lopez-Santos, C.; Puerto, D.; Siegel, J.; Macias-Montero, M.; Florian, C.; Gil-Rostra, J.; López-Flores, V.; Borras, A.; González-Elipe, A.R.; Solis, J. Anisotropic Resistivity Surfaces Produced in ITO Films by Laser-Induced Nanoscale Self-organization. *Adv. Opt. Mater.* **2021**, *9*, 2001086. [CrossRef]

37. Han, X.J.; Schober, H.R. Transport properties and Stokes-Einstein relation in a computer-simulated glass-forming $Cu_{33.3}Zr_{66.7}$ melt. *Phys. Rev. B* **2011**, *83*, 224201. [CrossRef]

Article

Straightforward Patterning of Functional Polymers by Sequential Nanosecond Pulsed Laser Irradiation

Edgar Gutiérrez-Fernández [1,†], Tiberio A. Ezquerra [1], Aurora Nogales [1] and Esther Rebollar [2,*]

[1] Instituto de Estructura de la Materia, IEM-CSIC, Serrano 121, 28006 Madrid, Spain; edgar.gutierrez@ehu.eus (E.G.-F.); t.ezquerra@csic.es (T.A.E.); aurora.nogales@csic.es (A.N.)

[2] Instituto de Química Física Rocasolano, IQFR-CSIC, Serrano 119, 28006 Madrid, Spain

* Correspondence: e.rebollar@csic.es

† Present address: POLYMAT and Polymer Science and Technology Department, Faculty of Chemistry, University of the Basque Country UPV/EHU, Av. de Tolosa 72, 20018 San Sebastián, Spain.

Abstract: Laser-based methods have demonstrated to be effective in the fabrication of surface micro- and nanostructures, which have a wide range of applications, such as cell culture, sensors or controlled wettability. One laser-based technique used for micro- and nanostructuring of surfaces is the formation of laser-induced periodic surface structures (LIPSS). LIPSS are formed upon repetitive irradiation at fluences well below the ablation threshold and in particular, linear structures are formed in the case of irradiation with linearly polarized laser beams. In this work, we report on the simple fabrication of a library of ordered nanostructures in a polymer surface by repeated irradiation using a nanosecond pulsed laser operating in the UV and visible region in order to obtain nanoscale-controlled functionality. By using a combination of pulses at different wavelengths and sequential irradiation with different polarization orientations, it is possible to obtain different geometries of nanostructures, in particular linear gratings, grids and arrays of nanodots. We use this experimental approach to nanostructure the semiconductor polymer poly(3-hexylthiophene) (P3HT) and the ferroelectric copolymer poly[(vinylidenefluoride-co-trifluoroethylene] (P(VDF-TrFE)) since nanogratings in semiconductor polymers, such as P3HT and nanodots, in ferroelectric systems are viewed as systems with potential applications in organic photovoltaics or non-volatile memories.

Keywords: laser-induced periodic surface structures; polymer nanostructures; ordered nanostructures; linear gratings; nanodots arrays; semiconductor polymer; ferroelectric polymer

Citation: Gutiérrez-Fernández, E.; Ezquerra, T.A.; Nogales, A.; Rebollar, E. Straightforward Patterning of Functional Polymers by Sequential Nanosecond Pulsed Laser Irradiation. *Nanomaterials* **2021**, *11*, 1123. https://doi.org/10.3390/nano11051123

Academic Editors: Jörn Bonse, Peter Simon and Jürgen Ihlemann

Received: 30 March 2021
Accepted: 23 April 2021
Published: 27 April 2021

Publisher's Note: MDPI stays neutral with regard to jurisdictional claims in published maps and institutional affiliations.

1. Introduction

More and more effort is being made to fabricate functional surfaces with an increased number of applications. In particular, the fabrication of structures in the micro- and nanoscale allows the design of materials with advanced surface properties. Examples of complex structures possessing specific functions and properties can be found in nature [1]. For instance, the control of wettability by the presence of hierarchical surface structures, ranging from micro- to the nanometer scale, may provide self-cleaning properties similar to the lotus leaf in nature [2,3]. Changing and controlling the wetting properties of a material by nanostructuring may allow new applications in the fields of microfluidics, nanofluidics, optofluidics, biomedicine, environmental science and self-cleaning, among others [4,5]. Fabrication of superhydrophobic surfaces by laser irradiation has been reported for different materials [6,7] by the chemical modification of the surface or micro-structuring. Hierarchical structures also play a role in the excellent adhesion of gecko feet [8], and the presence of anisotropic micro- and nanostructures in snake skin affect friction [9].

In order to obtain complex structures, different strategies have been developed based on chemical etching [10,11], lithographic techniques [12,13] or template-based methods [14,15]. Laser-based methods have demonstrated to be effective in the fabrication of surface micro- and nanostructures, which have found a wide range of applications [16]. For

299

example, cell orientation, proliferation and differentiation can be stimulated on biomaterials [17–19], bacterial adhesion can be reduced [20,21], wettability may be controlled [22–24] and tribological performance [25,26] and superconducting properties may be modified [27]. Some laser-based techniques used for micro- and nanostructuring of surfaces are Direct Laser Interference Patterning (DLIP) [22,28,29] and the formation of laser-induced periodic surface structures (LIPSS) [30–32]. LIPSS are formed upon repetitive irradiation at fluences well below the ablation threshold and, in particular, linear structures are formed in the case of irradiation with linearly polarized laser beams, while irradiation with circularly polarized lasers gives rise to the formation of circular nanostructures [30]. The period of the obtained structures is related to the laser wavelength and close to it when irradiation is carried out at normal incidence. During the last 50 years, LIPSS formation has gained attention due to its simplicity and robustness, since it is a single process step which can be carried out in ambient air and is fully compatible with industrial demands in terms of costs, reliability and productivity. Comparing LIPSS with the DLIP set up is simpler, since only one laser beam is employed, and smaller periods may be obtained when high spatial frequency LIPSS (HSFL) are formed [31,33].

Recent research has focused on the functionalization of different materials with structures ranging between a few tens of nanometers up to several micrometers [31,34–36]. Substrates nanostructured with LIPSS are proposed to be interesting for applications in different fields, such as optics, electronics, fluidics, sensing, mechanical engineering and biomedicine [16,25,31,37–39]. In the case of polymers, most of the research has been focused on the use of nanosecond laser pulses [30], although more recent studies have reported LIPSS formation upon irradiation with picosecond and femtosecond laser pulses [40–42]. Some of the applications proposed for the polymer substrates nanostructured in this way are sensors, photovoltaics, non-volatile memories, cell culture and antibacterial substrates [19,21,30,43–48].

Additionally, the combination of different laser techniques, such as direct laser writing and DLIP [49] or DLIP with mask imaging [50], may induce the formation of more complex structures, and also multi-pulse DLIP [51], laser micropatterning followed by LIPSS formation [52] or LIPSS simultaneously with diffraction patterns have been reported [53].

In this work, we show how to fabricate a library of ordered nanostructures in a polymer surface by repeated irradiation with a nanosecond laser beam. In particular, we prepared structures on two functional polymers: poly(3-hexylthiophene) (P3HT) and poly(vinylidenefluoride-co-trifluoroethylene) (P(VDF-TrFE)). Nanogratings in semiconductor polymers as P3HT and nanodots in ferroelectric systems are viewed as systems with potential applications in organic photovoltaics [44,54,55] or non-volatile memories [56–58]. In the same way that the formation of conventional LIPSS may be induced both in thin polymer films supported on different substrates [59] and in free-standing polymer films [26], the library or ordered structures reported here may be extended to this variety of substrates.

2. Materials and Methods

Poly(3-hexylthiophene) (P3HT) thin films were prepared using the spin processor Laurell WS-650 Series. P3HT (purchased from Ossila, Sheffield, UK, batch M102, Mw = 65,200 g/mol, regioregularity 95.7%) powder was dissolved in chloroform (99.98 purity, QUIMIPUR-Spain) with a concentration of 16 g/L. Conductive silicon wafers n-silicon (100, Arsenic dopant, ACM (Lannion, France)) were cut into pieces of 2 cm × 2 cm, cleaned by acetone and 2-propanol, and then dried under nitrogen flow. A fixed volume of 0.2 mL solution was spun coated on silicon wafer at 2400 rpm. Thickness of the polymer films, determined by AFM, was ca. 130 nm.

Additionally, bilayers of P3HT and poly(vinylidene fluoride-trifluoroethylene) P(VDF-TrFE) were prepared. P(VDF-TrFE), purchased from Piezotech S.A.S. (Pierre-Benite, France), with a molar concentration of the random copolymer 76:24 (VDF:TrFE content), Mw = 367,000 g/mol, Mw/Mn = 1.72, was dissolved in methylethylketone (Merck, Darmstadt, Germany) with a concentration of 5 g/L, stirring for 3 h at 70 °C. A fixed volume of 0.2 mL of the P(VDF-TrFE) solution was spun coated on top of the P3HT films. It was previously reported that P3HT is not soluble in methylethylketone [45]. Thickness of the P(VDF-TrFE) film, as determined by AFM, was around 20 nm.

Laser irradiation was carried out in ambient air under normal incidence, with the linearly polarized laser beam of a Q-switched Nd:YAG laser (Lotis TII LS-2131M (Minsk, Belarus), pulse duration 8 ns full width half-maximum) at a repetition rate of 10 Hz. Both the second (532 nm) and the fourth (266 nm) harmonics were used for irradiation. Laser fluence was determined by measuring the laser energy in front of the sample and considering an irradiated area of 5 mm. The total number of pulses and the laser fluence values were chosen on the basis of previous experiments in order to obtain optimal LIPSS [44,60]. Sequential irradiation by changing the laser beam polarization was carried out using a half-wave plate. Large areas were nanostructured by using a sample scanning process. The scanning speed, and consequently, the spatial overlap of successive pulses, was chosen to ensure the delivery of optimal number of pulses for LIPSS formation.

The morphology of the samples was inspected under ambient conditions using a Multimode 8 AFM (Bruker, Karlsruhe, Germany) with a Nanoscope V controller (Bruker, Karlsruhe, Germany). Images were collected in tapping mode using Tap300GHB-G probes (BudgetSensors, Sofia, Bulgaria) and an analysis was carried out using the Nanoscope Analysis software 1.50 (Bruker, Karlsruhe, Germany). Furthermore, in the case of bilayers, piezoresponse force microscopy (PFM) measurements were carried out using the same equipment in the piezoresponse mode. For this, conductive SCM-PIT (Bruker, Karlsruhe, Germany) tips were used. The PFM out-of-plane and in-plane signal was taken applying an AC voltage of 2 V.

Water contact angle (CA) measurements were carried out on the nanostructured films using a pocket goniometer PG2 (FIBRO system, Stockholm, Sweden). The static wetting CA was determined at room temperature and ambient humidity using deionized water.

Grazing incidence small- and wide-angle X-ray scattering (GISAXS and GIWAXS, respectively) were performed by using synchrotron radiation at BL11-NCD-SWEET beamline in ALBA (Cerdanyola del Vallès, Spain). The sample was placed with its surface horizontal and parallel to the X-ray beam and at a height which intercepted half of the beam intensity. Then, the sample was tilted in order to reach an incidence angle of 0.15° between the sample surface and the beam. GISAXS patterns were taken using a PILATUS 1M detector (Dectris, Baden-Daettwil, Switzerland) at 6.612 m from the sample, with exposure times of 5 s. GIWAXS patterns were taken using a LX255-HS detector from Rayonix (Evanston, IL, USA) located at 0.109 m, with exposure times of 10 s.

3. Results and Discussion

Several experimental parameters have been used to obtain different patterns by laser irradiation at the surface of P3HT films. In particular, number of laser irradiation steps, number of pulses, polarization orientation and laser wavelength have been varied to fabricate different kinds of nanostructures. For example, by performing two consecutive laser irradiations using the same wavelength and rotating the polarization vector 90°, a square-like pattern may be obtained (Figure 1a,b), while if the laser wavelength is changed in the second irradiation step, a rectangle-like pattern will be obtained (Figure 1c).

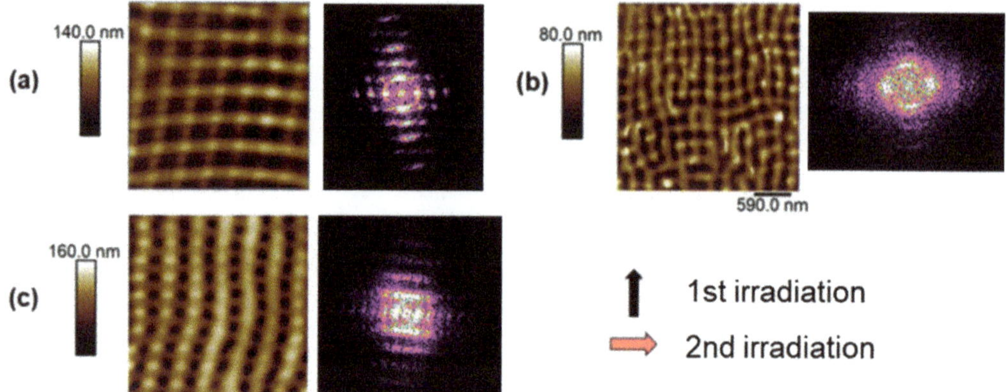

Figure 1. AFM images of combined LIPSS structures formed by sequential irradiation in P3HT. Each AFM image is complemented with its corresponding Fast Fourier Transform (FFT) pattern. (**a**) Sequential irradiation process at λ= 532 nm, 3500 pulses and λ = 532 nm, 100 pulses with the polarization rotated 90°, (**b**) λ = 266 nm, 3300 pulses and λ = 266 nm 300 pulses with the polarization rotated 90° and (**c**) 532 nm, 3600 pulses and λ = 266 nm 300 pulses with the polarization rotated 90°. Arrows indicate the direction of the laser polarization during each irradiation step.

During a typical process for the generation of gratings obtained by repeated nanosecond pulsed laser irradiation, the polarization direction of the laser is kept constant and the fluence and number of pulses are optimized for each material [30,53,61]. However, in our approach, a selected set of pulses is applied with a different polarization state using a half-wave plate, imposing a new geometry to the pre-existing grating-like one. Since our aim is avoiding ablation of the polymer, the fluence and number of pulses must be similar to the one for which optimum LIPSS gratings were obtained. Thus, we use for both sequential irradiations the laser fluence which gives rise to the formation of well-ordered LIPSS (fluence = 26 mJ·cm^{-2} for irradiations at λ = 532 nm and fluence = 13.4 mJ·cm^{-2} for irradiations at λ = 266 nm) and the total number of pulses is the one that induces the formation of LIPSS upon single irradiation [60].

Figure 1 features atomic force microscopy images of surface nanostructures, forming squares of around 500 nm size (Figure 1a) and of around 200 nm size (Figure 1b), and 500 nm × 200 nm rectangles (Figure 1c). In particular, the image shown in Figure 1a corresponds to a P3HT sample irradiated at 532 nm with 3500 pulses in the first irradiation and 100 pulses in the second irradiation with the polarization shifted 90° with respect to that of the first irradiation. Figure 1b displays the AFM pattern of the surface structures obtained by irradiating a P3HT sample at 266 nm with 3300 pulses in the first irradiation and 300 pulses with 90° shift in the laser polarization in the second. Finally, more complex patterns can be obtained by combining different polarization and different wavelengths. In Figure 1c, a rectangular pattern is formed by a first irradiation at a wavelength of 532 nm and 3600 pulses and afterwards at 266 nm with 300 pulses with a shifted 90° polarization. Besides the real space inspection of the order, possible by the AFM images, the degree of order in the nanostructures can be inferred from the Fourier transformed (FFT) images shown in Figure 1. It can be seen that the FFT image from the sample irradiated at 532 nm at both directions (Figure 1a) shows the intensity maxima along the vertical and horizontal directions. P3HT thin film, irradiated at 266 nm in both directions (Figure 1b), shows a lower order of the LIPSS. Its FFT image depicts two intensity maxima in the horizontal direction and one intensity maximum in the vertical direction. On the other hand, the AFM image from P3HT irradiated with both wavelengths (Figure 1c) shows a clear rectangular structure, corroborated by its FFT image, which shows the intensity maxima along the vertical direction more separated from the origin than the intensity maxima along the horizontal direction.

The quality of the ordered structures can be assessed by GISAXS by placing the sample with the X-ray beam parallel to the polarization of the first irradiation or parallel to the second irradiation. The results are shown also in Figure 2. The GISAXS patterns taken in both orientations are the characteristic ones for one-dimensional paracrystalline structures typical of LIPSS [44,62,63], exhibiting vertical diffraction maxima out of the meridian ($\omega \neq 0$). In the GISAXS cuts, it is observed that, in the case of squares of ~500 nm (Figure 2a), the array is more ordered along the direction of first irradiation, as revealed by the larger amount of diffraction maxima. The sample with squares of ~200 nm (Figure 2b) shows only one clear diffraction maximum in both directions. The low order in P3HT LIPSS at λ = 266 nm is clearly seen by AFM (Figure 1b), which has already been reported [60]. The sample with rectangles shows a high level of order in both directions (Figure 2c).

Other structures were obtained by using different irradiation conditions. Figure 3 shows AFM images of patterns produced by fixing the wavelength (λ = 532 nm) and polarization shift (90°) and varying the number of pulses, in a way that the total number of pulses is the optimal number to fabricate a conventional P3HT LIPSS (3600 pulses).

It can be seen in Figure 3 that with an equal number of pulses in both irradiation steps, the second polarization direction determines the main direction of the LIPSS. Even with a shorter number of pulses during the second irradiation step (Figure 3c–d), LIPSS are clearly better formed in the direction of the second polarization vector, although FFT patterns indicate that some ordering due to the first irradiation is still present, buried by more perfect LIPSS that appear parallel to the polarization of the second irradiation. Irradiating with 3300 pulses in the vertical direction followed by 300 pulses in the horizontal direction, LIPSS show good order in both directions (Figure 3d), as can be seen in the FFT image, which shows the intensity maxima along both orthogonal directions.

Moreover, more complex structures are fabricated by varying the polarization shifts, number of pulses and adding more irradiation steps. Figure 4 shows the AFM images of these patterns.

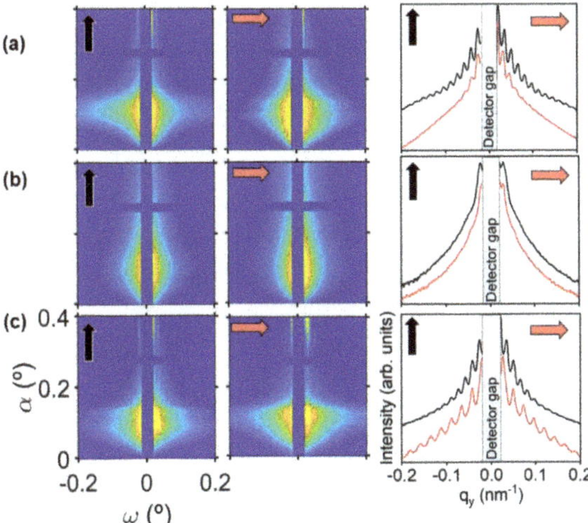

Figure 2. GISAXS patterns from P3HT irradiated samples with (**a**) squares of ~500 nm, (**b**) squares of ~200 nm and (**c**) rectangles of ~500 nm × ~200 nm. Left column: GISAXS patterns taken with the X-ray beam parallel to the first irradiation direction. Center column: GISAXS patterns taken with the X-ray beam parallel to the second irradiation direction. Right column: horizontal cuts from their corresponding GISAXS patterns in the same row, left column (black curves) and center column (red curves). Black arrows indicate the direction of the first irradiation step and red arrows the direction of the second irradiation step. GISAXS patterns taken at incident angle α = 0.4°. Intensity in logarithmic scale.

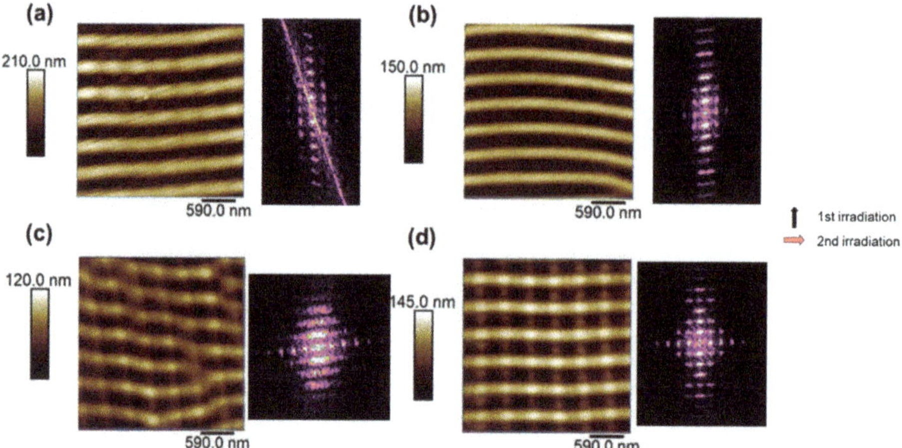

Figure 3. AFM images of combined LIPSS formed by sequential irradiation in P3HT at λ = 532 nm by varying the number of pulses of each irradiation: (**a**) 1800 + 1800 pulses, (**b**) 2400 +1200 pulses, (**c**) 3000 + 600 pulses and (**d**) 3300 + 300 pulses. In the right column, the Fourier Transform of each image is shown. Arrows indicate the direction of the laser polarization during each irradiation step.

Figure 4a shows the AFM image of a P3HT thin film irradiated first with 3600 pulses followed by 100 pulses polarized in the orthogonal direction and 100 pulses polarized 45° with respect to the initial polarization. It can be seen that LIPSS are formed preferentially in the original polarization direction with lower order at 45°, as revealed by several intensity maxima in FFT along the horizontal direction and one maximum along the direction of the polarization used in the third irradiation step. On the other hand, Figure 4b shows an AFM image of the sample irradiated with the same polarization shifts but with 3600 pulses, initially followed by 1800 pulses and then another 1800 pulses. The total number of pulses exceeds the optimum value, so LIPSS hardly preserved a preferential orientation and the corresponding FFT indicates a loss of order compared with Figure 4a. This shows that the polarization used during the last irradiation step will be a determinant to the final order of the sample using a number of pulses about an order of magnitude lower than the number of pulses used during the first irradiation step.

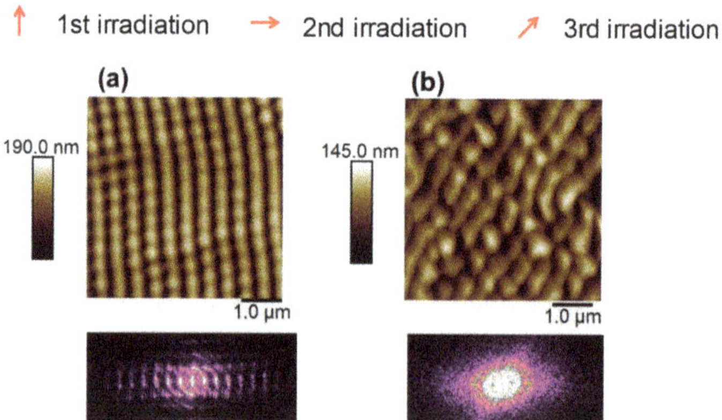

Figure 4. AFM images of combined LIPSS formed by sequential irradiation in P3HT at λ = 532 nm by varying the polarization vector of the laser: (**a**) 3600 + 100 + 100 pulses; (**b**) 3600 + 1800 + 1800 pulses. Next to each AFM figure, its corresponding FFT is shown.

Additionally, GIWAXS measurements were carried out to see changes in the inner crystalline structure of LIPSS, in particular whether the crystallinity degree and orientation is affected. Samples with squares of ~500 nm and ~200 nm and rectangles of ~500 nm × ~200 nm were selected and the corresponding GIWAXS patterns are presented in Figure 5.

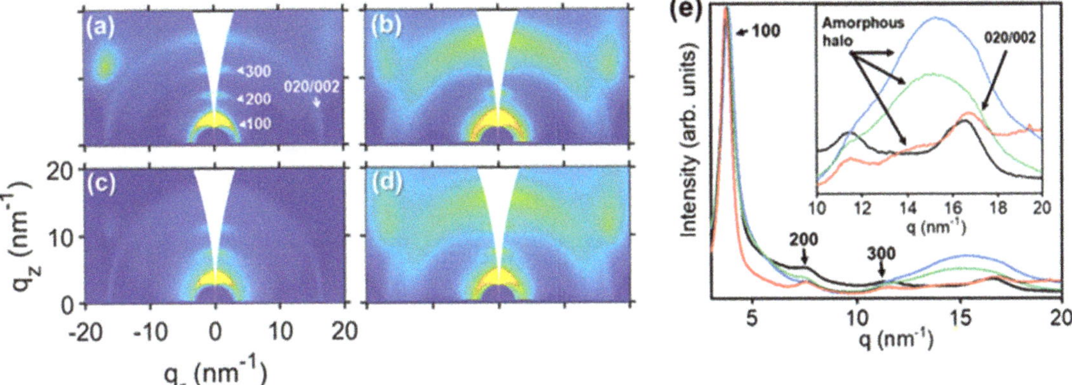

Figure 5. GIWAXS patterns of (**a**) P3HT thin film non-irradiated, and irradiated: (**b**) squares ~500 nm, (**c**) squares ~200 nm, (**d**) rectangles. (**e**) Azimuthal integrated intensity profiles of the GIWAXS pattern: non-irradiated thin film of P3HT (black curve), squares ~500 nm (red curve), squares ~200 nm (blue curve) and rectangles (green curve). GIWAXS of irradiated samples taken with X-ray beam parallel to first irradiation. Inset shows a magnification of the q-range region where the 020/002 reflection and amorphous halo of P3HT appear. Main reflections of crystalline P3HT labeled. Intensity of GIWAXS patterns in logarithmic scale.

GIWAXS patterns from Figure 5 show typical P3HT diffraction pattern with preferential "edge-on" orientation, with {h00} reflections oriented along the q_z axis. The 020/002 reflection and the halo from the amorphous domains are located around $q = 17 \ \text{nm}^{-1}$. It can be seen that GIWAXS patterns from irradiated samples (Figure 5b–d) present a broad amorphous halo, compared with the pattern from non-irradiated P3HT (Figure 5a).

Azimuthal integrations from GIWAXS patterns were done and are represented in Figure 6. It can be observed from the azimuthal integrations (Figure 5f) that the amorphous halo of P3HT increases in the irradiated samples in comparison to the non-irradiated one. Therefore, laser irradiation induces a reduction of the crystallinity degree, as previously reported, as a consequence of rapid heating and cooling processes involved during LIPSS formation [60]. Interestingly enough, the amorphous halo is more prominent in the profiles from samples which have been irradiated at 266 nm in at least one of the irradiation steps: the sample with squares ~200 nm (Figure 5e, blue curve) and the sample with rectangles (Figure 5e, green curve).

Wettability of the nanostructures surfaces was inspected by measuring the water contact angle and the obtained values are cited in Table 1. Initial surface of P3HT is hydrophobic, in agreement with previous results reported in literature [64]. It has been previously reported that laser micro/nanostructuring and in particular LIPSS formation may provoke changes in the wettability of materials due to the morphological changes, to the chemical changes induced upon irradiation or to a combination of both factors. In the present case, after irradiation, differences are not significant, although it seems that CA slightly increases after irradiation at 532 nm and slightly decreases after irradiation at 266 nm. According to the homogeneous wetting model for water, Wenzel's model [65], the contact between water and the surface of the sample is not altered by the presence of air, and it may explain the variation of the contact angle as a function of 'r,' which is given by the relation between the total surface of solid in the solid-liquid interface, and the projection of the total surface of solid in the interface: r = (total surface)/(projected surface), in such a way that the CA of the nanostructured sample is related to the original CA by: cos(CA*) = r·cos(CA). Since LIPSS formation increases 'r,' if the original sample is hydrophilic, it will become more hydrophilic, and if hydrophobic, as is the present case, it will become more hydrophobic. Changes in the observed CA may then be explained considering only the morphological changes induced by laser irradiation in the case of 532 nm. As previously reported by some of us [60], an analysis of irradiated P3HT by Near Edge X-ray Absorption Fine Structure (NEXAFS) and Raman spectroscopies revealed good chemical stability after LIPSS formation under laser irradiation conditions similar to the ones used here (same fluence and higher number of pulses). However, in the case of irradiation at 266 nm, we have previously reported that additional modifications are induced upon irradiation, in particular, some photooxidation [66], which could explain the slight decrease of the CA.

Table 1. Water contact angle of nanostructured P3HT.

1st Irradiation	2nd Irradiation	3rd Irradiation	Water Contact Angle (°)
Non-irradiated	-	-	96 ± 2
532 nm, 3600 p	-	-	100 ± 3
532 nm, 3500 p	532 nm, 100 p	-	96 ± 1
532 nm, 3300 p	532 nm, 300 p	-	96 ± 3
532 nm, 3000 p	532 nm, 600 p	-	101 ± 4
532 nm, 2400 p	532 nm, 1200 p	-	98 ± 3
532 nm, 1800 p	532 nm, 1800 p	-	98 ± 8
532 nm, 3600 p	532 nm, 100 p	532 nm, 100 p	101 ± 2
266 nm, 3600 p	-	-	86 ± 4
266 nm, 3300 p	266 nm, 300 p	-	72 ± 5
532 nm, 3600 p	266 nm, 300 p	-	89 ± 3

The process of obtaining large patterned areas consists of mounting the sample in a translation stage. In this way, areas of several centimeters can be patterned within a few minutes by scanning the sample with the laser beam. For this purpose, the scanning speed, and consequently, the spatial overlap of successive pulses, was chosen to ensure the delivery of optimal number of pulses for LIPSS formation, previously determined by normal single spot irradiation. Nanostructuring in large areas conferred iridescence to the polymer surface. Figure 6 shows a 1 cm × 1 cm piece of P3HT structured with 532 nm squares on a silicon wafer. Pictures are taken by illuminating it with white light at different incident angles and changes in the structural color can be observed. Supplementary Movie 1, recorded with a cellphone, shows the different colors observed as a function of the illumination angle.

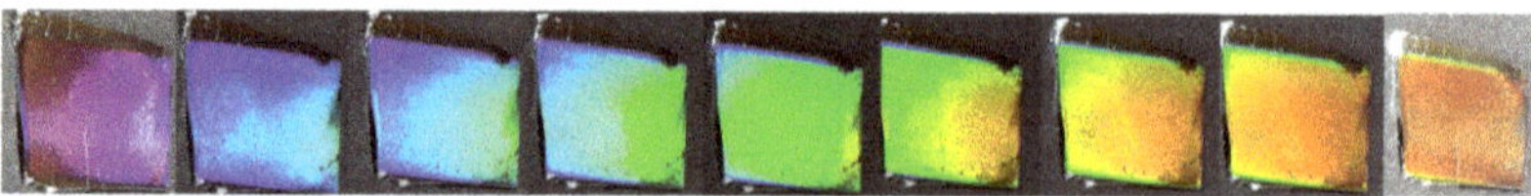

Figure 6. Optical images, obtained under different angles of white light illumination, of 1×1 cm^2 nanostructured samples obtained by using the sequential nanosecond pulsed laser irradiation.

This kind of gratings fabricated on functional polymer materials allows nanoscale-controlled functionality. A square pattern was obtained by sequential irradiation at 532 nm (3500 pulses plus 100 pulses after a 90° polarization shift) in a functional polymer bilayer P3HT (bottom)/P(VDF-TrFE) (up). It has been proven in the past that this configuration is suitable in order to obtain nanostructures by laser irradiation on ferroelectric polymers, besides the fact that they are non-absorbers at the used laser wavelength [45]. Figure 7 shows the morphology of the structured bilayer and its piezoelectric nature as revealed by PFM. These results demonstrate that it is possible to obtain an ordered square array of ferroelectric dots. The preserved ferroelectricity in the nanostructured polymer surface is demonstrated by the existence of hysteresis in the out of plane phase signal obtained by piezoresponse force microscopy, although GIWAXS results show an almost completely disappearance of the P(VDF-TrFE) Bragg peak when irradiated (Figure 8).

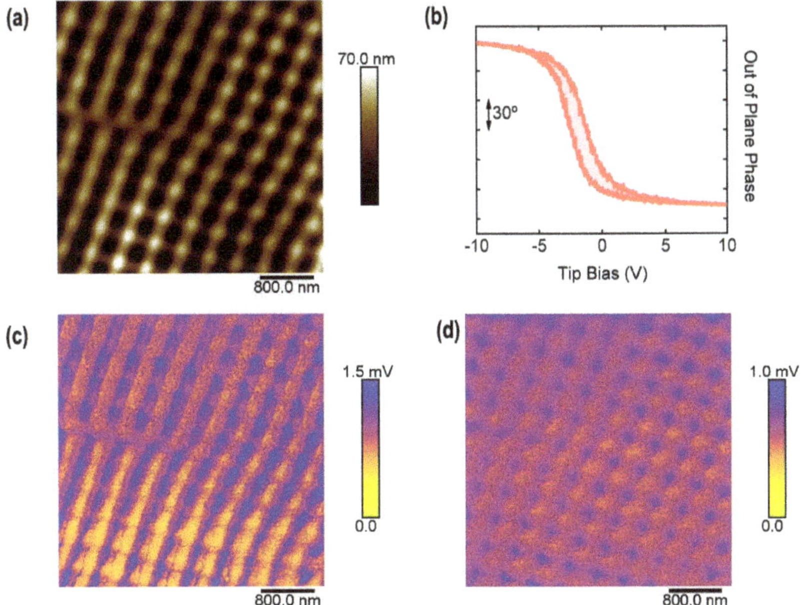

Figure 7. AFM height (**a**) and PFM results (**b–d**) of a patterned functional polymer surface exhibiting ferroelectricity. (**b**) Hysteresis of the out-of-plane PFM phase as a function of a DC bias applied to the conducting AFM tip. (**c**) Out-of-plane amplitude mapping of the patterned surface. (**d**) In-plane amplitude mapping of the patterned surface.

Figure 8a shows the reflection associated with the (110/200) planes of the ferroelectric phase of P(VDF-TrFE) highly oriented in the q_z axis, whereas the GIWAXS pattern from the irradiated bilayer does not show this orientation (Figure 8b). In Figure 8c, it can be seen that the (110/200) reflections from P(VDF-TrFE) along the q_z axis is clearly detected in the profile from non-irradiated bilayer (black curve) and it is slightly detected in the profile from the irradiated bilayer (red curve). Looking at Figure 8d, the peak from P(VDF-TrFE) is highly reduced between the profile from the non-irradiated (black curve) and the irradiated bilayer (red curve). Therefore, there is a clear loss of crystallinity in P(VDF-TrFE) when it is irradiated, as well as in irradiated P3HT (Figure 5). However, since PFM results (Figure 7) show piezoelectric behavior, it is supposed that the irradiated P(VDF-TrFE) in the bilayer keeps some degree of crystallinity. Radial integration in the q-range of the (110/002) reflection of P(VDF-TrFE) (Figure 8e) reveals indeed a complete disorientation of a crystalline order that was originally highly oriented in the out-of-plane axis, corresponding with the azimuthal angle $\chi = 90°$, as can be seen in Figure 8a. Therefore, the crystallinity of irradiated P(VDF-TrFE) is not completely removed, but it loses the original preferential orientation that it presents in the non-irradiated bilayer.

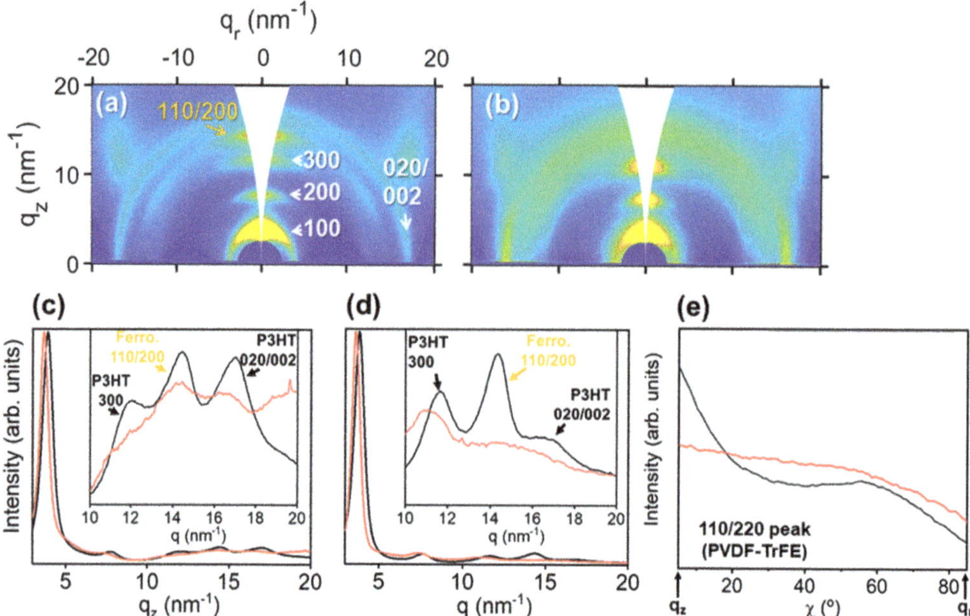

Figure 8. GIWAXS patterns of (**a**) a bilayer of P3HT and P(VDF-TrFE) non-irradiated, and (**b**) bilayer irradiated with two orthogonal polarizations at 532 nm (3500 + 100 pulses). In the right half of (**a**), reflections from P3HT are labeled, and in the left half, the P(VDF-TrFE) reflection is labeled. (**c**) Integration along the q_z axis of the non-irradiated bilayer (black curve) and irradiated bilayer (red curve). (**d**) Integration along the whole azimuthal range of the non-irradiated bilayer (black curve) and irradiated bilayer (red curve). Insets in (**c,d**) are magnifications of the q-range, where the reflection from P(VDF-TrFE) appears. Reflections from P3HT and P(VDF-TrFE) (ferro.) are labeled in black and orange, respectively. Intensity of GIWAXS patterns in logarithmic scale. (**e**) Radial integration of GIWAXS patterns in the q-range to delimit the 110/200 reflection of P(VDF-TrFE).

4. Conclusions

In summary, we reported on the simple fabrication of a library of ordered nanostructures by repeated irradiation using a nanosecond pulsed laser operating in the UV and visible region in order to obtain nanoscale-controlled functionality. As an example, we use this experimental approach to nanostructure a ferroelectric polymer so that an ordered square array of ferroelectric dots is obtained.

Supplementary Materials: The following are available online at https://www.mdpi.com/article/10.3390/nano11051123/s1, Video S1: color changes at different angles of white light illumination of nanostructured samples.

Author Contributions: Conceptualization, A.N. and E.R.; methodology, A.N. and E.R.; investigation, E.G.-F., T.A.E., A.N. and E.R.; writing—original draft preparation, E.G.-F., A.N. and E.R.; writing—review and editing, E.G.-F., T.A.E., A.N. and E.R.; visualization, E.G.-F., A.N. and E.R.; supervision, A.N. and E.R.; project administration, A.N. and E.R.; funding acquisition, T.A.E., A.N. and E.R. All authors have read and agreed to the published version of the manuscript.

Funding: This research was funded by the Spanish State Research Agency (AEI) through projects PID2019-107514GB-I00/AEI/10.13039/501100011033 and PID2019-106125GB-I00/AEI/10.13039/501100011033.

Acknowledgments: The GISAXS and GIWAXS experiments were performed at NCD-SWEET beamline at ALBA synchrotron with the collaboration of ALBA staff.

Conflicts of Interest: The authors declare no conflict of interest.

References

1. Bar-Cohen, Y. *Biomimetics: Biologically Inspired Technologies*; CRC Press: Boca Ratón, FL, USA, 2005.
2. Barthlott, W.; Mail, M.; Bhushan, B.; Koch, K. Plant Surfaces: Structures and Functions for Biomimetic Innovations. *Nano-Micro Lett.* **2017**, *9*, 23. [CrossRef]
3. Chu, Z.; Seeger, S. Superamphiphobic surfaces. *Chem. Soc. Rev.* **2014**, *43*, 2784–2798. [CrossRef] [PubMed]
4. Xia, D.; Johnson, L.M.; López, G.P. Anisotropic Wetting Surfaces with One-Dimesional and Directional Structures: Fabrication Approaches, Wetting Properties and Potential Applications. *Adv. Mater.* **2012**, *24*, 1287–1302. [CrossRef]
5. Wang, Z.; Elimelech, M.; Lin, S. Environmental Applications of Interfacial Materials with Special Wettability. *Environ. Sci. Technol.* **2016**, *50*, 2132–2150. [CrossRef]
6. Žemaitis, A.; Mimidis, A.; Papadopoulos, A.; Gečys, P.; Račiukaitis, G.; Stratakis, E.; Gedvilas, M. Controlling the wettability of stainless steel from highly-hydrophilic to super-hydrophobic by femtosecond laser-induced ripples and nanospikes. *RSC Adv.* **2020**, *10*, 37956–37961. [CrossRef]
7. Yong, J.; Yang, Q.; Guo, C.; Chen, F.; Hou, X. A review of femtosecond laser-structured superhydrophobic or underwater superoleophobic porous surfaces/materials for efficient oil/water separation. *RSC Adv.* **2019**, *9*, 12470–12495. [CrossRef]
8. Autumn, K.; Liang, Y.A.; Hsieh, S.T.; Zesch, W.; Chan, W.P.; Kenny, T.W.; Fearing, R.; Full, R.J. Adhesive force of a single gecko foot-hair. *Nature* **2000**, *405*, 681–685. [CrossRef] [PubMed]
9. Filippov, A.E.; Gorb, S.N. Modelling of the frictional behaviour of the snake skin covered by anisotropic surface nanostructures. *Sci. Rep.* **2016**, *6*, 23539. [CrossRef]
10. Wang, S.; Feng, L.; Jiang, L. One-Step Solution-Immersion Process for the Fabrication of Stable Bionic Superhydrophobic Surfaces. *Adv. Mater.* **2006**, *18*, 767–770. [CrossRef]
11. Pérez-Díaz, O.; Quiroga-González, E. Silicon conical structures by metal assisted chemical etching. *Micromachines* **2020**, *11*, 402. [CrossRef]
12. Kim, T.; Baek, C.H.; Suh, K.Y.; Seo, S.; Lee, H.H. Optical Lithography with Printed Metal Mask and a Simple Superhydrophobic Surface. *Small* **2008**, *4*, 182–185. [CrossRef]
13. Menumerov, E.; Golze, S.D.; Hughes, R.A.; Neretina, S. Arrays of highly complex noble metal nanostructures using nanoimprint lithography in combination with liquid-phase epitaxy. *Nanoscale* **2018**, *10*, 18186–18194. [CrossRef]
14. Lai, Y.; Lin, Z.; Huang, J.; Sun, L.; Chen, Z.; Lin, C. Controllable construction of ZnO/TiO$_2$ patterning nanostructures by superhydrophilic/superhydrophobic templates. *New J. Chem.* **2010**, *34*, 44–51. [CrossRef]
15. Chu, C.; Liu, Z.; Sun, Q.; Lin, P.; Guo, C.; Sheng, X.; Dong, Y.; Chen, F. Electrodeposition growth of ZnO nanorods on a TiO$_2$ nanotube template prepared by two-step anodization. *Xiyou Jinshu Cailiao Yu Gongcheng/Rare Met. Mater. Eng.* **2014**, *43*, 1246–1249.
16. Vorobyev, A.Y.; Guo, C. Direct femtosecond laser surface nano/microstructuring and its applications. *Laser Photon. Rev.* **2013**, *7*, 385–407. [CrossRef]
17. Zwahr, C.; Günther, D.; Brinkmann, T.; Gulow, N.; Oswald, S.; Grosse Holthaus, M.; Lasagni, A.F. Laser Surface Pattering of Titanium for Improving the Biological Performance of Dental Implants. *Adv. Healthc. Mater.* **2017**, *6*, 1600858. [CrossRef] [PubMed]
18. Rebollar, E.; Frischauf, I.; Olbrich, M.; Peterbauer, T.; Hering, S.; Preiner, J.; Hinterdorfer, P.; Romanin, C.; Heitz, J. Proliferation of aligned mammalian cells on laser-nanostructured polystyrene. *Biomaterials* **2008**, *29*. [CrossRef] [PubMed]
19. Rebollar, E.; Pérez, S.; Hernández, M.; Domingo, C.; Martín, M.; Ezquerra, T.A.; García-Ruiz, J.P.; Castillejo, M. Physicochemical modifications accompanying UV laser induced surface structures on poly(ethylene terephthalate) and their effect on adhesion of mesenchymal cells. *Phys. Chem. Chem. Phys.* **2014**, *16*, 17551–17559. [CrossRef]
20. Schwibbert, K.; Menzel, F.; Epperlein, N.; Bonse, J.; Krüger, J. Bacterial adhesion on femtosecond laser-modified polyethylene. *Materials* **2019**, *12*, 3107. [CrossRef]
21. Fajstavr, D.; Neznalová, K.; Kasálková, N.S.; Rimpelová, S.; Kubičíková, K.; Švorčík, V.; Slepička, P. Nanostructured polystyrene doped with acetylsalicylic acid and its antibacterial properties. *Materials* **2020**, *13*, 3609. [CrossRef]
22. Alamri, S.; Aguilar-Morales, A.I.; Lasagni, A.F. Controlling the wettability of polycarbonate substrates by producing hierarchical structures using Direct Laser Interference Patterning. *Eur. Polym. J.* **2018**, *99*, 27–37. [CrossRef]
23. Liu, M.; Li, M.-T.; Xu, S.; Yang, H.; Sun, H.-B. Bioinspired Superhydrophobic Surfaces via Laser-Structuring. *Front. Chem.* **2020**, *8*. [CrossRef]
24. Allahyari, E.; Nivas, J.J.; Oscurato, S.L.; Salvatore, M.; Ausanio, G.; Vecchione, A.; Fittipaldi, R.; Maddalena, P.; Bruzzese, R.; Amoruso, S. Laser surface texturing of copper and variation of the wetting response with the laser pulse fluence. *Appl. Surf. Sci.* **2019**, *470*, 817–824. [CrossRef]
25. Kirner, S.V.; Slachciak, N.; Elert, A.M.; Griepentrog, M.; Fischer, D.; Hertwig, A.; Sahre, M.; Dörfel, I.; Sturm, H.; Pentzien, S.; et al. Tribological performance of titanium samples oxidized by fs-laser radiation, thermal heating, or electrochemical anodization. *Appl. Phys. A Mater. Sci. Process.* **2018**, *124*. [CrossRef]

26. Rodríguez-Beltrán, R.I.; Martínez-Tong, D.E.; Reyes-Contreras, A.; Paszkiewicz, S.; Szymczyk, A.; Ezquerra, T.A.; Moreno, P.; Rebollar, E. Laterally-resolved mechanical and tribological properties of laser-structured polymer nanocomposites. *Polymer* **2019**, *168*. [CrossRef]

27. Cubero, A.; Martínez, E.; Angurel, L.A.; de la Fuente, G.F.; Navarro, R.; Legall, H.; Krüger, J.; Bonse, J. Effects of laser-induced periodic surface structures on the superconducting properties of Niobium. *Appl. Surf. Sci.* **2020**, *508*. [CrossRef]

28. Roessler, F.; Lasagni, A.F. Protecting Sub-Micrometer Surface Features in Polymers from Mechanical Damage Using Hierarchical Patterns. *J. Laser Micro/Nanoeng.* **2018**, *13*. [CrossRef]

29. Fraggelakis, F.; Tsibidis, G.D.; Stratakis, E. Tailoring submicrometer periodic surface structures via ultrashort pulsed direct laser interference patterning. *Phys. Rev. B* **2021**, *103*. [CrossRef]

30. Rebollar, E.; Castillejo, M.; Ezquerra, T.A. Laser induced periodic surface structures on polymer films: From fundamentals to applications. *Eur. Polym. J.* **2015**, *73*, 162–174. [CrossRef]

31. Bonse, J. Quo vadis LIPSS?—Recent and future trends on laser-induced periodic surface structures. *Nanomaterials* **2020**, *10*, 1950. [CrossRef] [PubMed]

32. Bonse, J.; Gräf, S. Maxwell Meets Marangoni—A Review of Theories on Laser-Induced Periodic Surface Structures. *Laser Photon. Rev.* **2020**, *14*, 2000215. [CrossRef]

33. Soldera, M.; Fortuna, F.; Teutoburg-Weiss, S.; Milles, S.; Taretto, K.; Fabián Lasagni, A. Comparison of Structural Colors Achieved by Laser-Induced Periodic Surface Structures and Direct Laser Interference Patterning. *J. Laser Micro/Nanoeng.* **2020**, *15*, 97–103.

34. Bonse, J.; Kirner, S.V.; Krüger, J. Laser-Induced Periodic Surface Structures (LIPSS). In *Handbook of Laser Micro- and Nano-Engineering*; Sugioka, K., Ed.; Springer International Publishing: Cham, Switzerland, 2020; pp. 1–59. ISBN 978-3-319-69537-2.

35. Gräf, S. Formation of laser-induced periodic surface structures on different materials: Fundamentals, properties and applications. *Adv. Opt. Technol.* **2020**, *9*, 11–39. [CrossRef]

36. Florian, C.; Kirner, S.V.; Krüger, J.; Bonse, J. Surface functionalization by laser-induced periodic surface structures. *J. Laser Appl.* **2020**, *32*, 22063. [CrossRef]

37. Stratakis, E.; Bonse, J.; Heitz, J.; Siegel, J.; Tsibidis, G.D.; Skoulas, E.; Papadopoulos, A.; Mimidis, A.; Joel, A.-C.; Comanns, P.; et al. Laser engineering of biomimetic surfaces. *Mater. Sci. Eng. R Rep.* **2020**, *141*, 100562. [CrossRef]

38. Vercillo, V.; Tonnicchia, S.; Romano, J.-M.; García-Girón, A.; Aguilar-Morales, A.I.; Alamri, S.; Dimov, S.S.; Kunze, T.; Lasagni, A.F.; Bonaccurso, E. Design Rules for Laser-Treated Icephobic Metallic Surfaces for Aeronautic Applications. *Adv. Funct. Mater.* **2020**, *30*, 1910268. [CrossRef]

39. Cunha, A.; Elie, A.-M.; Plawinski, L.; Serro, A.P.; do Rego, A.M.B.; Almeida, A.; Urdaci, M.C.; Durrieu, M.-C.; Vilar, R. Femtosecond laser surface texturing of titanium as a method to reduce the adhesion of Staphylococcus aureus and biofilm formation. *Appl. Surf. Sci.* **2016**, *360*, 485–493. [CrossRef]

40. Mezera, M.; Bonse, J.; Römer, G.R.B.E. Influence of Bulk Temperature on Laser-Induced Periodic Surface Structures on Polycarbonate. *Polymers* **2019**, *11*, 1947. [CrossRef] [PubMed]

41. Mezera, M.; van Drongelen, M.; Römer, G.R.B.E. Laser-Induced Periodic Surface Structures (LIPSS) on Polymers Processed with Picosecond Laser Pulses. *J. Laser Micro/Nanoeng.* **2018**, *13*, 105–116. [CrossRef]

42. Prada-Rodrigo, J.; Rodríguez-Beltrán, R.I.; Paszkiewicz, S.; Szymczyk, A.; Ezquerra, T.A.; Moreno, P.; Rebollar, E. Laser-Induced Periodic Surface Structuring of Poly(trimethylene terephthalate) Films Containing Tungsten Disulfide Nanotubes. *Polymers* **2020**, *12*, 1090. [CrossRef]

43. Sánchez, E.H.; Normile, P.S.; De Toro, J.A.; Caballero, R.; Canales-Vázquez, J.; Rebollar, E.; Castillejo, M.; Colino, J.M. Flexible, multifunctional nanoribbon arrays of palladium nanoparticles for transparent conduction and hydrogen detection. *Appl. Surf. Sci.* **2019**, *470*. [CrossRef]

44. Cui, J.; Rodríguez-Rodríguez, A.; Hernández, M.; García-Gutiérrez, M.-C.; Nogales, A.; Castillejo, M.; Moseguí González, D.; Müller-Buschbaum, P.; Ezquerra, T.A.; Rebollar, E. Laser-induced periodic surface structures on P3HT and on its photovoltaic blend with PC$_{71}$BM. *ACS Appl. Mater. Interfaces* **2016**, *8*. [CrossRef]

45. Martínez-Tong, D.E.; Rodríguez-Rodríguez, Á.; Nogales, A.; García-Gutiérrez, M.-C.; Pérez-Murano, F.; Llobet, J.; Ezquerra, T.A.; Rebollar, E. Laser Fabrication of Polymer Ferroelectric Nanostructures for Nonvolatile Organic Memory Devices. *ACS Appl. Mater. Interfaces* **2015**, *7*. [CrossRef]

46. Rebollar, E.; Sanz, M.; Pérez, S.; Hernández, M.; Martín-Fabiani, I.; Rueda, D.R.; Ezquerra, T.A.; Domingo, C.; Castillejo, M. Gold coatings on polymer laser induced periodic surface structures: Assessment as substrates for surface-enhanced Raman scattering. *Phys. Chem. Chem. Phys.* **2012**, *14*. [CrossRef] [PubMed]

47. Svorcik, V.; Nedela, O.; Slepicka, P.; Lyutakov, O.; Slepickova Kasalkova, N.; Kolska, Z. Construction and Properties of Ripples on Polymers for Sensor Applications. *Manuf. Technol. J.* **2018**, *18*, 851–855. [CrossRef]

48. Slepička, P.; Siegel, J.; Lyutakov, O.; Slepičková Kasálková, N.; Kolská, Z.; Bačáková, L.; Švorčík, V. Polymer nanostructures for bioapplications induced by laser treatment. *Biotechnol. Adv.* **2018**, *36*, 839–855. [CrossRef] [PubMed]

49. Cardoso, J.T.; Aguilar-Morales, A.I.; Alamri, S.; Huerta-Murillo, D.; Cordovilla, F.; Lasagni, A.F.; Ocaña, J.L. Superhydrophobicity on hierarchical periodic surface structures fabricated via direct laser writing and direct laser interference patterning on an aluminium alloy. *Opt. Lasers Eng.* **2018**, *111*, 193–200. [CrossRef]

50. Lechthaler, B.; Fox, T.; Slawik, S.; Mücklich, F. Direct laser interference patterning combined with mask imaging. *Opt. Laser Technol.* **2020**, *123*. [CrossRef]

51. El-Khoury, M.; Alamri, S.; Voisiat, B.; Kunze, T.; Lasagni, A.F. Fabrication of hierarchical surface textures using multi-pulse direct laser interference patterning with nanosecond pulses. *Mater. Lett.* **2020**, *258*. [CrossRef]
52. Martínez-Calderon, M.; Rodríguez, A.; Dias-Ponte, A.; Morant-Miñana, M.C.; Gómez-Aranzadi, M.; Olaizola, S.M. Femtosecond laser fabrication of highly hydrophobic stainless steel surface with hierarchical structures fabricated by combining ordered microstructures and LIPSS. *Appl. Surf. Sci.* **2016**, *374*, 81–89. [CrossRef]
53. Nedĕla, O.; Slepička, P.; Sajdl, P.; Veselý, M.; Švorčík, V. Surface analysis of ripple pattern on PS and PEN induced with ring-shaped mask due to KrF laser treatment. *Surf. Interface Anal.* **2017**, *49*, 25–33. [CrossRef]
54. Yang, Y.; Mielczarek, K.; Aryal, M.; Zakhidov, A.; Hu, W. Effects of nanostructure geometry on nanoimprinted polymer photovoltaics. *Nanoscale* **2014**, *6*, 7576–7584. [CrossRef]
55. Yang, Y.; Mielczarek, K.; Zakhidov, A.; Hu, W. Large Molecular Weight Polymer Solar Cells with Strong Chain Alignment Created by Nanoimprint Lithography. *ACS Appl. Mater. Interfaces* **2016**, *8*, 7300–7307. [CrossRef] [PubMed]
56. Fang, H.; Yan, Q.; Geng, C.; Chan, N.Y.; Au, K.; Yao, J.; Ng, S.M.; Leung, C.W.; Li, Q.; Guo, D.; et al. Facile fabrication of highly ordered poly(vinylidene fluoride-trifluoroethylene) nanodot arrays for organic ferroelectric memory. *J. Appl. Phys.* **2016**, *119*, 14104. [CrossRef]
57. Chen, X.-Z.; Li, Q.; Chen, X.; Guo, X.; Ge, H.-X.; Liu, Y.; Shen, Q.-D. Nano-Imprinted Ferroelectric Polymer Nanodot Arrays for High Density Data Storage. *Adv. Funct. Mater.* **2013**, *23*, 3124–3129. [CrossRef]
58. Son, J.Y.; Ryu, S.; Park, Y.-C.; Lim, Y.-T.; Shin, Y.-S.; Shin, Y.-H.; Jang, H.M. A Nonvolatile Memory Device Made of a Ferroelectric Polymer Gate Nanodot and a Single-Walled Carbon Nanotube. *ACS Nano* **2010**, *4*, 7315–7320. [CrossRef]
59. Cui, J.; Nogales, A.; Ezquerra, T.A.; Rebollar, E. Influence of substrate and film thickness on polymer LIPSS formation. *Appl. Surf. Sci.* **2017**, *394*. [CrossRef]
60. Rodríguez-Rodríguez, A.; Rebollar, E.; Soccio, M.; Ezquerra, T.A.; Rueda, D.R.; Garcia-Ramos, J.V.; Castillejo, M.; Garcia-Gutierrez, M.-C. Laser-Induced Periodic Surface Structures on Conjugated Polymers: Poly(3-hexylthiophene). *Macromolecules* **2015**, *48*. [CrossRef]
61. Csete, M.; Bor, Z. Laser-induced periodic surface structure formation on polyethylene-terephthalate. *Appl. Surf. Sci.* **1998**, *133*, 5–16. [CrossRef]
62. Rebollar, E.; Pérez, S.; Hernández, J.J.; Martín-Fabiani, I.; Rueda, D.R.; Ezquerra, T.A.; Castillejo, M. Assessment and formation mechanism of laser-induced periodic surface structures on polymer spin-coated films in real and reciprocal space. *Langmuir* **2011**, *27*. [CrossRef] [PubMed]
63. Rueda, D.R.; Martín-Fabiani, I.; Soccio, M.; Alayo, N.; Pérez-Murano, F.; Rebollar, E.; García-Gutiérrez, M.C.; Castillejo, M.; Ezquerra, T.A. Grazing-incidence small-angle X-ray scattering of soft and hard nanofabricated gratings. *J. Appl. Crystallogr.* **2012**, *45*. [CrossRef]
64. Gao, L.; Hou, S.; Wang, Z.; Gao, Z.; Yu, X.; Yu, J. One-Step Coating Processed Phototransistors Enabled by Phase Separation of Semiconductor and Dielectric Blend Film. *Micromachines* **2019**, *10*, 716. [CrossRef] [PubMed]
65. Wenzel, R.N. Resistance of solid surfaces to wetting by water. *Ind. Eng. Chem.* **1936**, *28*, 988–994. [CrossRef]
66. Rodríguez-Rodríguez, Á.; Gutiérrez-Fernández, E.; García-Gutiérrez, M.-C.; Nogales, A.; Ezquerra, T.A.; Rebollar, E. Synergistic effect of fullerenes on the laser-induced periodic surface structuring of poly(3-hexyl thiophene). *Polymers* **2019**, *11*, 190. [CrossRef] [PubMed]

nanomaterials

Article

Extreme Sub-Wavelength Structure Formation from Mid-IR Femtosecond Laser Interaction with Silicon

Kevin Werner [1,†] and Enam Chowdhury [1,2,3,*]

1   Department of Physics, The Ohio State University, Columbus, OH 43210, USA;
     kevin.werner@baesystems.com
2   Department of Material Science and Engineering, The Ohio State University, Columbus, OH 43210, USA
3   Department of Electrical and Computer Engineering, The Ohio State University, Columbus, OH 43210, USA
*   Correspondence: chowdhury.24@osu.edu
†   Current address: BAE Systems, 130 Daniel Webster Hwy, MER15-1813, Merrimack, NH 03054, USA.

Abstract: Mid-infrared (MIR) wavelengths (2–10 μm) open up a new paradigm for femtosecond laser–solid interactions. On a fundamental level, compared to the ubiquitous near-IR (NIR) or visible (VIS) laser interactions, MIR photon energies render semiconductors to behave like high bandgap materials, while driving conduction band electrons harder due to the λ^2 scaling of the ponderomotive energy. From an applications perspective, many VIS/NIR opaque materials are transparent for MIR. This allows sub-surface modifications for waveguide writing while simultaneously extending interactions to higher order processes. Here, we present the formation of an extreme sub-wavelength structure formation ($\sim\lambda/100$) on a single crystal silicon surface by a 3600 nm MIR femtosecond laser with a pulse duration of 200 fs. The 50–100 nm linear structures were aligned parallel to the laser polarization direction with a quasi-periodicity of 700 nm. The dependence of the structure on the native oxide, laser pulse number, and polarization were studied. The properties of the structures were studied using scanning electron microscopy (SEM), atomic force microscopy (AFM), cross-sectional transmission electron-microscopy (CS-TEM), electron diffraction (ED), and energy-dispersive X-ray spectroscopy (EDX). As traditional models for the formation of laser induced periodic surface structure do not explain this structure formation, new theoretical efforts are needed.

Keywords: laser induced periodic surface structure (LIPSS); mid-IR; femtosecond laser; surface structuring; nano-structure; ultrafast melting; laser induced damage; ATG instability; surface engineering

check for
updates

Citation: Werner, K.; Chowdhury, E. Extreme Sub-Wavelength Structure Formation from Mid-IR Femtosecond Laser Interaction with Silicon. *Nanomaterials* **2021**, *11*, 1192. https://doi.org/10.3390/nano11051192

Academic Editors: Peter Simon, Jürgen Ihlemann and Jörn Bonse

Received: 1 April 2021
Accepted: 28 April 2021
Published: 30 April 2021

Publisher's Note: MDPI stays neutral with regard to jurisdictional claims in published maps and institutional affiliations.

1. Introduction

Strong field interaction with solids in the mid-infrared (MIR) regime of light provides an exciting platform in contrast with that of visible (VIS) and near-infrared (NIR) regimes [1,2]: first, the reduction of photon energy allows the interaction to be more field dominated than photon dominated, where the material valence band to conduction band electronic transition leaning away from multiphoton processes towards tunneling process; second, traditional semiconductors and other 'small bandgap' materials, which are highly absorptive in the VIS-NIR regime, become transparent in the MIR, potentially changing ultrafast absorption dynamics; third, cycle averaged energy (a.k.a. ponderomotive energy, $U_p = e^2 E_0^2 \lambda^2 / 4m_{cb}$, where λ is the laser wavelength, E_0 is the peak laser electric field strength, and m_{cb} is the effective mass in the conduction band) of the free carriers increases significantly with longer wavelength, changing free carrier absorption and collisional ionization; fourthly, with decreasing plasma critical density as wavelength increases, semiconductors may become ideal plasmonic platforms, creating exotic metamaterials and surfaces [3]. With the vast array of important applications available with controlled laser induced structure formation demonstrated with shorter wavelengths, e.g., solar cells [4], black silicon [5], waveguide fabrication [6], surface-enhanced Raman scattering [7], colorization [8], fabrication of hydrophobic surfaces [9], and many others, there is a lack of both experimental studies and theoretical considerations on this topic in the MIR regime.

With transparency extending to $\lambda = 5$ µm, silicon is an important MIR material with, by far, the widest industrial platform to date. It has also been used as a fundamental platform for the study of laser induced periodic structure (LIPSS) formation [10] and its many applications [11]. Multipulse LID effects have been studied extensively on silicon with NIR wavelengths [12]. One of the first extensive femtosecond LIPSS studies on Si revealed fine ripples, also known as high-spatial frequency (HSFL) LIPSS (period $\lambda/4$), along with coarse ripples, or low spatial frequency LIPSS (LSFL) (period 0.8λ) in the NIR regime ($\lambda = 800$ nm, pulse duration = 100 fs, and 0° incidence, AOI) on p-doped (100) silicon [13], under vacuum with oxide layer etched. HSFL was observed to be aligned to the electric field polarization, indicating a surface scattered wave (SSW) mechanism [14], while LSFL was perpendicular to the electric field polarization, pointing towards a surface plasmon polariton (SPP) mechanism [15]. Under circular polarization, straight ripple patterns can still form but require a larger number of pulses [16].

LSFL has also been observed in various works on Si in the NIR regime (e.g., [17–19], without any evidence for HSFL formation. It has also been shown that LIPSS formation happens on the same timescale as material removal, suggesting that material is ejected [19]. An earlier study in the short wave IR (SWIR) regime (1300–2100 nm) showed that, beyond NIR, multi-pulse femtosecond interaction with Si produces not only traditional LSFL in the central region of the damage spot, but also a different type of HSFL feature with a period as small as $\lambda/6$ [20]. LIPSS formation on silicon has also been studied using ultraviolet (UV) wavelengths. A recent paper reports nano-island and HSFL formation on lightly doped n-type Si (100) at 0° AOI with 390 nm wavelength, and 150 fs pulse duration under rough vacuum. The authors observe within a single shot: nano-islands, bifurcated islands, HSFL, and LSFL, with increasing local fluence, respectively. This indicates a universal origin for HSFL and LSFL [21].

Determinations of the multi-shot laser-induced damage threshold (LIDT) of silicon with NIR wavelengths have also been well studied. Experiments have considered variations in the pulse duration [22,23], as well as number of pulses [24,25].

While studies of this sort using MIR wavelengths on silicon are generally lacking, LIPSS formation has been studied on other materials with MIR wavelengths. For example, studies in the picosecond regime on GaP and CaF_2 have observed damage initiated from surface defects, including LSFL [26], with $\lambda = 4.7$ µm laser wavelength and an LIPSS period of 2.5 µm. Several studies have been performed on Germanium with MIR wavelengths [27–29]. One such study observed HSFL both within a central damage spot and within a peripheral ring [29], where both types of HSFL were parallel to the laser polarization and the peripheral HSFL had double the spatial frequency as compared to central HSFL. The central frequency HSFL was about 0.27 times the wavelength for $\lambda = 3.6$ µm. Under cross-sectional transmission electron microscope (TEM), a thin 50 nm amorphous layer was found throughout a cross section of the HSFL. A very recent study of LIPSS formation on Si with MIR wavelength [30] (range $\lambda = 2.5$–4.5 µm) reveals HSFL ($\lambda/4$) parallel and near-subwavelength LSFL perpendicular to MIR laser polarization, consistent with observation from past NIR experimental observations. In our present study, we present a systematic study of femtosecond laser surface structuring at a MIR wavelength, and show formation of completely new classes of surface structure formation, one extreme sub-wavelength (using linearly polarized pulses), and another one chiral (using circularly polarized pulses).

2. Materials and Methods

A home-built Ti:Sapphire ultrashort pulse laser [31] is converted to MIR wavelengths via the Extreme Mid-InfraRed (EMIR) optical parametric amplifier (OPA) [3,32] with central wavelength $\lambda = 3.6$ µm, pulsed repetition frequency (PRF) of 500 Hz, and pulse duration of $\tau \approx 200$ fs full-width at half-maximum (FWHM).

Using our previously-described laser-induced damage (LID) setup [1], high purity (>1000 Ω-cm) FZ mono-crystalline silicon (111) and (100) wafers were exposed to a maximum of 10,000 pulses per site with a maximum average peak fluence of 1.6 J/cm^2, a focal diameter of 24 µm FWHM, at an angle of incidence (AOI) of 45 degrees (*p*-polarization). Details of the focal spot characterization method and positioning of the sample surface with respect to the focal plane are described in detail in reference [1]. The (111) wafer is oriented such that the laser pulse electric field polarization was parallel to the (100) direction. In this work, the reflex objective used for sample irradiation with MWIR pulses in reference [1] was replaced with an $f = 200$ mm CaF$_2$ lens. Each site was exposed to a predetermined number of pulses via a combination of the GRAY laser's external Pockels Cell pulse-picker and a synchronized mechanical shutter. The average fluence of these pulses was controlled by a variable attenuator [1]. For some tests, a quarter waveplate (QWP) was inserted to study the effect of circular polarization.

All experiments were performed in air with air-exposed samples; a few nm thick native-oxide layers were expected to be present on the wafer surface. To determine the importance of the native oxide layer, some of the silicon (111) wafers were etched by a 10% hydrofluoric acid (HF) buffer solution. We estimate a maximum oxide-layer growth of <1 atomic layer based on a timed 8 min transfer from etching to irradiation and the average rate of oxide layer growth on silicon at room temperature ($\leq$0.02 Å/min) [33].

Post-exposure examination of the samples was performed ex-situ via scanning electron microscopy (SEM), atomic force microscopy (AFM), cross-sectional transmission electron-microscopy (CS-TEM) [1], electron diffraction (ED), and energy-dispersive X-ray spectroscopy (EDX). CS-TEM, ED, and EDX were all performed on a single representative site in a representative region of the sample (500 pulses, 547 mJ/cm^2 average peak fluence, (111) wafer, etched oxide layer).

3. Results and Discussion

LIDT fluence was determined for varying number of pulses. Here, we define damage as a permanent change induced by the laser as detectable by SEM.

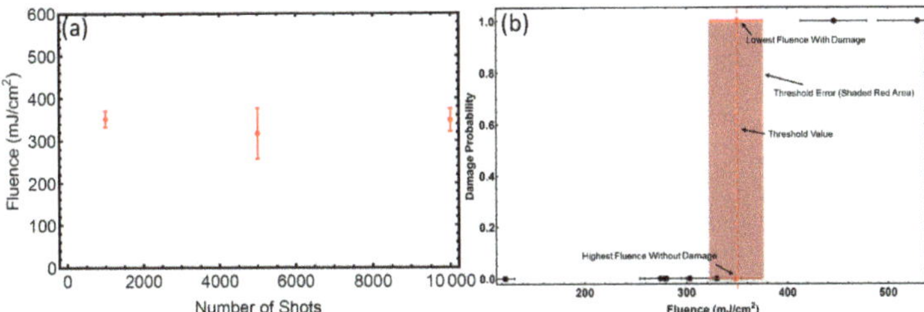

Figure 1. LIDT of silicon (111) vs. number of shots. (**a**) LIDT fluence vs. number of shots for Si; (**b**) example of damage threshold determination method for 10,000 shots. The LIDT here is considered to be the average of the highest fluence with zero damage probability and the lowest fluence with 100% damage probability.

The results of the LIDT study are shown in Figure 1. The S-on-1 multi-shot damage thresholds between 1000–10,000 shots were found to be shot number independent at around 350 mJ/cm^2. This is consistent with a well-known phenomenon, where the material LIDT steadily exhibits reduction with increasing number of pulses so that F(infinity)/F(1) approaches an asymptotic value around F(1000) i.e., the 1000-on-1 damage threshold fluence [34]. The LIDT were determined by damage probability as described in Figure 1b. The LIDT error bar is considered to be within the range of possible fluence values for the highest fluence without damage and lowest fluence with 100% damage.

Figure 2 shows SEM for a fixed fluence with 10,000, 5000, and 1000 shots. With 10,000 shots, central LSFL and peripheral HSFL are observed. In both cases, the structures are elongated in a direction perpendicular to the electric field polarization.

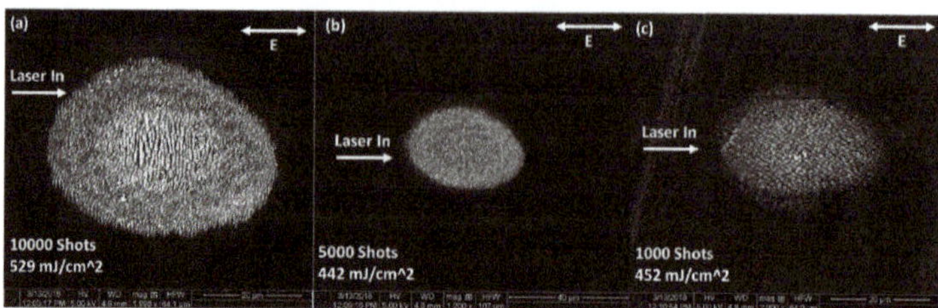

Figure 2. SEM micrographs of multi-pulse LID in silicon (111) with a high number of shots. From parts (**a–c**), the number of shots is decreasing while the fluence is mainly constant (up to fluctuations in the average pulse energy). Number of shots and fluence are located in the bottom left of the figure. In each figure, the input direction of the laser and polarization direction are indicated (as projected onto the sample surface).

Decreasing the number of shots to 5000; the LIPSS structures tend to disappear, resulting in chaotic damage features. Looking at 1000 shot damage sites, a new kind of structure begins to form in the peripheral region, predominantly on the side of the sample opposite to the laser input direction (i.e., on the right side of Figure 2c. These nano-structures will be the topic of focus for this study.

We found that the nanostructure formation works best with 500 pulse accumulated damage sites (see Figure 3). Two different types of nanostructures are observed. First, very narrow (<50 nm) regularly spaced (500 nm) nano-trenches extend out to the peripheral. The trenches are very straight and are parallel to the electric field polarization orientation. There appears to be a "rim" on either side of the trenches. Outside of the rim, there is a zone which has a contrast difference on the SEM. Drawing from the results of Ref. [1], this contrast change could be due to either subsurface melting or a weak ablation. Moving towards the center, the trenches bifurcate and become wider. Eventually they overlap and combine, resulting in apparent chaotic damage without a discernible structure. Debris can be seen around this area which may have come from the trenches.

The second type of observed nanostructure is a series of nanoscale outgrowths or clusters of "nano-spheroids", as shown in Figure 3b,e,f. They are always found aligned to the trenches, consisting of multiple spheroids, and appear to be molten material that has erupted from the trenches. They are not as regularly spaced as the trenches, but they occur roughly every μm. The exact shape of the nano-spheroid clusters (NSC) vary, but they are found to be generally round, form in clusters which are around 200 nm long, with the width of an individual spheroid about 50 nm. Since these observations were from Si sample surface with native oxide layers present, an obvious question emerged: what is the role of the native oxide layer in the formation of these two types of nanostructures?

To test the effect of the absence of a native oxide layer, we used an etched silicon (111) sample (see the Methods section). The results are presented in Figure 4. The morphology in this case appears qualitatively different from those in Figure 3. Focus first on the trench structures, and it is observed that the width is 20–40 nm near an NSC and less than 10 nm away from them. The nanotrench width still increases towards the center of the damage spot, eventually merging with one another. The trenches in this case remain extremely straight "distinct" single lines over a much longer distance (nearly 10 μm) before bifurcating and merging into chaotic damage. The spacing between trenches is still about 500 nm at the peripheral. An AFM depth profile taken along the trenches show that the trenches may grow up to 40 nm deeper toward the center of the damage spot.

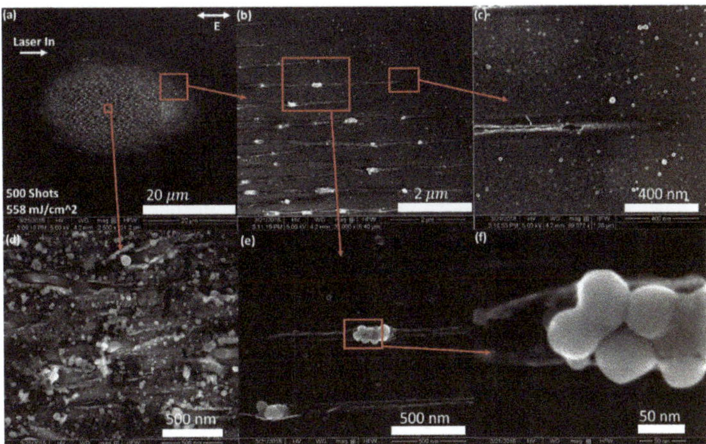

Figure 3. SEM images of multi-shot LID on silicon (111) with 500 shots. (**a**) shows an overall SEM image of the LID site. The fluence is indicated in the figure, as well as the direction of laser propagation and polarization, both projected onto the sample surface; (**b,d**) are zoom-ins of the red boxed areas in (**a**), as indicated by the red arrows; (**c**) is a zoom-in of a red boxed area in (**b**) as indicated by the red arrow; (**e**) is a zoom in of the larger red boxed area of (**b**), and an even larger magnification of (**e**) is shown in (**f**).

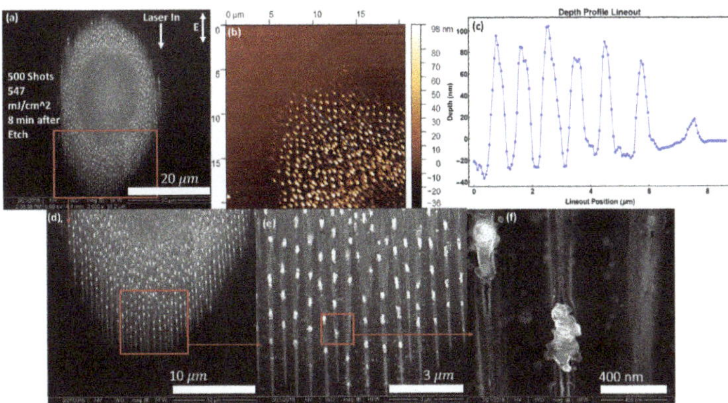

Figure 4. SEM and AFM images of multi-shot LID on silicon (111) with 500 shots, with oxide layer etched 8 min prior to exposure. (**a**) indicates the fluence used and the direction of laser input with polarization both projected onto the sample surface; (**b**) shows an AFM depth profile of the same sample area as in (**d**), but rotate by approx. 135 degrees clockwise. Samples with structures that may mimic AFM artifacts are purposely rotated to avoid scanning along or perpendicular to nanocracks; (**c**) lineout along the blue dotted line in (**b**); (**d**) shows a zoom-in of the boxed area in (**a**); (**e,f**) show further zoom-ins of this area indicated by the red boxes and arrows.

The nano-outgrowths also exhibit differences in the absence of the native oxide layer. While they are still spaced about 1 μm apart, their spacing is now more well defined and regular spacing occurs over the entire distance for which the trenches remain single lines. The shape of the nano-outgrowths is also different. Rather than distinct roughly sphere shaped clusters, we now see a single chaotic eruption of material without a well-defined shape. Furthermore, a closer look at Figure 4f shows that the debris in the area are also shaped differently. Instead of very round balls of debris, we see what appear to be disorderly clumps without definitive shapes.

Under the same conditions (oxide layer etched, Si (111)), but with a lower fluence, the morphology changes (see Figure 5). In this case, the entire damage site looks more like the peripheral regions from the previous two figures. The trenches are still strongly aligned to the laser polarization and become wider towards the center. In addition, Figure 5c shows that the end of the trenches are raised relative to the original surface.

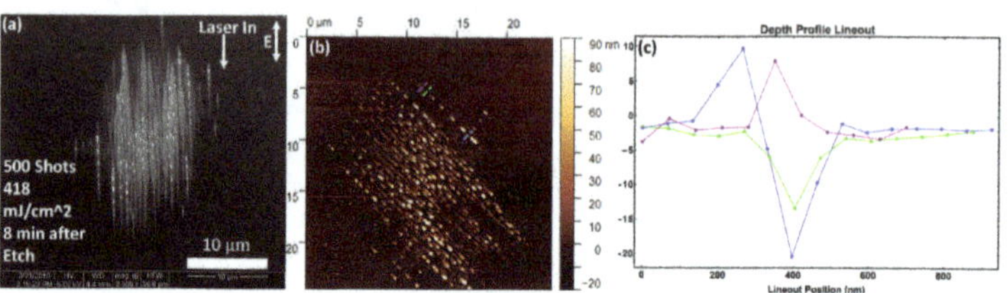

Figure 5. SEM and AFM images of multi-shot LID on silicon (111) with 500 shots, with oxide layer etched 8 min prior to exposure at a low fluence. (**a**) SEM image, indicates the fluence, laser input direction, and polarization projected onto the sample surface; (**b**) AFM depth profile, blue, green, and red dotted lines indicate lineouts in (**c**).

Figure 6 shows the effect of switching to a (100) sample under similar laser conditions. Note that this sample was not etched to remove the oxide layer. Here, perhaps, there is some sign of a diagonal ordering of the NSCs across distinct nano-trenches. The sample was oriented such that the vertical and horizontal edges in Figure 6 are aligned to the (110) direction such that the diagonal ordering (white dotted line) was aligned to the (100) direction. The trenches seemed to bifurcate more readily along this diagonal direction. The trenches also seemed to have lost a well-defined ridge. The spacing between NSC eruptions had decreased to about 500 nm. These clusters of nano-spheroid outgrowths are less consistent. They showed a larger number of spheroids per clump.

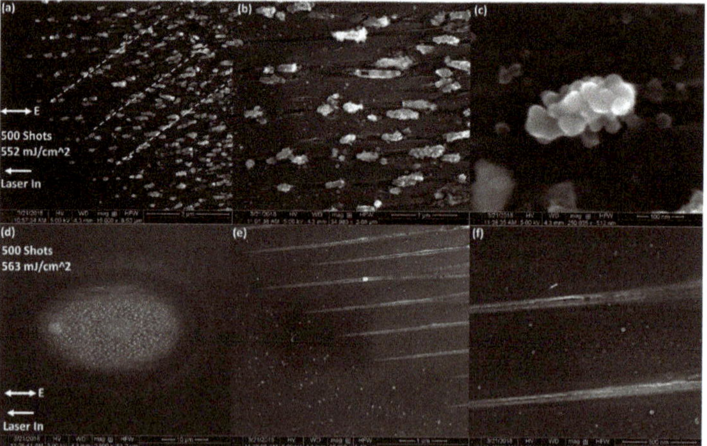

Figure 6. SEM images of multi-shot LID on silicon (100) with 500 shots. The laser input direction and polarization direction are indicated as projected onto the sample plane. The fluence is indicated on the bottom left for figures (**a–c**) and for figures (**d–f**); (**a–c**) show a damage site with progressively increasing magnification. The white dotted lines in (**a**) overlay ordering along a diagonal; (**d–f**) show a higher fluence damage site; (**d,f**) show increased magnification of the peripheral nano-trenches in (**d**).

Above a certain fluence (see Figure 6d–f), we found that the formation of the NSCs on the peripheral was suppressed while the trenches remain. Furthermore, at this higher fluence, the trenches have regained their ridges.

In order to obtain some insights into how the nanostructures formed, we looked at a higher fluence and varied the number of shots. A higher fluence was chosen so that the lower shot number exposures would still exhibit damage. Due to the Gaussian nature of the focal fluence profile on the surface, this allows us to explore features' formation at several different fluences within the same site based on distances from the center of the site.

The results of the shot number study are shown in Figure 7. At the site with 8 shot accumulation, we observe a damage area reminiscent of ultrafast melting/amorphization [24]. After 20 shots, one can observe what would eventually become the trenches, beginning to grow from the bottom rim of the site. The length of the trenches continues to grow until about 100 shots. At 100 shot accumulation and beyond, one observes a 'breaking' of a top layer, revealing the wavelength scale LSFL growing underneath. Similar LSFL periods have been recently observed in Ref. [30]. 2D Fourier analysis of the sites from the last row ($N = 200, 250$, and 500 shots) do not reveal sharp periodicity. However, few distinct diffuse period peaks appear, and the strongest is a peak representing an average LSFL period of 1.4 µm, with a range from 0.6–2 µm. A weaker peak oriented the same way (perpendicular to the laser polarization direction) shows a smaller period of 0.37 µm. These smaller features could arise in two steps, first bifurcation of largest LSFLs resulting in 0.7 µm periods, and then a second bifurcation of these to obtain the smallest periodicities. The bifurcation mechanism may be similar to that observed in another semiconductor [35]. The nano-trenches forming in the peripheral region result in a periodicity of 0.5 µm, as observed in previous figures. From sites with 125 to 500 shots, NSCs on nano-trenches begin to form in an increasing number corresponding to an increasing number of pulses per site.

Based on our observations so far, several characteristics begin to emerge pertaining to the formation of nano-trenches: first, they form via multi-pulse irradiation; second, their formation fluence is lower than multipulse ablation threshold, and the formation fluence range is relatively narrow; third, they do not form due to SPP driven mechanisms; fourth, it is unlikely that the surface scattered wave model applies to their formation mechanism, as their features appear qualitatively different from SSW HSFLs, which are also parallel to the laser polarization direction; fifth, the formation mechanism, albeit not needing the existence of native oxide, is nevertheless affected by the presence of native oxide on the surface, and perhaps the crystal orientation. We also note that some of the nano-structures presented here seem to resemble some of those presented in ref. [20], especially Figures 3 and 5 there (generated by $\lambda = 2100$ nm) in comparison to our Figure 5. In both theirs and our cases, the photon energies were below the Si bandgap, theirs within two photon absorption regime, where ours were beyond three photon absorption regime. They used unetched (native oxide present on surface) doped n- and p-type Si samples, with dopant concentrations of $\sim 10^{15}$ cm^{-3}, whereas, in our case, impurity concentrations were $\sim 10^{12}$ cm^{-3}. Laser illumination was at normal incidence in their case, where we used 45 degree AOI at p-polarization, lowering Fresnel reflection in our case. Still, a comparison can be drawn between their "bump" period along a "line" of ~ 1.1 µm (Figure 4 bottom panel), and ours from Figure 5c showing a period of ~ 1 µm, which do not preserve wavelength scaling. On the other hand, for the "periods" of lines parallel to laser polarization direction, they showed two types, the first one presented in the Figure 4 top panel, with a period of of ~ 900 nm, or $\lambda/2.3$ ("lines with bumps"), and the second ones at Figure 5 with spatial period of ~ 370 nm ("lines without bumps"), or $\lambda/5.7$. In our case, lines ("nanocracks") with or without bumps, parallel to laser polarization directions exhibit periods between 400–600 nm, i.e., between $\lambda/9$–$\lambda/6$.

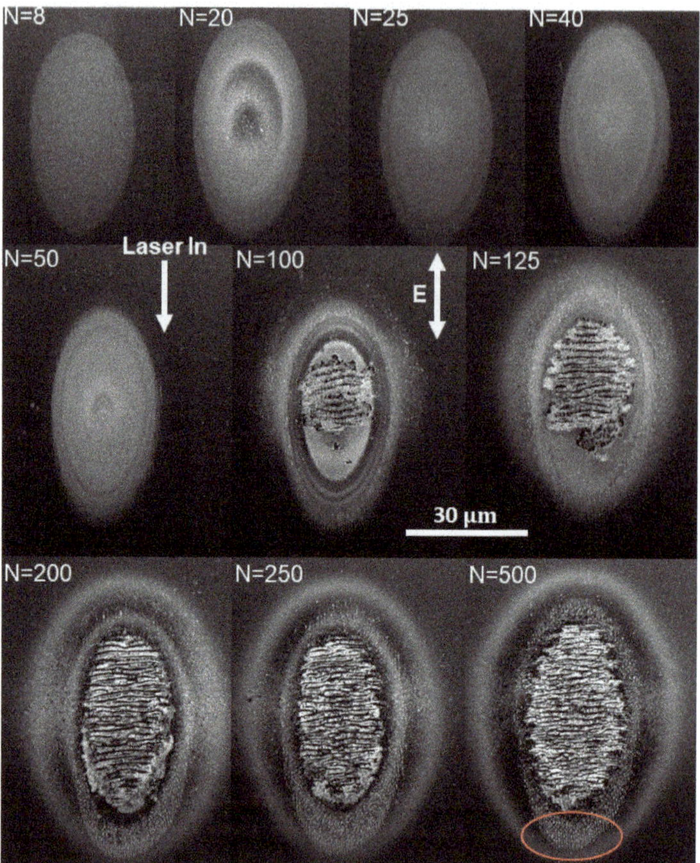

Figure 7. SEM images of multi-shot LID on silicon (111) with a fixed fluence and varying number of shots. The fluence used ranged from 0.8–0.9 J/cm^2 . The laser input and polarization directions are indicated as projected onto the sample surface. A universal length scale bar is indicated towards the middle of the figure. Each damage spot has the number of shots (N) indicated in the top left corner. The red circle in the bottom-right $N = 500$ shots image shows the location where the nanostructures most strongly form with 500 shots.

Next, focused circularly polarized MIR laser pulses were incident on (111) silicon by inserting a MIR quarter wave plate in the beamline before the focusing element. The results were quite surprising, with features qualitatively different than that from linear polarization case. SEM images of the damage sites generated by circularly polarized pulses is shown in Figure 8. Here, a higher fluence was used, since the damage threshold for circular polarization case was found to be generally higher.

A wavy, chiral pattern is observed in the damage sites. These waves appear to exhibit a contrast change in the SEM consistent with ultrafast melting and amorphization near the peripheral regions. The features narrow on the peripheral region. Trench like features can be seen towards the central region of damage, which exhibits multi-zone damage. The trenches seem to be aligned to the peripheral melting structures. A central region of damage seems to be surrounded by a ring of chaotic damage with no discernable structure aligned to the trenches. Lower fluence shots exhibit a similar wavy structure but without the trenches. The chiral nature of these surface features seem to imprint the handedness of polarization state (rotating clockwise) on these bright surface peripheral features, and may point towards an effect of strongly driven free carriers near the surface by the MIR electric field.

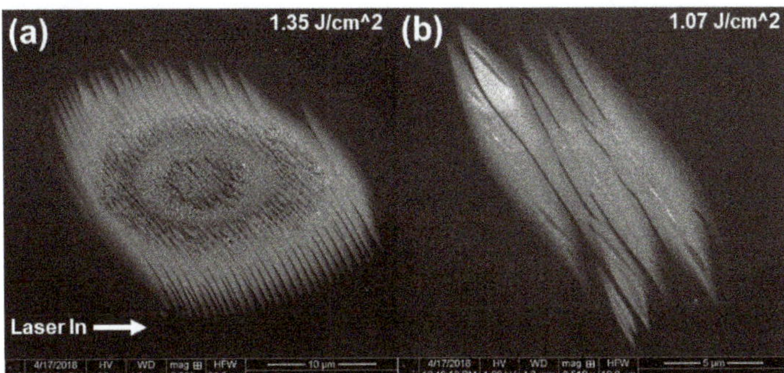

Figure 8. SEM images of multi-shot LID on silicon (111) with 500 shots and circular polarization. The fluence for each shot is indicated in the top right corner and the laser input direction is indicated in (**a**).

Cross-sectional transmission electron microscopy (TEM) was used to investigate one of the etched sample damage areas on silicon (111) with 500 shots and a fluence of 547 mJ/cm^2. The results are shown in Figure 9. This figure is meant to familiarize the reader with the sample under low magnification. Uppercase letters mark certain locations of interest. Before moving forward, it is important to understand that this figure shows a cross section of the LID spot near the peripheral, along several holes and trenches. The largest features in the figure are the protective layers (Pt and Au) which appear as very dark, and the sample bulk, which has characteristic bend contours. Only a very thin region at the interface between the bulk and the protective layers has been modified due to the laser, barely visible in this image. This is why we present higher magnification TEM images next. Our objective is to investigate the structure of the sample underneath the surface: determine which parts are crystalline and amorphous.

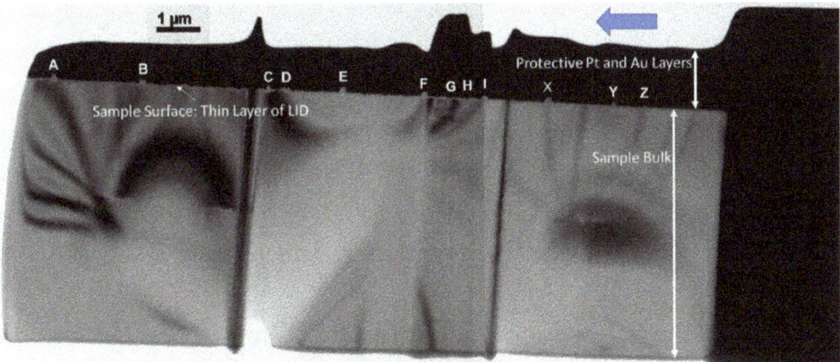

Figure 9. Cross-sectional TEM of a single, multi-shot LID site on silicon (111) with 500 shots, *p*-polarization. The sample oxide layer was etched 8 min prior to being shot by a train of pulses with an average fluence of 547 mJ/cm^2. The large blue arrow points from the peripheral of the LID spot towards the center. This sample has undergone FIB processing as described in the Methods section. The letters indicate various points of interest. The different layers involved in the cross section are labeled in the figure. Additionally, a capital letter labels several areas of interest. They include raised features (NSCs or nano-spheroidal clusters) A–F,I,X,Y,Z and a few trenches G,H.

Figure 10 shows a couple of the NSCs in higher magnification. The bright white areas correspond to amorphous material and gray corresponds to crystalline material, as evidenced by the ED scans in parts b and c. The TEM analysis shows that, away from NSCs, the laser modified surface amorphization is approximately 40 nm deep. Sometimes, black areas can be seen within the sample, attributed to voids or absence of material.

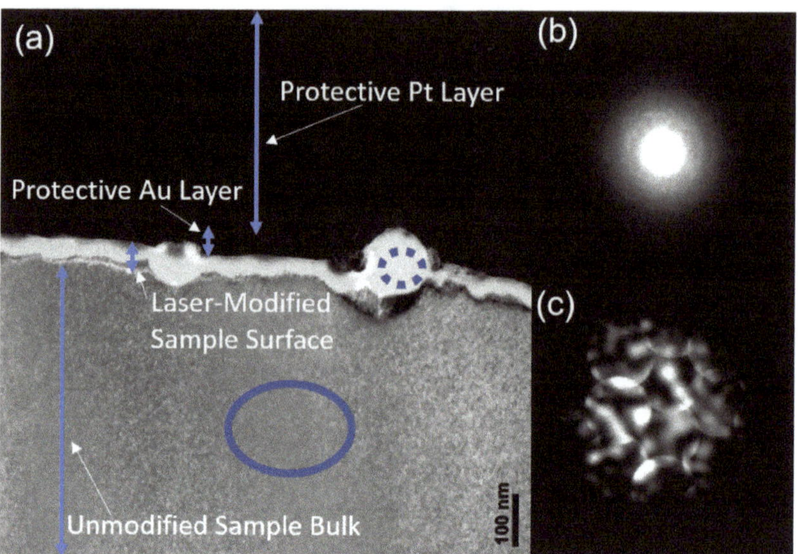

Figure 10. Electron diffraction and cross-sectional TEM measurements of multi-pulse LID on silicon. The different layers in the image are labeled in (**a**); (**b**) shows the result of electron diffractometry performed in areas like the blue dotted circle, i.e., on a nano-outgrowth; (**c**) shows the result of electron diffractometry performed in areas like the solid blue circle. It shows a strong crystalline ordering. In contrast, a strong amorphous signal is observed in (**b**). This means that, in the sample, the brighter white areas correspond to amorphized material while the gray/dark areas in the bulk correspond to monocrystalline silicon material.

Figure 11 summarizes the results of the TEM study. At locations A and B, a bisection of a single NSC is captured. Clearly, the spheroids are amorphous and they sometimes have a void beneath them. At location C, a partially detached NSC can be seen. At location D, it seems an NSC has completely been detached, leaving behind a ridge-like feature. Locations A,B,C,D all have an NSC and an increase in the amorphization thickness underneath the nano-outgrowth or NSC. Location G and H show 20–50 nm trenches. Clearly, the amorphous layer is thicker beneath the trenches, but rather than an NSC, there is an absence of material (which points towards crack formation). Location I shows another NSC. A slight contrast change can be seen beneath the NSC, which could indicate that the material there is at a lower density, i.e., the NSC may have been starting to detach as it cooled.

Next, we will explore the elemental make-up of the NSCs and surrounding material. In order to investigate this, we performed EDX analysis on the NSC at location Y.

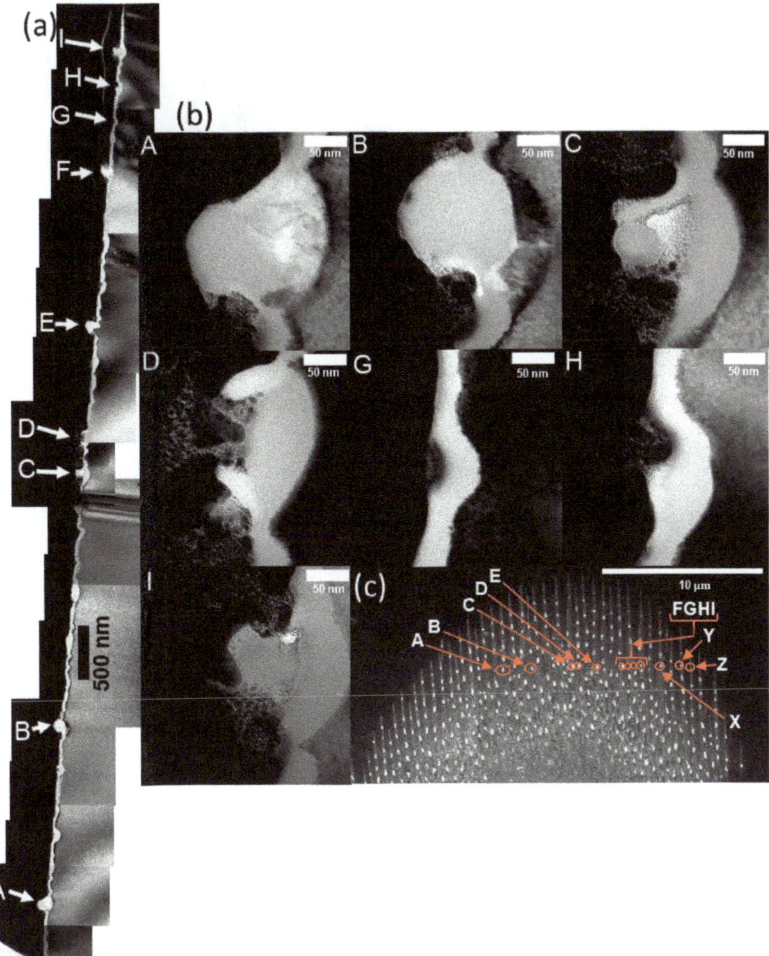

Figure 11. High resolution cross-sectional TEM measurements of multi-pulse nanostructure formation on silicon. (**a**) TEM images of a wide area have been stitched together. Areas of interest are marked by an upper-case letter. Area A is near the end of the sample (see Figures 5 and 6); (**b**) several of the features from (**a**) are shown magnified. They are labeled by the corresponding capitol letter locator in the top left corner; (**c**) the sample location for each feature is labeled in an overhead SEM of the site taken prior to any FIB/TEM processing.

The results of the EDX study are shown in Figure 12. The peak value of four different elements is displayed for each location by the intensity of the pixel at that location. The gold from the protective layer, which clusters together, can be seen clearly above the surface. Pt clearly has been deposited above the sample everywhere besides where there is gold. The silicon signal is strongly distinguishable, although near the top of the NSC, it seems some gold has penetrated. Finally, significant oxygen is only seen near the undercut of the NSC.

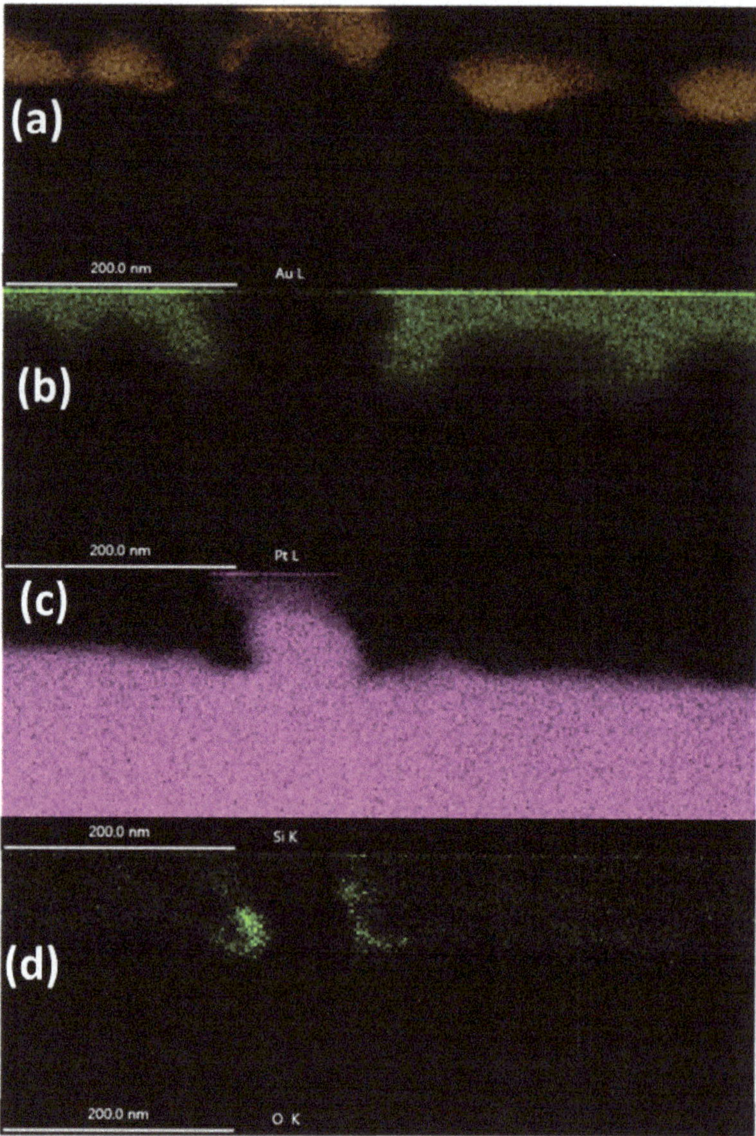

Figure 12. Spatially resolved Energy-dispersive X-ray spectroscopy (EDX) of an NSC outgrowth. The EDX is taken at location Y of the cross-sectional TEM sample. The peak signal count value is plotted for Au (**a**), Pt (**b**), Si (**c**), and O (**d**).

A proprietary algorithm used by the EDX software is able to plot spatially resolved EDX of several elements simultaneously while allowing for the colors to be mixed if both elements are present at the same location. This is shown in Figure 13.

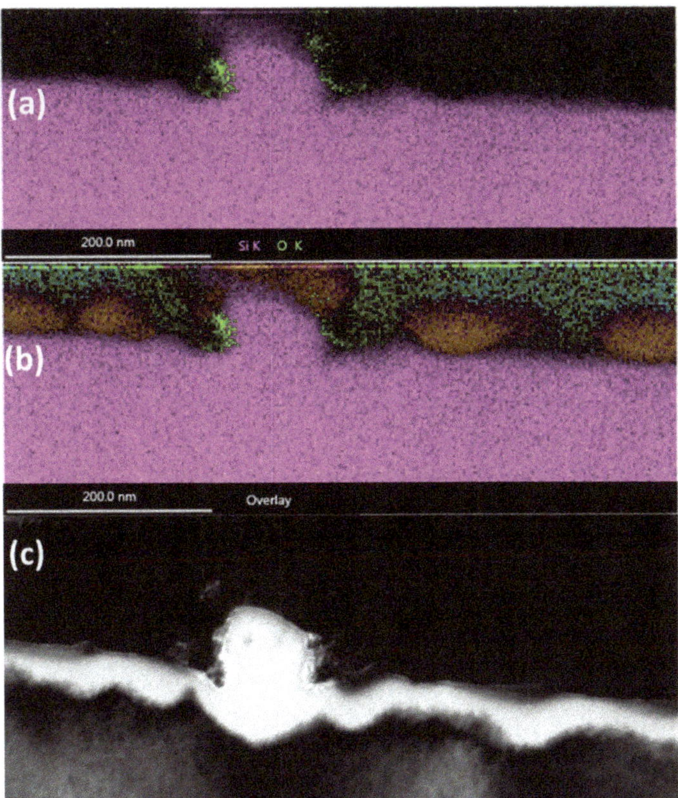

Figure 13. Spatially resolved Energy-dispersive X-ray spectroscopy (EDX) of an NSC outgrowth combining elemental signals. (**a**) silicon and oxygen peaks plotted; (**b**) Silicon, Copper, Gallium, Platinum, Gold, Carbon, and Oxygen are all plotted together, as color coded by the words used to describe them here; (**c**) cross-sectional TEM of the area which has undergone EDX in (**a**,**b**).

Figure 13 clearly indicates that the NSC outgrowth is not made of a silicon-oxide. Instead, it is made of pure silicon. Furthermore, the top part of the NSC was contaminated by gold during FIB preparation. Near the undercut of the NSC, higher levels of oxygen are indeed present. The very thin, brighter white area in Figure 13c is characteristic of an oxide layer. Carbon is expected to be present on the sample since it was exposed to air. Gallium is present from the gallium FIB. The Cu signal is from scattering off of the copper sample mount.

Based on our analysis of these unique extreme sub-wavelength surface structures (smallest nano-trench size 20 nm $= \lambda/180$) in single crystal Si generated from multi-pulse femtosecond MIR pulses, it is clear that the formation mechanism of these nanotrenches do not correspond to traditional LIPSS forming mechanisms. As it is clear from Figures 3–5 that these straight-line like structures are formed by cracks on the surface, one may consider a very well known ATG instability theory, named after pioneering works of Asaro, Tiller [36], and Grinfeld [37]. The ATG model has been successfully used to explain many different types of periodic surface structure generation caused by induced stress in the surface layer [38–40]. The basic mechanism of such a stress induced periodic surface structure formation (SIPSS) in creating self organized nano-islands [41] and nano-ripples [42] is explained briefly thus: a flat surface under stress can relax its strain energy by undulating the surface via surface diffusion, which may also result in crack formation. For SIPSS, the stress is introduced by lattice mismatch between the surface solid solution layer and the substrate, whereas, in the case of the nano-trenches in discussion here, the stress is essentially localized 'thermal stress' on the surface layer introduced by the laser. The strain energy is then relaxed via a process similar to the ATG instability, thus creating periodic trench/crack structures on the surface. There are two previous observations that support the above-mentioned hypothesis. First, these structures can only be generated after multiple laser shots. This can be explained by the fact that multiple laser shots induce more defect states, and disorder into the material, thereby increasing the strain energy. The sign of laser energy deposition and extreme disordering at the surface is also apparent due to the formation of the ~40 nm deep amorphous Si (a-Si) layer at the top (see Figure 13c), which contains all the nano-cracks and NSCs. On top of that, a-Si is in a metastable state, and thus always subject to local structural rearrangements when perturbed by external energy input, in an effort to minimize free energy. With a lower melting point of ~1200 °C, a-Si also exhibits a significantly larger increase in absorption co-efficient in comparison to crystalline Si [43]. The heat generated by the laser (via free carrier absorption on the surface by the laser E-field parallel to the surface) can also enhance mass transfer via surface diffusion in the top layer. Mass transfer and diffusion towards the surface are also corroborated by the nano-spheroid clusters of pure a-Si (see Figure 12) appearing to be 'bursting' out of the trenches/cracks in near-periodic fashion. The amorphous nature of Si NSCs also points towards ultrafast melting to locally reach temperatures above the Si melting point of 1410 °C, followed by rapid cooling to prevent recrystallization. Meanwhile, a recent work on bifurcation of LIPSS on GaAs [35] shows that LIPSS bifurcation crack formation due to stress on the surface created by diffusion of defects towards surface do not occur until hours after the laser treatment, whereas the plasmon should immediately die away upon laser cut-off. This time-dependent phenomenon may be explained by slow kinetics of surface diffusion at low temperatures, which may take seconds to hours for the strain energy to relax. We plan to study the formation time scale of such structures in a future effort.

4. Conclusions

In summary, we have observed a new kind of 'LIPSS' generated by strong MIR field interaction with single crystal Si with several distinctive features compared to traditional HSFL and LSFL. These 'straight-line' Nano-cracks/trenches were aligned to the laser polarization forming a near-periodic structure on the peripheral of the damage spot, with a spatial period similar to HSFL formed on germanium with MIR wavelengths [29]. However, unlike the case for common HSFL, the size/width of the individual features which make up the periodic surface structure is much smaller than both the spatial periodicity of the lines and the laser wavelength (as small as $\lambda/180$), and this size depends strongly on distance from the central damage site. Furthermore, nano-spheroidal cluster outgrowths of molten silicon have been captured re-solidified in their final amorphous state; in some cases, these nano-spheroidal clusters or NSCs, were in the middle of escaping the material. To the best of our knowledge, nothing like this has been previously reported. This phenomenon occurs regardless of the presence of a native oxide layer, although the morphology is slightly different. The morphology of the structures also was found to be affected by the crystal orientation of the sample. Silicon oxide was not detectable within the nano-cracks or NSCs, although an increased presence of oxygen was found in the undercut of the NSCs. Further theoretical models will be needed to understand how these structures form. With more study, better control over the formation of these structures could lead to extremely straight, nanometer sized extreme-sub-wavelength (ESW) nanostructure device formation.

Author Contributions: Conceptualization, K.W. and E.C.; methodology, K.W. and E.C.; software, K.W. and E.C.; validation, K.W. and E.C.; formal analysis, K.W. and E.C.; investigation, K.W. and E.C.; resources, K.W. and E.C.; data curation, K.W. and E.C.; writing—original draft preparation, K.W. and E.C.; writing—review and editing, K.W. and E.C.; visualization, K.W. and E.C.; supervision, E.C.; project administration, E.C.; funding acquisition, E.C. All authors have read and agreed to the published version of the manuscript.

Funding: This material is based upon work supported by the Air Force Office of Scientific Research (AFOSR) grants (FA9550-16-1-0069), (FA9550-20-1-0278) and AFOSR multidisciplinary research program of the university Research initiative (MURI) grant (FA9550-16-1-0013). The authors would like to further acknowledge instrument support from the Nanosystems Lab, and the Center for Electron Microscopy and Analysis (CEMAS) at the Ohio State University for SEM, EDS, and TEM analysis.

Institutional Review Board Statement: Not applicable.

Informed Consent Statement: Not applicable.

Data Availability Statement: Data can be made available by the authors upon request.

Conflicts of Interest: The authors declare no conflict of interest. The funders had no role in the design of the study; in the collection, analyses, or interpretation of data; in the writing of the manuscript, or in the decision to publish the results.

References

1. Werner, K.; Gruzdev, V.; Talisa, N.; Kafka, K.; Austin, D.; Liebig, C.; Chowdhury, E. Single-shot multi-stage damage and ablation of silicon by femtosecond mid-infrared laser pulses. *Sci. Rep.* **2019**, *9*, 19993. [CrossRef]
2. Soref, R. Mid-infrared photonics in silicon and germanium. *Nat. Photonics* **2010**, *4*, 495–497. [CrossRef]
3. Shcherbakov, M.; Werner, K.; Fan, Z.; Talisa, N.; Chowdhury, E.; Shvets, G. Photon acceleration and tunable broadband harmonics generation in nonlinear time-dependent metasurfaces. *Nat. Commun.* **2019**, *10*, 1345. [CrossRef]
4. Bian, Q.; Yu, X.; Zhao, B.; Chang, Z.; Lei, S. Femtosecond laser ablation of indium tin-oxide narrow grooves for thin film solar cells. *Opt. Laser Technol.* **2013**, *45*, 395–401. [CrossRef]
5. Torres, R.; Vervisch, V.; Halbwax, M.; Sarnet, T.; Delaporte, P.; Sentis, M.; Ferreira, J.; Barakel, D.; Bastide, S.; Torregrosa, F.; et al. Femtosecond laser texturization for improvement of photovoltaic cells: Black silicon. *J. Optoelectron. Adv. Mater.* **2010**, *12*, 621–625.
6. Li, J.; Ho, S.; Haque, M.; Herman, P.R. Nanograting Bragg responses of femtosecond laser written optical waveguides in fused silica glass. *Opt. Mater. Express* **2012**, *2*, 1562. [CrossRef]
7. Han, W.; Jiang, L.; Li, X.; Liu, P.; Xu, L.; Lu, Y. Continuous modulations of femtosecond laser-induced periodic surface structures and scanned line-widths on silicon by polarization changes. *Opt. Express* **2013**, *21*, 15505–15513. [CrossRef] [PubMed]
8. Vorobyev, A.Y.; Guo, C. Colorizing metals with femtosecond laser pulses. *Appl. Phys. Lett.* **2008**, *92*, 041914. [CrossRef]
9. Zorba, V.; Stratakis, E.; Barberoglou, M.; Spanakis, E.; Tzanetakis, P.; Anastasiadis, S.H.; Fotakis, C. Biomimetic artificial surfaces quantitatively reproduce the water repellency of a lotus leaf. *Adv. Mater.* **2008**, *20*, 4049–4054. [CrossRef]
10. Sokolowski-Tinten, K.; Barty, A.; Boutet, S.; Shymanovich, U.; Chapman, H.; Bogan, M.; Marchesini, S.; Hau-Riege, S.; Stojanovic, N.; Bonse, J.; et al. Short-pulse laser induced transient structure formation and ablation studied with time-resolved coherent XUV-scattering. *MRS Online Proc. Libr.* **2009**, *1230*, 503. [CrossRef]
11. Bonse, J.; Höhm, S.; Kirner, S.; Rosenfeld, A.; Krüger, J. Laser-induced Periodic Surface Structures—A Scientific Evergreen. *IEEE J. Sel. Top. Quantum Electron.* **2017**, *23*, 9000615. [CrossRef]
12. Tull, B.R.; Carey, J.E.; Mazur, E.; McDonald, J.P.; Yalisove, S.M. Silicon surface morphologies after femtosecond laser irradiation. *MRS Bull.* **2006**, *31*, 626–633. [CrossRef]
13. Costache, F.; Kouteva-Arguirova, S.; Reif, J. Sub-damage-threshold femtosecond laser ablation from crystalline Si: surface nanostructures and phase transformation. *Appl. Phys. A* **2004**, *79*, 1429–1432. [CrossRef]
14. Driel, H.V.; Sipe, J.; Young, J. Laser-induced periodic surface structure on solids: A universal phenomenon. *Phys. Rev. Lett.* **1982**, *49*, 1955. [CrossRef]
15. Huang, M.; Zhao, F.; Cheng, Y.; Xu, N.; Xu, Z. Origin of laser-induced near-subwavelength ripples: Interference between surface plasmons and incident laser. *ACS Nano* **2009**, *3*, 4062–4070. [CrossRef]
16. Tran, D.V.; Zheng, H.Y.; Lam, Y.C.; Murukeshan, V.M.; Chai, J.C.; Hardt, D.E. Femtosecond laser-induced damage morphologies of crystalline silicon by sub-threshold pulses. *Opt. Lasers Eng.* **2005**, *43*, 977–986. [CrossRef]
17. Bonse, J.; Krüger, J. Pulse number dependence of laser-induced periodic surface structures for femtosecond laser irradiation of silicon. *J. Appl. Phys.* **2010**, *108*, 034903. [CrossRef]
18. Shaheen, M.E.; Gagnon, J.E.; Fryer, B.J. Femtosecond laser ablation behavior of gold, crystalline silicon, and fused silica: A comparative study. *Laser Phys.* **2014**, *24*, 106102. [CrossRef]
19. Murphy, R.D.; Torralva, B.; Adams, D.P.; Yalisove, S.M. Pump-probe imaging of laser-induced periodic surface structures after ultrafast irradiation of Si. *Appl. Phys. Lett.* **2013**, *103*, 141104. [CrossRef]
20. Crawford, T.H.; Haugen, H.K. Sub-wavelength surface structures on silicon irradiated by femtosecond laser pulses at 1300 and 2100 nm wavelengths. *Appl. Surf. Sci.* **2007**, *253*, 4970–4977. [CrossRef]
21. Cahyadi, R.S.; Torralva, B.; Yalisove, S.M. femtosecond laser irradiation High spatial frequency periodic structures formation on silicon using near UV femtosecond laser irradiation. *Appl. Phys. Lett.* **2018**, *112*, 032105. [CrossRef]
22. Izawa, Y.; Setuhara, Y.; Hashida, M.; Fujita, M.; Izawa, Y. Ablation and amorphization of crystalline Si by femtosecond and picosecond laser irradiation. *Jpn. J. Appl. Phys. Part 1 Regul. Pap. Short Notes Rev. Pap.* **2006**, *45*, 5793–5794. [CrossRef]

23. Sudani, N.; Venkatakrishnan, K.; Tan, B. Experimental study of the effect of pulsewidth on ablation of silicon substrate using mega hertz repetition rate femtosecond laser. *Mater. Manuf. Process.* **2011**, *26*, 661–667. [CrossRef]

24. Bonse, J.; Baudach, S.; Krüger, J.; Kautek, W.; Lenzner, M. Femtosecond laser ablation of silicon–modification thresholds and morphology. *Appl. Phys. A* **2002**, *74*, 19–25. [CrossRef]

25. Armbruster, O.; Naghilou, A.; Kitzler, M.; Kautek, W. Spot size and pulse number dependence of femtosecond laser ablation thresholds of silicon and stainless steel. *Appl. Surf. Sci.* **2017**, *396*, 1736–1740. [CrossRef]

26. Lee, H. Picosecond mid-IR laser induced surface damage on Gallium Phosphate (GaP) and Calcium Fluoride (CaF_2) . *J. Mech. Sci. Technol.* **2007**, *21*, 1077–1082. [CrossRef]

27. Malik, R.; Mills, B.; Price, J.H.V.; Petrovich, M.; Moktadir, Z.; Li, Z.; Rutt, H.N. Determination of the mid-IR femtosecond surface-damage threshold of germanium. *Appl. Phys. A* **2012**, *113*, 127–133. [CrossRef]

28. Austin, D.R.; Kafka, K.R.P.; Trendafilov, S.; Shvets, G.; Li, H.; Yi, A.Y.; Szafruga, U.B.; Wang, Z.; Lai, Y.H.; Blaga, C.I.; et al Laser induced periodic surface structure formation in germanium by strong field mid IR laser solid interaction at oblique incidence. *Opt. Express* **2015**, *23*, 19522. [CrossRef] [PubMed]

29. Austin, D.R.; Kafka, K.R.P.; Lai, Y.H.; Wang, Z.; Zhang, K.; Li, H.; Blaga, C.I.; Yi, A.Y.; DiMauro, L.F.; Chowdhury, E.A. High spatial frequency laser induced periodic surface structure formation in germanium under strong mid-IR fields. *J. Appl. Phys.* **2016**, *120*, 143103. [CrossRef]

30. Kudryashov, S.I.; Pflug, T.; Busleev, N.I.; Olbrich, M.; Horn, A.; Kovalev, M.S.; Stsepuro, N.G. Topological transition from deeply sub- to near-wavelength ripples during multi-shot mid-IR femtosecond laser exposure of a silicon surface. *Opt. Mater. Express* **2021**, *11*, 1. [CrossRef]

31. Kafka, K.; Austin, D.; Li, H.; Yi, A.; Cheng, J.; Chowdhury, E. Time-resolved measurement of single pulse femtosecond laser-induced periodic surface structure formation induced by a pre-fabricated surface groove. *Opt. Express* **2015**, *23*, 19432–19441. [CrossRef]

32. Werner, K.; Hastings, M.; Schweinsberg, A.; Wilmer, B.; Austin, D.; Wolfe, C.; Kolesik, M.; Ensley, T.; Vanderhoef, L.; Valenzuela, A.; et al. Ultrafast mid-infrared high harmonic and supercontinuum generation with n_2 characterizaion in zinc selenide. *Opt. Express* **2019**, *27*, 2867–2885. [CrossRef]

33. Morita, M.; Ohmi, T.; Hasegawa, E.; Kawakami, M.; Ohwada, M. Growth of native oxide on a silicon surface. *J. Appl. Phys.* **1990**, *68*, 1272–1281. [CrossRef]

34. Mero, M.; Clapp, B.; Jasapara, J.C.; Rudolph, W.; Ristau, D.; Starke, K.; Krüger, J.; Martin, S.; Kautek, W. On the damage behavior of dielectric films when illuminated with multiple femtosecond laser pulses. *Opt. Eng.* **2005**, *44*, 051107. [CrossRef]

35. Abere, M.J.; Torralva, B.; Yalisove, S.M. Periodic surface structure bifurcation induced by ultrafast laser generated point defect diffusion in GaAs point defect diffusion in GaAs. *Appl. Phys. Lett.* **2016**, *108*, 153110. [CrossRef]

36. Asaro, R.J.; Tiller, W.A. Interface morphology development during stress corrosion cracking: Part I. Via surface diffusion. *Metall. Trans.* **1972**, *3*, 1789–1796. [CrossRef]

37. Grinfeld, M.A. Instability of the interface of a nonhydrostatically loaded elastic solid and melt. *Sov. Phys. Dokl.* **1986**, *31*, 831.

38. Leonard, D.; Krishnamurthy, M.; Reaves, C.M.; Denbaars, S.P.; Petroff, P.M. Direct formation of quantum-sized dots from uniform coherent islands of InGaAs on GaAs surfaces. *Appl. Phys. Lett.* **1993**, *63*, 3203. [CrossRef]

39. Drucker, J. Self-Assembling Ge(Si)/Si(100) Quantum Dots. *IEEE J. Quantum Electron.* **2002**, *38*, 975–987. [CrossRef]

40. Rauscher, M.D.; Boyne, A.; Dregia, S.A.; Akbar, S.A. Self-assembly of pseudoperiodic arrays of nanoislands on YSZ-(001). *Adv. Mater.* **2008**, *20*, 1699–1705. [CrossRef]

41. Ansari, H.M.; Dixit, V.; Zimmerman, L.B.; Rauscher, M.D.; Dregia, S.A.; Akbar, S.A. Self assembly of nanoislands on YSZ-(001) surface: A mechanistic approach toward a robust process. *Nano Lett.* **2013**, *13*, 2116–2121. [CrossRef] [PubMed]

42. Ansari, H.M.; Niu, Z.; Ge, C.; Dregia, S.A.; Akbar, S.A. Spontaneous Rippling and Subsequent Polymer Molding on Yttria-Stabilized Zirconia (110) Surfaces. *ACS Nano* **2017**, *11*, 2257–2265. [CrossRef] [PubMed]

43. Hoyland, J.D.; Sands, D. Temperature dependent refractive index of amorphous silicon determined by time-resolved reflectivity during low fluence excimer laser heating. *J. Appl. Phys.* **2006**, *99*, 063516. [CrossRef]

MDPI
St. Alban-Anlage 66
4052 Basel
Switzerland
Tel. +41 61 683 77 34
Fax +41 61 302 89 18
www.mdpi.com

Nanomaterials Editorial Office
E-mail: nanomaterials@mdpi.com
www.mdpi.com/journal/nanomaterials